Schauder Bases:
Behaviour and Stability

π Pitman Monographs and
Surveys in Pure and Applied Mathematics 42

Schauder Bases: Behaviour and Stability

P K Kamthan & M Gupta
Indian Institute of Technology, Kanpur

Longman
Scientific &
Technical
Copublished in the United States with
John Wiley & Sons, Inc., New York

Longman Scientific & Technical
Longman Group UK Limited
Longman House, Burnt Mill, Harlow
Essex CM20 2JE, England
and Associated Companies throughout the world.

Copublished in the United States with
John Wiley & Sons, Inc., 605 Third Avenue, New York, NY 10158

First published 1988

AMS Subject Classifications: (main) 46A02, 46A35, 46A45
(subsidiary) 46A05, 46A12, 47B10

ISSN 0269-3666

British Library Cataloguing in Publication Data
Kamthan, P.K.
Schauder bases : behaviour and stability.
— (Pitman monographs and surveys in
pure and applied mathematics, ISSN 0269-
3666 ; 42)
1. Linear Topical Spaces
I. Title II. Gupta, M.
515.7′3 QA322
ISBN 0-582-01482-4

Library of Congress Cataloging-in-Publication Data
Gupta, Manjul, 1951–
Schauder bases.

(Pitman monographs and surveys in pure and applied
mathematics, ISSN 0269-3666 ; 42)
Bibliography: p.
Includes index.
1. Schauder bases. I. Kamtham, P. K., 1938–
II. Title. III. Series.
QA322.G87 1988 515.7′3 87-31211
ISBN 0-470-21029-X (USA only)

Printed and Bound in Great Britain at The Bath Press, Avon

Contents

Preface

This monograph is the third and last part of a project on Schauder bases (S.b.) in topological vector spaces (TVS) initiated by the authors in 1972, the first two parts having already appeared in the form of [129] and [130].

As in the earlier work, the aim of the authors is twofold: to collect systematically more or less all the significant work on S.b. theory in locally convex spaces (l.c. TVS) which we could not cover in [130] and communicate this to graduates and advanced undergraduates with the sole intention of generating interest in this field and, secondly, to provide for experts a single source of reference containing information on the heuristic development of the subject matter treated in this book. As such, while we have been didactic in our approach to the subject matter, we have included as well several deep results with some obvious problems on the basis theory in locally convex spaces.

Whereas the previous work [130] on S.b. theory dealt exclusively with its few initial topics, including the geometrical interpretation with occasional briefing on types of bases, the present work is primarily aimed at providing a comprehensive and an up-to-date account of two major aspects of S.b. theory in l.c. TVS. These are the behaviour of an S.b. $\{x_n;f_n\}$ depending upon the convergence of the underlying infinite series and the restrictions put in a natural way on $\{x_n\}$, $\{f_n\}$ or on both and, secondly, the stability criteria of a given S.b. Both of these aspects are treated in the last three Parts of the book.

As remarked above, the present work on Schauder bases is exclusively studied in the setting of l.c. TVS, except for a few in normed spaces which seem to be unavoidable, showing thereby the importance of bases in such spaces. Nevertheless, the vast importance of the basis theory in non-normable l.c. TVS has been sufficiently explored in several chapters; for instance, discussing the different notions of an absolute base in normed spaces is a futile exercise. Thus, we expect an average reader to have a working knowledge of locally convex spaces, convergence of infinite series in l.c. TVS and sequence spaces. However, to acquire an advanced knowledge of the subject matter of this project, the reader requires further knowledge and the prerequisites are given in chapter 1. We apologise that, all the results in chapter 1 are given without proofs - after all this material is considered to be a prelude to the overall subject matter of this monograph.

A preliminary discussion on the behaviour of bases is provided in chapter 2, where one can find listed almost all the different types of S.b., including their brief history. Examples and counter-examples are also given here to enhance the importance of one type of S.b. over another.

Each l.c. TVS having an S.b. may be regarded as a sequence space (s.s.) and a basic study in this direction is carried out in chapter 3 where one also finds a glimpse of the relationship of an S.b. with an infinite matrix, resulting in new bases from the old ones.

Chapters 4 to 12 provide a rigorous picture of the various types of bases discussed in chapter 2. Several applications of these different notions of types of bases in the structural study of the underlying spaces have also been included in these chapters.

Part III is exclusively devoted to the second aspect of the basis theory. Whereas chapter 13 begins with the fundamentals of this aspect, chapter 14 contains results on the similarity of bases. With this background, chapters 15 and 16 present stability theorems on bases in different types of spaces.

We now turn to Part IV which may be considered as partly a study on behaviour plus a small amount of the similarity of Schauder bases. However, the nuclearity of one or another kind of the space in question plays a vital role. While chapters 17 and 18 deal with the relationship of λ-bases with the λ-nuclearity of the underlying space, an exclusive study on Dragilev bases (a restrictive kind of absolute bases) is carried out in chapters 19 and 20 of this work. The last chapter is concerned with the existence problem of conditional bases in non-nuclear spaces.

Our job will remain incomplete if we fail to acknowledge our gratitude and sincere thanks to quite a few colleagues and friends, along with our family members whose help and support in one way or the other at times of need have gone a long way toward the completion of the present project. At the outset, we put on record our sincere thanks to Professor S. Sampath, former Director of our Institute, who promptly supported the financing of this project for preparing the final typed draft of the book.

We heartily thank Professor Dr. Gottfried Köthe of Frankfurt, West Germany and Professor William, H. Ruckle of Clemson, U.S.A. with whom the authors had fruitful discussions on some parts of the manuscript. Our thanks are also due to Professor W. Wojtyński of Warsaw, Poland for his correspondence concerning the subject matter of chapter 21.

Finally, our thanks are also due to the staff at Longman for their excellent cooperation and to Ms. Terri Moss for her expert typing of the final manuscript.

PKK and MG

Kanpur

List of abbreviations

a.s.s.	associated sequence space
a.s.S.b.	almost shrinking Schauder base
b.c.	base constants
b.c.b.	balanced, convex and bounded
b.m-S.b.	bounded multiplier - Schauder base
b.m.c.	bounded multiplier convergent
b.s.	biorthogonal sequence
bl. ptb.	block perturbation
c.ab-S.b.	completely abnormal Schauder base
c-H	countably-Hilbert
c.l.o.	continuous linear operator
c.n-S.b.	completely normal Schauder base
CSBS	complemented Schauder basic sequence
c.u.e.	continuous unique extension
e.m-S.b.	extended monotone Schauder base
e-nz.S.b.	equi-normalized Schauder base
e-S.b.	e-Schauder base
f.nz-S.b.	fully normalized Schauder base
i.i.	isometric isomorphism
K-F	Köthe-Fréchet
l.c.	locally convex
l.c. TVS	locally convex topological vector spaces

l.c.o.	linear continuous and open
l.i.	linearly independent
M.f.	Minkowski functional
m.K.m.	monotone Köthe matrix
m-S.b.	monotone Schauder base
n-S.b.	normal Schauder base
nz.	normalized
nz-bases	normalized bases
nz-S.b.	normalized Schauder base
q.r.b.s	quasi-regular biorthogonal sequence
s.a.c.f.	sequence of associated coordinate functionals
S.b.	Schauder base(s)
S. basic	Schauder basic
s.c.	subseries convergent
s.o.e.	seminorm of equicontinuity
s.s.	sequence space
s-S.b.	subseries Schauder base
s.s.s.	symmetric sequence space
s-sy.S.b.	s-symmetric Schauder base
s-shrinking	semishrinking
s. γ-complete	semi γ-complete
sy.S.b.	symmetric base
t.b.	topological base
t.basic	topologically basic
t.i.	topological isomorphism
TVS	topological vector space(s)
P-W	Paley-Wiener
u.b.c.	unconditional base constants
u.c.	unconditionally convergent

u.e-S.b.	uniformly equicontinuous Schauder base
u-S.b.	unconditional Schauder base
u-sy.S.b.	u-symmetric Schauder base
ud.b.	unordered bounded
ud.c.	unordered convergent
ud.C.	unordered Cauchy
VS	vector space
v.v.	vector valued
w.e-S.b.	weakly e-Schauder bases
w.u.S.b.	weakly uniform Schauder base
w.λ-u.C.	weakly λ-unconditionally Cauchy

PART I

1 Preliminaries

1.1 INTRODUCTION AND NOTATION

This book is primarily concerned with the theory of Schauder bases in locally convex topological vector spaces (l.c. TVS). However, we will occasionally have to deal with this theory in still more general spaces, namely, topological vector spaces (TVS). For obvious reasons with regard to convergence problems, all such spaces will hereafter be assumed to be Hausdorff. In order to appreciate the material of this book, the reader is expected to have a working knowledge of the theories of infinite series, nuclear spaces and sequence spaces.

In order to facilitate the reading of the material of subsequent chapters, we will briefly touch upon all these four aspects in functional analysis, indicating only their salient features and giving more stress to the less common results that are to be used subsequently. We will also mention a few results on absolutely summing operators. No proofs of the results will, however, be provided in this chapter. To begin with, let us single out some frequently used notation:

$\mathbb{R}$ = Set of all reals equipped with the usual topology

$\mathbb{C}$ = Set of all complex numbers endowed with its usual topology

$\mathbb{K}$ = $\mathbb{R}$ or $\mathbb{C}$

X = Nontrivial vector space (VS) over $\mathbb{K}$

X' = Algebraic dual of X

$\mathbb{N}$ = $\{1,2,\ldots,n,\ldots\}$

$\mathbb{N}_0$ = $\{0,1,\ldots,n,\ldots\}$

I = Collection of all increasing subsequences of $\mathbb{N}$

I_∞ = Collection of all increasing subsequences of $\mathbb{N}$ with infinite complements in $\mathbb{N}$

Φ = Family of all finite subsets of $\mathbb{N}$ ordered by set-theoretic inclusion

$\#(A)$ = Cardinality of a set

dim X = Dimension of X

$\aleph_0$ = $\#(\mathbb{N})$

$\aleph$ = $\#([0,1])$

sp$\{A\}$ = Space generated by $A \subset X$

P = Set of all permutations of $\mathbb{N}$

Λ = A directed set

1.2 TOPOLOGICAL VECTOR SPACES

We will not linger here to introduce, discuss elementary facts of or even recall standard results for either l.c. TVS or TVS, for these can be traced out in any of the standard texts listed in [80], [140], [143], [190], [201], [205] and [233]. For notation and terminology, our reference is [80]. The expected background for the reader is also summarized in chapter 1 of [129] and [130].

Throughout we use the notation (X,T) to mean an arbitrary (Hausdorff) TVS (resp. l.c. TVS) where T is the linear topology (resp. locally convex topology, abbreviated hereafter as l.c. *topology*) generated by the family D_T of all T-continuous F-seminorms or pseudonorm functions at the origin (resp. the family of all T-continuous seminorms). Corresponding to a TVS (X,T) (resp. an l.c. TVS (X,T)), we use the symbol $\mathcal{B}_T$ or $\mathcal{B}_X$ to denote a fundamental system of neighbourhoods at the origin in X consisting of all balanced,

T-closed (resp. and convex) subsets of X. Given a TVS (X,T) (resp. l.c. TVS (X,T)) with D_T as above, an arbitrary family D of F-seminorms (resp. seminorms) on X is said to be equivalent if the linear topology (resp. l.c. topology) S generated by D is equivalent to T; that is, $S \approx T$. When T is metrizable, the countable family D is said to be *admissible* if it is equivalent and its members form an increasing sequence of seminorms.

For a subset A of a TVS (X,T), we write $[A] \equiv [A]^T$ for the T-closure of sp{A} and $\overline{\Gamma(A)}^T$ for the balanced, T-closed and convex hull of A; also $X^* \equiv (X,T)^*$ is reserved for the topological dual of $X \equiv (X,T)$.

Quotients and normed subspaces

Corresponding to an l.c. TVS (X,T), we write p_u for the *Minkowski functional* (M.f.) of u in $\mathcal{B}_X$. Let $X_u = X/\ker p_u$ and $K_u : X \to X_u$, the usual quotient map, where X_u is endowed with the norm $\hat{p}_u$ given by $\hat{p}_u(x_u) = \inf\{p_u(x + y) : y \in \ker p_u\}$; $x_u = x + \ker p_u$, $x \in X$. K_u is a linear continuous and open (l.c.o.) map from (X,T) onto $(X_u, \hat{p}_u)$.

For v and u in $\mathcal{B}_X$, let us write $v < u$ to mean that $v \subset \alpha u$ for some $\alpha > 0$. Then there is a natural map $K_u^v : X_v \to X_u$ with $K_u = K_u^v \circ K_v$.

With u,v as above ($v < u$), the set $v(u) \equiv K_u[v] = \{x_u : x \in v\}$ is clearly bounded in the normed space $(X_u, \hat{p}_u)$. Observe that $K_u[u]$ is the closed unit ball of X_u and so $K_u[v] < K_u[u]$.

We will generally write $\hat{X} \equiv (\hat{X},\hat{T})$ for the completion of an l.c. TVS (X,T) and if (Y,S) is another l.c. TVS such that $R : (X,T) \to (Y,S)$ is a *continuous linear operator* (c.l.o.) from (X,T) to (Y,S), then $\hat{R}$ is a *continuous unique extension* (c.u.e.) of R from $\hat{X}$ to $\hat{Y}$. With this notation, the c.u.e. $\hat{K}_u^v$ of K_u^v is a c.l.o. from the Banach space $\hat{X}_v \equiv (\hat{X}_v, \hat{p}_v)$ onto the Banach space $\hat{X}_u \equiv (\hat{X}_u, \hat{p}_u)$.

For u in $\mathcal{B}_X$ (X is an l.c. TVS), let us write $X^*(u^o)$ for $sp\{u^o\}$, u^o being the polar of u in X^*. Then $(X^*(u^o), q_{u^o})$

is easily seen to be a Banach space, where the norm q_{u^o} is given by $q_{u^o}(f) = \sup \{|f(x)| : x \in u\}$.

For two TVS (X,T) and (Y,S), we use the notation $(X,T) \simeq (Y,S)$ to mean that these spaces are *topologically isomorphic* and if we want to emphasize the underlying *topological isomorphism* (t.i.) R between (X,T) and (Y,S), the same is expressed as $(X,T) \simeq (Y,S)$ under R. In case X and Y are normed spaces and R is an *isometric isomorphism* (i.i.) from X onto Y, then $X \simeq Y$ is emphasized as $X \overset{ism}{\simeq} Y$ under R.

Returning to the foregoing notation, we have

$$(X^*(u^o), q_{u^o}) \overset{ism}{\simeq} ((X_u)^*, \hat{p}_u^*) \quad (1.2.1)$$

under R; $Rf = \hat{f}$, $\hat{f}(x_u) = f(x)$, where $\hat{p}_u^*$ is the dual norm on $X_u^* \equiv (X_u)^*$ obtained with respect to p_u.

NOTE. Sometimes it will be convenient to use different notation for certain symbols used above. For p in D_T((X,T) is an l.c. TVS), let $u_p = \{x \in X : p(x) \leq 1\}$. Then we write X_p, K_p and $\hat{p}$ for X_{u_p}, K_{u_p} and $\hat{p}_{u_p}$ respectively; also, let us write $\hat{A} \equiv \hat{A}^p$ for $K_p[A]$ and $\hat{x} \equiv \hat{x}^p$ for $K_p x$, where $x \in X$ and $A \subset X$.

Corresponding to a balanced, convex and bounded (b.c.b.) subset B of an l.c. TVS (X,T), let X_B denote the *Grothendieck space* defined by $X_B = \cup\{nB : n \geq 1\}$ and if $p_B(x) = \inf \{\alpha > 0 : x \in \alpha B\}$, then (X_B, p_B) is a normed space. If B is also *sequentially complete* (abbreviated hereafter as *ω-complete*), then (X_B, p_B) is a Banach space. Further, the letter $\mathcal{D}$ or $\mathcal{D}_X$ is reserved for the family of all subsets of an l.c. TVS (X,T) such that each member of $\mathcal{D}$ is b.c.b. and T-closed. A subfamily $\mathcal{D}_o$ of $\mathcal{D}$ is called *fundamental* if for each B in $\mathcal{D}$ there exists D in $\mathcal{D}_o$ with $B \subset D$.

Nets and sequences

Nets and sequences in a TVS (X,T) will be denoted by $\{x_\alpha\} \equiv \{x_\alpha : \alpha \in \Lambda\}$ and $\{x_n\} \equiv \{x_n : n \geq 1\}$ and if x in X is a convergent point of $\{x_\alpha\}$ (resp. $\{x_n\}$), these convergence limits will be expressed by $x_\alpha \to x$ or $T\text{-}\lim_\alpha x_\alpha = x$ (resp. $x_n \to x$ in T) and if there is no confusion of the underlying topology T, the letter T will be dropped. Further, we shall write that a family $\{x_\alpha\}$ is l.i. to mean it is *linearly independent*.

DEFINITION 1.2.2 A sequence $\{x_n\}$ in (X,T) is called (i) *regular* if for some p in D_T and $\varepsilon > 0$, $p(x_n) \geq \varepsilon$ for $n \geq 1$; (ii) *irregular* if $\alpha_n x_n \to 0$ for each $\{\alpha_n\}$ in $\mathbb{K}$; (iii) *normalized* if $\{x_n\}$ is bounded and regular; (iv) *normal* if for some $\{\alpha_n\}$ in $\mathbb{K}$ ($\alpha_n \neq 0$, $n \geq 1$), the sequence $\{\alpha_n x_n\}$ is normalized; and (v) ω-*linearly independent* (ω-l.i.) if the T-convergence of $\sum_{n\geq1} \alpha_n x_n$ to 0 yields $\alpha_n = 0$, $n \geq 1$.

Continuity of linear maps

Given two l.c. TVS (X,T) and (Y,S), let us write $\mathcal{L}(X,Y)$ for the collection of all c.l.o. from X into Y.

The following two results are reproduced from [140], p. 104 and [134].

PROPOSITION 1.2.3 If (X,T) is barrelled, then any subfamily H of $\mathcal{L}(X,Y)$ such that $\{f(x) : f \in H\}$ is bounded in (Y,S) for each x in X, is equicontinuous.

PROPOSITION 1.2.4 Let $\langle X,Y\rangle$ be a dual system of VS X and Y; let X be equipped with the l.c. topology $\sigma(X,Y)$, the *weak topology*. Then any subfamily H of $\mathcal{L}(X,X)$ such that $\{Rx : R \in H\}$ is $\beta(X,Y)$-bounded for x in X, is $\beta(X,Y)$-equicontinuous, $\beta(X,Y)$ being the *strong topology*.

An l.c. TVS (X,T) is called *Pták* or *fully complete* if each subspace M of X^* is $\sigma(X^*,X)$-closed whenever $M \cap A$ is $\sigma(X^*,X)$-closed for any balanced, convex $\sigma(X^*,X)$-closed and equicontinuous subset A of X^* (cf. [80]).

The next two results (often called different forms of the closed graph theorem) are recalled from [190] and [97], respectively.

THEOREM 1.2.5 Let X be barrelled and Y a Pták space. A linear map $R : X \to Y$ is continuous whenever its graph is closed.

THEOREM 1.2.6 Let (X,T) be a Mackey space (that is, (X,T) is an l.c. TVS with $T = \tau(X,X^*)$) and Y a separable Fréchet space. Then a linear map $R : X \to Y$ is continuous whenever its graph is closed.

Extensions of barrelledness

Corresponding to an l.c. TVS (X,T), let T^+ denote the finest l.c. topology on X having the same convergent sequences as T, then $(X,T^+)^* = X^+$, where X^+ is the sequential dual of (X,T) (cf. [229]). Also, let X^b denote the subspace of X' such that each member in X^b is bounded on bounded subsets of (X,T).

DEFINITION 1.2.7 An l.c. TVS (X,T) is called (i) a *Mazur space* if $X^* = X^+$; (ii) an S-*space* if $(X^*,\sigma(X^*,X))$ is ω-complete; (iii) W-*space* if $\sigma(X^*,X)$ and $\beta(X^*,X)$-bounded subsets of X^* are the same; (iv) σ-*barrelled* (resp. σ-*infrabarrelled*) if every countable $\sigma(X^*,X)$ (resp. $\beta(X^*,X)$)-bounded subset of X^* is equicontinuous; and (v) ω-*barrelled* if every $\sigma(X^*,X)$-convergent sequence is equicontinuous.

Following [141], we have

THEOREM 1.2.8 The following statements are equivalent:

(i) $(X^*,\sigma(X^*,X))$ is barrelled, (ii) $\sigma(X^*,X) \approx \beta(X^*,X)$, (iii) $X^+ = X^b = X'$ and (iv) each $\sigma(X,X^*)$-bounded subset A of X is finite dimensional.

The next is quoted from [229].

PROPOSITION 1.2.9 Each ω-barrelled space is a W-space. If (X,T) is an ω-complete l.c. TVS, then $(X^*, \tau(X^*,X))$ is ω-barrelled; conversely if (X,T) is a metrizable l.c. TVS and $(X^*,\tau(X^*,X))$ is ω-barrelled, then (X,T) is complete, where $\tau(X^*,X)$ is the *Mackey topology* on X^*.

T-limited subsets of X^+

DEFINITION 1.2.10 Let (X,T) be an l.c. TVS. A subset K of X^+ is called T-*limited* if for every null sequence $\{x_n\}$ in (X,T)

$$\lim_{n\to\infty} \sup_{f\in K} |f(x_n)| = 0.$$

Several basic facts about T-limited sets and their relation to the construction of T^+ are given in [229], let us single out

PROPOSITION 1.2.11 For an l.c. TVS (X,T), T^+ is the topology of uniform convergence on all T-limited subsets of X^+.

DEFINITION 1.2.12 Corresponding to an l.c. TVS (X,T), let G_c^T (resp. G_n^T) denote the family of all subsets A of X with the property that every sequence in A has a T-Cauchy subsequence (resp. the family of all sequences which are T-convergent to zero).

Let us write T_c (resp. T_n) for the l.c. topology on X^+ generated by the polars of members of G_c^T (resp. G_n^T). After [65], Exercise 2,2), p. 214, we have

PROPOSITION 1.2.13 Let (X,T) be an l.c. TVS and $K \subset X^+$. Then the following statements are equivalent: (i) K is T-limited, (ii) K is T_c-precompact and (iii) K is T_n-precompact.

NOTE. Clearly the preceding result remains valid if T is replaced by $\sigma(X,X^*)$.

Other useful concepts and results

The next result is given in [162].

LEMMA 1.2.14 Let M be a set, (X,T) a TVS and $\{A_n\}$, A transformations from M into (X,T). A subset B of M is transformed into a precompact subset A[B], provided (i) $A_n[B]$ is precompact for $n \geq 1$ and (ii) $A_n x \to Ax$ uniformly in $x \in B$.

If (X,T) and (Y,S) are two l.c. TVS, then the adjoint $R^* : Y^* \to X^*$ of a linear operator $R : X \to Y$ exists provided R is $\sigma(X,X^*) - \sigma(Y,Y^*)$ continuous and conversely. Note that $\langle Rx,g\rangle = \langle x,R^*g\rangle$; $x \in X$, $g \in Y^*$.

PROPOSITION 1.2.15 If R is T - S continuous, then R is also $\sigma(X,X^*) - \sigma(Y,Y^*)$ continuous, and R is $\sigma(X,X^*)-\sigma(Y,Y^*)$ continuous if and only if R is $\tau(X,X^*)-\tau(Y,Y^*)$ (or S) continuous.

Corresponding to any subspace Y of an l.c. TVS (X,T), let $Y^\perp$ denote the *annihilator* of Y. Following [218], p. 65, we have

LEMMA 1.2.16 Let $\langle X,Y\rangle$ be a dual system. Let A (resp. L) be a balanced convex subset (resp. subspace) of X. Then $(A+L)^0 = A^0 \cap L^\perp$; if A and L are also $\sigma(X,Y)$-closed then $(A \cap L)^0 = \overline{(A^0 + L^\perp)}$.

The proof of the next Auerbach lemma is given in [178], p. 136.

LEMMA 1.2.17 Let X and Y be normed spaces with closed unit balls u and v respectively. Then for any finite dimensional continuous linear operator $R : X \to Y$ with $\dim R[X] = n$, we have

$$Rx = \sum_{i=1}^{n} \alpha_i \, f^i(x) y^i,$$

where $f^i \in u^o$, $y^i \in v$ and $|\alpha_i| \leq \|R\|$ for $1 \leq i \leq n$.

For arbitrary l.c. TVS (X,T) and (Y,S), we will also need

DEFINITION 1.2.18 A linear operator $R : (X,T) \to (Y,S)$ is called *roughly bounded* if R transforms bounded subsets of X into bounded subsets of Y.

PROPOSITION 1.2.19 Let (X,T) be bornological and (Y,S) an arbitrary l.c. TVS, then each roughly bounded linear map $R : (X,T) \to (Y,S)$ is continuous.

DEFINITION 1.2.20 An l.c. TVS (X,T) is called *stable* if $X \times X \simeq X$.

Next, we reproduce from [43], the following

LEMMA 1.2.21 Let $\beta^{-1} : \mathbb{N} \times \mathbb{N} \to \mathbb{N}$ be a bijection and $\{\pi_k\}$ be a sequence of injections; $\pi_k : \mathbb{N} \to \mathbb{N}$ such that for each n in $\mathbb{N}$, there exist k, m in $\mathbb{N}$ such that $\pi_k(m) = n$. Then there exist an injection $\gamma: \mathbb{N} \to \mathbb{N} \times \mathbb{N}$ and π in P such that (i) $\beta^{-1}\gamma : \mathbb{N} \to \mathbb{N}$ is strictly increasing; (ii) $\beta^{-1} \gamma(n) \geq n$, $n \geq 1$; (iii) if $\pi_{\gamma_1(n)} (\gamma_2(n)) = \pi_k(m)$, then $\beta^{-1} \gamma(n) \leq \beta^{-1}(k,m)$ $[\gamma(n) \equiv (\gamma_1(n),\gamma_2(n))]$; and (iv) $\pi(n) = \pi_{\gamma_1(n)}(\gamma_2(n))$, $n \geq 1$.

1.3 SEQUENCE SPACES

To appreciate the limit and scope of Schauder bases in l.c. TVS, we have to rely heavily on the theory of sequence spaces and our major reference in this direction is [129] (cf. also [27] and [143]). We will only recall hereafter certain relevant and useful results and pay specific attention to the terminological aspect of this theory.

Whenever we use the symbol ω as a space, it will, henceforth, be understood as the sequence space of all $\mathbb{K}$-valued sequences $x \equiv \{x_n\}$: $x_n \in \mathbb{K}$, for $n \geq 1$. Usual coordinatewise addition and scalar multiplication force ω to be a VS over $\mathbb{K}$. Let $e^n = \{0,\ldots,0,1,0,\ldots,\}$, 1 being placed at the n-th coordinate, $n \geq 1$. Also, let $e = \{1,1,\ldots\}$. e^n is called the n-th *unit vector* whereas e is called the *unity* of ω. Put $\phi = \mathrm{sp}\,\{e^n : n \geq 1\}$. Any subspace λ of ω with $\phi \subset \lambda$ is called a *sequence space* (abbreviated hereafter as s.s.). For ready reference, we reproduce as Table 1.3.3, a table containing several frequently-used sequence spaces, their natural norms or F-norms and their topological, α- and β-duals. Given an s.s. λ , its α-*dual* (= Köthe dual) and β-*dual* are defined as:

$$\lambda^\beta = \{y \in \omega: q_y(x) \equiv |\sum_{n\geq 1} x_n y_n| < \infty, \forall x \in \lambda\} \ ; \qquad (1.3.1)$$

$$\lambda^\alpha = \{y \in \omega: p_y(x) \equiv \sum_{n\geq 1} |x_n y_n| < \infty, \forall x \in \lambda\} \ . \qquad (1.3.2)$$

We will, however, prefer to write $\lambda^\times$ for λ^α .

TABLE 1.3.3

No.	Sequence Space λ; its Natural Norm/ F-Norm	λ^*	$\lambda^\times$	λ^β
1	ϕ, $\|x\|_\infty = \sup \lvert x_n\rvert$	ℓ^1	ω	ω
2	ω, $\lvert x\rvert_\omega = \sum_{n\geq 1} \frac{1}{2^n} \frac{\lvert x_n\rvert}{1+\lvert x_n\rvert}$	ϕ	ϕ	ϕ
3	$c_o = \{x \in \omega: x_n \to 0\}$, $\|x\|_\infty = \sup \lvert x_n\rvert$	ℓ^1	ℓ^1	ℓ^1
4	$\ell^p = \{x \in \omega: \|x\|_p \equiv [\sum_{n\geq 1} \lvert x_n\rvert^p]^{1/p}, 1\leq p<\infty\}$	ℓ^q	ℓ^q	ℓ^q
		$(p^{-1}+q^{-1}=1)$		
5	$\ell^p = \{x \in \omega: \lvert x\rvert_p \equiv \sum_{n\geq 1} \lvert x_n\rvert^p < \infty, 0<p<1\}$	ℓ^∞	ℓ^∞	ℓ^∞
6	$\ell^\infty = \{x \in \omega: \|x\|_\infty \equiv \sup \lvert x_n\rvert < \infty\}$	$ba(\mathbb{N}, \Phi_\infty)$	ℓ^1	ℓ^1
7	$m_o = \{x \in \omega$: the set $\{x_n\}$ is finite$\}$, $\|x\|_\infty = \sup \lvert x_n\rvert$	$ba(\mathbb{N}, \Phi_\infty)$	ℓ^1	ℓ^1
8	$k = \{x \in \omega: x_n$ is eventually constant$\}$, $\|x\|_\infty$	ℓ^1	ℓ^1	cs
9	$\delta = \{x \in \omega: \lvert x_n\rvert^{1/n} \to 0\}$, $\lvert x\rvert_\delta = \sup \lvert x_n\rvert^{1/n}$	d	d	d
10	$d = \{x \in \omega: \limsup \lvert x_n\rvert^{1/n} < \infty\}$		δ	δ
11	$cs = \{x \in \omega: \{\sum_{i=1}^n x_i\}$ converges$\}$, $\|x\|_{cs} = \sup \lvert \sum_{i=1}^n x_i\rvert$	bv	ℓ^1	bv
12	$bs = \{x \in \omega: \{\sum_{i=1}^n x_i\}$ is bounded$\}$, $\|x\|_{bs} = \sup \lvert \sum_{i=1}^n x_i\rvert$	$ba(\mathbb{N}, \Phi_\infty)$	ℓ^1	bv_o
13	$bv = \{x \in \omega: \{\sum_{i=1}^n \lvert x_{i+1}-x_i\rvert\}$ converges$\}$, $\|x\|_{bv} = \sum_{n\geq 1} \lvert x_{n+1}-x_n\rvert + \lim_{n\to\infty} \lvert x_n\rvert$	$bs \oplus \mathbb{K}$	ℓ^1	cs
14	$bv_o = \{x \in bv : x_n \to 0\}$, $\|x\|_{bv_o} = \sum_{n\geq 1} \lvert x_{n+1}-x_n\rvert$	bs	ℓ^1	bs

NOTE. In Table 1.3.3, $ba(\mathbb{N},\Phi_\infty)$ denotes the collection of all charges ($\mathbb{K}$- valued finitely additive set fuctions F on the ring Φ_∞ of all subsets of $\mathbb{N}$ with $|F[A]| < \infty$ for every A in Φ_∞).

Given an s.s. λ, if μ is a subspace of λ^β with $\phi \subset \mu$, then $<\lambda,\mu>$ forms a dual system under the bilinear form $(x,y) \to \sum_{n\geq 1} x_n y_n$, where $x \in \lambda$ and $y \in \mu$. Thus, we have all familiar polar topologies on λ as well as on μ. Besides, there is another natural polar topology $\eta(\lambda,\mu)$ on λ, called *normal topology*, when $\mu \subset \lambda^\times$. This is generated by $\{p_y : y \in \mu\}$.

Let us call an x in an s.s. λ *positive* (*strictly positive*) and write it as $x \geq 0$ ($x > 0$) provided $x_n \geq 0$ ($x_n > 0$) for $n \geq 1$; further, let $\lambda_+ = \{x \in \lambda: x \geq 0\}$. Clearly $\eta(\lambda,\mu)$ is also generated by $\{p_y : y \in \mu_+\}$.

DEFINITION 1.3.4 An s.s. λ is called (i) *monotone* if $m_0\lambda \subset \lambda$; *normal* if $\ell^\infty\lambda \subset \lambda$; (iii) *perfect* if $\lambda = \lambda^{\times\times}$; and (iv) *symmetric* if $x_\sigma \equiv \{x_{\sigma(n)}\} \in \lambda$ for each x in λ and σ in P.

PROPOSITION 1.3.5 For a monotone s.s., $\lambda^\times = \lambda^\beta$.

Given an s.s. λ and $J \in I$, let

$$\lambda_J = \{\{x_i\} : \exists\{y_i\} \in \lambda \ni x_i = y_{n_i}, \quad \forall n_i \in J\}$$

and call λ_J, the J-*stepspace* of λ and let x_J denote an arbitrary element of λ_J. The *canonical preimage* of an x_J is the sequence $\bar{x}_J$ such that the latter agrees with the former on indices in J and is zero elsewhere. The s.s. $\bar{\lambda}_J = \{\bar{x}_J : x_J \in \lambda_J\}$ is called the canonical preimage of λ_J.

For x in ω, the *close-up* of x is defined to be a sequence x' where

$$x'_i = \begin{cases} \text{the i-th nonzero term of x, if such a term exists;} \\ 0, \text{ otherwise.} \end{cases}$$

Frequently used results

Following [129], p.82-83, 136 and [143], p. 409, 413, we have the following two results where μ is a normal subspace of $\lambda^\times$

PROPOSITION 1.3.6 For an s.s., $\sigma(\lambda,\mu) \subset \eta(\lambda,\mu) \subset \tau(\lambda,\mu)$.

THEOREM 1.3.7 An s.s. λ is $\sigma(\lambda,\mu)$-ω-complete (resp. $\eta(\lambda,\mu)$-complete) if and only if $\lambda = \mu^\times$.

Also, we have (cf. [129], p. 77)

PROPOSITION 1.3.8 The sequential convergences in λ relative to $\sigma(\lambda,\mu)$ and $\eta(\lambda,\mu)$ are the same, where μ is a normal subspace of $\lambda^\times$.

For an x in an arbitrary s.s. λ, let $x^{(n)} = \{x_1,\ldots,x_n, 0,0,\ldots\}$; $x^{(n)}$ is called the n-th *section* of x. Let λ be an s.s., μ a subspace of λ^β and x an arbitrary element of λ, then

$$x^{(n)} \to x \quad \text{in } \sigma(\lambda,\mu), \tag{1.3.9}$$

and if $\mu \subset \lambda^\times$, then we also have

$$x^{(n)} \to x \text{ in } \eta(\lambda,\mu) \tag{1.3.10}$$

The following extended form of Köthe's fundamental theorem [143] is extremely useful (cf. also [129], p. 189).

THEOREM 1.3.11 If λ is monotone, then $x^{(n)} \to x$ in $\tau(\lambda,\lambda^\times)$.

DEFINITION 1.3.12 An s.s. λ which is also a TVS or an l.c. TVS with topology T will be referred to as an s.s. TVS or an s.s. l.c. TVS. An s.s. TVS (λ,T) is called a K-*space* if each $p_i : \lambda \to \mathbb{K}$, $p_i(x) = x_i$ is continuous. A K-space (λ,T) is called an AK-*space* if $x^{(n)} \to x$ in T for each x in λ.

We have ([199], [129] (p. 59-60), [57])

THEOREM 1.3.13 Let (λ,T) be an AK-s.s. l.c. TVS. Then λ^* can be identified as $\lambda_s = \{\{f(e^n)\} : f \in \lambda^*\}$ and $\lambda^* \subset \lambda^\beta$; if (λ,T) is also barrelled, then $\lambda^* = \lambda^\beta$ and additionally, if λ is monotone as well, then $\lambda^* = \lambda^\times$.

Köthe spaces

A subset P of ω is called a *Köthe set* provided (i) $a \geq 0$ for each a in P, (ii) for each n, there exists a in P with $a_n > 0$ and (iii) for each a,b in P there exists c in P with $a,b \leq c$.

Throughout this section, we let $\Psi: [0,\infty) \to [0,\infty]$ such that $\Psi(x+y) \leq \Psi(x) + \Psi(y)$, $\Psi(x) < \Psi(y)$ for $x < y$ and $\Psi(0) = 0$. Let P be a Köthe set and define

$$\Lambda(p;\Psi) = \{x \in \omega: p_{a,\Psi}(x) \equiv \sum_{n\geq 1} \Psi(|x_n|)a_n < \infty, \forall a \in P\};$$

$\Lambda(P;\Psi)$, equipped with the l.c. topology $T_{P,\Psi}$ generated by $\{p_{a,\Psi} : a \in P\}$, is called a *Köthe space*. If $\Psi = I$, the identity map of $[0,\infty)$, then we write $\Lambda(P;I)$ as $\Lambda(P)$, $p_{a,I}$ as p_a and $T_{P,\Psi}$ as T_P.

PROPOSITION 1.3.14 $(\Lambda(P),T_p)$ is complete.

DEFINITION 1.3.15 A Köthe set P satisfying (i) each a in P satisfies $a > 0$ and nondecreasing (resp. nonincreasing) and (ii) for each a in P, there is b in P with $a_n^2 \leq b_n$ (resp. $a_n \leq b_n^2$) for $n \geq 1$, is called a Köthe set of *infinite* (resp. *finite*) *type* and the corresponding Köthe space $\Lambda(P)$ is called a smooth *sequence space* (s.s) of infinite (resp. *finite*) *type*. These spaces are called respectively G_∞ (resp. G_1)-*spaces* and are written as $\Lambda_\infty(P)$ (resp. $\Lambda_1(P)$).

In the special case, when $\alpha \in \omega$ is chosen so as to satisfy $0 \leq \alpha_1 \leq \ldots \leq \alpha_n \to \infty$ with n, and P is taken to be

$\{\{k^{\alpha_n}\} : 0 < R < \infty\}$ (resp. $\{\{R^{\alpha_n}\} : 0 < R < 1\}$), the G_∞-space $\Lambda_\infty(P)$ (resp. the G_1-space $\Lambda_1(P)$) is generally denoted by $\Lambda_\infty(\alpha)$ (resp. $\Lambda_1(\alpha)$) and is called *∞-power series space (power series space of infinite type)* (resp. *1-power series space (power series space of finite type)*). The topology for $\Lambda_\infty(\alpha)$ (resp. $\Lambda_1(\alpha)$) will be denoted by T_∞ (resp. T_1).

The spaces $\Lambda_1(\alpha)$ and $\Lambda_\infty(\alpha)$ are related by

PROPOSITION 1.3.16 We have (i) $\Lambda_\infty(\alpha) \,.\, \Lambda_\infty^\times(\alpha) = \Lambda_\infty(\alpha)$ and (ii) $\Lambda_1(\alpha) \,.\, \Lambda_1^\times(\alpha) = \Lambda_1^\times(\alpha)$.

NOTE. Observe that

$$\Lambda_1(\alpha) = \{x \in \omega: p_k(x) \equiv \sum_{n\geqq 1} |x_n| \left(\frac{k}{k+1}\right)^{\alpha_n} < \infty;\ k = 1,2,\ldots\};$$

$$\Lambda_\infty(\alpha) = \{x \in \omega: p_k(x) \equiv \sum_{n\geqq 1} |x_n| k^{\alpha_n} < \infty : k = 1,2,\ldots\}.$$

Concerning the stability of $\Lambda(P)$, we have ([223], p. 186).

PROPOSITION 1.3.17 A nuclear G_∞-space $(\Lambda(P),T_p)$ is stable if and only if for each a in P, there exists b in P so that $\{a_{2n}/b_n\} \in \ell^\infty$.

The following notion is essentially due to Köthe [142], [145], cf. also [86], [110].

DEFINITION 1.3.18 Let (λ,T) be an s.s. l.c. TVS with $\sigma(\lambda,\lambda^\times) \subset T \subset \tau(\lambda,\lambda^\times)$. Then λ is called *simple* if for each bounded subset A of λ there exists y in λ such that $|x_n| \leqq |y_n|$ for all x in A and $n \geqq 1$.

1.4 NUCLEAR SPACES

A map R in $\mathcal{L}(X,Y)$ [cf. section 1.2] is called *nuclear* provided there exists an equicontinuous sequence $\{f_n\}$ in X^*, a bounded sequence $\{y_n\}$ in Y and α in ℓ^1 such that

$$Rx = \sum_{n\geq 1} \alpha_n f_n(x)\, y_n, \quad \forall x \in X.$$

An l.c. TVS (X,T) is called *nuclear* if for each u in $\mathcal{B}_X$ there exists v in $\mathcal{B}_X$ with $v < u$ such that $\hat{K}_u^v$ is nuclear.

NOTE. A closely related concept is that of quasi-nuclearity. If X and Y are normed spaces, then R in $\mathcal{L}(X,Y)$ is *quasi-nuclear* provided there exists α in ℓ^1 and an equicontinuous sequence $\{f_n\}$ in X^* such that $\|Rx\| \leq \sum_{n\geq 1} |\alpha_n f_n(x)|$ for each x in X. An l.c. TVS (X,T) is *quasi-nuclear* if for each u in $\mathcal{B}_X$ there exists v in $\mathcal{B}_X$ with $v < u$ such that K_u^v is quasi-nuclear; this in turn implies the existence of α in ℓ^1 and an equicontinuous sequence $\{\hat{f}_n\}$ in $(X_v)^*$ so that

$$(*) \quad \hat{p}_u(K_u^v(x_v)) \leq \sum_{n\geq 1} |\alpha_n|\, |\hat{f}_n(x_v)|.$$

Define f_n in X^* $(n \geq 1)$ by $f_n(x) = \hat{f}_n(x_v)$, then $|f_n(x)| \leq p_v(x)$ for $n \geq 1$ and x in X and let

$$(x,y)_u = \sum_{n\geq 1} |\alpha_n|\, f_n(x)\, \overline{f_n(y)} \;; \quad \forall x,y \in X.$$

Clearly $(x,y)_u$ defines an inner product on $X \times X$. If $\|x\|_u^2 = (x,x)_u$, then $\|x\|_u \leq \sqrt{A}\, p_v(x)$, where $A = \sum_{n\geq 1} |\alpha_n|$ and depends upon u. Further, from (*)

$$p_u(x) = \hat{p}_u(K_u^v(x_v)) \leq [\sum_{n\geq 1} |\alpha_n|]^{1/2} [\sum_{n\geq 1} |\alpha_n|\,|f_n(x)|^2]^{1/2}$$

$$= \sqrt{A}\;\|x\|_u.$$

Therefore

$$p_u(x) \leq \sqrt{A}\,\|x\|_u \leq A\, p_v(x), \quad \forall x \in X.$$

Thus, the topology T is equivalent to the l.c. topology generated by Hilbertian seminorms $\|\cdot\|_u$, $u \in \mathcal{B}_X$ and since each nuclear map is quasi-nuclear, the preceding conclusion also holds if (X,T) is nuclear. We may now sum up the discussion in the form of (cf. also [164])

PROPOSITION 1.4.1 Let (X,T) be an l.c. TVS such that it is quasi-nuclear. Then $\mathcal{B}_X$ may be replaced by $\mathcal{B}_X^H$ consisting of all balanced, convex and closed neighbourhoods u at the origin such that p_u is a Hilbertian seminorm; in particular, under the hypothesis, each $\hat{X}_u$ is a Hilbert space.

A map R in $\mathcal{L}(X,Y)$ is called *bounded* (resp. *precompact*, *compact*) if there exists u in $\mathcal{B}_X$ such that R[u] is bounded (resp. precompact, relatively compact) in Y and the class of all such operators will be denoted by $\mathcal{L}_b(X,Y)$ [resp. $\mathcal{L}_p(X,Y)$, $\mathcal{L}_c(X,Y)$]. Clearly, $\mathcal{L}_c(X,Y) \subset \mathcal{L}_p(X,Y) \subset \mathcal{L}_b(X,Y) \subset \mathcal{L}(X,Y)$. An l.c. TVS (X,T) is called *Schwartz* provided that for every u in $\mathcal{B}_X$ there exists v in $\mathcal{B}_X$ such that $\hat{K}_u^v$ is precompact.

To begin with, we quote from [178] the following characterization of nuclear spaces.

THEOREM 1.4.2 An l.c. TVS (X,T) is nuclear if and only if for each u in $\mathcal{B}_X$, there exists α in ℓ^1 and an equicontinuous sequence $\{f_n\}$ in X* such that

$$p_u(x) \leq \sum_{n \geq 1} |\alpha_n f_n(x)|, \quad \forall x \in X. \tag{1.4.3}$$

More interesting characterizations of nuclear and also those of Schwartz spaces are obtained with the help of Kolmogorov's diameters and related notions. Our basic references are [178] and [218].

For subsets A,B in a VS X with $A < B$ and a subspace L of X, let

$$\delta(A,B;L) = \inf\{\delta \geqq 0 : A \subset \delta B+L\}$$

and

$$\delta_n \equiv \delta_n(A,B) = \inf\{\delta(A,B;L) : \dim L < n\}.$$

Then δ_n is called the n-th *Kolmogorov diameter*, where $n = 1,2,\ldots$.

Some simple properties on Kolmogorov's diameters are collected in the following (cf. also [218])

PROPOSITION 1.4.4 Let A,B,C be subsets of a VS E, and R a linear map from E into another VS F. If $A \prec B \prec C$ and $\alpha,\beta > 0$, then (i) $\{\delta_n(A,B)\}$ is decreasing, (ii) $\alpha\, \delta_n(A,B) = \beta\, \delta_n(\alpha A,\beta B)$ and (iii) $\delta_{m+n}(A,C) \leqq \delta_m(A,B)\, \delta_n(B,C)$. If $A \prec B \subset C$, then

(iv) $\delta_n(R[A], R[B]) \leqq \delta_n(A,B)$;

(v) $\delta_n(A,C) \leqq \delta_n(A,B)$.

If $A \subset B \prec C$, then

(vi) $\delta_n(A,C) \leqq \delta_n(B,C)$

If X and Y are normed spaces and $R \in \mathcal{L}(X,Y)$, then $\delta_n(R)$ is defined by

$$\delta_n(R) = \delta_n(R[u],v), \tag{1.4.5}$$

where u and v are closed unit balls in X and Y respectively. After [218], p. 69, we have

PROPOSITION 1.4.6 Let (X,T) be an l.c. TVS and $u,v \in \mathcal{B}_X$ with $v \prec u$. Then

$$\delta_n(v,u) = \delta_n(K_u[v],\, K_u[u]) \equiv \delta_n(v(u)) \equiv \delta_n(K_u[v]).$$

The following well known result of Tikhomirov [224] is stated below, cf. [218], p. 58

PROPOSITION 1.4.7 Let L be a subspace of a normed space X with dim L = n. Then for every bounded subset A of X and $\varepsilon \geqq 0$

$$\varepsilon(u \cap L) \subset A \Rightarrow \delta_n(A) \equiv \delta_n(A,u) \geqq \varepsilon,$$

where u is the closed unit ball of X.

PROPOSITION 1.4.8 A bounded subset B of a normed space X is precompact if and only if $\delta_n(B;u) \to 0$.

For the above result, see [178], p. 146.

DEFINITION 1.4.9 The *diametral approximative dimension* $\Delta(X)$ of an l.c. TVS (X,T) is the collection of all sequences α in ω such that whenever $u \in \mathcal{B}_X$, there exists v in $\mathcal{B}_X$ with $v < u$ so that $\{\alpha_n \delta_n(v,u)\} \in c_o$. The *inverse diametral dimension* $\Delta^*(X)$ of an l.c. TVS (X,T) is the collection of sequneces $\alpha \in \omega$ satisfying the relation: there exists u in $\mathcal{B}_X$ such that for each v in $\mathcal{B}_X$ with $v < u$, one has $\{\alpha_n / \delta_n(v,u)\} \in c_o$.

After [209], p. 14 and [218], p. 75, one has

THEOREM 1.4.10 Let (X,T) be an l.c. TVS. Then the following statements are equivalent: (i) (X,T) is nuclear, (ii) there exists $\alpha > 0$ so that whenever there is u in $\mathcal{B}_X$, one finds v in $\mathcal{B}_X$ with $v < u$ and $\{\delta_n^\alpha(v,u)\} \in \ell^1$, (iii) for every $\beta > 0$ and u in $\mathcal{B}_X$, there exists v in $\mathcal{B}_X$ with $v < u$ and $\{\delta_n^\beta(v,u)\} \in \ell^1$, (iv) $\{n^\alpha\} \in \Delta(X)$ for some $\alpha > 0$ and (v) $\{n^\beta\} \in \Delta(X)$ for every $\beta > 0$.

Concerning Schwartz spaces, we quote the next two results from [219] and [52] (p. 20), [218], p. 73 respectively.

THEOREM 1.4.11 An l.c. TVS (X,T) is Schwartz if and only

if for every p in D_T, there exist q in D_T and $\{f_n\} \subset X^*$ so that $\sup\{|f_n(x)| : q(x) \leq 1\} \to 0$ and $p(x) \leq \sup\{|f_n(x)| : n \geq 1\}$ for each x in X.

THEOREM 1.4.12 For an l.c. TVS (X,T), the following statements are equivalent: (i) (X,T) is Schwartz, (ii) for every u in $\mathcal{B}_X$, there exists v in $\mathcal{B}_X$ with $v < u$ and $\{\delta_n(v,u)\} \in c_o$, (iii) $c_o \neq \Delta(X)$ and (iv) $\ell^\infty \subset \Delta(X)$.

The extensions of the next two results will be obtained in subsequent chapters. We have ([129] and [111]; cf. also [178])

THEOREM 1.4.13 An s.s. l.c. TVS $(\lambda, \eta(\lambda,\lambda^\times))$ is nuclear if and only if for each α in $\lambda^\times$ there exists β in $\lambda^\times$ so that $\alpha/\beta \in \ell^1$ $(0/0 = 0)$.

THEOREM 1.4.14 The Köthe space $(\Lambda(P),T_p)$ is nuclear if and only if for each a in P there exists b in P with $a/b \in \ell^1$, where $0/0 = 0$.

The preceding theorem is termed the *Grothendieck-Pietsch criterion* for the nuclearity of $\Lambda(P)$.

In particular, one derives

PROPOSITION 1.4.15 The space $(\Lambda_\infty(\alpha),T_\infty)$ (resp. $(\Lambda_1(\alpha),T_1)$) is nuclear if and only if there exists $R > 1$ (resp. for each R, $0 < R < 1$) we have

$$\sum_{n\geq 1} R^{-\alpha_n} < \infty.$$

After [142], p. 270, we have

LEMMA 1.4.16 Let an s.s.$\lambda \equiv (\lambda,\eta(\lambda,\lambda^\times))$ be perfect and nuclear. Then λ is simple.

After [43], p. 95, we recall

LEMMA 1.4.17 Let P be a Köthe set such that $\inf |a_n| > 0$ for some a in P. Then $\{|\alpha_n|^t\} \in \Lambda(P)$ for any $t > 0$ and α in $\Lambda(P)$. In addition, if P is a Köthe set of infinite type and $(\Lambda(P), T_p)$ is nuclear, then $\alpha \in \Lambda(P)$ provided $\{|\alpha_n|^t\} \in \Lambda(P)$ for any $t > 0$.

Approximation numbers related to operators

Let us recall the concept of the n-th *approximation number* $\alpha_n(R)$ of $R \in \mathcal{L}(X,Y)$ where X and Y are normed spaces:

$$\alpha_n(R) = \inf \{\|R-A\| : \dim A[X] < n\}$$

DEFINITION 1.4.18 For R in $\mathcal{L}(X,Y)$, if $\{\alpha_n(R)\} \in \lambda$, then R is said to be of λ-*type*, λ being an arbitrary s.s.

For future reference, let us recall the following results, see [178]; cf. also [82] especially for Proposition 1.4.19 (i) and (ii).

PROPOSITION 1.4.19 Let X and Y be normed spaces and $R \in \mathcal{L}(X,Y)$. Then

(i) $\delta_n(R) \leqq \alpha_n(R) \leqq n^{1/2}\, \delta_n(R)$.

If X and Y are Banach spaces and R is compact, then

(ii) $\alpha_n(R) = \alpha_n(R^*)$;

and if in addition, X and Y are Hilbert spaces, then

(iii) $\alpha_n(R) = \delta_n(R)$.

PROPOSITION 1.4.20 If X and Y are normed spaces, then each R in $\mathcal{L}(X,Y)$ of ℓ^1-type is nuclear.

THEOREM 1.4.21 Let X and Y be Hilbert spaces and

$R \in \mathcal{L}_c(X,Y)$. Then there exist orthonormal sequences $\{e_n\}$ in X, $\{h_n\}$ in Y and $\{\lambda_n\} \in c_o$ ($\lambda_n > 0$) such that

$$Rx = \sum_{n\geq 1} \lambda_n \langle x, e_n \rangle h_n, \quad \forall x \in X \tag{1.4.22}$$

and $\alpha_n(R) = \lambda_n$, $n \geq 1$.

1.5 INFINITE SERIES

We sum up hereafter relevant results on infinite series. Throughout, we confine ourselves to an arbitrary l.c. TVS (X,T) containing a formal series $\sum_{n\geq 1} x_n$; $x_n \in X$, $n \geq 1$ and follow [129] for the rest of this section.

DEFINITION 1.5.1 $\sum_{n\geq 1} x_n$ is said to be (i) *absolutely* or *pre-absolutely* convergent if $\sum_{n\geq 1} p(x_n) < \infty$ for each p in D_T, (ii) *bounded multiplier convergent* (b.m.c.) if $\sum_{n\geq 1} \alpha_n x_n$ is convergent for each α in ℓ^∞, (iii) *subseries convergent* (s.c.) if $\sum_{n\geq 1} \alpha_n x_n$ converges for each α in m_o, (iv) *unconditionally convergent* (u.c.) if $\sum_{n\geq 1} x_{\sigma(n)}$ converges for each σ in P to the same point in X, and (v) *unordered convergent* (ud.c.) if there exists x in X with

$$x = T - \lim_{\sigma \in \Phi} \sum_{i \in \sigma} x_i .$$

THEOREM 1.5.2 Let there be a formal infinite series in an ω-complete l.c. TVS (X,T). Then the notions (ii) to (v) of Definition 1.5.1 are all equivalent. Further, the series $\sum_{n\geq 1} x_n$ is u.c. if and only if

$$\sum_{n\geq 1} |f(x_n)| < \infty$$

uniformly in f belonging to each equicontinuous subset M of X*.

DEFINITION 1.5.3 A series $\sum_{n\geq 1} x_n$ in an l.c. TVS (X,T) is *unordered bounded* (ud. b.) (resp. *unordered Cauchy* (ud.C.)) if $\{S_\sigma : \sigma \in \Phi\}$ is bounded (resp. Cauchy) in (X,T), where $S_\sigma = \sum_{i\in\sigma} x_i$.

THEOREM 1.5.4 Let $\sum_{n\geq 1} x_n$ be a formal series in an l.c. TVS (X,T). Then this series is T-s.c. if and only if it is $\sigma(X,X^*)$-s.c. If (X,T) is also $\sigma(X,X^*)$-ω-complete, then it is $\sigma(X,X^*)$-u.c. if and only if it is T-u.c. if and only if it is T- ud.b.

THEOREM 1.5.5 Let T be a polar topology with respect to a dual pair <X,Y> of VS such that (X,T) is separable. If $\sum_{n\geq 1} y_n$ is $\sigma(X,Y)$ s.c. in X, then it is T-s.c.

PROPOSITION 1.5.6 $\sum_{n\geq 1} x_n$ in an l.c. TVS (X,T) is ud.b. if and only if for each equicontinuous subset M of X*, there exists $k_M > 0$ such that $\sum_{n\geq 1} |f(x_n)| \leq k_M$ for each $f \in M$ if and only if $\sum_{n\geq 1} |f(x_n)| < \infty$ for each f in X*.

NOTE. The preceding results of this section can be traced in [129] and are originally to be found in [162], [231] and [240].

We also need a variation of a result of [159]; cf. also [129], Proposition 3.8.4.

PROPOSITION 1.5.7 Let $\sum_{n\geq 1} x_n$ be ud.c. in an l.c. TVS (X,T). Then for each bounded subset B of ℓ^∞, the set

$$S(B;\Phi) = \{ \sum_{i\in\sigma} b_i x_i : b \in B, \sigma \in \Phi\}$$

is precompact in (X,T).

1.6 SUMMING OPERATORS

There are several inequalities of far reaching consequence for sequences in Banach spaces and they have all resulted from the discussion of p-absolutely summing operators. As such, let us begin with (cf. [153], p. 284)

DEFINITION 1.6.1 Let X,Y be two Banach spaces, $A \in \mathcal{L}(X,Y)$ and $1 \leqq p < \infty$. Then A is called *p-absolutely summing* if there exists a constant $K > 0$ such that for any choice of finite set $\{x_1,\ldots,x_n\}$ in X

$$\left[\sum_{i=1}^{n} \|Ax_i\|^p \right]^{1/p} \leqq K \sup \left\{ \left[\sum_{i=1}^{n} |f(x_i)|^p \right]^{1/p} : f \in X^*, \|f\| \leqq 1 \right\}.$$

NOTE. It is known that A is *p-absolutely* summing if and only if $\{\|Ax_n\|\} \in \ell^p$ whenever for any sequence $\{x_n\}$ in X, $\{f(x_n)\} \in \ell^p$ uniformly in f with $\|f\| \leqq 1$ ([238], p. 59). In particular, if $p = 1$, then A is 1-absolutely summing if and only if A converts each unconditionally convergent series in X into an absolutely convergent series in Y. Some useful examples of these operators are mentioned below in the form of

PROPOSITION 1.6.2 Each A in $\mathcal{L}(\ell^1,\ell^2)$ is 1-absolutely summing whereas each A in $\mathcal{L}(c_0,\ell^p)$ with $1 \leqq p \leqq 2$ is 2-absolutely summing.

The following classical theorem of Orlicz (cf. [172], p. 44) has been a source of motivation for the development of p-absolutely summing operators and allied topics, e.g. cotypes of spaces (cf. [238]).

THEOREM 1.6.3 If $\sum_{n\geq 1} f_n$ converges unconditionally in $(L^p, \|\cdot\|_p)$; $L^p = L^p[0,1]$, $1 \leq p < \infty$, then

$$\sum_{n\geq 1} \|f_n\|_p^2 < \infty, \ 1 \leq p \leq 2; \quad \sum_{n\geq 1} \|f_n\|_p^p < \infty, \ 2 \leq p < \infty.$$

This result is obviously true for ℓ^p and in particular, for $p = 1$, if $\sum_{n\geq 1} x^n$ converges unconditionally in $(\ell^1, \|\cdot\|_1)$, the operator $S : c_o \to \ell^1$, $S(\alpha) = \sum_{n\geq 1} \alpha_n x^n$ is bounded and so it is 2-absolutely summing (cf. Proposition 1.6.2). Thus we easily deduce the following result (cf. also [153], p. 295) for later use.

THEOREM 1.6.4 If $\sum_{n\geq 1} x^n$ converges unconditionally in $(\ell^1, \|\cdot\|_1)$, then there exists a universal constant K_G such that

$$[\sum_{n\geq 1} \|x^n\|_1^2]^{1/2} \leq K_G \sup_{n\geq 1, |\varepsilon_i|=1} \|\sum_{i=1}^{n} \varepsilon_i x^i\|_1$$

$$\leq K_G \sup_{n\geq 1, |\varepsilon_i|\leq 1} \|\sum_{i=1}^{n} \varepsilon_i x^i\|_1.$$

We also have (cf. [238], p. 118)

THEOREM 1.6.5 Given L^p, $1 \leq p \leq 2$, there exists $M_p > 0$ so that for any finite sequence $\{f_i : 1 \leq i \leq n\}$ in L^p,

$$[\sum_{i=1}^{n} \|f_i\|_p^2]^{1/2} \leq M_p [\int_0^1 \|\sum_{i=1}^{n} r_i(t) f_i\|_p^2 \, dt]^{1/2}$$

$$\leq M_p \sup_{\varepsilon_i = \pm 1} \|\sum_{i=1}^{n} \varepsilon_i f_i\|_p,$$

where $\{r_n\}$ denotes the well known sequence of Rademacher functions in $[0,1]$.

PART II

2 Types of bases – an introduction

2.1 BACKGROUND AND PURPOSE

In our earlier work [130], we had occasion to discover a few applications of basis theory, especially those relating to the structural study of topological vector spaces and characterization of continuous mappings and compact subsets of these spaces. There are still several applications in this direction and to dig out them, we first need to examine carefully the behaviour of several ingredients which constitute a base; indeed, a Schauder base.

The applications of basis theory, which we plan to present in this book, will be revealed in subsequent chapters, for example, establishing various topological structures, characterizing the nuclear and Schwartz properties and finding precise forms of continuous linear functionals on locally convex spaces.

In order to study the so-called behaviour of the ingredients forming a Schauder base, let us recall the rudiments and notation from basis theory, which we will follow hereafter in this book without making reference.

Throughout this work we intend (X,T) (unless we say to the contrary) to mean an arbitrary Hausdorff locally convex topological vector space, or, merely an Hausdorff locally convex space (l.c. TVS) over the field $\mathbb{K}$ of complex numbers $\mathbb{C}$ or real numbers $\mathbb{R}$, T being the corresponding Hausdorff locally convex (l.c.) topology on the vector space $X \supsetneq \{0\}$. The notations D_T and $\mathcal{B}_T$ stand as before, as discussed in chapter 1.

A sequence $\{x_n\}$ in (X,T) is said to be a *topological*

base (t.b.) if for each x in X, there exists a unique sequence $\{\alpha_n\}$ from $\mathbb{K}$ such that

$$x = T - \lim_{n\to\infty} \sum_{i=1}^{n} \alpha_i x_i \equiv \sum_{n\geq 1} \alpha_n x_n .$$

The sequence $\{\alpha_n\}$ in the above expansion depends on the particular choice of x and consequently we write $\alpha_n = f_n(x)$, $n \geq 1$. Observe that $f_n \in X'$ for each $n \geq 1$ and $f_n(x_m) = \delta_{mn}$, the Kronecker delta. The unique sequence $\{f_n\}$ related to the t.b. $\{x_n\}$ is often called the *sequence of associated coordinate functionals* (s.a.c.f.) corresponding to the base $\{x_n\}$ and we will frequently and interchangeably write a t.b. $\{x_n\}$ as $\{x_n;f_n\}$ provided we want to stress the s.a.c.f. $\{f_n\} \subset X'$ corresponding to a given $\{x_n\} \subset X \equiv (X,T)$. More important to us is the case when $\{f_n\} \subset X^*$; for, there are examples of bases $\{x_n\}$ for which $\{f_n\} \subset X^*$ and $\{f_n\} \not\subset X^*$. In view of this discussion, a t.b. $\{x_n;f_n\}$ for (X,T) is called a *Schauder base* (S.b.) provided $f_n \in X^*$ for each $n \geq 1$.

Corresponding to an S.b. $\{x_n;f_n\}$ for (X,T), let us introduce the *expansion operators* S_n and S_n^* $(n \geq 1)$ with $S_n : X \to X$, $S_n^* : X^* \to X^*$ and

$$S_n(x) = \sum_{i=1}^{n} f_i(x)x_i, \quad S_n^*(f) = \sum_{i=1}^{n} f(x_i)f_i \tag{2.1.1}$$

Observe that S_n^* is the adjoint of S_n.

Most of the work which is carried out in this book, depends upon Schauder bases rather than topological bases and there is a class of spaces for which every t.b. is an S.b; indeed, let us recall the following continuity theorem of Arsove [9] (cf. also [130], chapter 2 for further details).

THEOREM 2.1.2 Every t.b. for an F-space is an S.b.

Occasionally we will find the following criterion for an S.b. to be quite useful (cf. [121], and also [130], chapter 5 for more details).

THEOREM 2.1.3 Let $\{x_n\}$ be a sequence in (X,T) with $x_n \neq 0$, $n \geq 1$. Suppose that for each p in D_T there exist $M > 0$ and q in D_T so that

$$p\left(\sum_{i=1}^{m} \alpha_i x_i\right) \leq Mq\left(\sum_{i=1}^{n} \alpha_i x_i\right), \tag{2.1.4}$$

valid for all m,n in $\mathbb{N}$ with $m \leq n$ and all scalars $\alpha_1, \ldots, \alpha_n$ in $\mathbb{K}$. Then $\{x_n\}$ is an S.b. for $[x_n] \equiv [x_n]^T \subset X$. Conversely, if (X,T) is barrelled or contains a set of second category and $\{x_n\}$ is an S.b. for (X,T), then (2.1.4) holds good.

REMARK. A sequence $\{x_n\}$ in (X,T) is called a *topologically basic* (*t. basic*) (resp. *Schauder basic* (*S. basic*)) *sequence* if $\{x_n\}$ is a t.b. (resp. an S.b.) for the T-closure $[x_n]$ of $sp\{x_n\}$. Thus, in particular, if $\{x_n\}$ satisfies (2.1.4), then $\{x_n\}$ is S. basic in (X,T). By a *block sequence* (bl. s.) of an S.b. $\{x_n;f_n\}$ in (X,T), we mean a sequence $\{y_n\}$ of the form

$$y_n = \sum_{i=m_{n-1}+1}^{m_n} \alpha_i x_i, \quad m_0 = 0$$

where $\alpha \in \omega$, $\{m_n\} \in I$ and $y_n \neq 0$ for $n \geq 1$. A bl. s. of an S.b. is always S. basic (cf. [130], p. 139). Further, if $\{x_n\}$ is S. basic in (X,T) and $[x_n]$ is complemented in (X,T), then $\{x_n\}$ is called a *complemented Schauder basic sequence* (CSBS) in (X,T). If (X,T) is a Banach space, Theorem 2.1.3 is generally referred to as Nikol'skii's Theorem (cf. [170]).

Returning to the main discussion of this chapter, the behaviour of an S.b. $\{x_n;f_n\}$ can be broadly classified into three distinct categories which depend upon (I) the particular mode of convergence of the underlying infinite series, (II) the dual properties of $\{x_n\}$ and $\{f_n\}$ and (III) the topological properties of the sequence $\{S_n\}$. Accordingly, let us introduce

DEFINITION 2.1.5 An S.b. $\{x_n;f_n\}$ for an l.c. TVS (X,T) is said to be (i) *semiabsolute* if $\sum_{n\geq 1} |f_n(x)|p(x_n) < \infty$ for each x in X and p in D_T, (ii) *unconditional* if $\sum_{n\geq 1} f_n(x)x_n$ converges unconditionally in (X,T) for each x in X, (iii) *subseries* if $\sum_{n\geq 1} f_n(x)x_n$ is subseries convergent in (X,T) for each x in X, (iv) *bounded multiplier* if $\sum_{n\geq 1} f_n(x)x_n$ is bounded multiplier convergent in (X,T) and (v) *symmetric* if $\sum_{n\geq 1} f_{\pi(n)}(x)\, x_{\rho(n)}$ converges in (X,T) for each x in X and each pair (π,ρ) in $P \times P$, P being the family of all permutations on $\mathbb{N}$.

NOTE. An S.b. $\{x_n;f_n\}$ which is unconditional (resp. subseries, bounded multiplier) will henceforth be abbreviated as u-S.b. (resp. s-S.b., b.m.-S.b.)

Regarding (II), we have

DEFINITION 2.1.6 An S.b. $\{x_n;f_n\}$ for an l.c. TVS (X,T) is called (i) *shrinking* if $\{f_n\}$ is an S.b. for $(X^*,\beta(X^*,X))$, (ii) γ-*complete* or *boundedly complete if* $\sum_{n\geq 1} \alpha_n x_n$ converges whenever $\{\sum_{i=1}^{n} \alpha_i x_i\}$ is bounded in (X,T) for α in ω, (iii) *regular* if $\{x_n\}$ is regular in (X,T), (iv) *bounded* if $\{x_n\}$ is bounded in (X,T), (v) *normalized* if $\{x_n\}$ is bounded and regular in (X,T), (vi) *equicontinuous* if $\{f_n\}$ is equicontinuous on (X,T), (vii) *uniformly equicontinuous* provided that to every p in D_T there corresponds some q in D_T so that $p(x_n f_n(x)) \leq q(x)$ for each x in X and n in $\mathbb{N}$, (viii) *of type*

P if $\{x_n\}$ is regular and $\{\sum_{i=1}^{n} x_i\}$ is bounded in (X,T) and (ix) *of type* P* if $\{x_n\}$ is bounded in (X,T) and $\{\sum_{i=1}^{n} f_i\}$ is bounded in $(X^*,\beta(X^*,X))$.

Concerning (III), let us pass on to

DEFINITION 2.1.7 An S.b. $\{x_n;f_n\}$ for an l.c. TVS (X,T) is termed (i) *simple* if $\{S_n^*(f)\}$ is bounded in $(X^*,\beta(X^*,X))$ for each f in X^*, (ii) *monotone* if for each x in X and p in D_T, $p(S_n(x)) \leqq p(S_{n+1}(x))$ for all $n \geqq 1$ and (iii) *weakly e-Schauder* (resp. *e-Schauder*) if $\{S_n\}$ is σ-σ (resp. T-T) equicontinuous on X, $\sigma \equiv \sigma(X,X^*)$.

In the rest of Part 2, we will attempt to make an extensive study of these several notions of types of an S.b. introduced in Definitions 2.1.5 to 2.1.7. In particular, we will focus our attention on the necessity of each of these types, their possible interrelationship as well as their impact on the underlying space and the nature of the associated sequence spaces.

2.2 HISTORICAL BRIEFING

In order to achieve a comprehensive discussion of the present subject matter, let us briefly touch upon certain salient features of the development of these various types of an S.b. outlined above; we follow [182] (cf. also [59] and [66]).

The notions of bases contained in Definitions 2.1.5 to 2.1.7 were introduced by different analysts during their study of the structure of spaces having Schauder bases. However, a few of these concepts appear to be the consequence of abstracting natural properties enjoyed by bases in some familiar Banach and Fréchet spaces.

The idea of a semiabsolute base is essentially due to Karlin [139] who called it an absolutely convergent base and its inception may be traced back to the work of

Markushevich [156] and Urysohn [227]. Gelbaum ([60], [61]) and Fullerton [55] have also worked on absolutely convergent bases in Banach spaces. The work on absolutely convergent bases - now frequently called absolute bases - in Fréchet spaces is also implicit in the contribution by Iyer ([60], [61]) on spaces of entire functions, and possibly encouraged by this work of Iyer; Newns [169] in 1953 took up a brief but systematic study of absolute bases in Fréchet spaces. In order to extend the work on absolute bases from Fréchet to more general spaces, Kalton [99] introduced the idea of a semiabsolute base and other related notions. In [133], there appears the notion of a λ-semibase which envelops semiabsolute bases.

The notions of unconditional, subseries and bounded multiplier bases in Banach spaces are due to Karlin [139]. These three concepts of an S.b. are the same in Banach spaces. However, Karlin preferred to call unconditional bases 'absolute bases'. Subsequent extensions of these three types of an S.b. are to be found in the work of McArthur and his school (cf. [160] and also [44]).

Symmetric bases in Banach spaces first appeared in the work of Singer [210] and later on this notion was considered by Ruckle [197] in Fréchet spaces (cf. also [23]). Subsequently Garling [58] gave a very systematic account of these bases in the general setting of locally convex spaces and much later, the same study was again investigated in [75].

James [87] introduced shrinking bases to study the reflexivity of Banach spaces, whereas boundedly complete bases in Banach spaces were considered by Dunford and Morse in [47]. Jame's study of reflexivity of Banach spaces was further carried over by Singer [211] and a few years later, Retherford [184] extended these notions of bases and results of James and Singer to the setting of locally convex spaces. Tumarkin [226] has rightly called shrinking bases 'stretching bases'. Further extensions of shrinking bases to $\mathfrak{S}$-uniform bases in locally convex spaces were carried out in [123].

Bounded, regular and normalized bases in Banach spaces are implicit in the work of Bessaga and Pelczynski [17], [18] and so is the case in [188] of a regular base in an arbitrary l.c. TVS. Kalton [96] has explicitly used the terminology of bounded, regular and normalized bases in an arbitrary l.c. TVS. Simple bases are also due to Kalton [96] who, in fact, considered simple Schauder decompositions instead of Schauder bases. Dragilev [39] has also used the term 'regular base' for a class of absolute bases.

Uniformly equicontinuous bases appear to have been used first by Pietsch [176] in connection with his study of 'absoluter basis'. However, he calls these bases merely as equicontinuous bases, whereas Terzioğlu [220] maintains the use of uniformly equicontinuous bases in his study on nuclear spaces.

P and P* bases were once again introduced by Singer [211] to study the reflexivity of Banach spaces and their subsequent extensions to locally convex spaces are to be found in [184].

Monotone bases in Banach spaces were considered by James [87] and Kozlov [146] independently around the year 1951. Russo [202], [203] made an extensive study of e-Schauder and monotone Schauder decompositions in topological vector spaces, whereas the notion of weakly e-Schauder bases is essentially due to Cook [24]. In [105], there is given a concept of another type of an S.b. apparently looking weaker than a monotone base.

2.3 ILLUSTRATIVE EXAMPLES

Before we take up a detailed study on different types of S.b., we consider several examples exhibiting the importance of each one of the notions of an S.b. contained in Definitions 2. .5 to 2.1.7.

Semiabsolute character

Let us start with an unfamiliar example of a Fréchet space having a semiabsolute base. In fact, we have (cf. [116])

EXAMPLE 2.3.1 Let X_1 be the class of all functions $f : \mathbb{C} \to \mathbb{C}$ with

$$f(s) = \sum_{n\geq 1} a_n \exp(s\lambda_n), \ a_n \equiv a_n(f) \tag{2.3.2}$$

where $\{\lambda_n\}$ is a fixed sequence of reals satisfying $0 = \lambda_1 < \ldots < \lambda_n \to \infty$, $\log \lambda_n/n \to K < \infty$ and $\limsup \log|a_n|/\lambda_n \to -\infty$. Each f in X_1 is called an entire function represented by Dirichlet series (cf. [155]). Fix $\rho > 0$ and $A > 0$ and let $X \equiv X(\rho, A)$ be a subspace of X_1 such that each f in X satisfies the inequality $\limsup \log M(\sigma,f)/\exp(\rho\sigma) \leq A$ as $\sigma = Rl(s) \to \infty$, where $M(\sigma;f) = \sup\{|f(\sigma + it)| : |t| < \infty\}$. It is known (cf. [100], [155], [189]) that an f in X_1 is indeed in X if and only if

$$(+) \qquad \limsup_{n\to\infty} \lambda_n^{1/\rho} |a_n|^{1/\lambda_n} \leq (Ae\rho)^{1/\rho},$$

where $e = \sum_{n\geq 0} 1/n!$. Equip X with the natural l.c. topology T generated by the family $\{\|\cdot;A+\delta\| : \delta > 0\}$ of norms on X, with

$$\|f;A+\delta\| = \sum_{n\geq 1} |a_n| \left\{\frac{\lambda_n}{(A+\rho)e\rho}\right\}^{\lambda_n/\rho}.$$

Following [81] or [101], we can easily show that (X,T) is a Fréchet space. Define $e_n \in X$ by $e_n(s) = \exp(s\lambda_n)$; $n \geq 1$, $s \in \mathbb{C}$. The following inequality

$$\left\| \sum_{i=1}^{m} \alpha_i e_i ; A + \delta \right\| \leqq \left\| \sum_{i=1}^{n} \alpha_i e_i ; A + \delta \right\|$$

is easily verified for any choice of $\delta > 0$, integers m, n with $n \geqq m \geqq 1$ and scalars $\alpha_1,\ldots,\alpha_n$. Hence by Theorem 2.1.3, $\{e_n\}$ is an S.b. for (X,T), and let $\{\phi_n\}$ be the s.a.c.f. corresponding to $\{e_n\}$. Hence, for f in X given by (2.3.2), $\phi_n(f) = a_n$ and so $\{\phi_n\}$ satisfies the inequality (+) with a_n replaced by ϕ_n. Thus

$$\sum_{n\geqq 1} |\phi_n(f)| \, \|e_n ; A+\delta\| < \infty ; \ \forall \ \delta > 0,$$

thereby yielding the semiabsolute character of the S.b. $\{e_n ; \phi_n\}$ for (X,T).

NOTE. Examples similar to the preceding one and which have semiabsolute bases may also be traced in [113], [114] and [117].

There are reasons which allow us to introduce topological bases as infinite matrices. For instance, an infinite matrix (x_{mn}) in an l.c. TVS (X,T) is called a topological (resp. Schauder) base if for each x in X there exists a unique infinite matrix (f_{mn}) in X' (resp. X*) such that

$$x = T\text{-}\lim_{N\to\infty} \sum_{0\leqq m+n\leqq N} f_{mn}(x) x_{mn} \equiv \sum_{m+n\geqq 0} f_{mn}(x) x_{mn}.$$

Several notions and results related to an S.b. as a sequence can be suitably modified so as to be valid for an S.b. when represented as an infinite matrix. For instance, an S.b. $\{x_{mn}, f_{mn}\}$ for an l.c. TVS (X,T) is *semiabsolute* if for each x in X and p in D_T,

$$\sum_{m+n\geqq 0} |f_{mn}(x)| p(x_{mn}) \equiv T\text{-}\lim_{N\to\infty} \sum_{0\leqq m+n\leqq N} |f_{mn}(x)| p(x_{mn}) < \infty .$$

The following example (cf. [120]) illustrates the foregoing situation and provides another space having a semi-absolute base.

EXAMPLE 2.3.3 Let there be two numbers ρ_1 and ρ_2 satisfying $0 < \rho_1, \rho_2 < \infty$. Denote by $X \equiv X(\rho_1,\rho_2)$ the class of all functions $f : \mathbb{C}^2 \to \mathbb{C}$ with

$$(*) \qquad f(z_1,z_2) = \sum_{m+n\geq 0}\sum a_{mn} z_1^m z_2^n,$$

where (a_{mn}) is the unique infinite matrix determining f and satisfies the condition: to each $\delta > 0$, there corresponds N in $\mathbb{N}$ such that

$$(**) \qquad |a_{mn}| \leq m^{-m/(\rho_1+\delta)} n^{-n/(\rho_2+\delta)}, \quad \forall\ m+n \geq N.$$

The space X coincides with the family of all entire functions f of two complex variables, considered earlier by Fred Gross [54], having order point at best equal to (ρ_1,ρ_2); that is, for every $\varepsilon > 0$, one has $M(f;r_1,r_2) \leq \exp\,(r_1^{\rho_1+\varepsilon} + r_2^{\rho_2+\varepsilon})$ valid for all large r_1 and r_2, where $M(f;r_1,r_2) = \sup\,\{|f(z_1,z_2)|: |z_i| \leq r_i;\ i = 1,2\}$. Endow X with the l.c. topology T generated by the family $\{\|\cdot;\rho_1+\delta,\ \rho_2+\delta\|\}$ of norms on X, where

$$\|f;\rho_1+\delta,\rho_2+\delta\| = |a_{oo}| + \sum_{m+n\geq 1}\sum |a_{mn}| m^{m/(\rho_1+\delta)} n^{n/(\rho_2+\delta)}.$$

It is known (cf. [120]) that (X,T) is a Fréchet space. The functions e_{mn} $(m,n \geq 0)$ defined by $e_{mn}(z_1,z_2) = z_1^m z_2^n$ belong to X. For any matrix (a_{mn}), let

$$V_N = \sum_{0\leq m+n\leq N}\sum a_{mn}e_{mn};$$

then $V_N \in X$ for all integers $N \geq 0$. Observe that for all $N \geq M \geq 0$

$$\|V_M;\rho_1 + \delta,\ \rho_2 + \delta\| \leq \|V_N;\rho_1 + \delta,\rho_2 + \delta\|.$$

Therefore, by the matrix analogue of Theorem 2.1.3, the matrix (e_{mn}) forms an S.b. for (X,T) and let $\{\phi_{mn}\}$ denote the matrix associated with $\{e_{mn}\}$ so that $\phi_{mn} \in X^*$ and

$$f = \sum_{m+n\geq 0}\sum \phi_{mn}(f)\ e_{mn}.$$

Using (**) above, it is seen that $(e_{mn};\phi_{mn})$ is a semiabsolute base for (X,T).

<u>NOTE</u>. For examples similar to the preceding one, one may refer to [68], [103], [118] and [119].

<u>NOTE</u>. For further examples of spaces of analytic functions having semiabsolute bases, one can refer to [3], [5], [6] and [107] and several references given therein.

The next two examples present bases which are not semiabsolute

<u>EXAMPLE 2.3.4</u> (i) $\{e^n;e^n\}$ is an S.b. for $(c_o, \|\cdot\|_\infty)$ and since $\{1/n\} \notin \ell^1$, it is not a semiabsolute base. (ii) Next we consider the S.b. $\{e^n;e^n\}$ for $(\ell^p, \|\cdot\|_p)$, $p > 1$. We can find q with $1 < q < p$. Then $x \in \ell^p$, where $x_n = n^{-q/p}$. But $\sum_{n\geq 1} |<x,e^n>|\ \|e^n\|_p = \sum_{n\geq 1} n^{-q/p} = \infty$ and so $\{e^n;e^n\}$ is not a semiabsolute base for ℓ^p.

<u>NOTE</u>. We refer to chapter 8 of [130] for various examples and counter-examples on bounded multiplier, subseries and unconditional bases.

A collective approach

The following proposition provides a set of examples and counter examples of the various types of S.b. described earlier.

PROPOSITION 2.3.5 Let λ be an s.s. equipped with the topology $\sigma(\lambda,\mu)$, μ being a subspace of λ^{β}. Then $\{e^n;e^n\}$ is an S.b. for $(\lambda,\sigma(\lambda,\mu))$. Moreover, the following statements are also valid.

(i) Let $\mu \subset \lambda^{\times}$. Then $\{e^n;e^n\}$ is unconditional as well as semiabsolute for λ.

(ii) $\{e^n;e^n\}$ is subseries (resp. bounded multiplier) for λ if and only if λ is monotone (resp. normal).

(iii) If $\{e^n;e^n\}$ is symmetric for λ, then λ is a symmetric s.s.

(iv) $\{e^n;e^n\}$ is shrinking for λ if and only if $y^{(n)} \to y$ in $\beta(\mu,\lambda)$ for each y in μ.

(v) Let μ be normal and $\mu \subset \lambda^{\times}$. Then $\{e^n;e^n\}$ is γ-complete for λ if and only if $\lambda = \mu^{\times}$.

(vi) Let $\mu = \lambda^{\times}$. Then $\{e^n;e^n\}$ is regular for λ if and only if $\ell^{\infty} \subset \mu$.

(vii) $\{e^n;e^n\}$ is bounded for λ if and only if $\mu \subset \ell^{\infty}$.

(viii) Let $\mu = \lambda^{\times}$. Then $\{e^n;e^n\}$ is normalized for λ if and only if $\mu = \ell^{\infty}$.

(ix) Let $\mu = \lambda^{\times}$. Then $\{e^n;e^n\}$ is of type P* for λ if and only if $\mu = \ell^{\infty}$.

(x) Let $\mu = \lambda^{\times}$. Then $\{e^n;e^n\}$ is not of type P for λ.

(xi) $\{e^n;e^n\}$ is e-Schauder or equivalently weakly e-Schauder for $\lambda^{\times}$ if and only if $\mu = \phi$.

PROOF. The first part is a consequence of the definition of the dual system $\langle\lambda,\mu\rangle$ with $\mu \subset \lambda^{\beta}$. The proofs of (i),

(iii) and (iv) are straightforward, whereas for (ii), see Proposition 8.2.2 of [130].

(v) Let $\lambda^\times = \mu$ and $\{\sum_{i=1}^{n} \alpha_i e^i\}$ be $\sigma(\lambda,\mu)$-bounded for α in ω. For each β in μ, we can determine γ in μ so that $\gamma_n = \theta_n\beta_n$, where $|\theta_n| = 1$ and $|\alpha_n\beta_n| = \theta_n\alpha_n\beta_n$ for $n \geqq 1$. There exists $M \equiv M(\gamma) > 0$ such that

$$\sum_{i=1}^{n} |\alpha_i\beta_i| = |< \sum_{i=1}^{n} \alpha_i e^i, \gamma >| \leqq M, \ \forall\ n \geqq 1.$$

Hence $\alpha \in \mu^\times = \lambda$.

For converse, observe that for x in $\mu^\times$, $\{\sum_{i=1}^{n} x_i e^i\}$ is bounded in $(\lambda,\sigma(\lambda,\mu))$ and so $x \in \lambda$.

(vi) The regularity of $\{e^n\}$ yields a $\delta > 0$ and y in μ with $|<e^n,y>| \geqq \delta$, $n \geqq 1$. Thus $\lambda \subset \ell^1$. The converse is trivial.

(vii) This is obvious.

(viii) This follows from (vi) and (vii).

(ix) Assume that $\mu = \ell^\infty$. By (vii), $\{e^n\}$ is $\sigma(\lambda,\mu)$-bounded. Next, let B be an arbitrary $\sigma(\lambda,\mu)$-bounded subset of λ. Then B is also $\sigma(\ell^1,\ell^\infty)$-bounded subset of ℓ^1. But $\eta(\ell^1,\ell^\infty)$ is compatible for the dual system $<\ell^1,\ell^\infty>$; thus for some $K > 0$,

$$\sup_{x \in B} \sum_{n \geqq 1} |x_n| \leqq K$$

$$\Rightarrow |<x, \sum_{i=1}^{n} e^i>| \leqq K;\ \forall n \geqq 1,\ x \in B.$$

Hence $\{e^n;e^n\}$ is of type P*.

Conversely, by (vii), $\mu \subset \ell^\infty$. It is enough to show that $\lambda \subset \ell^1$. If $x \in \lambda$, then $|x|^{(n)} = \{|x_1|,\ldots,|x_n|,0,0,\ldots\} \in \lambda$, $n \geqq 1$. Since $\{\sum_{i=1}^{n} |x_i| e^i\}$ is $\sigma(\lambda,\mu)$-bounded,

$$\sup_{m,n\geq 1} \left|\left< \sum_{i=1}^{m} |x_i| e^i, \sum_{j=1}^{n} e^j \right>\right| < \infty .$$

Therefore $x \in \ell^1$.

(x) Suppose on the contrary that $\{e^n;e^n\}$ is of type P. Since $\{e^{(n)}\}$ is $\sigma(\lambda,\mu)$-bounded, $\mu \subset \ell^1$. Also $\{e^n\}$ is $\sigma(\lambda,\mu)$-regular; hence from (vi), $\ell^\infty \subset \ell^1$. But this is absurd.

(xi) Cf. [129], Proposition 2.3.24, p. 65. □

NOTE. A part of Proposition 2.3.5 was announced in [44] and its proof was subsequently worked out in [46]. This result is incorporated in [74] where a similar theorem is proved for Schauder decompositions.

Further examples

Proposition 2.3.5 does justify the introduction of some of the types of an S.b. mentioned in Definitions 2.1.5 to 2.1.7, yet its discussion concerning items (i), (iii), (iv) and (x) is incomplete. We refer to chapter 8 of [130] and chapter 7 for examples of an S.b. which are, respectively, not unconditional and symmetric. Examples and counter-examples relating to shrinking bases are also given in chapter 8 of [130]. Thus, in order to complete the discussion on the preceding proposition, especially relating to (x), we have

EXAMPLE 2.3.6 The S.b. $\{e^n;e^n\}$ for $(c_0,\tau(c_0,\ell^1))$ is of type P. Indeed, $\|e^n\|_\infty = \|e^{(n)}\|_\infty = 1$, $n \geq 1$.

We, however, postpone the construction of examples and counter-examples on equicontinuous bases to chapter 10 and in the meantime, let us deal with the uniformly equicontinuous bases.

EXAMPLE 2.3.7 This is the space $(\phi,\sigma(\phi,\omega))$ for which we show that its S.b. $\{e^n;e^n\}$ is not uniformly equicontinuous.

For otherwise, there would exist some z in ω so that

$$(*) \quad |x_i| \leq |\sum_{j\geq 1} x_j z_j|; \ \forall i \geq 1, \ x \in \phi$$

In order to see that (*) holds, we have to have $z \neq 0$. If m is the only index such that $z_m \neq 0$, then for $x = \frac{1}{2} e^m + z_m e^{m+1}$ in ϕ we infer from (*) that $|z_m| \leq \frac{1}{2} |z_m|$ which is absurd. Hence, there are more indices than one for which the corresponding coordinates of z are nonvanishing. Accordingly, we may choose a finite set $F \subset \mathbb{N}$ so that #(F) is even and $z_j \neq 0$ for j in F. Define x in ϕ with $x_j = 0$ for $j \notin F$ such that $x_j = z_j^{-1}$ for half of the indices j in F and $x_j = -z_j^{-1}$ otherwise. Therefore, by (*) again, $|z_j|^{-1} \leq 0$, $j \in F$. This is again absurd. Consequently the inequality (*) is no more valid for any z in ω.

<u>EXAMPLE 2.3.8</u> The S.b. $\{e^n;e^n\}$ for $(\phi,\eta(\phi,\omega))$ is uniformly equicontinuous. In fact, for y in ω, $p_y(\langle x,e^i\rangle e^i) \leq p_y(x)$ for all $i \geq 1$ and x in ϕ.

The next example is of a simple base.

<u>EXAMPLE 2.3.9</u> Consider the S.b. $\{e^n;e^n\}$ for $(\ell^1,\tau(\ell^1,\ell^\infty))$. Choose arbitrary y in ℓ^∞ and a $\sigma(\ell^1,\ell^\infty)$-bounded subset B. Then for x in B,

$$|\langle x, \sum_{i=1}^{n} y_i e^i\rangle| \leq \|y\|_\infty \sum_{i=1}^{n} |x_i|$$

$$\Rightarrow \sup_{x\in B} \sup_{n\geq 1} |\langle x, \sum_{i=1}^{n} y_i e^i\rangle| < \infty ;$$

cf. proof of Proposition 2.3.5 (ix).

Kalton ([97], p. 379) observes that bases in most of the spaces of interest are simple and following him, we offer the following example of an S.b. which is not simple.

EXAMPLE 2.3.10 Consider the S.b. $\{e^n; e^n\}$ for the space $(\phi, \sigma(\phi, \lambda))$, where

$$\lambda = \{\alpha \in \omega : \sup_n |\sum_{i=1}^{2n} \alpha_i|, \sup_n \frac{1}{n} |\sum_{i=1}^{2n+1} \alpha_i| < \infty\}$$

Define α in λ by $\alpha_i = n$ if $i = 2n-1$ and $\alpha_i = -n$ if $i = 2n$, where $n \geq 1$. In order to prove that $\{S_n^*(\alpha)\}$ is not $\beta(\lambda, \phi)$-bounded, let us consider the $\sigma(\phi, \lambda)$-bounded subset $A = \{e^{(2n)}\}$ of ϕ. Then $\langle e^{(2n)}, S_{2n-1}^*(\alpha)\rangle = n$.

Next we have

EXAMPLE 2.3.11 Here we take up the S.b. $\{e^n; e^n\}$ for $(\omega, \sigma(\omega, \phi))$ and show that this base is not monotone. For, otherwise, choosing y in ϕ with $y_i \neq 0$ $(1 \leq i \leq m)$; $y_i = 0$, $i > m$, where m is even, we find

$$(*) \qquad |\sum_{i=1}^{m-1} x_i y_i| \leq |\sum_{i=1}^{m} x_i y_i|, \ \forall\, x \in \omega .$$

However, (*) is negated by taking x in ω as follows:

$x_i = (-1)^i y_i^{-1}$, $1 \leq i \leq m$ and x_i is arbitrary for $i > m$.

Finally, we come to

EXAMPLE 2.3.12 The S.b. $\{e^n; e^n\}$ for $(\phi, \eta(\phi, \omega))$ is clearly monotone.

3 Associated sequence spaces

3.1 INTRODUCTION

The theory of sequence spaces has now emerged as a very powerful tool in establishing and strengthening many important results in several branches of functional analysis. In particular, there is a close relationship between the sequence space theory and the theory of Schauder bases. Indeed, an l.c. TVS X containing an S.b. may be regarded as a sequence space and hence it seems plausible to transfer many of the known properties of sequence spaces to locally convex spaces equipped with an S.b. Furthermore, we may also enrich the sequence space theory by considering certain concrete results from the theory of locally convex spaces having an S.b.

Our aim in successive chapters will be to find the behaviour of sequence spaces resulting from the presence of different types of S.b. in locally convex spaces. Attempts will also be made to study the type of an S.b. by suitably restricting the resulting sequence space.

3.2 CONSTRUCTION OF SEQUENCE SPACES

In this section, we confine ourselves to a general situation to recover a sequence space from a given arbitrary nonzero sequence of vectors in an arbitrary TVS; we begin with the following result (cf. [78], [195] and [196]):

PROPOSITION 3.2.1 Let (X,T) be a TVS containing a sequence $\{x_n\}$ with $x_n \neq 0$, $n \geq 1$. Suppose that

$$\eta \equiv \eta_T \equiv \eta_{T,\{x_n\}} = \{\alpha \in \omega : \sum_{n\geq 1} \alpha_n x_n \text{ converges in } X\}. \tag{3.2.2}$$

For each v in $\mathcal{B}_T$, let $v^* = \{\alpha \in \eta : \sum_{i=1}^{n} \alpha_i x_i \in v$, for all $n \geq 1\}$ and $\mathcal{B}^* = \{v^* : v \in \mathcal{B}_T\}$. If T^* is the linear topology on η, generated by $\mathcal{B}^*$, then (η,T^*) is metrizable (resp. complete) provided (X,T) is metrizable (resp. complete). If (X,T) is an l.c. TVS, then so is (η,T^*). Finally, if p_v is the Minkowski functional on X corresponding to v in $\mathcal{B}_T$, then the Minkowski functional $p^*_{v^*}$ relative to v^* is given by

$$p^*_{v^*}(\alpha) = \sup_{n\geq 1} p_v(\sum_{i=1}^{n} \alpha_i x_i), \quad \forall \alpha \in \eta.$$

PROOF. It is convenient to start from the last part. So, let $\alpha \in \eta$ and $v \in \mathcal{B}_T$, then for some $A > 0$, $\sum_{i=1}^{n} \alpha_i x_i \in Av$ for all $n \geq 1$. Hence $p^*_{v^*}(\alpha) \geq \sup p_v(\sum_{i=1}^{n} \alpha_i x_i)$. On the other hand, let $B = \sup p_v(\sum_{i=1}^{n} \alpha_i x_i)$. If $B = 0$, then clearly $p^*_{v^*}(\alpha) = 0$. So, let $B > 0$. Then $\alpha \in Bv^*$, giving thereby $p^*_{v^*}(\alpha) \leq B$. This proves the last part. The other parts now being obvious, we have only to show the completeness of (η,T^*) in case (X,T) is given to be complete.

Consider, therefore, a T^*-Cauchy set $\{\alpha^p\}$ in η. Fix $\varepsilon > 0$ and v in $\mathcal{B}_T$. Then there exists an index r with

$$p_v(\sum_{i=1}^{n} (\alpha_i^p - \alpha_i^q)x_i) \leq \varepsilon, \; \forall\, n \geq 1; \; p,q \geq r.$$

It follows that for some α in ω, $\alpha_i^p \to \alpha_i$. It is now a routine exercise to conclude that $\alpha \in \eta$ and $p^*_{v^*}(\alpha^p-\alpha) \leq \varepsilon$ for all $p \geq r$. □

In order to establish a topological relation between η and X, let us prove

PROPOSITION 3.2.3 Let (X,T) be a TVS containing a sequence $\{x_n\}$ with $x_n \neq 0$, $n \geq 1$. Then the natural map $F : (\eta,T^*) \to (X,T)$, $F(\alpha) = \sum_{n\geq 1} \alpha_n x_n$ is a continuous linear operator. If (X,T) is an F-space, $\{x_n\}$ is ω-linearly independent in (X,T) and $Y = F[\eta]$ is a closed subspace of X, then $(\eta,T^*) \simeq (Y,T|Y)$ under F.

PROOF. Observe that for each v in $\mathcal{B}_T$,

$$p_v(F(\alpha)) \leq \sup_{n\geq 1} \left(\sum_{i=1}^{n} \alpha_i x_i\right);$$

in particular, F is a continuous linear operator. ω-linear independence of $\{x_n\}$ forces F to be a 1-1 map. Now the result is a consequence of Proposition 3.2.1 and the closed graph theorem. □

Next we have

PROPOSITION 3.2.4 The s.s. $\eta \equiv (\eta,T^*)$ defined earlier is an AK-space.

PROOF. Let $\alpha^p \to \alpha$ in (η,T^*). If $v \in \mathcal{B}_T$, then for $i \geq 1$,

$$p_v((\alpha_i^p-\alpha_i)x_i) \leq 2p_{v^*}^*(\alpha^p-\alpha), \quad \forall p.$$

Since $x_i \neq 0$ for each $i \geq 1$, we can choose $v \equiv v(i)$ in $\mathcal{B}_T$ with $p_v(x_i) \neq 0$. Hence $\alpha_i^p \to \alpha_i$ for every $i \geq 1$, yielding thereby the K-character of (η,T^*). Also, for α in η and v in $\mathcal{B}_T$,

$$p_{v^*}^*(\alpha^{(n)}-\alpha) = \sup_{m>n} p_v\left(\sum_{i=n+1}^{m} \alpha_i x_i\right)$$

$$\to 0,$$

as $n \to \infty$ by the construction of η. Thus $\alpha^{(n)} \to \alpha$ in (η,T^*) for each α in η. □

PROPOSITION 3.2.5 For $i \geq 1$, let $P_i : (\eta,T^*) \to \mathbb{K}$ with $P_i(\alpha) = \alpha_i$. Then $\{e^n;P_n\}$ is an S.b. for (η,T^*).

PROOF. Since (η,T^*) is a K-space, each linear functional P_i belongs to $\eta^* \equiv (\eta,T^*)^*$. But then the AK-character of (η,T^*) forces $\{e^n,P_n\}$ to be an S.b. for this space. □

We now derive the useful

PROPOSITION 3.2.6 Let (X,T) be an l.c. TVS having a sequence $\{x_n\}$ with $x_n \neq 0$, $n \geq 1$. Then $\{e^n;e^n\}$ is an S.b. for the l.c. s.s. (η,T^*).

PROOF. Since (η,T^*) is an AK-space, η^* can be identified with $\{\{f(e^n)\} : f \in \eta^*\}$ (cf. Theorem 1.3.13). In particular, $P_i \equiv \{P_i(e^n): n \geq 1\} = e^i$ and now use Proposition 3.2.5. □

We also have

THEOREM 3.2.7 Let $\{x_n;f_n\}$ be a t.b. for an F-space (X,T). Then $(\eta,T^*) \simeq (X,T)$; in particular, $\{x_n;f_n\}$ is an S.b. for (X,T).

PROOF. Indeed, invoking both the notation and statement of Proposition 3.2.3, we find that $F[\eta] = X$ and F is a topological isomorphism from (η,T^*) onto (X,T). The last part follows by using the continuity of F^{-1} and the K-character of (η,T^*). □

REMARK. The last part of the foregoing theorem is the continuity theorem of a t.b. for F-spaces and for a different proof of this result, we refer to Theorem 2.2.12 of [130] and references given therein.

3.3 WEAK BASES AND SEQUENCE SPACES

In order to get further information on sequence spaces associated with strictly nonzero sequences and at the same time keeping pace with the ultimate aim of this chapter, we consider and study those sequence spaces which are developed on account of weak Schauder bases. Accordingly, we consider an arbitrary l.c. TVS (X,T) having a $\sigma(X,X^*)$-S.b. $\{x_n;f_n\}$ and suppose that

$$\delta = \{\{f_n(x)\}:x \in X\} \equiv \{\alpha \in \omega: \sum_{n\geq 1} \alpha_n x_n \text{ is } \sigma(X,X^*)\text{-convergent in } X\}. \quad (3.3.1)$$

Since $x_n \neq 0$ for $n \geq 1$, we also have the sequence space η associated with $\{x_n\} \equiv \{x_n;f_n\}$. Clearly $\eta \subset \delta$.

The natural maps $F_1 : \delta \to X$, $F_1(\alpha) = x = \sum_{n\geq 1} \alpha_n x_n$ (convergence in the weak topology) and $F_2 : \eta \to X$, $F_2(\alpha) = x = \sum_{n\geq 1} \alpha_n x_n$ (convergence in the original topology T) are respectively bijection and injection. Clearly $\eta = \delta$ if and only if $\{x_n;f_n\}$ is an S.b. for (X,T).

We may also consider two more natural sequence spaces corresponding to a $\sigma(X,X^*)$ - S.b. for an l.c. TVS (X,T), namely,

$$\mu = \{\{f(x_n)\} : f \in X^*\} = \{\alpha \in \omega: \sum_{n\geq 1} \alpha_n f_n \text{ is } \sigma(X^*,X)\text{-convergent in } X^*\} \quad (3.3.2)$$

and

$$\nu = \{\alpha \in \omega: \sum_{n\geq 1} \alpha_n f_n \text{ converges in } (X^*,\beta(X^*,X))\}. \quad (3.3.3)$$

As before, the natural maps $G_1 : X^* \to \mu$, $G_1(f) = \{f(x_n)\}$ and $G_2 : \nu \to X^*$ are bijection and injection maps respectively.

Let us single out the spaces δ and μ which form a natural dual system $\langle\delta,\mu\rangle$ under the bilinear form $\langle\alpha,\beta\rangle$ with

$$\langle\alpha,\beta\rangle \equiv \langle\{f_n(x)\},\{f(x_n)\}\rangle = \sum_{n\geq 1} f_n(x)f(x_n) = f(x). \quad (3.3.4)$$

It follows that F_1, F_1^{-1} are continuous in their respective weak topologies and if F_1^* is the adjoint of $F_1 (F_1^* : X^* \to \delta^* = \mu)$, then $F_1^* = G_1$ as well as the maps G_1, G_1^{-1} are continuous in their respective weak topologies. Summing up the discussion, we have the following important observation in the form of

THEOREM 3.3.5 Let $\{x_n;f_n\}$ be a $\sigma(X,X^*)$-S.b. for an l.c. TVS (X,T). Then $(\delta,\sigma(\delta,\mu)) \simeq (X,\sigma(X,X^*))$ under F_1 and $(X^*,\sigma(X^*,X)) \simeq (\mu,\sigma(\mu,\delta))$ under $F_1^* = G_1$.

NOTE. There does not seem to be a meaningful relationship between η^* and ν unless we restrict both (X,T) and $\{x_n;f_n\}$; a possible dialogue between these spaces may be traced back in chapter 9.

The sequence spaces δ, μ and their duals

Since $\langle\delta,\mu\rangle$ forms a dual system and $\mu \subset \delta^\beta$, the former may be expected to yield some results on the structure of (X,T) with the help of sequence space theory and vice-versa. In order to derive more advantageous results in this direction, the situation would, however, be more pleasant when $\mu = \delta^\beta$ or $\delta^\times$, the latter one being still helpful. Concerning $\mu = \delta^\beta$ and also $\delta = \mu^\beta$, we have

PROPOSITION 3.3.6 Let $\{x_n;f_n\}$ be a $\sigma(X,X^*)$-S.b. for an l.c. TVS (X,T). If $(X,\sigma(X,X^*))$ (resp. $(X^*,\sigma(X^*,X))$) is ω-complete, then $\delta = \mu^\beta$ (resp. $\mu = \delta^\beta$).

PROOF. If $y \in \mu^\beta$, then $\{y^{(n)}\}$ is Cauchy in $(\delta,\sigma(\delta,\mu))$ and so $y^{(n)} \to y \in \delta$ by Theorem 3.3.5. Thus $\delta = \mu^\beta$. The other part similarly follows. □

The ω-completeness in the above proposition cannot be ignored to get the desired result, for instance we have

EXAMPLE 3.3.7 Consider the S.b. $\{e^n;e^n\}$ for $(\ell^1,\sigma(\ell^1,c_o))$. Here $\delta = \ell^1$ and $\mu = c_o$. As $\delta^\beta = \ell^\infty$, we get $\mu \subsetneq \delta^\beta$. Observe that $(c_o,\sigma(c_o,\ell^1))$ is not ω-complete (cf. [129], p. 118).

On the other hand, we also have

EXAMPLE 3.3.8 Here we take the S.b. $\{e^n;e^n\}$ for $(k,\sigma(k,\ell^1))$ and so $\delta = k$, $\mu = \ell^1$. But $\mu^\beta = \ell^\infty$ and so $\delta \subsetneq \mu^\beta$. Since k is not perfect (not even monotone), the space in question is not ω-complete (cf. Theorem 1.3.7).

Contrary to the relations $\delta \subset \mu^\beta$ and $\mu \subset \delta^\beta$, in general we do not have the inclusions $\delta \subset \mu^\times$ or $\mu \subset \delta^\times$. For instance, let us confine ourselves to

EXAMPLE 3.3.9 This is the space $(k,\sigma(k,k^\beta))$ having the S.b. $\{e^n;e^n\}$. Then $\mu = cs \not\subset \ell^1 = \delta^\times$.

Similarly we may construct

EXAMPLE 3.3.10 Let us take up the space $(cs, \sigma(cs,k))$ along with its S.b. $\{e^n;e^n\}$. Then $\delta = cs \not\subset \ell^1 = \mu^\times$.

The last two examples are a simple consequence of the following straightforward:

PROPOSITION 3.3.11 For a sequence space λ with $\lambda^\times \neq \lambda^\beta$, consider the spaces (i) $(\lambda,\sigma(\lambda,\lambda^\beta))$ and (ii) $(\lambda^\beta,\sigma(\lambda^\beta,\lambda))$. Then $\mu \not\subset \delta^\times$ and $\delta \not\subset \mu^\times$.

REMARK 3.3.12. Let us consider the sequence space η when $\{x_n\}$ is an S.b. More precisely, let $\{x_n;f_n\}$ be an S.b. for an l.c. TVS (X,T). Then $\eta = \delta$ and the topology T^* on η is given by $\{P_p : p \in D_T\}$, where $P_p(\{f_n(x)\}) = \sup\{p(S_n(x)) : n \geq 1\}$. We may also transfer the topology T as T^{**} on η in a natural way. In fact, for p in D_T, let $N_p(\{f_n(x)\}) = p(x)$ and denote by T^{**} the topology on η generated by $\{N_p : p \in D_T\}$. Clearly $T^{**} \subset T^*$. If (X,T) is also barrelled, then $\{S_n\}$ is equicontinuous and so $T^* \approx T^{**}$. Next observe that (η,T^{**}) is an AK-space. Further, $\sigma(\eta,\phi) \subset T^{**}$; indeed, for $i \geq 1$ there exists p_i in D_T with $q_{e^i}(\{f_n(x)\}) = |f_i(x)| \leq p_i(x)$ and note that

$\sigma(\eta,\phi)$ is generated by $\{q_{e_i} : i \geq 1\}$. Also $(\eta,T^{**}) \simeq (X,T)$ under F (cf. Proposition 3.2.3) and hence $X^* \leftrightarrow \eta^* \equiv(\eta,T^{**})^*$ with respect to the adjoint F^*. But $X^* \leftrightarrow \mu$ and consequently η^* can be identified with $\{\{f(x_n)\}: f \in X^*\}$.

Finally, we make a note that whenever $\{x_n;f_n\}$ is an S.b. for an l.c. TVS (X,T), we will write δ for η from chapter 4 onward and call it occasionally the *associated sequence space* (a.s.s.) or else the *sequence space associated with an* S.b.

3.4 MATRICES AND NEW BASES FROM OLD ONES

Let $\{x_n;f_n\}$ be an S.b. for a TVS (X,T) and $A = [a_{ij}]$ an infinite matrix. Define

$$y_j = \sum_{i\geq 1} a_{ij}\, x_i \equiv T - \lim_{n\to\infty} \sum_{i=1}^{n} a_{ij}\, x_i. \tag{3.4.1}$$

Applying some results of this chapter, we explore necessary and sufficient conditions which ensure the t.b. or S.b. character of $\{y_n\}$ for (X,T). The consideration of this problem appears to have originated from the theory of matrix transformations from sequence spaces to themselves (cf. [129]) and we follow [196] for the rest of this section.

LEMMA 3.4.2 Let there be an infinite matrix $A = [a_{ij}]$. Suppose that $\{x_n;f_n\}$ is an S.b. for a TVS (X,T) and let $\{y_n\}$ be the sequence defined by (3.4.1). Assume further the existence of a sequence $\{\beta_n\}$ in $\mathbb{K}$ such that $\sum_{j\geq 1} \beta_j y_j$ converges to y in (X,T). Then $\sum_{j\geq 1} a_{ij}\beta_j$ converges for each $i \geq 1$ and

$$\sum_{j\geq 1} \beta_j y_j = \sum_{i\geq 1} q_i x_i \text{ with } q_i = \sum_{j\geq 1} a_{ij}\beta_j.$$

In particular, if $y_n \neq 0$ for each $n \geq 1$ and $\eta_1 \equiv \eta_{T,\{x_n\}}$ and

$\eta_2 \equiv \eta_{T,\{y_n\}}$, then A is a matrix transformation from η_2 into η_1.

PROOF. We have

$$f_i\left(\sum_{j\geq 1}\beta_j y_j\right) = \sum_{j\geq 1} \beta_j a_{ij}, \ \forall\ i \geq 1.$$

But

$$y = \sum_{j\geq 1} \beta_j y_j = \sum_{i\geq 1} f_i(y)x_i = \sum_{i\geq 1} \left(\sum_{j\geq 1} \beta_j a_{ij}\right)x_i,$$

and we are done. □

Throughout this section we write (X,T) for an arbitrary TVS containing an S.b. $\{x_n;f_n\}$ and $\{y_n\}$ for the sequence defined in (3.4.1). To achieve the objective of this section, we assume that $y_j \neq 0$ for $j \geq 1$ and η_1 and η_2 are the sequence spaces defined in Lemma 3.4.2. For the Minkowski functional p_V on X with v in $\mathcal{B}_T$, let

$$p_V^*(\alpha) = \sup_{n\geq 1} p_V\left(\sum_{i=1}^n \alpha_i x_i\right);\ q_V^*(\beta) = \sup_{n\geq 1} p_V\left(\sum_{i=1}^n \beta_i y_i\right),$$

where $\alpha \in \eta_1$ and $\beta \in \eta_2$. In what follows, we consider η_1 (resp. η_2) equipped with the linear topology T_1^* (resp. T_2^*) generated by $\{p_V^* : v \in \mathcal{B}_T\}$ (resp. $\{q_V^* : v \in \mathcal{B}_T\}$).

Let $Y = \{y \in X : y = \sum_{j\geq 1} \beta_j y_j,\ \beta \in \eta_2\}$ and $G : \eta_2 \to Y \subset X$ with $G(\beta) = y = \sum_{j\geq 1} \beta_j y_j$. Similarly, let $F : \eta_1 \to X$, $F(\alpha) = \sum_{i\geq 1} \alpha_i x_i$, $\alpha \in \eta_1$. Then F and G are continuous and onto (Proposition 3.2.3) linear operators, the former being one-to-one also. We may thus summarize the discussion by the following diagram

$$\eta_2 \xrightarrow{\ G\ } Y \subset X \xleftrightarrow{\ F\ } \eta_1. \qquad (3.4.3)$$

In addition, if (X,T) is an F-space also, then F is a topological isomorphism (Proposition 3.2.3) and (3.4.3) takes the following form:

$$\eta_2 \xrightarrow{G} Y \subset X \overset{F}{\simeq} \eta_1 . \tag{3.4.4}$$

Concerning the t.b. character of $\{y_n\}$, we have

THEOREM 3.4.5 Let (X,T) be a TVS containing an S.b. $\{x_n;f_n\}$. Suppose that $A = [a_{ij}]$ is an infinite matrix and $\{y_j\}$ is the sequence defined by (3.4.1) with $y_j \neq 0$ for $j \geq 1$. Then $\{y_j\}$ is a t.b. for (X,T) if and only if A is 1-1 and onto from η_2 to η_1.

PROOF. We prove here

(i) X = Y if and only if A is onto, where

$Y = \{y \in X : y = \sum_{j\geq 1} \beta_j y_j, \ \beta \in \eta_2\}$; and

(ii) $\{y_j\}$ is ω-linearly independent in X if and only if A is 1-1.

Then the statements (i) and (ii) together yield the required result.

(i) To prove this statement, suppose that Y = X. Consider $\{f_i(x)\}$ in η_1 for some x in X. Then $x = \sum_{j\geq 1} \beta_j y_j$ and so $\beta = \{\beta_j\} \in \eta_2$. By Lemma 3.4.2, $(A\beta)_i$ exists and $(A\beta)_i = q_i = f_i(x)$ for each $i \geq 1$. Thus A is onto η_2. Similarly, if A is onto, then Y = X.

(ii) Let $\{y_j\}$ be ω-linearly independent. Consider $\beta \in \eta_2$ such that $A\beta = 0$. Then by Lemma 3.4.2

$$(A\beta)_i = \sum_{j\geq 1} a_{ij}\beta_j = q_i = 0, \ i \geq 1$$

$$\Rightarrow \sum_{j\geq 1} \beta_j y_j = 0 \Rightarrow \beta_j = 0, \ \forall \ j \geq 1 \ \Rightarrow \beta = 0.$$

Hence A is 1-1.

Conversely, if A is 1-1 and $\sum_{j\geq 1} \beta_j y_j = 0$, then by Lemma 3.4.2, $\sum_{i\geq 1} q_i x_i = 0$ where $q_i = (A\beta)_i = \sum_{j\geq 1} a_{ij}\beta_j$, $i \geq 1$. As $\{x_i\}$ is an S.b. for (X,T), $q_i = 0$, for each $i \geq 1$ and so $A\beta = 0 \Rightarrow \beta = 0$, i.e. $\beta_j = 0$, for each $j \geq 1$. □

Regarding the S.b. character of $\{y_n\}$, we have

<u>THEOREM 3.4.6</u> Let (X,T), $\{x_n;f_n\}$, $\{y_j\}$ and A be as given in Theorem 3.4.5. If $\{y_j\}$ is an S.b. for (X,T), then there is an infinite matrix $B = [b_{ij}]$ so that $\{b_{ij} : j \geq 1\} \in \eta_1^\beta$ for $i \geq 1$, $\sum_{j\geq 1} b_{ij} f_j(x)$ converges for each x in X, $i \geq 1$ with $\{\sum_{j\geq 1} b_{ij} f_j(x) : i \geq 1\} \in \eta_2$ for all x in X and $AB = BA = [\delta_{ij}]$; in particular, A has an inverse matrix $B : \eta_1 \to \eta_2$. Conversely, if (X,T) is also a Fréchet space and there exists a matrix satisfying the above conditions, then $\{y_j\}$ is an S.b. for (X,T).

<u>PROOF</u>. Let $\{y_j\}$ be an S.b. for (X,T) and let $\{g_j\}$ be the s.a.c.f. corresponding to $\{y_j\}$. Put $b_{ij} = g_i(x_j)$ and suppose that $B = [b_{ij}]$. Since

$$\eta_1 = \{\{f_i(x)\} : x \in X\}; \ \eta_2 = \{\{g_i(x) : x \in X\},$$

and each x in X has a representation

$$x = \sum_{j\geq 1} f_j(x)x_j,$$

the series $\sum_{j\geq 1} b_{ij}f_j(x)$ converges to $g_i(x)$ for all x in X and $i \geq 1$. This proves the first and second parts. Further, for the equalities $AB = BA = [\delta_{ij}]$, let us note that

$$\sum_{k\geq 1} a_{ik}\, b_{kj} = \sum_{k\geq 1} f_i(y_k)g_k(x_j) = f_i(x_j) = \delta_{ij}$$

and

$$\sum_{k\geq 1} b_{ik}\, a_{kj} = \sum_{k\geq 1} g_i(x_k)f_k(y_j) = g_i(y_j) = \delta_{ij}.$$

Also, B is a matrix transformation from η_1 to η_2, since

$$(B(\{f_i(x)\}))_n = \sum_{j\geq 1} a_{nj} f_j(x), \ \forall\ n \geq 1.$$

In order to prove the converse part, let us observe that (η_1, T_1^*) and (η_2, T_2^*) are AK-FK spaces (cf. Propositions 3.2.1 and 3.2.4). Hence from Theorem 1.3.13, the spaces $(\eta_1^\beta, \sigma(\eta_1^\beta, \eta_1))$ and $(\eta_2^\beta, \sigma(\eta_2^\beta, \eta_2))$ are ω-complete. But $A : \eta_2 \to \eta_1$ and $B : \eta_1 \to \eta_2$ are matrix transformations and hence A (resp. B) is $\sigma(\eta_2, \eta_2^\beta)$-$\sigma(\eta_1, \eta_1^\beta)$ (resp. $\sigma(\eta_1, \eta_1^\beta)$ - $\sigma(\eta_2, \eta_2^\beta)$) continuous (cf. [129], proof of Proposition 4.3.2). Therefore using Proposition 1.2.15, we conclude that A (resp. B) is T_2^* - T_1^* (resp. T_1^*-T_2^*) continuous. The considerations of continuity of A and B, S.b. character of $\{e^n; e^n\}$ for (η_1, T_1^*) and (η_2, T_2^*) as well as the equality $A(B(e^n)) = B(A(e^n)) = e^n$, $n \geq 1$ force A to be 1-1 from η_2 onto η_1. Now make use of Theorem 3.4.5 to reach the desired conclusion. □

Another result on the characterization of $\{y_n\}$ as an S.b. is contained in

<u>THEOREM 3.4.7</u> Suppose that (X,T) is a Fréchet space having an S.b. $\{x_n; f_n\}$. Let $\{y_j\}$ be the strictly nonzero sequence in X associated with an infinite matrix $A = [a_{ij}]$ as defined earlier. Recall the sequence space η_1 and η_2 related to $\{x_n\}$ and $\{y_n\}$ respectively. Then $\{y_j\}$ is an S.b. for (X,T) if and only if $A^\perp$ is 1-1 from η_1^β onto η_2^β,

$A^{\perp}$ being the transpose of A with $A^{\perp} = [a_{ji}]$.

PROOF. At the outset, let us observe that $A : \eta_2 \to \eta_1$ is a matrix transformation and since $(\eta_2, \sigma(\eta_2, \eta_2^{\beta}))$ is ω-complete, A is $\sigma(\eta_2, \eta_2^{\beta})$-$\sigma(\eta_1, \eta_1^{\beta})$ continuous. Consequently from the existence properties of adjoints (cf. [90]), the adjoint $A^* : \eta_1^{\beta} \to \eta_2^{\beta}$ exists and is $\sigma(\eta_1^{\beta}, \eta_1)$ - $\sigma(\eta_2^{\beta}, \eta_2)$ continuous. However, A^* can be identified with $A^{\perp}$; see, for instance, a similar discussion on p. 207 in [129]. Hence $A : \eta_2 \to \eta_1$ is 1-1 and onto if and only if $A^{\perp} : \eta_1^{\beta} \to \eta_2^{\beta}$ is 1-1 and onto; for (η_1, T_1^*) and (η_2, T_2^*) are AK-FK spaces and A is T_2^*-T_1^* continuous.

To complete the proof, now make use of Theorem 3.4.5 and the property that each t.b. in a Fréchet space is an S.b. □

4 λ-bases

4.1 MOTIVATION

The presence of an absolute S.b. in an l.c. TVS has often revealed a number of interesting properties of the space in question. Clasically, there is a known result of Karlin [139] that a Banach space having an absolute S.b. is isometrically isomorphic to ℓ^1. According to Newns [169], the topological dual of a Fréchet space having an absolute S.b. can be explicitly characterized in terms of this S.b. Interestingly enough, Dynin and Mitiagin [49] and Wojtynski [235] characterized the nuclearity of a Fréchet space equipped with an S.b. in terms of the absolute character of the latter (cf. chapter 12). Nuclear spaces possessing absolute S.b. have many other applications too.

Besides, the development of and growing interest in λ-nuclear spaces have made it necessary to have an insight once again into the structure of an absolute S.b. which might ultimately help develop this class of spaces. This was accomplished to some extent in [138].

Keeping in view the above discussion, this chapter, therefore, presents a preliminary and an unified study on several ingredients forming an absolute S.b., which are generally lumped together in a Banach or a Fréchet space, thereby giving rise to the so-called notion of an "absolute" base for either of these spaces.

4.2 SEVERAL NOTIONS OF λ-BASES

Throughout this chapter, λ stands for an arbitrary s.s. equipped with its normal topology $\eta(\lambda,\lambda^{\times})$ and (X,T) denotes an arbitrary l.c. TVS containing an S.b. $\{x_n;f_n\}$. Further, we also use the notation Δ and ν for the following sequence spaces

$$\Delta = \cap \{\frac{\lambda}{p^*} : p \in D_T\}, \; p^* = \{p(x_n)\};$$
$$\nu = \{\alpha \in \omega : \sum_{i=1}^{n} \alpha_i x_i \text{ is T-Cauchy}\}. \tag{4.2.1}$$

Let us also recall the sequence space $\delta = \eta$ from chapter 3. Then we have (cf. [133])

DEFINITION 4.2.2 $\{x_n;f_n\}$ is called a (i) *semi λ-base* if for each p in D_T, the mapping $\Psi_p : X \to \lambda$ is well defined, where for x in X, $\Psi_p(x) = \{p(x_n)f_n(x)\}$, (ii) *pre λ-base* if for each p in D_T, the mapping $\phi_p : \nu \to \lambda$ is well defined, where for α in ν, $\phi_p(\alpha) = \{\alpha_n\, p(x_n)\}$, (iii) *$\lambda$-base* if it is a semi λ-base and $\Delta \subset \delta$, (iv) *λ-pre Köthe base* if it is a pre λ-base and the map Ψ_p is T- $\eta(\lambda,\lambda^\times)$ continuous for each p in D_T, (v) *λ-Köthe base* if it is a λ-base and λ-pre Köthe base and (vi) *fully λ-base* if it is a semi λ-base and each Ψ_p is T-$\eta(\lambda,\lambda^\times)$ continuous.

NOTE. When $\lambda = \ell^1$, the term "ℓ^1-base" is usually referred to as "absolute base", and in this special case Definitions 4.2.2, (i) to (v) were introduced in [99], whereas Definition 4.2.2 (vi) is to be found in [204], p. 36.

EXERCISE 4.2.3 Prove that $\{x_n;f_n\}$ is a (1) semi λ-base if and only if $\delta \subset \Delta$, (2) pre λ-base if and only if $\nu \subset \Delta$, (3) λ-base if and only if $\delta = \Delta$, (4) λ-pre Köthe base if and only if $\nu \subset \Delta$ and each seminorm $Q_{p,\alpha}$ on X given by

$$Q_{p,\alpha}(x) = \sum_{n\geq 1} |\alpha_n f_n(x)|\, p(x_n); \; \alpha \in \lambda^\times, \; p \in D_T \tag{4.2.4}$$

is T-continuous, (5) λ-Köthe base if and only if $\delta = \Delta$, $\nu \subset \Delta$ and each $Q_{p,\alpha}$ is T-continuous; and (6) fully λ-base if and only if $\delta \subset \Delta$ and each $Q_{p,\alpha}$ is T-continuous.

Before we justify the introduction of Definition 4.2.2 by examples and counter-examples, let us observe that some of these notions of types of an S.b. related to the sequence space λ are themselves interrelated as shown in the following implication diagram:

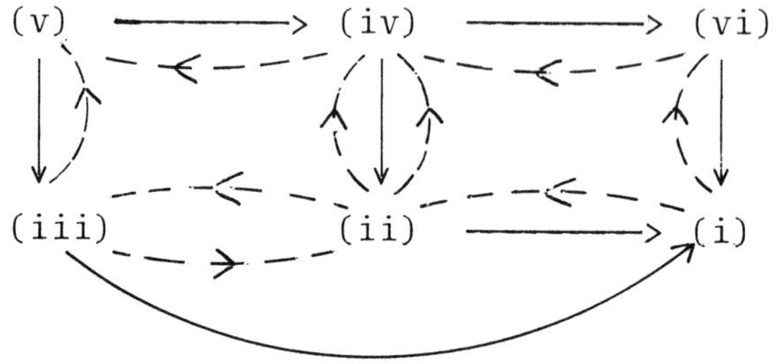

(4.2.5)

Here a dotted arrow means that the corresponding implication is not necessarily true.

Semi λ-bases

The concept of semi λ-bases is independent of other types of an S.b. mentioned above and its justification is provided in the following

PROPOSITION 4.2.6 (a) Let $c_0 \subset \lambda$, then $\{x_n;f_n\}$ is a semi λ-base for (X,T); (b) $\{e^n;e^n\}$ is a semi λ-base for an arbitrary l.c.s.s. $(\mu,\sigma(\mu,\mu^\times))$ (resp. $(\mu,\eta(\mu,\mu^\times))$) if and only if $\mu\cdot\mu^\times \subset \lambda$.

PROOF. (a) Observe that for x in X and p in D_T, $f_i(x)p(x_i) \to 0$ as $i \to \infty$. Hence $\{f_i(x)p(x_i)\} \in c_0 \subset \lambda$.

(b) Here $\{e^n;e^n\}$ is a semi λ-base for $(\mu,\sigma(\mu,\mu^\times))$ if and only if for every x in μ and y in $\mu^\times$, $\{\langle x,e^n\rangle q_y(e^n)\} \in \lambda$. The other part similarly follows. □

NOTE. Part (a) of the above proposition is a modification of a result proved in [35] for s.s. $\lambda \supset \ell^\infty$; whereas part (b) is amply demonstrated by the following (cf. [35], p. 513).

EXAMPLE 4.2.7 Consider the power series space $(\Lambda_\infty(\alpha), T_\infty)$; that is, the smooth s.s. $\Lambda_\infty(\alpha)$ of infinite type equipped with natural metrizable topology T_∞ so that it is a nuclear space; also it is a Fréchet space (Proposition 1.3.14). $\Lambda_\infty(\alpha)$ is clearly monotone and $(\Lambda_\infty(\alpha), T_\infty)$ is an AK-space. Thus $\eta(\Lambda_\infty(\alpha), \Lambda_\infty^\times(\alpha)) \subset \tau(\Lambda_\infty(\alpha), \Lambda_\infty^*(\alpha)) \approx T_\infty \subset \eta(\Lambda_\infty(\alpha), \Lambda_\infty^\times(\alpha))$. By Proposition 1.3.16, $\Lambda_\infty(\alpha) \cdot \Lambda_\infty^\times(\alpha) \subset \Lambda_\infty(\alpha)$ and $\Lambda_\infty^\times(\alpha)$. $\Lambda_\infty^{\times\times}(\alpha) \subset \Lambda_\infty^\times(\alpha)$; thus $\{e^n;e^n\}$ is a semi $\Lambda_\infty(\alpha)$- and $\Lambda_\infty^\times(\alpha)$-base for $(\Lambda_\infty(\alpha), T_\infty)$. Similarly $\{e^n;e^n\}$ is a semi $\Lambda_1(\alpha)$- and $\Lambda_1^\times(\alpha)$-base for the power series $(\Lambda_1(\alpha), T_1)$.

On the other hand, we have (cf. [136])

EXERCISE 4.2.8 Show that $\{e^n;e^n\}$ is not a semi ℓ^1-base either for $(k, \sigma(k, k^\beta))$ or for $(cs, \sigma(cs, k))$.

4.3 COUNTER-EXAMPLES

As mentioned in the preceding section, the concept of a semi λ-base is solely independent of other notions introduced in Definition 4.2.2 and the manner in which they are related is shown in (4.2.5). The importance of these notions is enhanced only by constructing counter-examples which should exhibit the nonequivalent character of these different types of an S.b. in the general setting of locally convex spaces and that is precisely what we are going to do in this section. However, the examples are related to the case when $\lambda = \ell^1$ and we follow [126], [127] and [136]. The general case $\lambda \neq \ell^1$ is yet to be investigated.

Observe that in what follows, each of the S.b. constructed below is a semi ℓ^1-base.

EXAMPLE 4.3.1 The S.b. $\{e^n;e^n\}$ for $(\ell^1,\sigma(\ell^1,k))$ is not a pre ℓ^1-base. For, if $\alpha_n = (-1)^{n+1}/n$; $n \geq 1$, then $\{\sum_{i=1}^{n} \alpha_i e^i\}$ is $\sigma(\ell^1,k)$-Cauchy but $\alpha \notin \ell^1$.

EXAMPLE 4.3.2 Consider the space $(c_o,\sigma(c_o,\ell^1))$ having the S.b. $\{e^n;e^n\}$. If $\{\sum_{i=1}^{n} \alpha_i e^i\}$ is $\sigma(c_o,\ell^1)$-Cauchy, then $\sum_{n\geq 1} \alpha_n y_n$ converges for each y in ℓ^1 and so $\alpha \in (\ell^1)^\beta = \ell^\infty$. Consequently, $\{\alpha_n\langle e^n,y\rangle\} \varepsilon\ \ell^1$ for each y in ℓ^1 and so $\{e^n;e^n\}$ is a pre ℓ^1-base and hence a semi ℓ^1-base. On the other hand $e \in \Delta$, $\notin c_o$; thus $\Delta \not\subset \delta$. Hence $\{e^n;e^n\}$ is a pre ℓ^1-base without being an ℓ^1-base.

EXAMPLE 4.3.3 This is once again the S.b. $\{e^n;e^n\}$ of Example 4.3.1. Here for α in Δ, $\{\alpha_n y_n\} \in \ell^1$ for each y in k. Hence $\alpha \in k^\times = \ell^1$; that is $\Delta \subset \delta = \ell^1$ and so $\Delta = \delta$. Thus $\{e^n;e^n\}$ is an ℓ^1-base for $(\ell^1,\sigma(\ell^1,k))$.

Next we present an example (cf. [136]) of an ℓ^1-pre Köthe base which is not an ℓ^1-Köthe base.

EXAMPLE 4.3.4 This is the S.b. $\{e^n;e^n\}$ for $(cs,\eta(cs,\ell^1))$. Consider α in ω so that $\{\sum_{i=1}^{n} \alpha_i e^i\}$ is Cauchy in $(cs,\eta(cs,\ell^1))$. Then $\alpha \in \ell^\infty$ and consequently $\{\alpha_n\langle e^n,\beta\rangle\} \in \ell^1$ for each β in ℓ^1. This shows that $\{e^n;e^n\}$ is a pre ℓ^1-base and hence also a semi ℓ^1-base. Further, for x in cs, α in ℓ^1 and β in ℓ^∞, $p_\beta(\{\langle x,e^n\rangle p_\alpha(e^n)\}) \leq \sup|\beta_n|\ p_\alpha(x)$, where p_α, etc. denote seminorms for $\eta(cs,\ell^1)$. This shows that $\{e^n;e^n\}$ is an ℓ^1-pre Köthe base. On the other hand, $e \in \Delta$ in this case but $e \notin \delta = cs$ and so $\{e^n,e^n\}$ is not an ℓ^1-base. Therefore $\{e^n;e^n\}$is not an ℓ^1-Köthe base for $(cs,\ \eta(cs,\ell^1))$.

EXAMPLE 4.3.5 The S.b. $\{e^n;e^n\}$ for $(\ell^2,\ \eta(\ell^2,\ell^2))$ is an ℓ^1-Köthe base. Here if $\alpha \in \Delta$, then $\{\alpha_n\beta_n\} \in \ell^1$ for each β in ℓ^2 and so $\alpha \in \ell^2 = \delta$. Thus $\{e^n;e^n\}$ is an ℓ^1-base.

Further, for α in ω with $\{\sum_{i=1}^{n} \alpha_i e^i\}$ being Cauchy in $(\ell^2, \eta(\ell^2, \ell^2))$, we find that $\alpha \in \ell^2$ (Theorem 1.3.7). Hence from Cauchy's inequality, $\{\alpha_n p_\beta(e^n)\} \in \ell^1$ for each β in ℓ^2 and this shows that $\{e^n; e^n\}$ is a pre ℓ^1-base for $(\ell^2, \eta(\ell^2, \ell^2))$. As in the preceding example, one can easily show that for each α in ℓ^2, the map $\Psi_{p_\alpha} : \ell^2 \to \ell^1$ is $\eta(\ell^2, \ell^2)$-$\eta(\ell^1, \ell^\infty)$ continuous and this completes the required construction.

We have already encountered one or two examples of fully ℓ^1-bases. On the other hand, let us offer an example of a S.b. which is a pre ℓ^1-base but not a fully ℓ^1-base.

EXAMPLE 4.3.6 This is the space $(\phi, \sigma(\phi, \omega))$ containing the S.b. $\{e^n; e^n\}$. Here one may use Theorem 1.3.7 to conclude the pre ℓ^1-character of $\{e^n; e^n\}$. Since $\sigma(\phi, \omega) \subsetneq \eta(\phi, \omega)$ (cf. [129], p. 106) $\{e^n; e^n\}$ cannot be a fully ℓ^1-base.

EXERCISE 4.3.7 Show that $\{e^n; e^n\}$ is a pre ℓ^1-base but not an ℓ^1-base for $(\phi, \sigma(\phi, \ell^1))$.

4.4 INTERRELATIONSHIP

Although the preceding section reveals that several notions relating to types of an S.b. as enunciated in Definition 4.2.2 are, in general, not equivalent, yet these are the same when λ satisfies a certain property, namely, the (K)-property and the space (X,T) is a ω-complete barrelled space. In particular, all the concepts relating to different types of "ℓ^1-bases" are equivalent in a Fréchet space. More precisely we have the following two results (cf. [133]).

PROPOSITION 4.4.1 Let $\{x_n; f_n\}$ be an S.b. for an l.c. TVS (X,T) and λ be an arbitrary s.s. Recall (i) to (vi) of Definition 4.2.2.

If (X,T) is ω-complete, then

(a) (i) $\Leftrightarrow$ (ii);

(b) (iv) $\Leftrightarrow$ (vi),

and if in addition λ satisfies the (K)-property, then

(c) (i) $\Leftrightarrow$ (iii).

Finally, if $\{x_n;f_n\}$ is a λ-pre Köthe base with λ satisfying the (K)-property, then

(d) (v) $\Leftrightarrow$ (X,T) is complete.

PROOF. The proof of (a), (b) and (c) is simple and so it is omitted (see (d) for the proof of (c), (i) $\Rightarrow$ (iii)).

(d) Let (X,T) be complete and hence it is ω-complete. If $\alpha \in \Delta$, then $\{\alpha_n\, p(x_n)\} \in \ell^1$ for each p in D_T (by the (K)-property of λ). Thus $\alpha \in \delta$; that is, $\Delta = \delta$.

Suppose that $\{x_n;f_n\}$ is a λ-Köthe base. Put $P = \{\{p(x_n)\beta_n\} : p \in D_T \text{ and } \beta \in \lambda_+^\times\}$, then P is a Köthe set. Since $\{p(x_n)f_n(x)\} \in \lambda$ for each p in D_T and x in X,

$$\sum_{n\geq 1} p(x_n)\ \beta_n|f_n(x)| < \infty;\ \forall x \in X,\ p \in D_T,\ \beta \in \lambda_+^\times. \tag{4.4.2}$$

Thus the linear map $\psi\colon X \to \Lambda(P)$, $\Psi(x) = \{f_n(x)\}$ is well defined and it is clearly injective. Next, we have some γ in $\lambda_+^\times$ such that

$$k_\gamma \equiv \inf_{n\geq 1} \gamma_n > 0. \tag{4.4.3}$$

Therefore, if $\alpha \in \Lambda(P)$, then

$$\sum_{n\geq 1} |\alpha_n|p(x_n) \leq k_\gamma^{-1} \sum_{n\geq 1} p(x_n)\gamma_n|\alpha_n| < \infty.$$

Since $\{x_n;f_n\}$ is a pre λ-base, $\{\alpha_n\ p(x_n)\} \in \lambda$ for each p in

D_T. Thus $\alpha \in \Delta = \delta$ (since $\{x_n;f_n\}$ is also a λ-base) and $\Psi(x) = \alpha = \{f_n(x)\}$. This shows that Ψ is bijective.

Next, observe that T_p is generated by $\{Q^*_{p,\beta}: p \in D_T, \beta \in \lambda^\times_+\}$, where for α in $\Lambda(P)$

$$Q^*_{p,\beta}(\alpha) = \sum_{n \geq 1} p(x_n)\ \beta_n |\alpha_n|. \tag{4.4.4}$$

Making use of the continuity of $Q_{p,\beta}$ (cf. (4.2.4)), we conclude the T-T_p continuity of Ψ. Also, for each p in D_T, $p(\psi^{-1}(\alpha)) \leq k_\gamma^{-1}\ Q^{P_*}_{p,\gamma}(\alpha)$ by using (4.4.3) and (4.4.4). Hence $(X,T) \simeq (\Lambda(P),T_p)$. Now make use of Proposition 1.3.14 to conclude the completeness of (X,T). □

<u>PROPOSITION 4.4.5</u> Let (X,T) be a Mackey S-space containing an S.b. $\{x_n;f_n\}$ and λ an arbitrary s.s. Then, recalling Definition 4.2.2, we have (a) (i) $\Leftrightarrow$ (vi), (b) (ii) $\Leftrightarrow$ (iv), and (c) (i) $\Rightarrow$ (ii) for $\lambda = \ell^1$.

<u>PROOF</u>. (a) It suffices to prove that (i) $\Rightarrow$ (vi). At the outset, we recall Proposition 3.3.6 to conclude that $\mu = \delta^\beta$. To prove the required result, it is enough to show that for each p in D_T, the map Ψ_p is $\sigma(X,X^*)$ - $\sigma(\lambda,\lambda^\beta)$ continuous. For, granting this temporarily, we find that Ψ_p is T-$\tau(\lambda,\lambda^\times)$ continuous (Proposition 1.2.15); that is, Ψ_p is T-$\eta(\lambda,\lambda^\times)$ continuous. Returning to the required proof, let $b \in \lambda^\beta$. Then $\{p(x_n)b_n\} \in \delta^\beta = \mu$ and so $p(x_n)b_n = f(x_n)$ for some f in X^* and all $n \geq 1$. Thus for each x in X,

$$\sum_{n\geq 1} p(x_n)b_n f_n(x) = f(x)$$

$$\Rightarrow q_b(\Psi_p(x)) = |f(x)|.$$

(b) The implication (ii) $\Rightarrow$ (iv) is contained in (a), (i) $\Rightarrow$ (vi).

(c) From (a) it follows that T is also generated by $\{Q_p : p \in D_T\}$ where Q_p is $Q_{p,e}$ as introduced in (4.2.4) (cf. Proposition 4.5.5 for a more general result). Hence for $\alpha \in \nu$, $p \in D_T$ and $\varepsilon > 0$, there exists N in $\mathbb{N}$ such that $\sum_{i=m}^{n} |\alpha_i| p(x_i) < \varepsilon$, for $m,n \geq N$ and so $\sum_{i\geq 1} |\alpha_i| \, p(x_i) < \infty$. □

NOTE. Proposition 4.4.5 extends a result proved in [34], p. 645; for minor variations of this proposition, see [35], p. 511-512.

4.5 FULLY λ-BASES AND APPLICATIONS

Fully λ-bases enjoy a comparatively privileged place among other notions introduced in Definition 4.2.2. Indeed, there are a number of applications which essentially follow by the presence of a fully λ-base in an l.c. TVS and in all such cases, the use of the (K)-property of λ is inevitable. In particular, this includes the case when $\lambda = \ell^1$ which is not $\eta(\ell^1, \ell^\infty)$-nuclear.

Before we cite some applications of fully λ-bases, let us pass on to an analytical characterization of these bases in the form of

PROPOSITION 4.5.1 Let $(\lambda, \eta(\lambda, \lambda^\times))$ be nuclear with λ being perfect. Then an S.b. $\{x_n; f_n\}$ for an l.c. TVS (X,T) is a fully λ-base if and only if for each p in D_T and β in $\lambda^\times$, there corresponds some q_β in D_T such that

$$\sup_{n\geq 1} |\beta_n f_n(x) p(x_n)| \leq q_\beta(x), \ \forall \ x \in X. \tag{4.5.2}$$

PROOF. Suppose that (4.5.2) is satisfied. By Theorem 1.4.13, we find γ in $\lambda^\times$ with $M \equiv \sum_{n\geq 1} |\beta_n/\gamma_n| < \infty$. Thus for each x in X, p in D_T and β in $\lambda^\times$,

$$\sum_{n\geq 1} |\beta_n f_n(x)| \, p(x_n) \leq M \, q_\gamma(x).$$

Thus $\{f_n(x)p(x_n)\} \in \lambda^{\times\times} = \lambda$ and Ψ_p is continuous. □

The proof of the following result is similar to the preceding one and so it is omitted.

PROPOSITION 4.5.3 Let an s.s. λ be nuclear with respect to $\eta(\lambda,\lambda^{\times})$. Then a semi λ-base $\{x_n,f_n\}$ for an l.c. TVS (X,T) is a fully λ-base if and only if (4.5.2) is satisfied.

Applications

In order to present some applications of the presence of a fully λ-base in an arbitrary l.c. TVS, we make the following assumptions in this subsection without further reference: (X,T) is an arbitrary l.c. TVS containing a fully λ-base with λ satisfying the (K)-*property*, namely, for some β in $\lambda_+^{\times}$,

$$k_\beta \equiv \inf \beta_n > 0. \tag{4.5.4}$$

On account of (4.5.4) and the fully λ-base character of $\{x_n.f_n\}$, the following is trivially verified

PROPOSITION 4.5.5 The topology T is also generated by $\{Q_{p,\alpha} : p \in D_T, \alpha \in \lambda^{\times}\}$, where $Q_{p,\alpha}$ is a seminorm on X defined by (4.2.4).

At the outset, let us observe that there are many spaces where weak sequential convergence is not the same as the original sequential convergence, e.g. consider ℓ^2. On the other hand, we have

THEOREM 4.5.6 Cauchy and convergent sequences in (X,T) and $(X,\sigma(X,X^*))$ are necessarily the same.

PROOF. Let $z_m \to 0$ in $\sigma(X,X^*)$. Since Ψ_p is T-$\eta(\lambda,\lambda^{\times})$ continuous for each p in D_T, Ψ_p^* exists from $\lambda^{\times}$ to X^*. Hence $\langle\Psi_p(z_m),\alpha\rangle = \langle z_m,\Psi_p^*(\alpha)\rangle \to 0$ for each α in $\lambda^{\times}$ and p in D_T.

Thus $\Psi_p(z_m) \to 0$ in $\sigma(\lambda,\lambda^\times)$ and hence in $\eta(\lambda,\lambda^\times)$ for each p (cf. Proposition 1.3.8). Therefore, $Q_{p,\alpha}(z_m) \to 0$ for each α in $\lambda^\times$ and p in D_T; consequently $z_m \to 0$ in T by the preceding proposition. The other part follows similarly. □

REMARK. If μ is a normal subspace of $\lambda^\times$, λ being an arbitrary s.s., then $\eta(\lambda,\mu)$ is compatible with the dual system $\langle\lambda,\mu\rangle$; cf. chapter 1. Hence we derive the following well-known result ([27], Chapter 10; [129], p. 77 and [225], p. 838).

COROLLARY 4.5.7 If λ is an arbitrary s.s., then for any normal subspace μ of $\lambda^\times$, Cauchy and convergent sequences in λ relative to $\sigma(\lambda,\mu)$ and $\eta(\lambda,\mu)$ are the same

THEOREM 4.5.8 An f in X' belongs to X* if and only if

$$f(x) = \sum_{n \geq 1} \alpha_n f_n(x), \ \forall\, x \in X \tag{4.5.9}$$

for some α in ω such that $\{\alpha_n/p(x_n)\} \in \lambda^\times$ for a suitable p in D_T, where 0/0 has to be regarded as 0.

PROOF. If f satisfies (4.5.9), then $|f(x)| \leq Q_{p,\beta}(x)$, where $\beta_n = \alpha_n/p(x_n)$, $n \geq 1$. Hence $f \in X^*$ by Proposition 4.5.5.

Conversely, let $f \in X^*$ and so $|f(x)| \leq p(x)$ for some p in D_T. If $\alpha_n = f(x_n)$, then

$$f(x) = \sum_{n\geq 1} \alpha_n f_n(x).$$

For showing $\{\alpha_n/p(x_n)\}$ to be a member of $\lambda^\times$, let $\gamma \in \lambda$. Then, using (4.5.4), we find that $\sum_{n\geq 1} |(\alpha_n/p(x_n))\gamma_n| \leq$

$k_\beta^{-1} \sum_{n\geq 1} |\gamma_n \beta_n| < \infty$. □

Next, we have (cf. [73])

PROPOSITION 4.5.10 There exists a Köthe set P such that $(X,T) \simeq (\delta, T_P|\delta)$, where δ is a dense subspace of $(\Lambda(P), T_P)$; in particular, if (X,T) is ω-complete, then $(X,T) \simeq (\Lambda(P), T_P)$.

PROOF. Let $P = \{\{p(x_n)\gamma_n\} : p \in D_T, \gamma \in \lambda_+^\times\}\}$. Since $\Psi_p : X \to \lambda$ is well defined for each p in D_T, $\delta \subset \Lambda(P)$ and the map $\Psi: X \to \delta$, $\Psi(x) = \{f_n(x)\}$ is a bijective linear map. The continuity of Ψ follows by the fully λ-base character of $\{x_n;f_n\}$. On the other hand, if $p \in D_T$ and $x \in X$,

$$p(\Psi^{-1}(\{f_n(x)\})) \leq \frac{1}{k_\beta} Q^*_{p,\beta} (\{f_n(x)\}).$$

Thus $(X,T) \simeq (\delta, T_p|\delta)$. To prove that $\bar{\delta} = \Lambda(P)$, let $\bar{\delta} \subsetneq \Lambda(P)$. Hence there exists α in $\Lambda(P)$ with $\alpha \notin \bar{\delta}$. By the Hahn-Banach theorem, we find a continuous linear functional f on $\Lambda(P)$ so that $\langle\alpha, f\rangle = 1$ and $\langle\beta, f\rangle = 0$ for every β in δ. The last equality yields $\langle e^n, f\rangle = 0$ for all $n \geq 1$ and so $\langle\alpha, f\rangle = 0$, a contradiction. Finally, let (X,T) be ω-complete and $\delta \subsetneq \bar{\delta}$. Then there exists α in $\Lambda(P)$ with $\alpha \notin \delta$. But $k_\beta \sum_{n\geq 1} |\alpha_n| p(x_n) < \infty$ for every p in D_T and so $\alpha = \{f_n(x)\}$ for some x in X. Therefore $\delta = \bar{\delta} = \Lambda(P)$ and the result follows from the first part. □

An immediate consequence of the above result is the following corollary, essentially due to Karlin [139].

COROLLARY 4.5.11 If (X,T) is a normed space containing an S.b. $\{x_n;f_n\}$ which is a fully ℓ^1-base, then $(X,T) \simeq (Y, \|\cdot\|_1)$, where Y is a dense subspace of ℓ^1 and if (X,T) is also ω-complete, then $(X,T) \simeq (\ell^1, \|\cdot\|_1)$.

PROPOSITION 4.5.12 Let (X,T) be ω-complete and let $(\lambda, \eta(\lambda, \lambda^\times))$ be nuclear. Then $(X,T) \simeq (\Lambda(P), T_p)$ for some Köthe space $\Lambda(P)$ and is nuclear.

PROOF. Invoking both the notation and proof of Proposition 4.5.10, it suffices to establish the nuclearity of

$(\Lambda(P), T_p)$. For each α in $\lambda_+^\times$, we find β in $\lambda_+^\times$ so that $\sum_{n\geq 1} \alpha_n/\beta_n < \infty$. The result now follows from Theorem 1.4.13 by observing that both $\{\alpha_n\, p(x_n)\}$ and $\{\beta_n\, p(x_n)\}$ belong to p. □

NOTE. The results of this subsection envelop results concerning ℓ^1-bases proved earlier in [126], [169], [179] and [225].

Relationship with γ-complete bases

Unless we state to the contrary, our assumption about (X,T), $\{x_n;f_n\}$ and λ remains the same as that which we spelled out in the last subsection. We will further discuss γ-complete bases in chapter 9; however, we present here some results on 'λ-bases' which ultimately yield γ-complete bases.

PROPOSITION 4.5.13 If (X,T) is ω-complete, then $\{x_n;f_n\}$ is γ-complete.

PROOF. Let $\{\sum_{i=1}^{n} \alpha_i x_i\}$ be bounded in (X,T) for some choice of α in ω. Fix p in D_T and ν in $\lambda_+^\times$. By Proposition 4.5.5, there exists q in D_T such that

$$\sum_{i=1}^{n} |\alpha_i|\nu_i\, p(x_i) \leq q\left(\sum_{i=1}^{n} \alpha_i x_i\right), \quad \forall n \geq 1.$$

The desired result follows by making use of (4.5.4) and ω-completeness of (X,T). □

PROPOSITION 4.5.14 Let $\{x_n;f_n\}$ be a λ-base for an S-space (X,T) with λ being perfect. Then $\{x_n;f_n\}$ is γ-complete.

PROOF. Here δ is perfect (Proposition 5.2.4). Therefore by Propositions 3.3.5, 3.3.6 and 2.3.5(v), $\{x_n;f_n\}$ is γ-complete for $(X,\sigma(X,X^*))$ and hence for (X,T). □

REMARK. The S-character of (X,T) in the preceding result cannot be dropped, for we have

EXAMPLE 4.5.15 Recall the ℓ^1-base $\{e^n;e^n\}$ for the space $(\ell^1,\sigma(\ell^1,k))$ (cf. Example 4.3.3). Obviously this space cannot be an S-space. If $\alpha_i = (-1)^{i+1}/i$, $i \geq 1$, then $\{\sum_{i=1}^{n} \alpha_i e^i\}$ is $\sigma(\ell^1,k)$-bounded in ℓ^1. But this sequence does not converge in this space.

Further applications : linear operators

Let us begin with the following proposition which envelops an earlier result in [34], p. 646.

PROPOSITION 4.5.16 Every continuous linear operator G from (X,T) into a Banach space $(Y, \|\cdot\|)$ factors continuously through a step-space of the s.s. λ. If λ is monotone, then G factors continuously through λ.

PROOF. There exists p in D_T so that $\|G(x)\| \leq p(x)$ and since $\lambda \subset \ell^1$, $\{\|G(x_n)\|/p(x_n)\} \in \lambda^\times$. Let $J = \{i \in \mathbb{N}: G(x_i) \neq 0\}$ and arrange it in the increasing order of integers, say $J = \{n_i\}$. Write λ_J for the stepspace of λ corresponding to J and let T_J denote the linear map from λ onto λ_J, defined as $T_J(\alpha) = \alpha_J$. Then T_J is $(\lambda,\eta(\lambda,\lambda^\times)) - (\lambda_J,\eta(\lambda_J,\lambda_J^\times))$ continuous (cf. [129], p. 135). Define $H_p : \lambda_J \to Y$ by

$$H_p(\alpha) = \sum_{i\geq 1} \alpha_{n_i} \frac{G(x_{n_i})}{p(x_{n_i})}, \quad \alpha \in \lambda_J.$$

Since the series on the right is absolutely convergent in Y, H_p is well defined. Further, $\|H_p(\alpha)\| \leq p_\beta(\alpha)$, where

$\beta = \{\| G(x_{n_i}) \| / p(x_{n_i})\} \in \lambda_J^{\times}$ and p_β is a seminorm generating $\eta(\lambda_J, \lambda_J^{\times})$. Thus H_p is continuous from $(\lambda_J, \eta(\lambda_J, \lambda_J^{\times}))$ to $(Y, \|\cdot\|)$. Also, note that $H_p \circ T_J \circ \Psi_p = G$. Finally, for the second statement observe that the canonical preimage $\bar{\lambda}_J$ of λ_J is a subspace of λ and then proceed as above with $\bar{\lambda}_J$ in place of λ_J. □

Next, we pass on to (cf. [35])

PROPOSITION 4.5.17 Let λ be normal and simple and $(X^*, \beta(X^*, x))$ be ω-complete. Then every continuous linear operator $G : (X,T) \to (\lambda, \eta(\lambda, \lambda^{\times}))$ is given by $G(x) = \{g_n(x)\}$, where $g_n \in X^*$, $n \geq 1$ and for each bounded subset B of (X,T), $\{p_B(g_n)\} \in \lambda$ with $p_B(f) = \sup \{|f(x)| : x \in B\}$, $f \in X^*$.

PROOF. Since λ is normal and simple, $\{p(x_n)p_B(f_n)\} \in \lambda$ for any bounded subset B of (X,T) and p in D_T; here we use the continuity of Ψ_p. By the continuity of G

$$G(x) = \sum_{n\geq 1} f_n(x) G(x_n) = \sum_{n\geq 1} \alpha^n f_n(x), \ \forall\, x \in X$$

where $\alpha^n = G(x_n)$. In particular, $(G(x))_m = \sum_{n \geq 1} \alpha_m^n f_n(x)$ where $\alpha_m^n = (G(x_n))_m$.

Choose γ in $\lambda^{\times}$ satisfying (4.4.3). Hence from the continuity of G, there exists q_γ in D_T such that

$$\sum_{m\geq 1} |(G(x))_m| \gamma_m \leq q_\gamma(x), \ \forall\, x \in X.$$

$$\Rightarrow \quad k_\gamma |\alpha_m^n| \leq q_\gamma(x_n), \ \forall\, m,n \geq 1$$

$$\Rightarrow \quad k_\gamma^2 \sum_{n\geq 1} |\alpha_m^n|\, p_B(f_n) < \infty,$$

for each $m \geq 1$ and every bounded subset B of (X,T). Therefore, there exists g_m in X^*, for each $m \geq 1$ with

$$g_m = \sum_{n \geq 1} \alpha_m^n f_n$$

and this yields $g_m(x) = (G(x))_m$ for $m \geq 1$.

Now

$$p_B(g_m) \leq \sum_{n \geq 1} |\alpha_m^n| \, p_B(f_n)$$

and for any β in $\lambda^{\times}$, there exists q_β in D_T with

$$\sum_{m \geq 1} |\alpha_m^n| \, |\beta_m| \leq q_\beta(x_n),$$

Therefore $\sum_{m \geq 1} |\beta_m| p_B(g_m) < \infty$ and so $\{p_B(g_m)\} \in \lambda$, since λ is perfect ([129], p. 89). □

Illustrations

In the past, much attention has been paid by different authors to determining exact forms of continuous linear functionals on spaces of analytic functions of one or more variables represented by Taylor or Dirichlet series. A bulk of such results may be traced back in [31], [59], [62], [63], [68], [81], [84], [85], [101], [102], [103], [114], [116], [117], [118], [120], [147], [148], [156] and many references given therein. It appears that in most of these results concerning the characterization of continuous linear functionals, an implicit use has been made of the presence of fully ℓ^1-bases in the underlying spaces. This is amply justified by the following two examples, illustrating the use of Theorem 4.5.8 to yield the precise forms of continuous linear functionals on two known Fréchet spaces having fully ℓ^1-bases.

EXAMPLE 4.5.18 This is the space X of all entire functions represented by Taylor series and equipped with the compact-open topology T. If $\delta_n \in X$ is defined by

$\delta_n(z) = z^n$ $(n \geq 0)$, then $\{\delta_n : n \geq 0\}$ is an S.b. for (X,T). This is indeed verified either by the classical complex variable theory or by use of Theorem 2.1.3 (cf. also Example 2.3.3 along with its single variable analogue). Since $\{\delta_n\}$ is clearly a semi ℓ^1-base, it is a fully ℓ^1-base by Proposition 4.4.5. Hence each ϕ in X^* is precisely of the form:

$$\phi(f) = \sum_{n \geq 0} a_n c_n \text{ with } f = \sum_{n \geq 0} a_n \delta_n,$$

where for some $r > 0$, the sequence $\{c_n/r^n : n \geq 0\}$ is bounded in $\mathbb{C}$.

EXAMPLE 4.5.19 Let $A > 0$ and $0 = \lambda_1 < \ldots < \lambda_n \to \infty$ with $\log n/\lambda_n \to 0$ be given. Denote by X_A the class of functions $f : \mathbb{C} \to \mathbb{C}$ such that $f(s) = \sum_{n \geq 1} \alpha_n \exp(s\,\lambda_n)$, $s = \sigma + it \in \mathbb{C}$, where

$$\limsup_{n \to \infty} \frac{\log |\alpha_n|}{\lambda_n} \leq -A.$$

Then X_A represents the family of all functions given by Dirichlet series such that each function has the same abscissa of convergence and absolute convergence, being at least equal to A (cf. [157], p. 33). One of the natural topologies T_A on X_A is the one generated by the sequence $\{\|\cdot\ \sigma_n\|\}$ of norms on X_A given by

$$\|f;\sigma_n\| = \sum_{i \geq 1} |\alpha_i| \exp(\sigma_n \lambda_i),$$

where $\sigma_1 < \ldots < \sigma_n \to A$. As outlined in Example 2.3.2, the sequence $\{e_n\}$ is an S.b. for the Fréchet space (X_A, T_A) (cf. also [114]). $\{e_n\}$ is clearly a fully ℓ^1-base for (X_A, T_A). Hence each ϕ in X_A^* is exactly of the form

$$\phi(f) = \sum_{n\geq 1} \alpha_n c_n ; \ f(s) = \sum_{n\geq 1} \alpha_n \exp(s \lambda_n),$$

where $\{|c_n|/\exp(\sigma \lambda_n)\} \in \ell^\infty$ for some $\sigma = \mathrm{Rl}(s) < A$.

4.6 THE WEAK SEMI λ-BASE PROBLEM

When considering the importance of fully λ-bases, it is worthwhile to investigate the minimum structural restrictions on an l.c. TVS equipped with a weak semi λ-base so that the latter becomes a semi λ-base for the space in question. Besides, it is (comparatively) easier to establish the weak semi λ-base character of an S.b. than to justify the semi λ-base property of the S.b. Throughout this section, we assume λ to be a normal s.s.

First we have

PROPOSITION 4.6.1 Every semi λ-base $\{x_n;f_n\}$ for an l.c. TVS (X,T) is a weak semi λ-base; that is, $\{x_n;f_n\}$ is a semi λ-base for $(X,\sigma(X,X^*))$ and if (X,T) is also an S-space, then the map $\phi_f : X \to \lambda$, $\phi_f(x) = \{f_n(x)f(x_n)\}$ is $\sigma(X,X^*)-\sigma(\lambda,\lambda^\times)$ continuous.

PROOF. Since for each f in X^* there exists p in D_T with $|f(x)| \leq p(x)$ and λ is normal, the element $\phi(x) \in \lambda$ for each x in X and f in X^*. This also shows that

$$\sum_{n\geq 1} |f(x_n)f_n(x)\alpha_n| < \infty \ ; \forall x \in X, \ f \in X^*, \ \alpha \in \lambda^\times.$$

In particular, for f in X^* $\{f(x_n)\alpha_n\} \in \delta^\beta$ for every α in $\lambda^\times$. But $\delta^\beta = \mu$ (Proposition 3.3.6) and so for some g in X^*, $\langle\phi_f(x),\alpha\rangle = \langle x,g\rangle$ for all x in X. This proves the last part. □

EXERCISE 4.6.2 Prove that the map ϕ_f in the preceding proposition is $\tau(X,X^*)-\tau(\lambda,\lambda^\times)$ continuous.

Finally, we pass on to a result which makes a weak semi

λ-base a fully λ-base in the original topology. Indeed, we have (cf. [35])

<u>PROPOSITION 4.6.3</u> Let λ be simple also. If $\{x_n; f_n\}$ is a weak semi λ-base for a barrelled space (X,T) which is $\sigma(X,X^*)$-ω-complete, then $\{x_n; f_n\}$ is a fully λ-base for (X,T).

<u>PROOF</u>. By the weak basis theorem for barrelled spaces (cf. [130], p. 88), $\{x_n; f_n\}$ is an S.b. for (X,T). In view of Proposition 4.4.5, it suffices to show that $\{x_n; f_n\}$ is a semi λ-base.

At the outset, let us observe that T is generated by $\{p_B$: B is $\sigma(X^*,X)$-bounded in $X^*\}$, where $p_B(x) = \sup\{|f(x)| : f \in B\}$, since (X,T) is barrelled. Consequently we will show that for each x in X and B as defined above, $\{f_n(x)p_B(x_n)\} \in \lambda$.

For x in X, define $F_x : X^* \to \lambda$ by $F_x(f) = \{f_n(x)f(x_n)\}$. Since $\{x_n; f_n\}$ is a weak λ-base and λ is normal, F_x is well defined. Thus for each α in $\lambda^\times$, x in X and f in X^*,

$$\sum_{n\geq 1} |f_n(x)\alpha_n \; f(x_n)| < \infty \Rightarrow \{f_n(x)\alpha_n\} \in \mu^\times; \forall x \in X, \; \alpha \in \lambda^\times.$$

By Proposition 3.3.6, for each pair of x and α with x in X and α in $\lambda^\times$, there corresponds y in X (we keep x fixed here) so that $f_n(x)\alpha_n = f_n(y)$ for $n \geq 1$. Thus $\langle F_x(f),\alpha\rangle = \langle y,f\rangle$ for all f in X^* and consequently $F_x[B]$ is $\sigma(\lambda,\lambda^\times)$-bounded in λ for every $\sigma(X^*,X)$-bounded subset B of X^*. Now the simple character of λ yields an element β in λ with β depending upon B (and also on x in X) such that

$$\sup_{f\in B} |f_n(x)f(x_n)| \leq |\beta_n|, \forall n \geq 1.$$

Therefore $\{f_n(x)p_B(x_n) \in \lambda\}$. □

5 λ-sequence spaces and ℓ^p-bases

5.1 INTRODUCTION

This chapter deals with two more aspects of the theory of bases as related to an s.s. λ introduced in chapter 4. First of all, we dispose of the question of the structure of the sequence space δ (see (3.3.1)) associated with one or another S.b. of Definition 4.2.2. In this case, δ is called a λ-sequence space. On the other hand, we also take up the structure and applications of λ-bases when $\lambda = \ell^p$, $1 \leqq p \leqq \infty$.

5.2 ASSOCIATED SEQUENCE SPACES

In what follows in this section, (X,T) denotes an arbitrary l.c. TVS containing an S.b. $\{x_n,f_n\}$ and λ stands for an arbitrary s.s. We once again recall the sequence space δ of (3.3.1).

PROPOSITION 5.2.1 If λ satisfies the (K)-property (4.5.4) and $\{x_n;f_n\}$ is a semi λ-base, then $\{p(x_n)\} \in \delta^\times$ for each p in D_T.

PROOF. Indeed, we have

$$k_\gamma \sum_{n\geq 1} |f_n(x)|p(x_n) < \infty;\ \forall p \in D_T,\ x \in X$$

and so $\{p(x_n)\} \in \delta^\times$ for each p in D_T. □

In general, the hypothesis of semi λ-base on $\{x_n;f_n\}$ cannot be dropped. For consider

EXAMPLE 5.2.2 We recall the space $(k,\sigma(k,k^{\beta}))$ of Exercise 4.2.8. Here $\lambda = \ell^1$ and if $\alpha = \{(-1)^{n+1}/n\} \in k^{\beta}$, then $\{q_{\alpha}(e^n)\} = \alpha \notin \ell^1 = \delta^{\times}$, where $\delta = k$.

PROPOSITION 5.2.3 If (X,T) is ω-complete, λ has the (K)-property and $\{x_n;f_n\}$ is a semi λ-base, then δ is perfect.

PROOF. Let $\alpha \in \delta^{\times\times}$. Since $\{p(x_n)\} \in \delta^{\times}$ for every p in D_T,

$$\sum_{n\geq 1} p(\alpha_n x_n) < \infty, \ \forall\ p \in D_T.$$

Therefore, for some x in X, $x = \sum_{n\geq 1} \alpha_n x_n$, that is, $\alpha \in \delta$. □

PROPOSITION 5.2.4 If λ is perfect and $\{x_n;f_n\}$ is a λ-base, then δ is perfect.

PROOF. Let $\alpha \in \delta^{\times\times}$. Since $\delta \subset \Delta$,

$$\sum_{n\geq 1} |f_n(x)\beta_n|\ p(x_n) < \infty;\ \forall\ x \in X,\ \beta \in \lambda^{\times} \text{ and } p \in D_T,$$

and so

$$\{\beta_n\ p(x_n)\} \in \delta^{\times} \text{ for each } \beta \in \lambda^{\times} \text{ and } p \in D_T.$$

Hence

$$\sum_{n\geq 1} |\alpha_n\beta_n|\ p(x_n) < \infty, \quad \forall \beta \in \lambda^{\times} \text{ and } p \in D_T.$$

Therefore $\{\alpha_n\ p(x_n)\} \in \lambda^{\times\times} = \lambda$, for each $p \in D_T$. Hence $\alpha \in \delta$ □

REMARK. If $\{x_n;f_n\}$ is a semi λ-base and δ is perfect, it does not necessarily follow that either $\{x_n;f_n\}$ is a λ-base with λ-perfect or (X,T) is ω-complete; in other words, the converse of Proposition 5.2.3 or 5.2.4 is not true. For, we have

EXAMPLE 5.2.5 Consider $(\phi,\sigma(\phi,cs))$. Then $\{e^n;e^n\}$ is clearly a semi ℓ^1-base (indeed, it is a semi λ-base with $\lambda = \ell^1$). Also $\delta = \phi$ and so δ is perfect. Recall the s.s. Δ of (4.2.1) with $\lambda = \ell^1$ and $p = q_\alpha$, $\alpha \in cs$. Then $\{1/n^2\} \in \Delta$ but $\{1/n^2\} \notin \delta = \phi$ and so $\{e^n;e^n\}$ is not an ℓ^1-base for $(\phi,\sigma(\phi,cs))$. Also, $(\phi,\sigma(\phi,cs))$ is not ω-complete, for $\{e^{(n)}\}$ is $\sigma(\phi,cs)$-Cauchy but does not converge in $(\phi,\sigma(\phi,cs))$.

EXERCISE 5.2.6 Prove that $\{e^n;e^n\}$ is a semi ℓ^1-base but not an ℓ^1-base for $(\phi,\sigma(\phi,\ell^1))$. Here δ is perfect.

REMARK. If λ satisfies the (K)-property and $\{x_n;f_n\}$ is a λ-Köthe base, then δ is perfect (cf. Proposition 4.4.2(d)). However, presence of a λ-pre Köthe base is not good enough even to yield the weakest Köthe structure of δ; for we have

EXAMPLE 5.2.7 This is the same as Example 4.3.4. Here the S.b. $\{e^n;e^n\}$ is an ℓ^1-pre Köthe base whereas δ is not even monotone.

PROPOSITION 5.2.8 If (X,T) is an S-space with $\{x_n;f_n\}$ being a λ-base with λ perfect, then $(X,\sigma(X,X^*)) \simeq (\delta,\sigma(\delta,\delta^\times))$.

PROOF. Apply Propositions 3.3.5, 3.3.6 and 5.2.4. □

REMARK. The restriction on (X,T) in the preceding proposition cannot be dropped. For, consider

EXAMPLE 5.2.9 Recall the space of Example 3.3.7 and note that $\{e^n;e^n\}$ is an ℓ^1-base for $(\ell^1,\sigma(\ell^1,c_o))$. Here $X = \ell^1$, $X^* = c_o$ and $\delta = \ell^1$ and so $(X,\sigma(X,X^*)) \not\simeq (\delta,\sigma(\delta,\delta^\times))$.

5.3 λ-BASES

This section is exclusively devoted to an application of a result of the last section. First we recall a result from [130], p. 40.

PROPOSITION 5.3.1 Let there be a Mackey S-space (X,T) such that $(X,\sigma(X,X^*))$ is ω-complete. If $(X,\sigma(X,X^*))$ contains an S.b., then (X,T) is complete.

Now we have

PROPOSITION 5.3.2 Let (X,T) be a Mackey S-space containing a λ-base $\{x_n;f_n\}$, where λ is perfect. Then (X,T) is complete.

PROOF. By Propositions 5.2.4, 5.2.8 and Theorem 1.3.7, $(X,\sigma(X,X^*))$ is ω-complete. Now make use of Proposition 5.3.1 to conclude the proof. □

REMARK. The condition that (X,T) is a Mackey S-space in Proposition 5.3.2 cannot be dropped, for we have

EXAMPLE 5.3.3 This is the space of Example 5.2.9. It is known that $(\ell^1,\sigma(\ell^1,c_0))$ is not complete (e.g. [129], p. 121). In view of our earlier disucssion, we need to show only $\sigma(\ell^1,c_0) \subsetneq \tau(\ell^1,c_0)$. In this direction, let us observe that $c_0' \supsetneq c_0^* = (c_0, \|\cdot\|_\infty)^*$. For, $\{1/n\} \in c_0 \setminus \phi$ and so there exists $f \in c_0'$ with $f(\{1/n\}) = 1$ and $f(\alpha) = 0$, $\alpha \in \phi$. Clearly $f \notin c_0^*$. Next we note that $(\ell^1,\tau(\ell^1,c_0))$ is complete (cf. [130], p. 39). Thus, if $\sigma(\ell^1,c_0) \approx \tau(\ell^1,c_0)$, then $\ell^1 = c_0'$ (cf. [80], p. 189). However, this leads to a contradiction.

On the other hand, the absence of either of the conditions on (X,T) in the foregoing proposition does not necessarily prevent (X,T) from being complete. This is illustrated in the next two examples.

EXAMPLE 5.3.4 Consider any arbitrary perfect s.s. λ with $\eta(\lambda,\lambda^\times) \subsetneq \tau(\lambda,\lambda^\times)$. Here $\{e^n;e^n\}$ is clearly an ℓ^1-base for $(\lambda,\eta(\lambda,\lambda^\times))$. By Theorem 1.3.7, this space is complete and using the same theorem, we also conclude that $(\lambda^x,\sigma(\lambda^x,\lambda))$ is ω-complete.

EXAMPLE 5.3.5 Let us confine ourselves to the S.b. $\{e^n;e^n\}$ for the complete space $(\ell^1,\tau(\ell^1,c_0))$ (cf. [130], p. 39). Here the space is not an S-space. In order to complete the desired requirement from this example, we have to show that $\{e^n;e^n\}$ is an ℓ^1-base. In this direction, it suffices to prove the semi ℓ^1-base character of $\{e^n;e^n\}$ for $(\ell^1,\tau(\ell^1,c_0))$. So, let B be any $\sigma(c_0,\ell^1)$-compact subset of c_0. Then $\sup\{q_\alpha(e^n) : \alpha \in B, n \geq 1\} < \infty$ and this establishes the required conclusion.

5.4 ℓ^p-BASES AND APPLICATIONS

If a Banach space X contains an S.b. $\{x_n;f_n\}$ satisfying either of the criteria (i) or (vi) of Definition 4.2.2 with $\lambda = \ell^1$, then $\{x_n;f_n\}$ is generally referred to as an "absolute base" for X and as mentioned earlier, in this case X behaves like ℓ^1 (cf. [139] and also [158], p. 42). This observation is also contained in Proposition 4.5.10 (in fact, here $P = \{\{\|x_n\| \beta_n\} : \beta \in \ell^\infty_+\}$ and let Ψ_1 be an isomorphism from $(\Lambda(P),T_p)$ onto $(\ell^1,\eta(\ell^1,\ell^\infty))$ defined by $\Psi_1(\alpha) = \{\|x_n\| \alpha_n\}$, then $p_\beta(\Psi_1(\alpha)) = Q^*_{\|\cdot\|,\beta}(\alpha)$ and $\Psi_1 \circ \Psi$ is the required isomorphism from $(X, \|\cdot\|)$ onto $(\ell^1, \|\cdot\|_1)$).

Obviously we cannot expect such a result as mentioned above for Fréchet spaces having ℓ^1-bases. However, in such a situation we can still find subsets L of a given Fréchet space (X,T) having an ℓ^1-base so that the corresponding Grothendieck space X_L behaves like the space ℓ^1 (cf. [176]). We will show in this section that this observation is a special case of a more general result due to Schock [208].

We first prove

PROPOSITION 5.4.1 Let $\{x_n;f_n\}$ be an ℓ^p-base $(1 \leq p < \infty)$ for a Fréchet space (X,T) with $D_T = \{p_i\}$, $p_1 \leq p_2 \leq \cdots$. Then $T \approx T^*$, where T^* is generated by $D^* = \{q_i\}$ with

$$q_i(x) = [\sum_{n\geq 1} (|f_n(x)|p_i(x_n))^p]^{1/p};\ \forall x \in X,\ i \geq 1.$$

PROOF. On account of the semi ℓ^p-base character of $\{x_n;f_n\}$, each q_i exists and is clearly a seminorm on X. For every $\varepsilon > 0$ and x in X there exists N so that

$$q_i(x) < [\sum_{j=1}^{N} (|f_j(x)|p_i(x_j))^p]^{1/p} + \varepsilon,$$

and if there is a net $\{y_\alpha\}$in X with $y_\alpha \to x$, it follows that $q_i(x) \leq \liminf q_i(y_\alpha) + \varepsilon$. Thus each q_i is T-lower semi-continuous and so $T^* \subset T$.

In order to apply the open mapping theorem to infer $T \subset T^*$, it suffices to show that (X,T^*) is ω-complete. So, let $\{y_k\}$ be a T^*-Cauchy sequence in X. Hence to every $i \geq 1$ and $\varepsilon > 0$, we can determine M in $\mathbb{N}$ such that

$$q_i(y_k-y_j) = [\sum_{n\geq 1} (|f_n(y_k-y_j)|p_i(x_n))^p]^{1/p} \leq \varepsilon;\ \forall k,j \geq M.$$

This allows us to find α_n in $\mathbb{K}$ with $f_n(y_k) \to \alpha_n$, $n \geq 1$ and consequently

$$\sum_{n=1}^{N} (|f_n(y_k)-\alpha_n|p_i(x_n))^p \leq \varepsilon^p;\ \forall\ N \geq 1,\ k \geq M.$$

This inequality shows that $\alpha \in \Delta$ (here $\lambda = \ell^p$) and so $\alpha = \{f_n(x)\}$ and also $q_i(y_k-x) \to 0$. □

NOTE. It is clear that the above result is also true for F-spaces. (Complete the proof as an exercise).

Let us now return to the main result (cf. [208]) of this section in the form of

THEOREM 5.4.2 Let $\{x_n; f_n\}$ be an ℓ^p-base $(1 \leqq p < \infty)$ for a Fréchet space (X,T) with $D \equiv D_T = \{p_1 \leqq p_2 \ldots\}$. Then corresponding to each bounded, closed, balanced and convex subset B of (X,T), there exists a similar subset L of (X,T) such that $B \subset L$ and the Grothendieck space (X_L, p_L) is isometrically isomorphic to $(\ell^p, \|\cdot\|_p)$.

PROOF. It is clear that for each fixed $i \geqq 1$, $\{p_i(x_n)/p_n(x_n)\}$ is bounded. Therefore for each $i \geqq 1$, there exists $k_i > 0$ such that

$$(*) \quad p_i(x_n) \leqq k_i \, p_n(x_n), \ \forall \ n \geqq 1$$

and

$$(**) \quad q_i(x) = \Big[\sum_{n \geqq 1} (|f_n(x)| p_i(x_n))^p \Big]^{1/p} \leqq k_i, \ \forall \ x \in B;$$

cf. Proposition 5.4.1. Set

$$\sigma_n = \Big[\sum_{i \geqq 1} (p_i(x_n)/2^{i/p} k_i)^p \Big]^{1/p}, \ n \geqq 1.$$

Then, by (*), $2^{-i/p} k_i^{-1} p_i(x_n) \leqq \sigma_n < \infty$ for $i, n \geqq 1$ and the set

$$L = \{x \in X : \|x\|_L \equiv \Big[\sum_{n \geqq 1} (|f_n(x)| \sigma_n)^p \Big]^{1/p} \leqq 1\}$$

is clearly balanced, convex and closed. Further, for x in L, it easily follows that $q_i(x) \leqq 2^{i/p} k_i$ and so L is also bounded in (X,T). If $x \in B$, then

$$\sum_{n \geqq 1} |f_n(x)|^p \sigma_n^p = \sum_{i \geqq 1} 2^{-i} k_i^{-p} \sum_{n \geqq 1} (|f_n(x)| p_i(x_n))^p \leqq 1$$

and this shows that $B \subset L$.

Next, we find that $\|\cdot\|_L$ is also a norm on X_L along with p_L. Since L is a closed unit ball for the normed spaces

(X_L, p_L) and $(X_L, \|\cdot\|_L)$, $p_L(x) = \|x\|_L$ for x in X_L.

Finally, consider the natural map F from the Banach space $(X_L, \|\cdot\|_L)$ into $(\ell^p, \|\cdot\|_p)$ with $F(x) = \{f_n(x)\sigma_n\}$. If $a \in \ell^p$, then $\{\alpha_n/\sigma_n\} \in \Delta = \delta$ and hence we find x in X with

$$x = \sum_{n\geq 1} \alpha_n \sigma_n^{-1} x_n .$$

It is evident that $x \in X_L$ and $F(x) = \alpha$. Thus F is a bijective map and for x in X_L

$$\|F(x)\|_p = [\sum_{n\geq 1} (|f_n(x)|\sigma_n)^p]^{1/p} = \|x\|_L . \quad \square$$

NOTE. Let $\mathcal{F}$ be a subfamily of the family $\mathcal{D}_X$, where X is an ω-complete l.c. TVS such that for each L in $\mathcal{F}$, $X_L \simeq \lambda$, λ being a fixed Banach s.s., then we write $\mathcal{D}_\lambda$ or $\mathcal{D}_{X,\lambda}$ for $\mathcal{F}$. With this notation, Theorem 5.4.2 can now be rephrased as follows (cf. [176], [208]).

THEOREM 5.4.3 For each Fréchet space (X,T) having an ℓ^p-base $\{x_n;f_n\}$, $1 \leq p < \infty$, there exists a family $\mathcal{D}_p$ which is fundamental in (X,T).

Estimation of Kolmogorov's diameters

If $\{x_n;f_n\}$ is a semi ℓ^p-base for an l.c. TVS (X,T), then for each ν in D_T, the function $\nu_p : X \to \mathbb{R}$ is well defined, where

$$\nu_p(x) = [\sum_{n\geq 1} |f_n(x)|^p \nu^p(x_n)]^{1/p}, \; x \in X. \tag{5.4.4}$$

Let $u_{\nu_p} = \{x \in X : \nu_p(x) \leq 1\}$. Then we have (cf. also [53], [209])

PROPOSITION 5.4.5 Let $\{x_n;f_n\}$ be a semi ℓ^p-base for an l.c. TVS (X,T). Let for each μ in D_T, there exist ν in D_T with $\mu(x_i) \leq \nu(x_i)$, $i \geq 1$, then

$$\delta_n(u_{\nu_p}, u_{\mu_p}) \leq \sup_{i \geq n} \{\mu(x_i)/\nu(x_i)\}, \; n \geq 1. \tag{5.4.6}$$

PROOF If $\nu(x_i) = 0$, then $\mu(x_i) = 0$ and we make the convention that $0/0 = 0$ (this will be clear from the proof).

Let $x \in X$ with $\nu_p(x) \leq 1$. Then

$$\mu_p(x - \sum_{i=1}^{n-1} f_i(x)x_i)$$

$$= [\sum_{i \geq n} \{|f_i(x)|^p \; \nu^p(x_i)\}\{\mu(x_i)/\nu(x_i)\}^p]^{1/p}$$

$$\leq \sup_{i \geq n} \{\mu(x_i)/\nu(x_i)\} \; \nu_p(x)$$

$$\Rightarrow u_{\nu_p} \subset \sup_{i \geq n} \{\mu(x_i)/\nu(x_i)\} u_{\mu_p} + L_{n-1},$$

where $L_{n-1} = \mathrm{sp}\,\{x_1, \ldots, x_{n-1}\}$. □

COROLLARY 5.4.7 If $p = 1$, then with the above hypothesis of Proposition 5.4.5,

$$\delta_n(u_{\nu_1}, u_\mu) \leq \delta_n(u_{\nu_1}, u_{\mu_1}) \leq \sup_{i \geq n} \{\mu(x_i)/\nu(x_i)\}, \; n \geq 1. \tag{5.4.8}$$

Indeed, $\mu \leq \mu_1$, and so the first inequality follows.

As an application of Corollary 5.4.7, we derive

PROPOSITION 5.4.9 Let $\{x_n; f_n\}$ be a fully ℓ^1-base for an l.c. TVS (X,T). Let for each μ in D_T there exist ν in D_T with $\mu(x_i) \leq \nu(x_i)$, $i \geq 1$ such that

$$\lim_{n \to \infty} \frac{\mu(x_n)}{\nu(x_n)} = 0. \tag{5.4.10}$$

Then (X,T) is Schwartz.

PROOF In fact, let $\mu \in D_T$. There exists ν in D_T so that (5.4.8) is satisfied. Thus $\delta_n(u_{\nu_1}, u_{\mu_1}) \to 0$. There exists η in D_T with $\nu_1 \leqq \eta$ and so $u_\eta \subset u_{\nu_1}$; also $u_{\mu_1} \subset u_\mu$. Thus

$$u_\eta \subset u_{\nu_1} \subset u_{\mu_1} \subset u_\mu$$

$$\Rightarrow \delta_n(u_\eta, u_\mu) \leqq \delta_n(u_{\nu_1}, u_\mu) \leqq \delta_n(u_{\nu_1}, u_{\mu_1}),$$

by Proposition 1.4.4, (v) and (vi). The required result follows by making use of Theorem 1.4.12. □

NOTE. A partial converse of Proposition 5.4.9 will be discussed in Proposition 12.4.5.

Restrictions on ℓ^p-bases : p-Köthe bases

The notions of different types of "λ-bases" introduced in Definition 4.2.2 happen to be of further use provided we modify some of them for specific spaces λ, for instance, when $\lambda = \ell^p$, $p \in (0,1) \cup (1,\infty)$. In fact, the prevailing motivation stems from the fact that the natural topologies of the spaces ℓ^p $(0 < p < \infty)$ are different from their normal topologies except when $p = 1$. Accordingly, let us introduce

DEFINITION 5.4.11 An S.b. $\{x_n; f_n\}$ for an l.c. TVS (X,T) is said to be *semi* p-*Köthe* $(0 < p < \infty)$ if it is a semi ℓ^p-base and $T \approx T^p$, where T^p is generated by $\{\nu_p : \nu \in D_T\}$.

Further $\{x_n; f_n\}$ is called p-*Köthe* if it is semi p-Köthe and an ℓ^p-base.

NOTE. An ℓ^p-base $(1 \leqq p < \infty)$ in an l.c. TVS is called a 'p-absolute base' by Kalton ([99], p. 220) whereas he calls an S.b. a p-Köthe base provided it is an ℓ^p-base $(1 \leqq p < \infty)$ and $T \approx T^p$. Schock [208] calls p-Köthe bases 'p-absolute bases', $1 \leq p < \infty$. An S.b. which is an ℓ^p-base as well as a fully ℓ^p-base $(0 < p < 1)$ is easily seen to be a p-Köthe base, whereas the converse is trivially true. The case $p = \infty$ will be dealt with subsequently.

PROPOSITION 5.4.12 Let $\{x_n; f_n\}$ be a p-Köthe base $(0 < p < \infty)$ for an l.c. TVS (X,T). Let $P = \{\{\nu(x_n)\}: \nu \in D_T\}$ and

$$\Lambda_p(P) = \{\alpha \in \omega: Q_{\nu,p}(\alpha) \equiv [\sum_{n\geq 1} |\alpha_n \nu(x_n)|^p]^{1/p} < \infty, \ \forall \nu \in D_T\}.$$

Equip $\Lambda_p(P)$ with the topology T_p generated by $\{Q_{\nu,p}: \nu \in D_T\}$. Then $(X,T) \simeq (\Lambda_p(P), T_p)$ and (X,T) is complete.

PROOF. Since $\delta \subset \Lambda_p(P)$, the map $F : X \to \Lambda_p(P)$, $F(x) = \{f_n(x)\}$ is a well-defined injection. By the ℓ^p-base character of $\{x_n; f_n\}$, if $\alpha \in \Lambda_p(P)$, then α also belongs to δ. Hence F is a bijection and since $T \approx T^p$, $(X,T) \simeq (\Lambda_p(P), T_p)$ under F. The completeness part of $(\Lambda_p(P), T_p)$ is a routine exercise.

PROPOSITION 5.4.13 Let $\{x_n; f_n\}$ be a p-Köthe base $(1 \leq p < \infty)$ for an l.c. TVS (X,T). Then for each ν in D_T, X_ν is either finite dimensional or is topologically isomorphic to a dense subspace of ℓ^p, where X_{ν_p} is the quotient space of X with respect to $\nu_p^{-1}(0)$ normed through $\hat{\nu}_p(\hat{x}) = \inf\{\nu_p(x-y) : \nu_p(y) = 0\}$.

PROOF. There are two cases: $\nu_p(x_n) \neq 0$ only for finitely many indices n or infinitely many indices $n = n_k$, $k \geq 1$. In the former case X_{ν_p} is finite dimensional. In the second case, let $Y = \{\{f_{n_k}(x)\ \nu_p(x_{n_k})\} : x \in X\}$. By the Hahn-Banach theorem, it is easy to show that Y is a dense subspace of $(\ell^p, \|\cdot\|_p)$. The map $R : X_{\nu_p} \to Y$, $R(\hat{x}) = \{f_{n_k}(x)\nu_p(x_{n_k})\}$ establishes an isometric isomorphism between x_{ν_p} and Y, since $\|R\hat{x}\|_p = \nu_p(x) = \hat{\nu}_p(\hat{x})$. □

Another estimation of Kolmogorov's diameters

The presence of semi p-Köthe bases in an l.c. TVS is further exploited in estimating a lower bound of Kolmogorov's diameters. In fact, we have (cf. also [53], [209])

LEMMA 5.4.14 Let $\{x_n;f_n\}$ be a semi p-Köthe base for an l.c. TVS (X,T), $1 \leqq p < \infty$. Let for each μ in D_T, there exist ν in D_T with $\mu(x_i) \leqq \nu(x_i)$, $i \geqq 1$. Then for $\mu(x_{n_i}) \neq 0$, $1 \leqq i \leqq s$,

$$\delta_s(u_{\nu_p}, u_{\mu_p}) \geqq \inf_{1 \leqq i \leqq s} \{\mu(x_{n_i})/\nu(x_{n_i})\} \tag{5.4.15}$$

PROOF Define $R : X \to \ell^p$, $Rx = \{f_n(x)\ \mu(x_n)\}$. Let $L_s = \mathrm{sp}\ \{e^{n_i} : 1 \leqq i \leqq s\}$, then L_s is a s-dimensional subspace of ℓ^p. Set $u_p = \{\alpha \in \ell^p : \|\alpha\|_p \leqq 1\}$. Observe that $R[u_{\nu_p}]$ is bounded in $(\ell^p, \|\cdot\|_p)$, for if $x \in u_{\nu_p}$, then $\|Rx\|_p \leqq \nu_p(x) \leqq 1$. Also, if $A = \inf\ \{\mu(x_{n_i})/\nu(x_{n_i}) : 1 \leqq i \leqq s\}$, then

$$(*) \quad A(u_p \cap L_s) \subset R[u_{\nu_p}].$$

In fact, let $\alpha \in A(u_p \cap L_s)$. Then $\alpha = A\beta$, where $\|\beta\|_p \leqq 1$ and

$$\beta = \sum_{i=1}^{s} \beta_{n_i} e^{n_i}.$$

Define

$$y = \sum_{i=1}^{s} A \frac{\beta_{n_i}}{\mu(x_{n_i})} x_{n_i}.$$

Then $\nu_p(y) \leqq \| \beta \|_p \leqq 1$ and

$$Ry = \sum_{i=1}^{s} A\beta_{n_i} e^{n_i};$$

thus $\alpha = Ry$. This proves (*). Hence by Proposition 1.4.6

$$\delta_s(R[u_{\nu_p}], u_p) \geqq A.$$

Since

$$R[u_{\nu_p}] \subset R[u_{\mu_p}] \subset u_p,$$

we have

$$\delta_s(u_{\nu_p}, u_{\mu_p}) \geqq \delta_s(R[u_{\nu_p}], R[u_{\mu_p}]) \geqq \delta_s(R[u_{\nu_p}], u_p) \geqq A$$

and this completes the proof. □

COROLLARY 5.4.16 Under the hypothesis of the preceding lemma with $p = 1$ or if $\{x_n; f_n\}$ is a fully ℓ^1-base, suppose that for each μ in D_T, there exists ν in D_T with $\mu(x_i) \leqq \nu(x_i)$ for $i \geqq 1$. Then for $\mu(x_{n_i}) \neq 0$, $1 \leqq i \leqq s$,

$$\delta_s(u_{\nu_1}, u_{\mu_1}) \geqq \inf_{1 \leqq i \leqq s} \{\mu(x_{n_i})/\nu(x_{n_i})\}. \qquad (5.4.17)$$

5.5 p AND q-KÖTHE BASES

In this section, we exploit the presence of an S.b. which is both p- and q-Köthe in an arbitrary l.c. TVS and it turns out that the corresponding space is nuclear. To achieve the objective, let us first prove (cf. [131])

LEMMA 5.5.1 Let $\{x_n; f_n\}$ be an S.b. for an l.c. TVS (X,T). Suppose that there is some $r > 0$ so that for each ν in D_T, there corresponds μ in D_T with

$$\sum_{n\geq 1} \left[\frac{\nu(x_n)}{\mu(x_n)}\right]^r < \infty, \quad (\tfrac{0}{0} = 0). \tag{5.5.2}$$

Then to every $s > 0$ and ν in D_T, there corresponds μ in D_T such that

$$\sum_{n\geq 1} \left[\frac{\nu(x_n)}{\mu(x_n)}\right]^s < \infty. \tag{5.5.3}$$

PROOF At the outset, let us mention that if $a_n, b_n > 0$; $n = 1,\ldots,N$ and $1/\gamma = 1/\alpha + 1/\beta$ with α and β being finite positive numbers, then it is a simple consequence of the Hölder inequality, namely,

$$(*) \qquad \left[\sum_{n=1}^{N} (a_n b_n)^{\gamma}\right]^{1/\gamma} \leq \left[\sum_{n=1}^{N} a_n^{\alpha}\right]^{1/\alpha} \left[\sum_{n=1}^{N} b_n^{\beta}\right]^{1/\beta}.$$

Choose the least positive integer k so that $k \geq r/s$ and let $\nu_0 = \nu$ be chosen arbitrarily from D_T. Determine $\nu_1,\ldots,\nu_k$ in D_T so that

$$\sum_{n\geq 1} \left[\frac{\nu_{i-1}(x_n)}{\nu_i(x_n)}\right]^r \leq M < \infty; \ \forall i, \ 1 \leq i \leq k.$$

Fix any N in $\mathbb{N}$. Then from (*), we get

$$\left[\sum_{n=1}^{N} \left(\frac{\nu_0(x_n)}{\nu_k(x_n)}\right)^s\right]^{1/s} \leq \left[\sum_{n=1}^{N} \left(\frac{\nu_0(x_n)}{\nu_1(x_n)}\right)^r\right]^{1/r} .$$

$$\left[\sum_{n=1}^{N} \left(\frac{\nu_1(x_n)}{\nu_2(x_n)} \cdots \frac{\nu_{k-1}(x_n)}{\nu_k(x_n)}\right)^{m_1}\right]^{\frac{1}{m_1}},$$

where $1/s = 1/r + 1/m_1$. Using (*) repeatedly, we obtain

$$\left[\sum_{n=1}^{N}\left(\frac{\nu(x_n)}{\nu_k(x_n)}\right)^s\right]^{1/s} \leqq \left[\sum_{n=1}^{N}\left(\frac{\nu_o(x_n)}{\nu_1(x_n)}\right)^r\right]^{1/r} \cdots$$

$$\cdots \left[\sum_{n=1}^{N}\left(\frac{\nu_{k-2}(x_n)}{\nu_{k-1}(x_n)}\right)^r\right]^{1/r} \quad \left[\sum_{n=1}^{N}\left(\frac{\nu_{k-1}(x_n)}{\nu_k(x_n)}\right)^{m_{k-1}}\right]^{\frac{1}{m_{k-1}}},$$

where

$$\frac{1}{s} = \frac{k-1}{r} + \frac{1}{m_{k-1}}$$

and so $m_{k-1} \geqq r$. The required result now follows by letting $N \to \infty$ in the preceding inequality. □

Next, we have (cf. [99], [111]).

PROPOSITION 5.5.4 Let $\{x_n;f_n\}$ be an uniformly equicontinuous S.b. for an l.c. TVS (X,T). Suppose that (5.5.2) is satisfied with $r = 1$, then (X,T) is nuclear.

PROOF. Choose ν and then μ with $\nu \leqq \mu$ so that (5.5.2) is satisfied with $r = 1$. Let $g_n = \nu(x_n)f_n$ and $\mu^*(x) = \sup\{|f_n(x)|\mu(x_n) : n \geqq 1\}$. Then μ^* is a member of D_T and $|g_n(x)| \leqq \mu^*(x)\ \{\nu(x_n)/\mu(x_n)\}$, for $n \geqq 1$ and x in X. Thus

$$\sum_{n\geqq 1} \sup\{|g_n(x)| : \mu^*(x) \leqq 1\} < \infty$$

and

$$(*) \qquad \nu(x) \leqq \sum_{n\geqq 1} |g_n(x)|, \quad \forall\ x \in X.$$

The nuclearity of (X,T) now follows by an application of Theorem 1.4.2. □

NOTE. The converse of the above result will be discussed in chapter 12.

Now we have the main result [99].

THEOREM 5.5.5 Let an l.c. TVS (X,T) have an S.b. $\{x_n;f_n\}$ which is both p- and q-Köthe with $1 \leq p < q < \infty$. Then (X,T) is nuclear.

PROOF. By the hypothesis, T is generated by $D_p^* = \{\nu_p : \nu \in D_T\}$, or, equivalently by $D_q^* = \{\mu_q : \mu \in D_T\}$, where ν_p is given by (5.4.4) and μ_q is defined similarly. Thus, for every ν in D_T, there exists a μ in D_T such that $\nu_p(x) \leq \mu_q(x)$ for all x in X; that is,

$$\sum_{n\geq 1} (|f_n(x)|\nu(x_n))^p \leq [\sum_{n\geq 1} (|f_n(x)|\mu(x_n))^q]^{p/q}, \forall x \in X.$$

Let $\alpha \in \omega$ with $\alpha_n \geq 0$, $n \geq 1$ and suppose that

$$y^N = \sum_{i\geq 1} (\alpha_i^{(N)})^{1/p} x_i = \sum_{i=1}^{N} \alpha_i^{1/p} x_i,$$

then

$$\sum_{i=1}^{N} \alpha_i(\nu(x_i))^p \leq [\sum_{i=1}^{N} \alpha_i^{q/p} (\mu(x_i))^q]^{p/q}.$$

Choose s with $1/s + p/q = 1$ and let

$$\alpha_i = \frac{[\nu(x_i)]^{ps-p}}{[\mu(x_i)]^{ps}}, \quad i \geq 1.$$

Hence we get

$$\sum_{i=1}^{N} [\frac{\nu(x_i)}{\mu(x_i)}]^{ps} \leq 1,$$

and therefore

$$\sum_{n\geq 1} \left[\frac{\nu(x_n)}{\mu(x_n)}\right]^r < \infty ,$$

where $1/p - 1/q = 1/r$ with $r = ps > 1$. Consequently (5.5.3) follows with $s = 1$. The final result is obtained by an application of Proposition 5.5.4. □

5.6 ℓ^p- AND p-KÖTHE BASES WHEN $p = \infty$

Earlier we discussed ℓ^p- and p-Köthe bases when $1 \leq p < \infty$. A natural question arises about the concept of these bases as well as their applications when $p = \infty$. This section offers only preliminaries for this special case and we will take up this issue again when we deal with shrinking bases.

To begin with, let us observe that if we consider an ℓ^∞-base $\{x_n;f_n\}$ for an l.c. TVS (X,T), then it follows that whenever $\{\alpha_n\, p(x_n)\} \in \ell^\infty$ for α in ω and any p in D_T, we should have $\alpha_n = f_n(x)$ for x in X and all $n \geq 1$. Consequently $p(\alpha_n x_n) \to 0$ necessarily. However, this could be inconsistent in general. Thus, in order to overcome this difficulty, we may modify the definition of ℓ^∞-bases in the form of (cf. [99])

DEFINITION 5.6.1 An S.b. $\{x_n;f_n\}$ for an l.c. TVS (X,T) is said to be ∞-*absolute* provided $\{\alpha_n c_n\} \in \delta$ whenever $\{c_n\} \in c_o$ and $\{\alpha_n\, p(x_n)\} \in \ell^\infty$ for every p in D_T.

The definition of an ℓ^p-Köthe base or p-Köthe base can also be modified as follows (cf. [99]).

DEFINITION 5.6.2 An S.b. $\{x_n;f_n\}$for an l.c. TVS (X,T) is called ∞-*Köthe* if it is an ∞-absolute base and T is generated by the family $D_\infty = \{\nu_\infty : \nu \in D_T\}$, where for x in X,

$$\nu_\infty(x) = \sup_{n\geq 1} |f_n(x)|\, \nu(x_n). \tag{5.6.3}$$

Following [99], let us begin with

PROPOSITION 5.6.4 Let (X,T) be a Fréchet space with $D_T = \{\nu^1 \leq \nu^2 \leq \ldots\}$. Then every ∞-absolute base $\{x_n;f_n\}$ for (X,T) is ∞-Köthe.

PROOF. The proof is similar to that of Proposition 5.4.1. To begin with, let us observe that each ν_∞^n is lower semi-continuous on (X,T) and hence $T^* \subset T$, where T^* is generated by $D_\infty^* = \{\nu_\infty^n : n \geq 1\}$.

To complete the proof, it suffices to prove that (X,T^*) is complete. Let $\{y_n\}$ be T^*-Cauchy in X. Then there exists α in ω so that $f_n(y_m) \to \alpha_n$ for $n \geq 1$ and $f_n(y_m)\ \nu^k(x_n) \to \alpha_n\ \nu^k(x_n)$ uniformly in $n \geq 1$ and for each $k \geq 1$. Since $f_n(y_m)\nu^k(x_n) \to 0$ as $n \to \infty$, we deduce $\nu^k(\alpha_n x_n) \to 0$ for each $k \geq 1$. By the metrizable character of T, we can find $\beta \in c_o$ so that $\beta_n^{-1}\alpha_n x_n \to 0$ in (X,T) (cf. [201], p. 20). Thus for x in X, $\alpha_n = f_n(x)$, $n \geq 1$. Employing the usual arguments, it now readily follows that $y_m \to x$ in (X,T^*). □

PROPOSITION 5.6.5 Let $\{x_n;f_n\}$ be an ∞-absolute base for an ω-complete l.c. TVS (X,T). Then $\{f_n\} \equiv \{f_n;\Psi(x_n)\}$ is a semi ℓ^1-base for $(X^*,\beta(X^*,X))$, Ψ being the canonical embedding from X into X^{**}.

PROOF. For any bounded subset B of (X,T), let $p_B(f) = \sup\{|f(x)| : x \in B\}$ denote a seminorm belonging to D_β, where $\beta \equiv \beta(X^*,X)$. Put $\mu_n = p_B(f_n)$, $n \geq 1$.

Since (X,T) is a W-space, $\{x_n;f_n\}$ is simple and so the set $E = \{\sum_{i=1}^{n} f_i(x)x_i : x \in B,\ n \geq 1\}$ is T-bounded in X. Hence $\{\mu_n x_n\}$ is also bounded in (X,T). Consequently, if $\alpha \in c_o$, then for some x in X, $\alpha_n\mu_n = f_n(x)$ for $n \geq 1$. In other words, for each α in c_o, the series $\Sigma\ \alpha_n\mu_n x_n$ converges in (X,T) and hence it converges weakly. Thus for each f in X^*, $\{\mu_n\ f(x_n)\} \in c_o^\beta = \ell^1$; we have

$$(*) \qquad \sum_{n\geq 1} \mu_n |f(x_n)| < \infty, \quad \forall f \in X^*.$$

Hence, for f in X*,

$$p_B(\sum_{i=1}^{n} f(x_i)f_i - f) \leq \sum_{i>n} |f(x_i)| \mu_i \to 0 \text{ as } n \to \infty ,$$

and so $\{f_n;\Psi(x_n)\}$ is an S.b. for $(X^*,\beta(X^*,X))$; also (*) shows that this S.b. is a semi ℓ^1-base. □

NOTE. The above result is further strengthened in Theorem 9.5.15.

EXERCISE 5.6.6 (cf. [99]) let $\{x_n;f_n\}$ be an ∞-absolute base for an l.c. TVS (X,T). Prove that $\{x_n;f_n\}$ is unconditional for $(X,\sigma(X,X^*))$. If (X,T) is also $\sigma(X,X^*)$ ω-complete, then show that $\{x_n;f_n\}$ is bounded multiplier. [Hint: First establish the fact that for x in X and f in X*, $\{f_n(x)f(x_n)\} \in c_o^\beta$ and next use Theorem 1.5.4.]

Our next result corresponds to an extreme case of Theorem 5.4.3 when $p = \infty$ and is reproduced from [176].

THEOREM 5.6.7 For each weakly ω-complete Fréchet space (X,T) with $D_T = \{p_1 \leq p_2 \leq \dots\}$ and having an ∞-absolute base $\{x_n;f_n\}$, there corresponds a family $\mathcal{D}_{\ell^\infty}$ which is fundamental in (X,T).

PROOF. By Proposition 5.6.5, $\{f_n;\Psi(x_n)\}$ is a semi ℓ^1-base for $(X^*,\beta(X^*,X))$. Put $y_n = x_n/(p_n(x_n) + 1)$, $n \geq 1$. Then $\{y_n\}$ is bounded in (X,T). For each B in $\mathcal{D}_X$, let $\hat{B}$ denote the balanced closed convex hull of $B \cup \{y_n\}$. Then for each B in $\mathcal{D}_X$, $\hat{B} \in \mathcal{D}_X$ and $B \subset \hat{B}$.

For each B in $\mathcal{D}_X$, consider the seminorm p_B on X* relative to the topology $\beta(X^*,X)$. If

$$M_n^B = \{x \in X : |f_n(x)| \leq p_B(f_n)\},$$

then $B \subset M_n^B$ for $n \geq 1$ and so the set

$$M^B = \cap \{M_n^B : n \geq 1\}$$

contains B. Also $M^B \in \mathcal{D}_X$. Indeed, for x in M^B and any given f in X*,

$$|f(x)| \leq \sum_{n\geq 1} |f(x_n)|\, p_B(f_n) \equiv K_{B,f} < \infty .$$

Since M^B is already balanced, convex and closed, the requirement on M^B that it belongs to $\mathcal{D}_X$ is established. Thus, if A is bounded, then we can find some B in $\mathcal{D}_X$ so that $A \subset B \subset M^B \in \mathcal{D}_X$. Hence, to show that $\mathcal{D}_{\ell^\infty} = \{M^B : B \in \mathcal{D}_X\}$ is the required family, we need prove that $X_{M^B} \simeq \ell^\infty$ for each B in $\mathcal{D}_X$.

Indeed, if we write M for M_B, then the natural map $F : (X_M, \|\cdot\|_M) \to (\ell^\infty, \|\cdot\|_\infty)$, $F(x) = \{f_n(x)/p_B(f_n)\}$ is well defined, where $\|\cdot\|_M$ is the usual norm associated with the Banach space X_M. If $\alpha \in \ell^\infty$, then $\{\sum_{i=1}^{n} \alpha_i p_B(f_i) x_i\}$ is weakly Cauchy and so it converges to some x in X and $F(x) = \alpha$. Clearly $\|F(x)\|_\infty \leq \|x\|_M$. Therefore, F is a continuous linear bijection from $(X_M, \|\cdot\|_M)$ onto $(\ell^\infty, \|\cdot\|_\infty)$. □

In the following results, we derive some consequences of the presence of an ∞-Köthe base in an arbitrary l.c. TVS.

We have ([108])

<u>PROPOSITION 5.6.8</u> Let $\{x_n ; f_n\}$ be an ∞-Köthe base for an l.c. TVS (X,T). Suppose that for each ν in D_T, there exists μ in D_T such that

$$\frac{\nu(x_n)}{\mu(x_n)} \to 0 \text{ as } n \to \infty. \tag{5.6.9}$$

Then (X,T) is a Schwartz space.

PROOF. It is easy to see that to every ν in D_T, there exists μ in D_T so that $\{\nu(x_n)/\mu(x_n)\} \in \ell^\infty$. Thus the requirement (5.6.9) is stronger than that which would follow as a natural consequence of the presence of the ∞-Köthe base $\{x_n;f_n\}$ in (X,T).

Let $\eta \in D_T$ be fixed. There exists ν in D_T so that $\eta \leqq \nu_\infty$ and choose μ in D_T so that (5.6.9) is satisfied. Define $g_n = \nu(x_n)f_n$, $n \geqq 1$. Then $\sup \{|g_n(x)| : \mu_\infty(x) \leqq 1\} \leqq \nu(x_n)/\mu(x_n) \to 0$ as $n \to \infty$; also $\eta(x) \leqq \nu_\infty(x) = \sup \{|g_n(x)| : n \geqq 1\}$ for each x in X. The required proof is completed by applying Theorem 1.4.11. □

NOTE. A variation of the converse of the foregoing proposition will be taken up in chapter 12.

p-Köthe bases, $1 \leqq p \leqq \infty$

Having discussed the concepts of p-Köthe bases $(0 < p < \infty)$ and ∞-Köthe bases, we now take up the general case of p-Köthe bases when $1 \leqq p \leqq \infty$ and essentially exploit their presence in an arbitrary l.c. TVS. Let us begin with

PROPOSITION 5.6.10 Let $\{x_n;f_n\}$ be an S.b. which is both ℓ^1-Köthe and ∞-Köthe. Then for every p, $1 \leqq p \leqq \infty$, $\{x_n;f_n\}$ is p-Köthe.

PROOF. Note that an ℓ^1-Köthe base is the same as a 1-Köthe base. To prove the result, it suffices to consider the case when p is any given number with $1 < p < \infty$. With this choice of p, find q so that $1 = 1/p + 1/q$. Let $\nu \in D_T$. There exist μ and η in D_T so that $\nu_p(x) \leqq \nu_1(x) \leqq \mu_\infty(x) \leqq \eta(x)$ and

$$\sum_{n\geqq 1} |\alpha_n| \nu(x_n) \leqq \Big[\sum_{n\geqq 1} (|\alpha_n| \mu(x_n))^p\Big]^{1/p}\Big[\sum_{n\geqq 1}\left(\frac{\nu(x_n)}{\mu(x_n)}\right)^q\Big]^{1/q}.$$

In fact, $\nu_1(x) \leqq \mu_\infty(x)$ with $x = \sum_{i=1}^{N} x_i/\mu(x_i)$, $N \in \mathbb{N}$, implies that $\sum_{i=1}^{N} \nu(x_i)/\mu(x_i) \leq 1$ and so $\sum_{i\geq 1} (\nu(x_i)/\mu(x_i))^q < \infty$, for $q > 1$. Thus $\{x_n;f_n\}$ is an ℓ^p-base and $\nu(x) \leq k\, \mu_p(x)$, $\nu_p(x) \leq \eta(x)$, where $k = [\sum_{n\geq 1} (\nu(x_n)/\mu(x_n))^q]^{1/q} > 0$, is a constant. □

PROPOSITION 5.6.11 Let an l.c. TVS (X,T) contain an S.b. $\{x_n;f_n\}$ which is both p- and q-Köthe, $1 \leq p < q \leq \infty$. Then (X,T) is nuclear.

PROOF. In view of Theorem 5.5.5 and Proposition 5.6.10, it is enough to prove the result when $1 < p < \infty$ and $q = \infty$.

Let $\nu \in D_T$, then $\nu_p \in D_T$. Then there exists μ in D_T so that $\nu_p(x) \leq \mu_\infty(x)$, for every x in X. For any $N \geq 1$, let

$$y_N = \sum_{i=1}^{N} \frac{x_i}{\mu(x_i)}$$

then

$$[\sum_{i=1}^{N} (\frac{\nu(x_i)}{\mu(x_i)})^p]^{1/p} \leq 1, \ \forall\, N \geq 1.$$

Therefore, as before, the result is proved by appealing to Lemma 5.5.1 and Proposition 5.5.4. □

5.7 p-KÖTHE BASES, $0 < p < 1$

Concerning p-Köthe bases, the only case which we have yet to deal with is the one when $0 < p < 1$. In this direction we prove the following result of which an alternative proof is given in [193].

THEOREM 5.7.1 Each l.c. TVS (X,T) having a p-Köthe base $\{x_n;f_n\}$ with $0 < p < 1$, is nuclear.

PROOF. Let $\nu \in D_T$. Then there exists μ in D_T so that

$$[\sum_{n\geq 1} (|f_n(x)|\ \nu(x_n))^p]^{1/p} \leq \mu(x), \quad \forall x \in X;$$

and also for each x in X

$$\mu(x) \leq \sum_{n\geq 1} |f_n(x)|\ \mu(x_n) \leq [\sum_{n\geq 1} (|f_n(x)|\mu(x_n))^p]^{1/p} < \infty .$$

Therefore

$$(*) \quad [\sum_{n\geq 1} (|f_n(x)|\ \nu(x_n))^p]^{1/p} \leq \sum_{n\geq 1} |f_n(x)|\ \mu(x_n), \quad \forall x \in X.$$

Let $r = p/(1-p)$ and for each $N \geq 1$, define

$$y_N = \sum_{i=1}^{N} \frac{(\nu(x_i))^r}{(\mu(x_i))^{r+1}} x_i \in X.$$

Then $|f_i(y_N)|\ \nu(x_i) = (\nu(x_i)/\mu(x_i))^{r+1}$ for $1 \leq i \leq N$

and $|f_i(y_N)|\ \mu(x_i) = (\nu(x_i)/\mu(x_i))^r$, for $1 \leq i \leq N$.

Hence from (*)

$$[\sum_{i=1}^{N} \left(\frac{\nu(x_i)}{\mu(x_i)}\right)^r]^{(1-p)/p} \leq 1$$

$$\Rightarrow \sum_{n\geq 1} \left(\frac{\nu(x_n)}{\mu(x_n)}\right)^r < \infty.$$

It now remains to apply Lemma 5.5.1 and Proposition 5.5.4 to conclude the nuclearity of (X,T). □

6 Unconditional bases

6.1 INTRODUCTION

Unconditional bases (abbreviated hereafter as u-S.b.) form an integral part of the Schauder basis theory in locally convex spaces. A sufficient amount of attention has already been paid in [130] to establishing characterizations and the importance of the geometrical aspect of unconditional bases. In this chapter, we will deal with those aspects of unconditional bases which we could not cover in our earlier work, namely, the size of the sequence space (s.s.) δ associated with a u-S.b., the behaviour of a u-S.b. in compatible topologies, the relationship of a u-S.b. with different types of absolute bases and so on.

It is clear that each bounded multiplier base (b.m-S.b.) in an l.c. TVS (X,T) is a subseries base (s-S.b.), which is, in turn, a u-S.b.: if (X,T) is ω-complete, these three notions of bases are the same (for example, [129], chapter 3). Although the concept of a u-S.b. is, in general, the weakest of the three types of S.b. described above, for all practical purposes, we generally work in ω-complete spaces and thus, in such a case, we are dealing with either of these three notions of an S.b. interchangeably.

It will be useful to recall a characterization of these bases. Let $I = \bar{a}$, $\bar{b}$ or $\bar{e}$, where $\bar{a} = \{\alpha \in \omega: \alpha_n = 1 \text{ or } -1, n \geq 1\}$, $\bar{b} = \{\beta \in \omega: \beta_n = 0 \text{ or } 1, n \geq 1\}$ and

$\bar{e} = \{\varepsilon \in \omega: |\varepsilon_n| \leq 1, n \geq 1\}$.

The following two results of Weill [232] (cf. also [77]) may be found in [130].

THEOREM 6.1.1 Let there be an arbitrary sequence $\{x_n\}$ in an l.c. TVS (X,T) with $x_n \neq 0$, $n \geq 1$. Consider the condition

Given p in D_T, there exists q in D_T such that for arbitrary a in ω, m and n in $\mathbb{N}$ with $m \leq n$ (m and n can be ∞) and γ in I such that

$$p(\sum_{i=1}^{m} \gamma_i a_i x_i) \leq q(\sum_{i=1}^{n} a_i x_i). \tag{6.1.2}$$

(i) If (X,T) is ω-complete, then (6.1.2) implies that $\{x_n\}$ is a u-S.b. for $[x_n]$. (ii) If $\{x_n\}$ is a u-S.b. for (X,T) which is assumed to be barrelled, then (6.1.2) holds. (iii) If (X,T) is ω-complete and barrelled, and $\{x_n\}$ is a u-S.b. for (X,T), then the topology T is also generated by the family $\{P_p : p \in D_T\}$ of seminorms on X, where for x in X.

$$P_p(x) = \sup_{|\alpha_n| \leq 1} p(\sum_{n \geq 1} \alpha_n f_n(x) x_n). \tag{6.1.3}$$

Theorem 6.1.1 leads to

PROPOSITION 6.1.4 Let $\{x_n; f_n\}$ be a u-S.b. for a barrelled space (X,T). For an arbitrary sequence $\{y_n\}$ in X and a sequence $\{\sigma_n\}$ from the family Φ of all finite subsets of $\mathbb{N}$ with the property that $\sigma_m \cap \sigma_n = \emptyset$ if $m \neq n$, let

$$z_n = \sum_{i \in \sigma_n} f_i(y_n) x_i. \tag{6.1.5}$$

Then, to every p in D_T, there corresponds q in D_T such that for arbitrary a in ω, m and n in $\mathbb{N}$ with $m \leq n$ (m and n may be ∞) and γ in I, we have

$$p(\sum_{i=1}^{m} \gamma_i a_i z_i) \leq q(\sum_{i=1}^{n} a_i z_i). \tag{6.1.6}$$

In addition, if (X,T) is ω-complete and $z_i \neq 0$ for $i \geq 1$, then $\{z_n\}$ is a u-S.b. for $[z_n]$.

6.2 SIZE OF δ AND A CHARACTERIZATION

In this section, we present a characterization of unconditional bases which depends upon the l.c. topology T^{**} considered in Remark 3.3.12 on the s.s. δ defined in (3.3.1). The basic result is (cf. [58])

THEOREM 6.2.1 Let $\{x_n;f_n\}$ be an S.b. for an l.c. TVS (X,T) with $\delta \subset c$ but $\delta \not\subset c_0$. Then $\{x_n;f_n\}$ is a u-S.b. if and only if $T^{**} \subset S$, S being the topology on δ generated by the polars of all compact subsets of ℓ^1.

PROOF. Let $T^{**} \subset S$ and consider any x in X and σ in P. If $\tilde{S}$ denotes the l.c. topology on c generated by the polars of all compact subsets of ℓ^1, then $S = \tilde{S}|\delta$. Now $\{f_{\sigma(n)}(x)\} \in c$ and let A be any relatively compact subset of ℓ^1. Since for y in A

$$|\langle\{f_{\sigma(n)}(x)\} - \sum_{i=1}^{n} f_{\sigma(i)}(x)e^{\sigma(i)}, y\rangle|$$

$$= |\sum_{i>n} f_{\sigma(i)}(x)y_{\sigma(i)}|$$

$$\leq \|\{f_i(x)\}\|_\infty \sum_{i>n} |y_{\sigma(i)}|$$

and $F_\sigma:\ell^1 \to \ell^1$, $F_\sigma(\alpha) = \{\alpha_{\sigma(i)}\}$ is an isometric isomorphism, the well-known criterion for compact subsets of ℓ^1 (for example, [129], p. 108) yields the unconditional convergence of $\sum_{n\geq 1} f_n(x)e^n$ to $\{f_n(x)\}$ in $(c,\tilde{S})$ and hence in (δ,S). Thus $\{x_n;f_n\}$ is a u-S.b. for (X,T).

Conversely, let $\{x_n;f_n\}$ be a u-S.b. for (X,T). Observe that $\delta^* \equiv (\delta,T^{**})^*$ can be identified with $\{f(x_n): f \in X^*\}$. On account of the weak unconditional convergence of

$\sum_{n\geq 1} f_n(x)x_n$ for each x and $\delta \subset c$, $\not\subset c_o$, we conclude that $\delta^* \subset \ell^1$.

Now assume that $T^{**} \not\subset S$ and so there exists a T^{**}-equicontinuous subset A of δ^* such that A is not relatively compact in $(\ell^1, \|\cdot\|)$. Then from the characterization of compact subsets of ℓ^1 referred to above, there exist $\eta > 0$ and an increasing sequence $\{m_r\}$ in $\mathbb{N}$ such that

$$\sup_{r\leq n\leq m_r} \ \sup_{a\in A} \ \sum_{j\geq n} |a_j| \geq \eta \ , \ \forall \ r \geq 1. \qquad (6.2.2)$$

Let us write Φ_r for the collection of all finite subsets of the set of integers r, r+1,... . Then by (6.2.2), we can find an $\varepsilon > 0$ such that for given r in $\mathbb{N}$, there exist J_r in Φ_r and a^r in A for which

$$|\sum_{j\in J_r} a_j^r| \geq \frac{1}{4} \sum_{j\in J_r} |a_j^r| \geq \varepsilon \ .$$

Thus, there exists a sequence $\{J_i\}$ in $\Phi \equiv \Phi_1$ and a sequence $\{a^i\}$ in A such that

$$|\sum_{j\in J_i} a_j^i| \geq \frac{1}{4} \sum_{j\in J_i} |a_j^i| \geq \varepsilon \ ,$$

and $\sup\{j:j \in J_i\} \leq \inf\{j:j \in J_{i+1}\} - 2$, $i \geq 1$.

Let $k_i = \sum_{j=1}^{i} \#(J_j)$ and define σ in P with

$$\sigma(\{i:1 \leq i \leq k_1\}) = J_1, \ \sigma(\{j:k_{i-1} + i \leq j \leq k_i+i-1\}) = J_i,$$

for $i \geq 2$ and elsewhere σ is defined identically.

From the restriction on δ and the fact that $\phi \subset \delta$, there exist α in δ and a nonzero k with $\alpha_n \to k$ and $|\alpha_n| \leq |k|/8$ for $n \geq 1$.

Now

$$|< \sum_{j=k_{i-1}+i}^{k_i+i-1} \alpha_{\sigma(j)} e^{\sigma(j)}, a^i>| = |\sum_{j\in J_i} \alpha_j a_j^i|$$

$$\geq \frac{|k|}{4} \sum_{j\in J_i} |a_j^i| - \sum_{j\in J_i} |\alpha_j - k||a_j^i|$$

$$\geq \frac{1}{8}|k| \varepsilon.$$

Consequently the series $\sum_{n\geq 1} \alpha_{\sigma(n)} e^{\sigma(n)}$ is not T**-Cauchy and this shows that the series $\sum_{n\geq 1} f_{\sigma(n)}(x)x_{\sigma(n)}$ is not convergent, where $\alpha = \{f_n(x)\}$ for x in X and hence we arrive at a contradiction. □

6.3 COMPATIBLE TOPOLOGIES, STRUCTURE OF δ

The behaviour of a biorthogonal sequence $\{x_n;f_n\}$ for $<X,X^*>$ as an S.b. for X relative to l.c. compatible topologies on X was discussed at length in chapter 6 of [130], where X is an arbitrary l.c. TVS. As mentioned there, the solution to this problem is, in general, negative. On the other hand, the situation corresponding to a u-S.b. (resp. s-S.b., b.m-S.b.) is quite different and rather pleasant.

Before we come to the main results let us recall the following results relevant to the present study. Let λ be an arbitrary s.s. and μ be a subspace of λ^β. Then we have (cf. [130])

PROPOSITION 6.3.1 $\{e^n;e^n\}$ is a u-S.b. for $(\lambda,\sigma(\lambda,\lambda^\times))$. Further $\{e^n;e^n\}$ is a s-S.b. (resp. a b.m-S.b.) for $(\lambda,\sigma(\lambda,\mu))$ if and only if λ is monotone (resp. normal).

PROPOSITION 6.3.2 Let $\{x_n;f_n\}$ be an S.b. for an l.c. TVS

(X,T). Then (i) $\{x_n;f_n\}$ is an s-S.b. (resp. a b.m-S.b.) for (X,T) if and only if δ is monotone (resp. normal) and (ii) if $\{x_n;f_n\}$ is an s-S.b. for (X,T), then $\mu = \delta^\times$ if and only if (X,T) is an S-space.

Throughout this section, we write $X \equiv (X,T)$ for an arbitrary l.c. TVS and $\{x_n;f_n\}$ for a biorthogonal sequence for the dual system $\langle X,X^*\rangle$. Also, let T_1 and T_2 denote any two l.c. topologies on X compatible with $\langle X,X^*\rangle$.

To begin with, let us mention the following result which is a consequence of the well-known Orlicz-Pettis-Grothendieck theorem on series (Theorem 1.5.4).

<u>THEOREM 6.3.3</u> $\{x_n;f_n\}$ is a T_1-s-S.b. for X if and only if it is a T_2-s-S.b.

Similarly, following Theorem 1.5.4, one can prove

<u>THEOREM 6.3.4</u> Let $(X,\sigma(X,X^*))$ be ω-complete. Then $\{x_n;f_n\}$ is a T_1-u-S.b. for X if and only if it is a T_2-u-S.b.

Concerning bounded multiplier bases, we have ([45])

<u>THEOREM 6.3.5</u> $\{x_n;f_n\}$ is a T_1-b.m-S.b. for X if and only if it is a T_2-b.m-S.b.

<u>PROOF</u>. Let $\{x_n;f_n\}$ be a T_1-b.m-S.b. By using Theorem 3.3.5, we conclude that $\{e^n;e^n\}$ is a b.m-S.b. for $(\delta,\sigma(\delta,\mu))$; that is, δ is normal (Proposition 6.3.1). Hence from Theorem 1.3.11, $\alpha^{(n)} \to \alpha$ in $\tau(\delta,\delta^\times)$, for each α in δ. Since $\mu \subset \delta^\beta = \delta^\times$, $\tau(\delta,\mu) \subset \tau(\delta,\delta^\times)$. Thus

$$\alpha^{(n)} \to \alpha \quad \text{in } \tau(\delta,\mu), \quad \forall\, \alpha \in \delta . \tag{6.3.6}$$

We next show that $\{x_n;f_n\}$ is an S.b. for $(X,\tau(X,X^*))$. However, this is a simple consequence of Proposition 1.2.15 and (6.3.6), since F_1 is $\sigma(\delta,\mu)$-$\sigma(X,X^*)$ continuous (cf. Theorem 3.3.5).

Since $T_2 \subset \tau(X,X^*)$, $\{x_n;f_n\}$ is an S.b. for (X,T_2). Finally, let $x \in X$ and $b \in \ell^\infty$. Then $\sum_{n\geq 1} b_n f_n(x)x_n$ converges to y (say) in the topology $\sigma(X,X^*)$ and so $f_n(y) = b_n f_n(x)$, $n \geq 1$. Therefore

$$\sum_{n\geq 1} b_n f_n(x)x_n = \sum_{n\geq 1} f_n(y)x_n = y,$$

the convergence of the infinite series being considered relative to the topology T_2. □

EXERCISE 6.3.7 Prove Theorem 6.3.3 without recourse to Theorem 1.5.4 [Hint: Follow the proof of Theorem 6.3.5].

REMARKS. In the preceding three theorems, Theorem 6.3.4 is the only result where the space (X,T) bears a restriction of being $\sigma(X,X^*)$- ω-complete. However, this restriction cannot be overlooked. To realize its importance, it will be interesting, for example, to construct a weak u-S.b. $\{x_n;f_n\}$ for an l.c. TVS (X,T) which is not ω-complete relative to $\sigma(X,X^*)$ such that $\{x_n;f_n\}$ is an S.b. for (X,T) but not u-S.b. for (X,T). On the other hand, the absence of weak ω-completeness of an l.c. TVS (X,T) may even prevent a weak u-S.b. from becoming a T-S.b. For instance, consider

EXAMPLE 6.3.8 It is clear that $\{e^n;e^n\}$ is not an S.b. for $(c, \|\cdot\|_\infty)$, although it is a u-S.b. for $(c,\sigma(c,\ell^1))$ (cf. Proposition 6.3.1). Note that $(c,\sigma(c,\ell^1))$ is not ω-complete by Theorem 1.3.7.

EXERCISE 6.3.9 Let $\{x_n;f_n\}$ be a $\sigma(X,X^*)$-S.b. for an l.c. TVS (X,T). Prove that $\{x_n;f_n\}$ is an s-S.b. (resp. a b.m-S.b.) for (X,T) if and only if δ is monotone (resp. normal).

EXERCISE 6.3.10 Let $\{x_n;f_n\}$ be an s-S.b. (resp. a

b.m-S.b.) for $(X,\sigma(X,X^*))$, $X \equiv (X,T)$ being an arbitrary l.c. TVS. Prove that $\mu = \delta^X$ if and only if (X,T) is an S-space.

6.4 RELATIONSHIP WITH ABSOLUTE BASES

If $\{x_n;f_n\}$ is an ℓ^1-base or even a semi ℓ^1-base, then it is a u-S.b. (see also Exercise 5.6.6). What about the converse of this result? Even in Banach spaces, the converse may not hold good, for instance, consider the usual S.b. $\{e^n;e^n\}$ for $(c_o, \|\cdot\|_\infty)$ (cf. also [48], [158]). However, a solution to this converse problem is not known in general except in the case of nuclear Fréchet spaces (cf. [178] and [180], p. 180) and in a certain class of spaces which we will take up in this section. We essentially follow [99] for the rest of this section.

DEFINITION 6.4.1 Let $\{x_n;f_n\}$ be an S.b. for an l.c. TVS (X,T). For θ in ϕ, let

$$P_\theta\Big(\sum_{n\geq 1} f_n(x)x_n\Big) = \sum_{n=1}^{\ell} \theta_n f_n(x)x_n,$$

where $x \in X$ and ℓ is the length of θ. The S.b. $\{x_n;f_n\}$ is called u-*Schauder* if $\{P_\theta: \theta \in \phi, |\theta_n| \leq 1\}$ is equicontinuous on (X,T).

Concerning the relationship of a u-S.b. with a u-Schauder S.b., we have

PROPOSITION 6.4.2 Let $\{x_n;f_n\}$ be an S.b. for an l.c. TVS (X,T). (i) If (X,T) is barrelled and $\{x_n;f_n\}$ is a u-S.b., then it is u-Schauder, (ii) If (X,T) is ω-complete and $\{x_n;f_n\}$ is u-Schauder, then it is a u-S.b.

PROOF. (i) This is a consequence of Theorem 6.1.1 (ii).

(ii) If $p \in D_T$, there exists q in D_T so that $p(P_\theta(x)) \leq q(x)$; $\forall x \in X$, $\theta \in \phi$, $|\theta_n| \leq 1$.

Therefore, for each x in X, the set $\{\sum_{i=1}^{n} \varepsilon_i f_i(x)x_i : \varepsilon \in \bar{e}, n \in \mathbb{N}\}$ is bounded and consequently for each equicontinuous subset B of X*, we find an $M_B > 0$ so that (cf. [129], p. 165)

$$\sum_{n \geq 1} |f(f_n(x)x_n)| \leq M_B, \ \forall\ f \in B.$$

Now apply Theorem 1.5.2. □

REMARK. Let $\{x_n;f_n\}$ be a fully ℓ^1-base for an l.c. TVS (X,T). Then T is also generated by $\{Q_p : p \in D_T\}$, where Q_p is the same as $Q_{p,e}$ introduced in (4.2.4) with $e = \{1,1,\ldots\}$. For convenience, let us write X_{Q_p} for $X/Q_p^{-1}(0)$ normed through $\hat{Q}_p(\hat{x}) = \inf\{Q_p(x-y) : Q_p(y) = 0\}$; $\hat{x} \in X_{Q_p}$, $\hat{x} = \Psi_{Q_p}(x)$, $x \in X$, Ψ_{Q_p} being the usual quotient map from X onto X_{Q_p}. Further, for p in D_T, let $J_p = \{n \in \mathbb{N} : p(x_n) \neq 0\}$. Then $\hat{f}_n : X_{Q_p} \to \mathbb{K}$ given by $\hat{f}_n(\hat{x}) = f_n(x)$ with n in J_p, is well defined (if $x_1, x_2 \in \hat{x}$, then $Q_p(x_1-x_2) = 0$ and so $|f_n(x_1-x_2)| = 0$, $n \in J_p$). Since $|\hat{f}_n(\hat{x})| \leq (p(x_n))^{-1}\hat{Q}_p(\hat{x})$ for n in J_p, $\hat{f}_n \in X^*_{Q_p}$. Finally note that $\hat{f}_n(\hat{x}_m) = f_n(x_m) = \delta_{mn}$.

LEMMA 6.4.3 Let $\{x_n;f_n\}$ be a fully ℓ^1-base for an l.c. TVS (X,T). Then for each p in D_T, $\{\hat{x}_n, \hat{f}_n : n \in J_p\}$ is a fully ℓ^1-base for the normed space $(X_{Q_p}, \hat{Q}_p)$: in particular, $(X_{Q_p}, \hat{Q}_p)$ is either finite dimensional or is topologically isomorphic to a dense subspace of $(\ell^1, \|\cdot\|_1)$.

PROOF. For N in $\mathbb{N}$, let $\sigma_N = \{1,\ldots,N\} \cap J_p$; then

$$\hat{Q}_p(\hat{x} - \sum_{i\in\sigma_N} \hat{f}_i(\hat{x})\hat{x}_i) \leq Q_p(x - \sum_{i=1}^{N} f_i(x)x_i)$$

$$\to 0 \text{ as } N \to \infty.$$

Therefore

$$\hat{x} = \sum_{n\in J_p} \hat{f}_n(\hat{x})\hat{x}_n;$$

and also

$$\sum_{n\in J_p} |\hat{f}_n(\hat{x})|\hat{Q}_p(\hat{x}_n) = Q_p(x) = \hat{Q}_p(\hat{x}).$$

This proves the first part and the last part follows from Corollary 4.5.11. □

With this background, it is now possible to prove the main result of this section (cf. [99]), namely,

THEOREM 6.4.4 Let an l.c. TVS (X,T) contain an ℓ^1-Köthe base $\{x_n;f_n\}$. Then every u-Schauder base $\{y_n;g_n\}$ for (X,T) is an ℓ^1-Köthe base.

PROOF. At the outset, let us observe that (X,T) is complete (cf. Proposition 4.4.2(d)). Further, by Lemma 6.4.3, for each p in D_T, $(X_{Q_p},\hat{Q}_p)$ is either finite dimensional or is topologically isomorphic to a dense subspace of $(\ell^1, \|\cdot\|_1)$. Since the series $\sum_{n\geq 1} g_n(x)y_n$ is unconditionally convergent for each x in X, by Proposition 6.4.2 (ii), there exists a universal constant K (cf. Theorem 1.6.4) so that

$$\pi_p(x) \equiv [\sum_{n\geq 1} (|g_n(x)|Q_p(y_n))^2]^{1/2}$$

$$\leq K \sup_{n\geq 1} \sup_{|\varepsilon_i|\leq 1} Q_p(\sum_{i=1}^{n} \varepsilon_i g_i(x)y_i),$$

for each x in X. But $\{y_n; g_n\}$ is u-Schauder, therefore there exists q in D_T such that

$$\sup_{n\geq 1} \sup_{|\varepsilon_i|\leq 1} Q_p\left(\sum_{i=1}^{n} \varepsilon_i g_i(x) y_i\right) \leq q(x), \forall x \in X.$$

Hence $\pi_p \in D_T$ for each p in D_T.

As before, the quotient space $(X_{Q_{\pi_p}}, \hat{Q}_{\pi_p})$ is either finite dimensional or is topologically isomorphic to a dense subspace of $(\ell^1, \|\cdot\|_1)$, where p is an arbitrary member of D_T.

Next, consider the quotient space $(X_{\pi_p}, \hat{\pi}_p)$ for any p in D_T, formed with respect to the kernel of π_p Since $(\pi_p(x+y))^2 + (\pi_p(x-y))^2 = 2\{(\pi_p(x))^2 + (\pi_p(y))^2\}$, the space $(X_{\pi_p}, \hat{\pi}_p)$ is a pre-Hilbert space. Clearly the natural map defined by (since $\pi_p \leq Q_p$)

$$K_{\pi_p}^{Q_{\pi_p}} : X_{Q_{\pi_p}} \to X_{\pi_p}, \quad K_{\pi_p}^{Q_{\pi_p}}(\hat{x}_{Q_{\pi_p}}) = \hat{x}_{\pi_p}$$

is a continuous linear operator and let us denote by $\hat{K}_{\pi_p}^{Q_{\pi_p}}$ the continuous extension of this operator from the completion $\hat{X}_{Q_{\pi_p}}$ into the completion $\hat{X}_{\pi_p}$. Following the proof of Lemma 6.4.3, we conclude that $\hat{X}_{\pi_p}$ is a separable Hilbert space. Hence by Theorem 1.6.5, the operator $\hat{K}_{\pi_p}^{Q_{\pi_p}}$ is absolutely summing. In particular, because of the unconditional convergence of $\sum_{n\geq 1} g_n(x) y_n$ for each x in X, we find from Theorem 1.6.3 that

$$\sum_{n\geq 1} |g_n(x)|\pi_p(y_n) \leqq K \sup_{n\geq 1} \sup_{|\varepsilon_i|\leqq 1} Q_p(\sum_{i=1}^{n} \varepsilon_i g_i(x)y_i)$$

$$\leqq q(x), \ \forall \ x \in X.$$

Therefore, for each x in X, we have

$$\sum_{n\geq 1} |g_n(x)|p(y_n) \leqq \sum_{n\geq 1} |g_n(x)|Q_p(y_n) \leqq q(x),$$

since $Q_p(y_n) = \pi_p(y_n)$. This shows that $\{y_n;g_n\}$ is a fully ℓ^1-base and so by Proposition 4.4.2, $\{y_n;g_n\}$ is an ℓ^1-Köthe base. □

EXERCISE 6.4.5 Let an l.c. TVS (X,T) contain a fully ℓ^1-base, then for each unconditionally convergent series $\sum_{n\geq 1} y_n$ in (X,T) and p in D_T,

$$\sum_{n\geq 1} (Q_p(y_n))^2 < \infty.$$

[Hint: Make use of Lemma 6.4.3, Theorem 1.6.4 and Proposition 1.5.7.]

PROPOSITION 6.4.6 Let $\{x_n;f_n\}$ be a semi ℓ^1-base for an l.c. TVS (X,T). If $\sum_{n\geq 1} y_n$ is a subseries convergent series in (X,T), then for each p in D_T,

$$\sum_{n\geq 1} (p(y_n))^2 < \infty.$$

PROOF. If $x \in X$ and $p \in D_T$, then

$$Q_p(x - \sum_{i=1}^{n} f_i(x)x_i) \leqq \sum_{i>n} |f_i(x)|p(x_i) \to 0 \text{ as } n \to \infty;$$

$$Q_{Q_p}(x) \leqq Q_p(x),$$

and

$$Q_p(\sum_{i=m}^{n} \alpha_i x_i) < \varepsilon \Rightarrow \sum_{i=m}^{n} |\alpha_i| p(x_i) < \varepsilon.$$

Hence $\{x_n;f_n\}$ is an ℓ^1-pre Köthe base for (X,T_Q), where T_Q is the l.c. topology generated by $\{Q_p : p \in D_T\}$. It is easily seen that each Q_p is T-lower semicontinuous (cf. proof of Proposition 5.4.1) and hence T_Q is $\langle X,X^*\rangle$-polar. Hence by Theorem 1.5.5, $\sum_{n\geq 1} y_n$ is subseries convergent in (X,T_Q). Since $p \leq Q_p$,

$$\sum_{n\geq 1} (p(y_n))^2 \leq \sum_{n\geq 1} (Q_p(y_n))^2 < \infty,$$

by Exercise 6.4.5. □

It is clear that each semi ℓ^1-base in an ω-complete l.c. TVS is a u-S.b. The converse is generally not true. But we have

PROPOSITION 6.4.7 Let $\{x_n;f_n\}$ be an ℓ^1-base (resp. a semi ℓ^1-base) for a barrelled space (X,T). Then each u-S.b. $\{y_n;g_n\}$ for (X,T) is an ℓ^1-Köthe base (resp. a fully ℓ^1-base).

PROOF. According to Proposition 4.4.5, $\{x_n;f_n\}$ is an ℓ^1-Köthe base (resp. a fully ℓ^1-base), whereas $\{y_n;g_n\}$ is also an u-Schauder base by Proposition 6.4.2(i). Thus, in the first case, the desired conclusion is a consequence of Theorem 6.4.4. On the other hand, making use of Lemma 6.4.3 and proceeding on the lines of proof of Theorem 6.4.4, we find that $\{y_n;g_n\}$ is a fully ℓ^1-base in the second case. □

NOTE. Proposition 6.4.7 extends a result of Lindenstrauss and Pelczynski [153] that every u-S.b. of ℓ^1 is an ℓ^1-Köthe base.

REMARK. The hypothesis of barrelledness in Proposition 6.4.7 cannot be dropped. For, we have

EXAMPLE 6.4.8 Let us confine ourselves to the space $(\ell^2, \eta(\ell^2, \ell^2))$ for which $\{e^n; e^n\}$ is an ℓ^1-Köthe base (cf. Example 4.3.5). Clearly this space is not barrelled, for otherwise $\eta(\ell^2, \ell^2) \approx T_2$, the topology on ℓ^2 generated by the Hilbertian norm $\|\cdot\|_2$. However, this is not true (consider the convergence of $\{e^n\}$).

Now we proceed to construct a u-S.b. for $(\ell^2, \eta(\ell^2, \ell^2))$ such that it is not an ℓ^1-base. But for variation in the approach, this construction is due to Kalton [99]. Let $X_n = sp\{e^{2^n+1}, \ldots, e^{2^{n+1}}\}$ denote the 2^n-dimensional subspace of ℓ^2, $n \geqq 1$. In each X_n, we construct the so-called *Haar system* (see [50] and [174]) consisting of 2^n linearly independent elements defined as follows:

$$y^{n,1} = 2^{-n/2} \sum_{i=1}^{2^n} e^{2^n+i}; \quad y^{n,2^k+s} = 2^{-(n-k)/2} \sum_{i=1}^{2^n} \beta_i(k,s) e^{2^n+i},$$

where $1 \leqq s \leqq 2^k$, $0 \leqq k \leqq n-1$ and

$$\beta_i(k,s) = \begin{cases} 1, & (2s-2)2^{n-k-1} + 1 \leqq i \leqq (2s-1)2^{n-k-1} \\ -1, & (2s-1)2^{n-k-1} + 1 \leqq i \leqq 2s.2^{n-k-1} \\ 0, & \text{otherwise} \end{cases}$$

Suppose $z^1 = e^1$ and $z^2 = e^2$. If $i \geqq 3$, there is a unique integer $n \geqq 1$ with $2^n < i \leqq 2^{n+1}$ and $i = 2^n+m$, where $1 \leqq m \leqq 2^n$. For $m = 1$, let $z^i = y^{n,1}$ and if $2 \leqq m \leqq 2^n$, we find a unique integer $k \geqq 0$ so that $2^k < m \leqq 2^{k+1}$ and $m = 2^k+s$ with $1 \leqq s \leqq 2^k$, $0 \leqq k \leqq n-1$ and we put $z^i = y^{n,2^k+s}$.

It is clear that $y^{n,m}$ $(m=1,\ldots,2^n)$ is an orthonoraml base for $X_n \equiv (x_n, \|\cdot\|_2)$ and it follows that (see arguments below also) $\{z^i : i \geq 1\}$ is an orthonormal base for the Hilbert space $(\ell^2, \|\cdot\|_2)$. Hence for each b in ℓ^2

$$b = \sum_{i\geq 1} \langle b, z^i\rangle z^i \Longleftrightarrow \langle b, b\rangle = \sum_{i\geq 1} |\langle b, z^i\rangle|^2 < \infty,$$

and as $\eta(\ell^2, \ell^2)$ is inferior to the topology given by $\|\cdot\|_2$, $\{z^i; z^i\}$ is a u-S.b. for $(\ell^2, \eta(\ell^2, \ell^2))$. We proceed to show that $\{z^i; z^i\}$ is not a semi ℓ^1-base for $(\ell^2, \eta(\ell^2, \ell^2))$, for otherwise

$$\sum_{i\geq 1} |\langle b, z^i\rangle|\, p_f(z^i) < \infty; \quad \forall\, b, f \in \ell^2. \tag{6.4.9}$$

Therefore

$$\sum_{i\geq 1} (p_f(z^i))^2 < \infty, \quad \forall\, f \in \ell^2. \tag{6.4.10}$$

Indeed, if (6.4.10) were not true, then for some f in ℓ^2, we can find an increasing sequence $\{n_i\}$ with $n_o = 0$ so that

$$\sum_{j=n_{i-1}+1}^{n_i} (p_f(z^j))^2 \equiv M_i^2 \geq 1, \; i \geq 1.$$

If $b_j = p_f(z^j)/iM_i$ for $n_{i-1}+1 \leq j \leq n_i$, $i \geq 1$, then $b \in \ell^2$. Define α in ω so that $\alpha_1 = b_1$, $\alpha_2 = b_2$ and for $2^n < i \leq 2^{n+1}$, $n \geq 1$; $\alpha_{2^n+1}, \ldots, \alpha_{2^{n+1}}$ are solutions of 2^n equations given by

$$\langle \alpha, z^i\rangle = b_i, \quad 2^n < i \leq 2^{n+1}.$$

Then $\alpha \in \ell^2$ and

$$\sum_{i\geq 1} \langle \alpha, z^i\rangle p_f(z^i) = \sum_{i\geq 1} \sum_{j=n_{i-1}+1}^{n_i} b_j p_f(z^j) \geq \sum_{i\geq 1} \frac{1}{i}.$$

This, however, contradicts (6.4.9). Using (6.4.10), we easily conclude that

$$\sum_{n\geq 1} \sum_{m=1}^{2^n} (p_f(y^{n,m}))^2 < \infty, \forall f \in \ell^2. \qquad (6.4.11)$$

Define f in ω by $f_1 = 0$, $f_2 = 0$ and

$$f_k = \frac{1}{n.2^{n/2}} \; ; \; 2^n+1 \leq k \leq 2^{n+1}, \; n \geq 1.$$

Then

$$\sum_{k\geq 1} |f_k|^2 = \sum_{n\geq 1} \frac{1}{n^2} < \infty,$$

and so $f \in \ell^2$. Observe that

$$p_f(y^{n,1}) = \frac{1}{n}, \; p_f(y^{n,2}) = \frac{1}{n},$$

and for $2^k < m \leq 2^{k+1}$, $1 \leq k \leq n-1$

$$p_f(y^{n,m}) = 2^{-(n-k)/2} . 2^{n-k} \frac{2^{-n/2}}{n} = \frac{1}{n} 2^{-k/2}.$$

Hence

$$\sum_{m=1}^{2^n} (p_f(y^{n,m}))^2 = \frac{2}{n^2} + \sum_{k=1}^{n-1} \frac{1}{n^2} \sum_{m=2^k+1}^{2^{k+1}} 2^{-k}$$

$$= \frac{n+1}{n^2}, \; \forall n \geq 1$$

and this contradicts (6.4.11). Therefore $\{z^i;z^i\}$ cannot be a semi ℓ^1-base for $(\ell^2,\eta(\ell^2,\ell^2))$.

Unconditional convergence and p-Köthe bases

We next prove a result [99] which is similar to Exercise 6.4.5 or Proposition 6.4.6 for p-Köthe bases, $1 \leq p < \infty$.

PROPOSITION 6.4.12 Let an l.c. TVS (X,T) contain a p-Köthe base $(1 \leqq p < \infty)$ $\{x_n;f_n\}$. Suppose $\sum_{n\geq 1} y_n$ converges unconditionally in (X,T). Then for every ν in D_T, we have

(a) if $2 \leqq p < \infty$, then

$$\sum_{n\geq 1} [\nu(y_n)]^p < \infty$$

and

(b) if $1 \leqq p \leqq 2$, then

$$\sum_{n\geq 1} [\nu(y_n)]^2 < \infty.$$

PROOF. By Proposition 5.4.13, X_{ν_p} is either finite dimensional or topologically isomorphic to a dense subspace of ℓ^p. In the former case, (a) is trivially true. In the other case, observe that T is also generated by $\{\nu_p : \nu \in D_T\}$ and so the result follows by an application of Theorem 1.6.3. □

As an application of Proposition 6.4.12, we derive

THEOREM 6.4.13 Let a barrelled space have a p-Köthe base $\{x_n;f_n\}$ and a q-Köthe base $\{y_n;g_n\}$, where $2 < p < \infty$ and $1 \leqq q < p < \infty$. Then (X,T) is nuclear.

PROOF. Let $r = \max(2,q)$. Since (X,T) is complete (cf. Proposition 5.4.12), $\sum_{n\geq 1} f_n(x)x_n$ converges unconditionally for each x in X and so by Proposition 6.4.9

$$\sum_{n\geq 1} |f_n(x)|^r \nu^r(x_n) < \infty; \quad \forall x \in X, \quad \nu \in D_T.$$

The seminorms ν_r are clearly lower semicontinuous on (X,T)

and hence T-continuous. Since $\nu_p \leqq \nu_r$ (observe that $r < p$) and T is also determined by $\{\nu_p : \nu \in D_T\}$, $\{x_n;f_n\}$ is r-Köthe, the result follows by Theorem 5.5.5. □

7 Symmetric bases and sequence spaces

7.1 INTRODUCTION

Symmetric bases are closely related to symmetric sequence spaces, the latter being helpful in locating the size of the s.s. δ defined corresponding to symmetric bases. In order to appreciate the size of δ, a study of symmetric sequence spaces is carried out in section 7.2 of which a part is relevant to symmetric bases, whereas the rest is of independent interest. Thereafter, we touch upon some elementary characterizations of symmetric bases, postponing some advanced results to chapter 8.

In addition, we consider a few applications of the presence of a symmetric base in an arbitrary l.c. TVS such as the metrizability of the space in question, the Hamel base character of the s.a.c.f and so on. In the course of this discussion, we also discover an interesting characterization of symmetric bases in a Fréchet space. Also, we discuss a few notions related to symmetric bases and explore the interrelationship of the former with the latter.

Unless specified otherwise, throughout, (X,T) denotes an arbitrary l.c. TVS containing an S.b. $\{x_n;f_n\}$. Further, we abbreviate hereafter a symmetric base as an sy.S.b. and a symmetric sequence space as an s.s.s.

7.2 SYMMETRIC SEQUENCE SPACES

This section is devoted to a discussion of some general results on the location of an s.s.s. At the end of the section, we also prove a related lemma which is of independent interest.

At the outset, let us first of all recall the following result ([56]; cf. also [129], p. 91).

THEOREM 7.2.1 Let λ be an s.s.s. If λ is also (i) normal, then $\lambda \subset c_0$ or $\lambda = \ell^\infty$ or $\lambda = \omega$, and (ii) perfect with $\lambda \neq \phi, \omega$, then $\ell^1 \subset \lambda \subset \ell^\infty$.

In the rest of this section, we follow [75] and begin with

THEOREM 7.2.2 Let λ be an s.s.s. with $\lambda \not\subset \ell^\infty$. Then $\lambda^\beta = \phi$ and if, in addition, λ is also an AK-space relative to an l.c. topology T, then $\{e^n\}$ generates λ^* and $T \approx \sigma(\lambda,\phi)$.

PROOF. We have an $s \in \lambda \setminus \ell^\infty$. Let $\lambda^\beta \not\subset \phi$. Then there exists $t \in \lambda^\beta \setminus \phi$ and so we may determine $\{m_k\}$ and $\{n_k\}$ in I_∞ with $t_{n_k} \neq 0$ and $|s_{m_k}|\,|t_{n_k}| \geq 1$ for $k \geq 1$. Define π in P so that $\pi(n_k) = m_k$. Then

$$|s_{\pi(n_k)} t_{n_k}| \geq 1, \ \forall\, k \geq 1. \tag{7.2.3}$$

Since $s_\pi \in \lambda$, (7.2.3) contradicts the choice of t unless $t \in \phi$. So $\lambda^\beta \subset \phi$; thus $\lambda^\beta = \phi$. For the final part, we make use of Theorem 1.3.13 to conclude that $\lambda^* = \phi$ and so $\sigma(\lambda,\lambda^*) \approx \tau(\lambda,\lambda^*)$. This shows that $T \approx \sigma(\lambda,\phi)$. □

For convenience, we assume in the next two results that all sequences are taken from $\mathbb{R}$. The following theorem reflects a categorical study of an s.s.s. depending on the size of its Köthe dual and for other similar results, one may refer to [200].

To begin with, let us mention

LEMMA 7.2.4 Let $\sum_{n\geq 1} s_n$ be a non-absolutely convergent series and t be in ω with $0 < a \leq t_n \leq b$ for $n \geq 1$. Then for some π in P, $\sum_{n\geq 1} s_{\pi(n)} t_n$ diverges.

PROOF. It is a classical result of Riemann that absolute and unconditional convergence of a series are the same in every finite dimensional space. □

THEOREM 7.2.5 Let λ be an s.s.s. such that $\phi \subset \lambda \subset c_o$. Then $\lambda^\times = \lambda^\beta$.

PROOF. If $\lambda = \phi$, the result is trivially true and so let $\phi \subsetneqq \lambda$. Hence there exist s in λ and $\{n_i\}$ in I_∞ so that $s_{n_i} \neq 0$ for $i \geq 1$. Therefore $\lambda^\beta \subset \ell^\infty$, for otherwise there exists t in λ^β with $t \notin \ell^\infty$; however, this gives a contradiction as in the proof of Proposition 7.2.2. □

In order to prove the required result, we consider two cases: $e \in \lambda^\beta$ and $e \notin \lambda^\beta$. If $e \in \lambda^\beta$, by the symmetric character of λ, $\lambda \subset \ell^1$. Thus $\lambda^\beta = \lambda^\times = \ell^\infty$.

Now let $e \notin \lambda^\beta$ and consider two cases of interest, namely, (a) when $\lambda^\beta \not\subset c_o$ and (b) when $\lambda^\beta \subset c_o$.

(a) Here one finds some $t \in \lambda^\beta \setminus c_o$ with $1/2 \leqq t_{n_i} \leqq 3/2$, $i \geqq 1$ for some $\{n_i\} \in I_\infty$. Put $\{m_i\} = \mathbb{N} \setminus \{n_i\}$.

We may clearly assume that $\lambda \not\subset \ell^1$ and let $\alpha \in \lambda \setminus \ell^1$. Since $\lambda \subset c_o$, we can find $\{p_i\}$ in I_∞ so that $\sum_{i \geq 1} |\alpha_{p_i}| < \infty$. Then by Lemma 7.2.4, we can determine a permutation of $\mathbb{N} \setminus \{p_i\}$, say, $\{q_i\}$ such that $\sum_{i \geq 1} \alpha_{q_i} t_{n_i}$ diverges. Let π in P be such that π maps $\{n_i\}$ onto $\{q_i\}$ and $\{m_i\}$ onto $\{p_i\}$. Clearly, the series $\sum_{i \geq 1} \alpha_{\pi(n_i)} t_{n_i}$ diverges and $\sum_{i \geq 1} |\alpha_{\pi(m_i)} t_{m_i}| < \infty$. Hence $\sum_{i \geq 1} \alpha_{\pi(i)} t_i$ is divergent, yielding thereby a contradiction since $\alpha_\pi \in \lambda$. Thus $\lambda^\beta \setminus c_o = \emptyset$ and we are, therefore, lead to the next case (b).

(b) Here $\lambda^\beta \subset c_o$. To prove the required result, assume that $t \in \lambda^\beta, \notin \lambda^\times$. Hence for some s in λ, $\sum_{n \geq 1} |s_n t_n| = \infty$.

Clearly $s, t \notin \ell^1$. Since $\sum_{n\geq 1} s_n t_n$ is convergent, we find an $I \in \mathcal{I}_\infty$ so that $s_n t_n$ has the same sign (either > 0 or < 0) for n in I with $\sum_{n\in I} |s_n t_n| = \infty$ and $\sum_{n\in I^c} s_n t_n$ does not converge, where $I^c = \mathbb{N} \setminus I$, consequently $\sum_{n\in I^c} |s_n| = \sum_{n\in I^c} |t_n| = \infty$. Since s and t belong to c_o, there exist infinite subsets $I_1, J_1; I_2, J_2$ of I^c such that

$$I_1 = \{i \in I^c : s_i = 0 \text{ or } \sum_{i\in I_1} |s_i| < \infty\} ,$$

$$J_1 = I^c \setminus I_1$$

and

$$I_2 = \{i \in I^c : t_i = 0 \text{ or } \sum_{i\in I_2} |t_i| < \infty\},$$

$$J_2 = I^c \setminus I_2.$$

Define π in P with $\pi(i) = i$ for $i \in I$, $\pi(I_2) = J_1$ and $\pi(J_2) = I_1$. Since

$$\sum_{i\geq 1} s_{\pi(i)} t_i = \sum_{i\in I} s_{\pi(i)} t_i + \sum_{i\in I_2} s_{\pi(i)} t_i + \sum_{i\in J_2} s_{\pi(i)} t_i ,$$

the series $\sum_{i\geq 1} s_{\pi(i)} t_i$ does not converge. Hence $s_\pi \notin \lambda$, a contradiction. Thus $\lambda^\beta = \lambda^\times$ in this case. □

Lastly, we have

<u>PROPOSITION 7.2.6</u> Let $\alpha \in \ell^\infty \setminus \phi$ and $\beta \in \omega$ satisfy

$$\sum_{n\geq 1} |\alpha_n \beta_{\sigma(n)}| < \infty , \forall \sigma \in P.$$

Then

$$\sup_{\sigma\in P} \sum_{n\geq 1} |\alpha_n \beta_{\sigma(n)}| < \infty . \qquad (7.2.7)$$

PROOF. For x in ω, let

$$x^{\delta} = \{y \in \omega: \sum_{n\geq 1} |y_n x_{\sigma(n)}| < \infty, \quad \forall \sigma \in P\}$$

denote the *symmetric dual* of x. We recall the following known result ([198]; cf. also [129], p. 92-99) : for x,y in ω, one has (a) $x \in y^{\delta}$ if and only if $y \in x^{\delta}$, (b) $x^{\delta} = \phi$ if and only if $x \notin \ell^{\infty}$, (c) $x^{\delta} = \omega$ if and only if $x \in \phi$, (d) $x^{\delta} = \ell^{\infty}$ if and only if $x \in \ell^1 \setminus \phi$, (e) $x^{\delta} = \ell^1$ if $x \in \ell^{\infty} \setminus c_o$ and if (f) $x \in c_o \setminus \ell^1$ and $y \in c_o$, then

$$\sup_{\sigma \in P} \sum_{n\geq 1} |y_n x_{\sigma(n)}| < \infty.$$

Returning to the main proof, let us confine ourselves to three cases:

(i) $\alpha \in \ell^{\infty} \setminus c_o$, (ii) $\alpha \in c_o \setminus \ell^1$ and (iii) $\alpha \in \ell^1 \setminus \phi$.

(i) Here $\beta \in \alpha^{\delta} = \ell^1$ and so

$$\sup_{\sigma \in P} \sum_{n\geq 1} |\alpha_n \beta_{\sigma(n)}| \leqq \| \alpha \|_{\infty} \; \| \beta \|_1 .$$

(ii) Here $\alpha \in \beta^{\delta}$. If $\beta \notin \ell^{\infty}$, then $\alpha \in \phi$ (cf. (b)) and hence this is a contradiction. Now let $\beta \in \ell^{\infty} \setminus c_o$, then $\alpha \in \ell^1$, a contradiction again. Thus $\beta \in c_o$ and we get (7.2.7) by using (f).

(iii) In this case $\beta \in \ell^{\infty}$, see (d) and we are again done. □

7.3 CHARACTERIZATIONS AND A NON-SYMMETRIC BASE

In this section, we provide (cf. [75]) a few characterizatins of a symmetric base (sy. S.b.) and more rigorous ones will be taken up at a later stage (see Theorem 7.4.18 and chapter 8). These results (for instance, Theorem 7.3.6) easily make it possible to construct examples of an sy.S.b. Therefore, at the end we pass on to an example of an S.b.

which is not an sy.S.b.

We recall the terminology from section 7.1 and begin with

THEOREM 7.3.1 $\{x_n;f_n\}$ is an sy. S.b. for (X,T) if and only if the following is true:

$$\text{For each } \sigma \text{ in } P\text{, the b.s. } \{x_{\sigma(n)};f_{\sigma(n)}\} \text{ is an S.b. for } (X,T) \text{ equivalent to } \{x_n;f_n\}. \tag{7.3.2}$$

PROOF. Assume the truth of (7.3.2) and choose x in X and ρ,σ in P arbitrarily. Then $x = \sum_{n\geq 1} f_{\sigma(n)}(x)x_{\sigma(n)}$. Since $\{x_{\sigma(n)}\} \sim \{x_n\}$, $\sum_{n\geq 1} f_{\sigma(n)}(x)x_n$ converges. But $\{x_n\} \sim \{x_{\rho(n)}\}$ and so $\sum_{n\geq 1} f_{\sigma(n)}(x)x_{\rho(n)}$ converges.

Conversely, let $\{x_n;f_n\}$ be an sy.S.b. If $x \in X$, then $\sum_{n\geq 1} f_{\sigma(n)}(x)x_{\sigma(n)}$ converges for each σ in P; in particular, for a given σ in P,

$$x = \sum_{n\geq 1} f_{\sigma(n)}(x)x_{\sigma(n)},$$

and this proves the first part of (7.3.2). Let $\sum_{n\geq 1} \alpha_n x_n$ converge to x in (X,T). By the symmetric character of $\{x_n;f_n\}$, $\sum_{n\geq 1} f_n(x)x_{\sigma(n)}$ converges in (X,T) for each σ in P. Similar arguments yield the convergence of $\sum_{n\geq 1} \alpha_n x_n$ from that of $\sum_{n\geq 1} \alpha_n x_{\sigma(n)}$ for each σ in P. □

THEOREM 7.3.3 $\{x_n;f_n\}$ is an sy.S.b. for (X,T) if and only if

$$\text{For each } x \text{ in } X \text{ and } \sigma \text{ in } P\text{, the series } \sum_{n\geq 1} f_n(x)x_{\sigma(n)} \text{ converges in } (X,T). \tag{7.3.4}$$

PROOF. The necessity is obvious.

To prove the converse, observe that the S.b. $\{x_n;f_n\}$ is a u-S.b. Indeed, let $x \in X$ and σ in P be chosen arbitrarily. Then for some y in X,

$$y = \sum_{j\geq 1} f_j(x)x_{\rho(j)} \qquad (\rho = \sigma^{-1}).$$

Hence $f_i(y) = f_{\sigma(i)}(x)$ for $i \geq 1$ and so

$$\sum_{n\geq 1} f_{\sigma(n)}(x)x_{\sigma(n)} = \sum_{n\geq 1} f_n(y)x_{\sigma(n)}$$

converges in (X,T) for each σ in P.

Now, let $x \in X$ and ρ, σ in P be again chosen arbitrarily. For each $j \geq 1$, we find a unique i such that $\sigma(j) = \mu(i)$, where $\mu = \sigma\rho^{-1}$. Then

$$\sum_{i\geq 1} f_i(x)x_{\mu(i)} = z = \sum_{i\geq 1} f_{\sigma(i)}(z)x_{\sigma(i)},$$

and so $f_{\sigma(j)}(z) = f_{\rho(j)}(x)$, $j \geq 1$. Therefore $\sum_{n\geq 1} f_{\rho(n)}(x)x_{\sigma(n)}$ converges in (X,T). □

Concerning the characterization of an sy. S.b. $\{x_n;f_n\}$ in terms of its corresponding s.s. $\delta \equiv \delta_{\{x_n;f_n\}}$ we first prove (cf. [58]; [75])

PROPOSITION 7.3.5 If $\{x_n;f_n\}$ is an sy.S.b. for (X,T), then δ is symmetric. Conversely, if δ is symmetric and for each σ in P, the map $F_\sigma : X \to X$ is continuous, then $\{x_n;f_n\}$ is an sy.S.b., where

$$F_\sigma(x) = \sum_{n\geq 1} f_{\sigma(n)}(x)x_n.$$

PROOF. If $\{x_n;f_n\}$ is an sy.S.b., then δ is a symmetric s.s. by (7.3.2).

To prove the converse, let us observe that F_σ is well defined by the symmetric character of δ, for each σ in P. Choose x in X and π,σ in P arbitrarily. Put $\rho = \sigma^{-1}$. Then for $j \geq 1$, $F_\rho(x_j) = \sum_{n\geq 1} f_{\rho(n)}(x_j)x_n = x_{\sigma(j)}$. Hence

$$F_\rho\left(\sum_{i=1}^{n} f_{\pi(i)}(x)x_i\right) = \sum_{i=1}^{n} f_{\pi(i)}(x)x_{\sigma(i)},$$

and the continuity of F_ρ immediately yields the convergence of $\sum_{i\geq 1} f_{\pi(i)}(x)x_{\sigma(i)}$. □

We now derive the main

THEOREM 7.3.6 Let (X,T) be barrelled. Then $\{x_n;f_n\}$ is an sy.S.b. if and only if δ is symmetric.

PROOF. The barrelledness of (X,T) yields the continuity of F_σ for each σ in P and now apply Proposition 7.3.5. □

REMARK. The barrelledness of (X,T) in the foregoing theorem is indispensible for inferring the symmetric character of $\{x_n;f_n\}$ from the symmetric nature of δ, for consider

EXAMPLE 7.3.7 This is the space $(k,\sigma(k,cs))$ which is not barrelled (e.g. [129], p. 65) and for which $\{e^n;e^n\}$ is an S.b. However, it is not an sy.S.b.; indeed, the series $\sum_{i\geq 1} \langle e,e^i\rangle e^i = \sum_{i\geq 1} e^i$ is not even unconditionally convergent. For otherwise, $\{\sum_{i=1}^{n} (-1)^i/i\}$ would be an unconditionally Cauchy sequence corresponding to the seminorm defined by $\{(-1)^n/n\}$ in cs, and this in turn implies that $\sum_{n\geq 1} (1/n) < \infty$ which is absurd.

At the end of this section, we pass on to an example of a non-sy.S.b.

EXAMPLE 7.3.8 Let us recall the subset $\mathcal{O}$ of ω consisting of all sequences with coordinates x_i's as 0 or 1 such that $(\sum_{i=1}^{n} x_i)/n \to 0$ as $n \to \infty$. Write $\omega_0 = \{\alpha x : \alpha \in \mathcal{O},\ x \in \omega\}$.
It is known (cf. [129], p. 200) that $(\omega_0, \sigma(\omega_0, \phi))$ is a barrelled subspace of $(\omega, \sigma(\omega, \phi))$. Clearly $\{e^n; e^n\}$ is a u-S.b. We proceed to show that this base is not symmetric. In fact, define α in ω by $\alpha_i = 1$ for $i = 2^n + n$ $(n \geqq 1)$ and $\alpha_i = 0$ otherwise. If $m \in \mathbb{N}$, then for some $n \geqq 1$,

$$2^{n-1} + n-1 \leqq m \leqq 2^n + n,$$

and so

$$\frac{1}{m} \sum_{i=1}^{m} \alpha_i \leqq \frac{n}{2^{n-1} + n - 1} \to 0 \text{ as } n \to \infty.$$

Consequently $\alpha \in \mathcal{O}$. Let now $x = \alpha e$. Consider σ in P defined by

$$\sigma(i) = \begin{cases} 2^{\frac{i+1}{2}} + \frac{i+1}{2}, & i = 1,3,5,\ldots; \\ 1, & i = 2; \\ \frac{i}{2} + j-1, & 2^{j-1}+1 \leqq \frac{i}{2} \leqq 2^j,\ j \geqq 1,\ i = 4,6,8,\ldots. \end{cases}$$

Then

$$x_{\sigma(i)} = \begin{cases} 1, & i = 1,3,5,\ldots; \\ 0, & i = 2,4,6,\ldots\ . \end{cases}$$

Hence $x_\sigma \notin \omega_0$ and consequently

$$\sum_{n \geqq 1} x_{\sigma(n)} e^n$$

does not converge in $(\omega_0, \sigma(\omega_0, \phi))$.

NOTE. For another example of an S.b. which is bounded multiplier but not symmetric, we refer to Example 7.6.9.

7.4 APPLICATIONS AND FURTHER CHARACTERIZATIONS

Let $\{x_n;f_n\}$ be an S.b. for a Banach space $(X, \|\cdot\|)$. According to Singer ([210], [212]) $\{x_n;f_n\}$ is an sy.S.b. for $(X, \|\cdot\|)$ if and only if

$$\sup_{\sigma\in P} \sup_{\substack{|\alpha_i|\leq 1 \\ 1\leq n<\infty}} \left\| \sum_{i=1}^{n} \alpha_i f_i(x) x_{\sigma(i)} \right\| < \infty \ , \ \forall x \in X \qquad (7.4.1)$$

Now let $\{x_n;f_n\}$ be an S.b. for an l.c. TVS (X,T) and define $R_p(x)$ for each x in X and p in D_T as given below:

$$R_p(x) = \sup_{\sigma\in P} \sup_{\substack{|\alpha_i|\leq 1 \\ 1\leq n<\infty}} p\left(\sum_{i=1}^{n} \alpha_i f_i(x) x_{\sigma(i)} \right). \qquad (7.4.2)$$

In general, the symmetric character of $\{x_n,f_n\}$ cannot be expressed in terms of the finiteness of $R_p(x)$ given by (7.4.2), for each p in D_T and x in X. This will be demonstrated in Example 7.4.18.

In this section, we confine our attention to finding out a suitable characterization of an sy.S.b. in terms of the finiteness of $R_p(x)$ for each p in D_T and x in X. In the process of this investigation, we discover a few interesting applications of the presence of an sy.S.b. in an l.c. TVS. We follow [75] (cf. also [23]) for the rest of this section and as mentioned in [75], our approach is more sequence space oriented. Let us recall certain unexplained terms from chapter 3, especially Remarks 3.3.12 and begin with

PROPOSITION 7.4.3 Let $\{x_n;f_n\}$ be an sy.S.b. for a barrelled space (X,T). If $\{x_n\}$ is unbounded, then $(X,T) \simeq (\phi,\eta(\phi,\omega))$.

PROOF. There exists f in X* such that $\{f(x_n)\} \notin \ell^\infty$. If $\phi \subsetneqq \delta$, then following the proof of Theorem 7.2.2, we find x in X, $\{n_k\}$ in I_∞ and ρ in P such that

$$|f_{n_k}(x)\ f(x_{\rho(n_k)})| \geq 1, \ \forall\ k \geq 1. \tag{7.4.4}$$

Since $\{x_n;f_n\}$ is an sy.S.b., for each x in X, σ in P and f in X*, we have

$$\sum_{n\geq 1} |f_n(x) f(x_{\sigma(n)})| < \infty. \tag{7.4.5}$$

However, (7.4.5) contradicts (7.4.4). Hence $\delta = \phi$. By Proposition 3.2.4 and Theorem 1.3.13, $\delta^* = \delta^\beta = \omega$. But $T^{**} \approx \tau(\delta,\delta^*) \approx \eta(\phi,\omega)$ and so $(X,T) \simeq (\phi,\eta(\phi,\omega))$. □

COROLLARY 7.4.6 If a metrizable barrelled space (X,T) contains an sy. S.b. $\{x_n;f_n\}$, then $\{x_n\}$ is bounded in (X,T).

For otherwise, $(\phi,\eta(\phi,\omega)) \simeq (X,T)$ and so $(\phi,\eta(\phi,\omega))$ is a Baire space (cf. Theorem 1.3.7) having a countable dimension. This is, however, not true (e.g. [89], p. 88).

With some mild restriction on an sy.S.b., we can infer the metrizability of the space in question; for instance, we have

THEOREM 7.4.7 Let $\{x_n;f_n\}$ be an sy.S.b. such that $\{f_n(x_0)\} \notin \ell^\infty$ for some x_0 in X. Then $\{f_n\}$ is a Hamel base for X* and $T \approx \sigma(X,X^*)$.

PROOF. At the outset, let us observe that for each $i \geq 1$, we find p_i in D_T so that $|f_i(x)| \leq p_i(x)$ for all x in X. The collection $D_S = \{p_i\}$ generates a metrizable l.c. topology S on X such that $S \subset T$.

For each p in D_T, let $I_p = \{i \in \mathbb{N} : p(x_i) \neq 0\}$. If possible, let I_{p_0} be infinite and write $I_{p_0} = \{n_1,\ldots,n_k,\ldots\}$ in I_∞. We can determine $\{m_k\}$ in I_∞ so that

$$|f_{m_k}(x_o)| \geq \frac{1}{p_o(x_{n_k})}, \ \forall\ k \geq 1.$$

Define π in P with $\pi(n_k) = m_k$, $k \geq 1$. Then $|f_{\pi(n_k)}(x_o)|p_o(x_{n_k}) \geq 1$ for $k \geq 1$ and so $\sum_{n\geq 1} f_{\pi(n)}(x_o)x_n$ does not converge in (X,T), thereby contradicting the symmetric character of $\{x_n;f_n\}$. Hence for each p in D_T, there exists N_p in $\mathbb{N}$ so that $I_p \subset \{1,\ldots,N_p\}$.

If $f \in X^*$, then $|f(x)| \leq p(x)$ for some p in D_T and all x in X and so

$$f(x) = \sum_{n=1}^{N_p} f_n(x)f(x_n), \ \forall\ x \in X.$$

This shows that $\{f_n\}$ is a Hamel base for X^*.

Lastly, for x in X, p in D_T and $\varepsilon > 0$, we can find $N > N_p$ such that

$$p(x) - \varepsilon \leq p(\sum_{i=1}^{N} f_i(x)x_i),$$

and so

$$p(x) \leq (\sum_{i\in I_p} p(x_i)) \max_{i\in I_p} p_i(x), \ \forall x \in X.$$

Hence $T \subset S$. □

Alternative proof. Under the hypothesis, (δ,T^{**}) is a symmetric AK-space with $\delta \not\subset \ell^\infty$. Now apply Theorem 7.2.2. □

COROLLARY 7.4.8 Let $\{x_n;f_n\}$ be an sy.S.b. for an ω-complete l.c. TVS (X,T) such that $\{f_n(x_o)\} \notin \ell^\infty$ for some x_o in X. Then (X,T) is a Fréchet space and $\{x_n\}$ is bounded in (X,T).

REMARK. In the preceding theorem and its corollary, the

condition $\{f_n(x_o)\} \notin \ell^\infty$ cannot be dropped, for we have

EXAMPLE 7.4.9 This is the space $(\phi,\sigma(\phi,\omega))$ considered in Example 4.3.6. It is not metrizable, otherwise it would be a Baire space with $\dim(\phi) = \aleph_o$, a contradiction (e.g. [89], p. 88).

Next we have

THEOREM 7.4.10 Let $\{x_n;f_n\}$ be an sy.S.b. for an ω-complete l.c. TVS (X,T) such that $\{f_n(x_o)\} \notin \ell^\infty$ for some x_o in X. Then $(\delta,T^{**}) \simeq (\omega,\sigma(\omega,\phi))$.

PROOF. By Theorem 7.2.2, Corollary 7.4.8 and Proposition 3.3.6, $\mu = \phi$. Further, $\{x_n;f_n\}$ being a u-S.b. for (X,T) is a b.m-S.b. and therefore using Proposition 6.3.2, δ is normal. Thus by Theorem 7.2.1, $\delta = \omega$. Consequently, $T^{**} \approx \tau(\delta,\delta^*) \approx \tau(\omega,\phi) \approx \sigma(\omega,\phi)$. □

REMARK. None of the restrictions that $\{f_n(x_o)\} \notin \ell^\infty$ or (X,T) is an ω-complete space can be omitted in order to have the desired conclusion in Theorem 7.4.10. This is demonstrated by the following two examples.

EXAMPLE 7.4.11 Consider the space $(\phi,\eta(\phi,\omega))$ with its S.b. $\{e^n;e^n\}$. Since $\eta(\phi,\omega) = \beta(\phi,\omega)$ (cf. [129], p. 106 or [143], p. 408), the space in question is barrelled and by Proposition 7.3.6, $\{e^n;e^n\}$ is an sy. S.b. for $(\phi,\eta(\phi,\omega))$ which is also ω-complete (cf. Theorem 1.3.7). Note that $\{\langle\alpha,e^n\rangle\} \in \ell^\infty$ for each α in ϕ and $(\phi,\eta(\phi,\omega)) \not\simeq (\omega,\sigma(\omega,\phi))$.

EXAMPLE 7.4.12 Let $A = \{x \in \omega_{\mathbb{R}} : x_i$ is rational for $i \geq 1\}$ and consider the s.s.$\lambda = \mathrm{sp}\,\{A\}$ over the field $\mathbb{R}$. Then λ is a monotone s.s.s. with $\lambda \not\subset \ell^\infty$ and so by Theorem 7.2.2, $\lambda^\beta = \lambda^\times = \phi$. Since λ is monotone, $(\phi,\sigma(\phi,\lambda))$ is ω-complete (cf. [129], p. 188). Consequently, each $\sigma(\phi,\lambda)$-bounded subset of ϕ is $\beta(\phi,\lambda)$-bounded ([234], p. 159). Further,

$(\phi,\beta(\phi,\lambda))^* = \omega$ (cf. [234], p. 261). Thus every $\sigma(\phi,\lambda)$-bounded subset of ϕ is $\sigma(\phi,\omega)$-bounded and therefore finite-dimensional ([129], p. 104). Hence $(\lambda,\sigma(\lambda,\phi))$ is barrelled without being ω-complete (cf. Theorems 1.2.8 and 1.3.7). By Theorem 7.3.6, $\{e^n;e^n\}$ is an sy.S.b. for $(\lambda,\sigma(\lambda,\phi))$; however, $(\lambda,\sigma(\lambda,\phi) \not\approx (\omega,\sigma(\omega,\phi))$.

PROPOSITION 7.4.13 Let $\{x_n;f_n\}$ be an sy.S.b. for a Banach space $(X, \|\cdot\|)$. Then

$$0 < \inf_{n\geq 1} \|x_n\| \leq \sup_{n\geq 1} \|x_n\| < \infty.$$

PROOF. By Corollary 7.4.8, $\sup \|x_n\| < \infty$. Further, on account of Theorem 7.4.10, $\{f_n(x)\} \in \ell^\infty$ for each x in X, otherwise $\sigma(\omega,\phi)$ would be normable which is, however, not true. Hence by the Banach Steinhaus theorem, $|f_n(x)| \leq k\|x\|$ for all x in X and $n \geq 1$, k being a positive constant. Therefore, $\inf \|x_n\| \geq k^{-1}$. □

We now turn our attention to one of the main results of this section promised in the opening paragraphs, namely,

PROPOSITION 7.4.14 Let (X,T) be a Fréchet space with $(X,T) \not\approx (\omega,\sigma(\omega,\phi))$. If $\{x_n;f_n\}$ is an sy.S.b. for (X,T) and $D_T = \{p_k\}$, then for each x in X and $k \geq 1$,

$$R_k(x) \equiv \sup_{\sigma\in P} \sup_{\substack{|\alpha_i|\leq 1 \\ 1\leq n<\infty}} p_k\left(\sum_{i=1}^{n} \alpha_i f_i(x) x_{\sigma(i)}\right) < \infty. \quad (7.4.15)$$

PROOF. By Corollary 7.4.6 and Theorem 7.4.10, $\{f(x_n)\}$, $\{f_n(x)\} \in \ell^\infty$ for all x in X and f in X*.

Let us consider two mutually exclusive cases:

(i) $\{f_n(x)\} \in \phi$ and (ii) $\{f_n(x)\} \in \ell^\infty \setminus \phi$.

(i) If ℓ denotes the length of $\{f_n(x)\}$, then for f in X*

$$\sup_{\sigma \in P} \sup_{\substack{|\alpha_i| \leq 1 \\ 1 \leq n < \infty}} \left| f\left(\sum_{i=1}^{n} \alpha_i f_i(x) x_{\sigma(i)} \right) \right|$$

$$\leq \| \{ f(x_n) \} \|_\infty \sum_{i=1}^{\ell} |f_i(x)| < \infty ,$$

and this proves (7.4.15).

(ii) Since $\{x_n; f_n\}$ is an sy.S.b., we have (7.4.5) for each f in X* and σ in P. Thus using Proposition 7.2.6, we find for each x in X with $\{f_n(x)\} \in \ell^\infty \setminus \phi$ and f in X*,

$$\sup_{\sigma \in P} \sup_{\substack{|\alpha_i| \leq 1 \\ 1 \leq n < \infty}} \left| f\left(\sum_{i=1}^{n} \alpha_i f_i(x) x_{\sigma(i)} \right) \right| < \infty ,$$

and we have, again, finished. □

The converse of the preceding proposition is true even with a slightly weaker hypothesis. Indeed, we have

PROPOSITION 7.4.16 Let (X,T) be ω-complete and barrelled. Suppose that for each x in X, p in D_T and σ in P,

$$p_\sigma(x) \equiv \sup_{n \geq 1} p\left(\sum_{i=1}^{n} f_i(x) x_{\sigma(i)} \right) < \infty. \tag{7.4.17}$$

Then $\{x_n; f_n\}$ is an sy.S.b. for (X,T).

PROOF. Fix σ in P and define $G_{\sigma,n} : X \to X$ by

$$G_{\sigma,n}(x) = \sum_{i=1}^{n} f_i(x) x_{\sigma(i)}, \; n \geq 1.$$

By the barrel theorem, for each p in D_T, we find q in D_T so that $p(G_{\sigma,n}(x)) \leq q(x)$, for all n ≥ 1 and x in X. Thus, $\{ \sum_{i=1}^{n} f_i(x) x_{\sigma(i)} \}$ is Cauchy in (X,T) and hence it converges in (X,T). Now apply Theorem 7.3.3. □

With the help of Propositions 7.4.14 and 7.4.16, it is now possible to offer a variation of a result given in [23], namely,

<u>THEOREM 7.4.18</u> Let $\{x_n;f_n\}$ be an S.b. for a Fréchet space $(X,T) \not\approx (\omega,\sigma(\omega,\phi))$ and let $D_T = \{p_k\}$. Then $\{x_n;f_n\}$ is an sy.S.b. for (X,T) if and only if (7.4.15) is valid.

<u>REMARK</u>. The next example shows that the restriction $(X,T) \not\approx (\omega,\sigma(\omega,\phi))$ in Proposition 7.4.14 or Theorem 7.4.18 cannot be dropped.

<u>EXAMPLE 7.4.19</u> This is the space $(\omega,\sigma(\omega,\phi))$ with its usual S.b. $\{e^n;e^n\}$. Since ω is symmetric and the space in question is barrelled, $\{e^n;e^n\}$ is an sy.S.b. for $(\omega(\sigma(\omega,\phi))$ by Theorem 7.3.6. Next observe that for x in ω and σ in P,

$$\sup_{1\leqq n<\infty} |<\sum_{i=1}^{n} x_i e^{\sigma(i)},e^j>| = |x_{\pi(j)}|,$$

where $\pi = \sigma^{-1}$. Suppose that (7.4.15) is true for this space; then in particular

$$\sup_{\sigma\in P}\ \sup_{1\leqq n<\infty} <\sum_{i=1}^{n} ie^{\sigma(i)},e^j>| \leqq K_j < \infty,$$

for j in $\mathbb{N}$. Choose m in $\mathbb{N}$ so that $K_j < m$ and define σ in P with $\sigma(2m) = j$; $\sigma(i) = i$, $i \neq j$, $2m$. Then

$$m > K_j \geqq \sup_{1\leqq n<\infty} |<\sum_{i=1}^{n} ie^{\sigma(i)}, e^j>| = 2m$$

which is absurd.

<u>EXERCISE 7.4.20</u> Prove that in an ω-complete barrelled space (X,T), $\{x_n;f_n\}$ is an sy.S.b. if and only if for each x in X and σ in P, the set

$$\{\sum_{i=1}^{n} \alpha_i f_i(x) x_{\sigma(i)} : \alpha \in \omega, |\alpha_i| \leq 1; n \geq 1\}$$

is bounded in (X,T).

EXERCISE 7.4.21 If $\{x_n;f_n\}$ is an sy.S.b. for (X,T), prove that either $\{x_n\}$ is bounded or it is a Hamel base for X.

EXERCISE 7.4.22 Let (X,T) be barrelled and δ an s.s.s. Prove that either $\delta \not\subset \ell^\infty$ or $\delta \subset c_o$; in the former case, show that (δ, T^{**}) is a metrizable barrelled space (cf. Remarks 3.3.12 for the definition of T^{**}). [Hint: if $\delta \subset \ell^\infty$, then $\ell^1 \subset \delta^*$ and $\beta(\ell^\infty, \ell^1)|\delta \subset T^{**}$, giving $\delta \subset c_o$.]

7.5 SIZE OF δ

As pointed out in section 7.1, the study of symmetric sequence spaces is quite helpful in locating the size of these spaces. Consequently, if $\{x_n;f_n\}$ is an sy.S.b. for an l.c. TVS (X,T), we may possibly expect a better estimate of the size of the associated s.s. δ discussed in section 3.3 and this is what we plan to discuss in this section. More results in this direction will be discussed in the next chapter. So, we begin with ([75]; cf. also [95] and [221])

THEOREM 7.6.1 Let $\{x_n;f_n\}$ be an sy.S.b. for an ω-complete barrelled space (X,T). Then (i) $(X,T) \simeq (\omega, \eta(\omega,\phi))$ or $(X,T) \simeq (\phi, \eta(\phi,\omega))$ or $\phi \subsetneq \delta \subset c_o$ and in the last case, T is given by the family of seminorms $\{M_p : p \in D_T\}$, where for x in X,

$$M_p(x) = \sup_{\sigma \in P} \sup_{|\alpha_n| \leq 1} p(\sum_{n \geq 1} \alpha_n f_n(x) x_{\sigma(n)}), \qquad (7.5.2)$$

and (ii) if (X,T) is also nuclear, then $(X,T) \simeq (\omega, \eta(\omega,\phi))$ or $(X,T) \simeq (\phi, \eta(\phi,\omega))$.

PROOF. By Proposition 6.3.2, δ is normal.

Let $\delta = \ell^\infty$. Then $\delta^* = \delta^\beta = \ell^1$, $\delta^* \equiv (\delta, T^{**})^*$ (cf. Proposition 3.2.6 and Remark 3.3.12). Thus $T^{**} \approx \beta(\ell^\infty, \ell^1)$ which is, however, not true as $(\ell^\infty, \beta(\ell^\infty, \ell^1))$ is not separable. Therefore, by Theorem 7.2.1, either $\delta = \omega$ or $\delta \subset c_o$.

(i) Let $\delta = \omega$. Then $(X,T) \simeq (\omega, \beta(\omega,\phi)) \simeq (\omega, \eta(\omega,\phi))$. If $\delta = \phi$, then $(X,T) \simeq (\phi, \beta(\phi,\omega)) \simeq (\phi, \eta(\phi,\omega))$. Finally, let $\phi \subsetneq \delta \subset c_o$. Observe that for each σ in P and α in ω with $|\alpha_n| \leq 1$ for $n \geq 1$, the operator $u_{\alpha,\sigma} : X \to X$,

$$u_{\alpha,\sigma}(x) = \sum_{n\geq 1} \alpha_n f_n(x) x_{\sigma(n)}$$

is well-defined and continuous. Using Proposition 7.2.6 for those x with $\{f_n(x)\} \in \delta \setminus \phi$ and Exercise 7.4.21 for x with $\{f_n(x)\} \in \phi$, we get

$$\sup_{|\alpha_n|\leq 1} \sup_{\sigma \in P} \Big| \sum_{n\geq 1} \alpha_n f_n(x) f(x_{\sigma(n)}) \Big| < \infty; \ \forall\, x \in X,\ f \in X^*.$$

Consequently $\{u_{\alpha,\sigma} : \alpha \in \omega, |\alpha_n| \leq 1; \sigma \in P\}$ is equicontinuous and this proves the last part in (i).

(ii) Let us note that $\{x_n; f_n\}$ is a fully ℓ^1-base (Proposition 12.4.15) and so from Proposition 5.2.3, δ is perfect. Hence in view of Theorem 7.2.1 and part (i) of the present result, it suffices to confine our attention to when $\ell^1 \subset \delta \subset c_o$.

Since $\{x_n; f_n\}$ is a fully ℓ^1-base, for each p in D_T, there exists q in D_T so that

$$(*) \qquad p(x) \leq \sum_{n\geq 1} |f_n(x)| p(x_n) \leq q(x), \ \forall\, x \in X.$$

In particular, for each f in X^*, we find q in D_T with

$$\sum_{n\geq 1} |f_n(x) f(x_n)| \leq q(x), \ \forall\, x \in X$$

and as $\mu = \delta^\beta = \delta^\times$ (Proposition 3.3.6), we find that $\eta(\delta,\delta^\times) \subset T^{**}$. On the other hand, by using (*) and Proposition 5.2.1, we conclude that $\eta(\delta,\delta^\times) \approx T^{**}$. Thus, to every α in $\delta^\times$ there corresponds β in $\delta^\times$ so that $\alpha/\beta \in \ell^1$ (cf. Theorem 1.4.13). Consequently, $\delta^\times = \ell^1$ and so $\delta = \ell^\infty$. Hence $\delta = \phi$ or ω and apply (i). □

7.6 RELATED NOTIONS

Theorem 7.3.1 motivates us in a natural way to introduce the following

DEFINITION 7.6.1 An S.b. $\{x_n;f_n\}$ for an l.c. TVS (X,T) is called s-*symmetric* (s-sy.S.b.) [resp. u-*symmetric* (u-sy.S.b.)] if $\{x_n;f_n\}$ is an s-S.b. [resp. u-S.b.] and the S.b. $\{x_{n_i}\}$ for $[X_{n_i}]$ is equivalent to $\{x_n\}$ for each $\{n_i\}$ in I.

A characterization of an s-sy.S.b. is contained in

PROPOSITION 7.6.2 An S.b. $\{x_n;f_n\}$ for an l.c. TVS (X,T) is an s-sy.S.b. if and only if for each x in X and $\{m_i\}$, $\{n_i\}$ in I, the series $\sum_{i\geq 1} f_{m_i}(x)x_{n_i}$ converges. If (X,T) is ω-complete, then every s-sy.S.b. for (X,T) is an u-sy.S.b. and conversely.

PROOF. Let $\{x_n;f_n\}$ be an s-sy.S.b. and fix x in X and $\{m_i\}$, $\{n_i\}$ in I arbitrarily. Since $\sum_{i\geq 1} f_{m_i}(x)x_{m_i}$ converges and $\{x_{m_i}\} \sim \{x_i\} \sim \{x_{n_i}\}$, $\sum_{i\geq 1} f_{m_i}(x)x_{n_i}$ also converges.

Conversely, for each $\{m_i\}$ in I and x in X, $\sum_{i\geq 1} f_{m_i}(x)x_{m_i}$ converges and so $\{x_n;f_n\}$ is an s-S.b. But then, by taking $\{n_i\}$ as $\{i\}$, $\sum_{i\geq 1} f_{m_i}(x)x_i$ also converges. Also, $\sum_{i\geq 1} f_i(x)x_{m_i}$ converges. This proves that $\{x_{m_i}\} \sim \{x_i\}$. □

In order to prove the main result of this section, let us first pass on to

<u>PROPOSITION 7.6.3</u> Let $\{x_n;f_n\}$ be an S.b. for an ω-complete barrelled space (X,T) such that $R_p : X \to \mathbb{R}$ is well defined (cf. (7.4.2)) for each p in D_T. Then the l.c. topology S on X defined by seminorms R_p, $p \in D_T$ is equivalent to T.

Further, for any σ in P, p in D_T and any pair of finite sequences $\alpha_1,\ldots,\alpha_n$ and $\beta_1,\ldots,\beta_n$ with $|\alpha_n| = 1$, $1 \leq n < \infty$, we have

$$R_p\left(\sum_{i=1}^{n} \beta_i x_i\right) = R_p\left(\sum_{i=1}^{n} \alpha_i \beta_i x_{\sigma(i)}\right). \tag{7.6.4}$$

Also, for x in X, p in D_T, (ρ,σ) in $P \times P$ and α in ω with $|\alpha_i| = 1$, $i \geq 1$, one has

$$R_p(x) = R_p\left(\sum_{n\geq 1} \alpha_n f_{\rho(n)}(x) x_{\sigma(n)}\right). \tag{7.6.5}$$

<u>PROOF</u>. For σ in P, n in $\mathbb{N}$ and α in ω, define $F_{\sigma,n,\alpha} : X \to X$ by

$$F_{\sigma,n,\alpha}(x) = \sum_{i=1}^{n} \alpha_i f_i(x) x_{\sigma(i)}.$$

Then the family $\{F_{\sigma,n,\alpha} : \sigma \in P,\ n \in \mathbb{N},\ \alpha \in \omega$ with $|\alpha_n| \leq 1,\ n \geq 1\}$ being pointwise bounded on account of the finiteness of $R_p(x)$ for each x in X, is therefore equicontinuous. Hence for each p in D_T, there exists q in D_T such that $R_p(x) \leq q(x)$ for all x in X. Clearly $p(x) \leq R_p(x)$ for all x in X and each p in D_T. Thus $S \approx T$.

To prove the second part, let us observe that

$$R_p\left(\sum_{i=1}^{n} \alpha_i\beta_i x_{\sigma(i)}\right) = \sup_{\sigma\in P}\ \sup_{\substack{|\gamma_j|\leq 1\\ 1\leq j\leq n}} p\left(\sum_{i=1}^{j} \gamma_{\sigma(i)}\alpha_i\beta_i x_{\rho\sigma(i)}\right)$$

$$(*) \qquad \leq R_p\left(\sum_{i=1}^{n} \beta_i x_i\right).$$

Put $M_1 = [1,\ell]\setminus\sigma[1,n]$ and $M_2 = [1,\ell]\setminus[1,n]$, where $\ell = \max\{\sigma(i):1 \leq i \leq n\}$ and $[1,n] = \{1,2,\ldots,n\}$ etc. Then $\#(M_1) = \#(M_2)$ and let Θ be the one-one map from M_1 onto M_2. Define ρ in P by

$$\rho(i) = \begin{cases} \sigma^{-1}(i), & i \in \{\sigma(1),\ldots,\sigma(n)\}; \\ \Theta(i), & i \in M_1; \\ i, & i \geq \ell + 1. \end{cases}$$

Also, let $\gamma_i = 0$ for $i \in M_1$ and $= \alpha_{\sigma^{-1}(i)}\beta_{\sigma^{-1}(i)}$ for $i \in \{\sigma(1),\ldots,\sigma(n)\}$. Then

$$R_p(\sum_{i=1}^{n} \beta_i x_i) = R_p(\sum_{j=1}^{\ell} \bar{\alpha}_{\sigma^{-1}(j)} \gamma_j x_{\rho(j)})$$

$$\leq R_p(\sum_{j=1}^{\ell} \gamma_j x_j) = R_p(\sum_{i=1}^{n} \alpha_i \beta_i x_{\sigma(i)}),$$

by using (*). This proves (7.6.4).

To prove (7.6.5), let us observe in view of Proposition 7.4.16 that $\{x_n;f_n\}$ is an sy.S.b. and so for any x in X and ρ,σ in P, the series $\sum_{n\geq 1} f_{\rho(n)}(x)x_{\sigma(n)}$ converges unconditionally and so the series on the right of (7.6.5) converges. The above arguments also allow us to conclude that $\sum_{n\geq 1} \alpha_n f_{\rho(n)}(x)x_{\sigma(n)}$ converges unconditionally (cf. Theorem 1.5.2). Hence

$$R_p(\sum_{n\geq 1} \alpha_n f_{\rho(n)}(x)x_{\sigma(n)}) = \lim_{n\to\infty} R_p(\sum_{i=1}^{n} \alpha_{\rho^{-1}(i)} f_i(x)x_{\sigma\rho^{-1}(i)})$$

$$= \lim_{n\to\infty} R_p(\sum_{i=1}^{n} f_i(x)x_i),$$

by (7.6.4). This proves (7.6.5). □

<u>PROPOSITION 7.6.6</u> Let (X,T) be an ω-complete barrelled

space having an S.b. $\{x_n;f_n\}$ such that $R_p(x)$ is finite for each p in D_T and x in X. Then $\{x_n;f_n\}$ is a u-sy.S.b. or equivalently an s-sy.S.b.

PROOF. Choose β in ω and $\{n_i\}$ in I arbitrarily. Put $y = \sum_{i=m}^{r} \beta_i x_{n_i}$. Then for p in D_T, using (7.6.5) we have

$$R_p(y) = R_p(\sum_{j=m}^{r} \beta_j x_{\tau(n_j)}), \quad \tau = \sigma\rho^{-1}$$

$$= R_p(\sum_{j=m}^{r} \beta_j x_j),$$

by (7.6.4). Since $S \approx T$ (see Proposition 7.6.3), $\{x_i\} \sim \{x_{n_i}\}$.

Also $\{x_n;f_n\}$ is a u-S.b. (indeed, it is an sy.S.b.), therefore $\{x_n;f_n\}$ is a u-sy.S.b. □

THEOREM 7.6.7 Every sy.S.b. $\{x_n;f_n\}$ in a Fréchet space $(X,T) \not\approx (\omega,\sigma(\omega,\phi))$ is a u-sy.S.b. or equivalently an s-sy.S.b.

PROOF. We may write $D_T = \{p_k\}$. Then $R_k(x) \equiv R_{p_k}(x)$ is finite for each x in X and $k \geq 1$ by Proposition 7.4.14. Now apply Proposition 7.6.6. □

COROLLARY 7.6.8 Every sy.S.b. in a Banach space is a u-sy.S.b. or equivalently an s-sy.S.b.

REMARK. The converse in Corollary 7.6.8 is not necessarily true, for we have ([58])

EXAMPLE 7.6.9 For each $n = \{n_i\}$ in I, let a^n in ω be defined by

$$a_i^n = \begin{cases} j^{-1/2}, & i = n_j; \\ 0, & \text{otherwise.} \end{cases}$$

Put $A = \{a^n : n \in I\}$ and consider the following Banach s.s. given by

$$\lambda_A = \{x \in \omega : \|x\|_A \equiv \sup_{a \in A} \sum_{i \geq 1} |x_i a_i| < \infty\}.$$

By Theorem 1.3.13, $\lambda_A^* = \lambda_A^\beta$; in particular, $e^i \in \lambda_A^*$ with $|\langle x, e^i \rangle| \leq \|x\|_A$ for each x in λ_A and $i \geq 1$. Also, $\{e^n ; e^n\}$ is an s-S.b. for $(\lambda_A, \|\cdot\|_A)$ (cf. [215], p. 584-585). Since

$$\|\sum_{i=r}^{s} \beta_i e^{n_i}\|_A = \|\sum_{i=r}^{s} \beta_i e^i\|_A,$$

$\{e^n ; e^n\}$ is an s-sy.S.b.

We proceed to show that $\{e^n ; e^n\}$ is not an sy.S.b. for $(\lambda_A, \|\cdot\|_A)$

First, let us construct a bounded sequence $\{x^r : r \geq 1\}$ in λ_A. Indeed, let $x_i^r = (r+1-i)^{-1/2}$ for $1 \leq i \leq r$ and $x_i^r = 0$ for $i > r$. Then

$$\|x^r\|_A = \sum_{i=1}^{r} (i(r+1-i))^{-1/2}.$$

A direct verification shows that for $s > 1$,

$$\sum_{i=1}^{n} (i(r+1-i))^{-1/2} = \begin{cases} 2r^{-1/2} + 2 \sum_{i=2}^{[r/2]} (i(r+1-i))^{-1/2} + \frac{2}{r+1}, & r \text{ is odd;} \\ 2r^{-1/2} + 2 \sum_{i=2}^{r/2} (r(r+1-i))^{-1/2}, & r \text{ is even.} \end{cases}$$

Since $(i(r+1-i))^{-1/2}$ decreases in $[1, [r/2]]$,

$$\sum_{i=1}^{[r/2]-1} \int_i^{i+1} (u(r+1-u))^{-1/2} du \geq \sum_{i=2}^{[r/2]} (r(r+1-i))^{-1/2}.$$

Thus

$$\sum_{i=1}^{r} (i(r+1-i))^{-1/2} \leq 2r^{-1/2} + 2 \sum_{i=1}^{[r/2]-1} \int_i^{i+1} (u(r+1-u))^{-1/2} du + \frac{2}{r+1}$$

$$\leq 2r^{-1/2} + 2 \int_1^{(r+1)/2} (u(r+1-u))^{-1/2} du + \frac{2}{r+1}$$

$$\leq 3+2 \sin^{-1} \left(\frac{r-1}{r+1}\right) \leq 3+\pi, \quad \forall r \geq 1.$$

Hence $\{x^r\}$ is bounded.

To derive the required result, suppose on the contrary that $\{e^n;e^n\}$ is an sy.S.b. for $(\lambda_A, \|\cdot\|_A)$ and so the map $F_\sigma:\lambda_A \to \lambda_A$,

$$F_\sigma\left(\sum_{i\geq 1} x_i e^i\right) = \sum_{i\geq 1} x_{\sigma(i)} e^i$$

is continuous (cf. Theorem 7.3.6). Since (7.4.1) is satisfied (cf. Proposition 7.4.14), the family $\{F_\sigma:\sigma \in P\}$ is equicontinuous (use (7.6.5) for the pointwise boundedness of this family). Consequently $\{F_\sigma:\sigma \in P\}$ is bounded on bounded subsets. However, this is not true. In fact, define $\{\sigma_r : r \geq 1\}$ in P by $\sigma_r(i) = r+1-i$, $1 \leq i \leq r$ and $\sigma_r(i) = i$ otherwise. Then

$$F_{\sigma_r}(x^r) = \{x^r_{\sigma_r(1)},\ldots,x^r_{\sigma_r(r)},0,0\ldots\}$$

$$= \{1,2^{-1/2},\ldots,r^{-1/2},0,0,\ldots\},$$

and so

$$\|F_{\sigma_r}(x^r)\|_A = \sum_{i=1}^{r} i^{-1} \to \infty \quad \text{as } r \to \infty.$$

8 Advances in symmetric bases

8.1 EARLIER BACKGROUND

Throughout this chapter we write (X,T) for an arbitrary l.c TVS containing an S.b. $\{x_n;f_n\}$. If $\{x_n;f_n\}$ is a symmetric S.b. (sy.S.b.), then the underlying sequence space (s.s.) $\delta = \{\{f_n(x)\} : x \in X\}$ is necessarily symmetric. We already know that if δ is symmetric, then it does not follow that $\{x_n;f_n\}$ is an sy.S.b. Accordingly, it is natural to investigate conditions which will ensure the symmetric character of $\{x_n;f_n\}$, given that δ is a symmetric s.s. In this direction, we have already established one result, namely, Theorem 7.3.6. In this chapter, we further concentrate on this converse problem, placing more emphasis this time on the size of the symmetric sequence space (s.s.s.) δ. In addition, related conditions are also considered, which ascertain the symmetric character of an S.b. The importance of the results is illustrated by several examples and the material of this chapter is essentially based on a paper of Garling [58].

Finally, we recall the topology T^{**} on δ from Remarks 3.3.12 and also the notation Φ_r for the collection of all finite subsets of $[r,\infty] = \{i \in \mathbb{N} : r \leq i < \infty\}$ with $\Phi \equiv \Phi_1$ for $r = 1$. Also for p,q in $\mathbb{N}$ with $p < q$, let $[p,q] = \{p,p+1,\ldots,q\}$.

8.2 THE SYMMETRIC δ AND ITS SIZE

Our notation for (X,T), $\{x_n;f_n\}$ and δ stand as in section 8.1 and begin with

PROPOSITION 8.2.1 Let δ be an s.s.s. with $\delta \subset \ell^\infty$ but $\delta \not\subset c$. Then $T^{**} \subset S$, S being the l.c. topology on δ considered in Theorem 6.2.1.

PROOF. As S is the topology on δ, generated by the polars of all compact subsets of ℓ^1, it suffices to prove that every T^{**}-equicontinuous subset of δ^* $(=\{\{f(x_n)\}: f \in X^*\}$; cf. Remarks 3.3.12) is relatively compact in $(\ell^1, \|\cdot\|_1)$.

Let us first note that $\delta^* \subset \ell^1$; for otherwise if there exists $a \in \delta^* \setminus \ell^1$, the series $\sum_{i\geq 1} a_i$ is not unordered convergent and then the method given below with $A = \{a\}$ yields a contradiction.

Now let A be a T^{**}-equicontinuous subset of δ^* and suppose that A is not relatively compact in $(\ell^1, \|\cdot\|_1)$. Then as in the proof of Theorem 6.2.1, there exists $\varepsilon > 0$ such that

$$\sup_{J\in\Phi_r} \sup_{a\in A} \Big|\sum_{j\in J} a_j\Big| \geq 4\varepsilon, \ \forall\ r \geq 1.$$

Consequently we can construct sequences $\{m_i\}$ and $\{n_i\}$ in $\mathbb{N}$, a sequence $\{J_i\}$ in Φ and a sequence $\{a^i\}$ in A satisfying

(i) $m_i \leq n_i \leq m_{i+1}-2$, (ii) $J_i \subset [m_i,n_i]$ and

(iii) $\Big|\sum_{j\in J_i} a^i_j\Big| \geq 2\varepsilon$, where $i = 1,2,\ldots$.

By the hypothesis, there exists α in $\delta\setminus c$. Since $\alpha \in \ell^\infty$, we can find numbers μ,ν and sequences $\{p_n\}$, $\{q_n\}$ in I_∞ with $\{p_n\}$ and $\{q_n\}$ being nonoverlapping such that $\alpha_{p_n} \to \mu$, $\alpha_{q_n} \to \nu$ and $\mu \neq \nu$.

Consider σ in P so that $\sigma(p_n) = q_n$ and $\sigma(q_n) = p_n$. If $\beta_n = \alpha_{\sigma(n)}$, then $\beta \in \delta$ and so $\gamma = \mu\alpha - \nu\beta \in \delta$. Note that $\gamma_{p_n} \to \mu^2 - \nu^2 \equiv k \neq 0$ and $\gamma_{q_n} \to 0$.

Hence we can find sequences $\{l_i\}$ and $\{s_i\}$ such that

$$|\gamma_{p_n} - k| \leq \frac{\varepsilon|k|}{2\sum_{s=m_i}^{n_i} |a_r^i|} \quad , \forall\, n \geq l_i$$

and

$$|\gamma_{q_n}| \leq \frac{\varepsilon|k|}{2\sum_{r=m_i}^{n_i} |a_r^i|} \quad , \forall\, n \geq s_i.$$

Let $K_i = [m_i, n_i] \setminus J_i$. Then by choosing $\{l_i\}$ and $\{s_i\}$ suitably, we can find π in P so that

$$\pi[\bigcup_{i\geq 1} J_i] \subset \{p_i\};\ \pi[\bigcup_{i\geq 1} K_i] \subset \{q_i\};$$

and for each $i \geq 1$,

$$|\gamma_{\pi(j)} - k| \leq \frac{\varepsilon|k|}{2\sum_{r=m_i}^{n_i} |a_r^i|},\ j \in J_i : |\gamma_{\pi(j)}| \leq \frac{\varepsilon|k|}{2\sum_{r=m_i}^{n_i} |a_r^i|}, j \in K_i$$

Thus

$$|\langle \sum_{j=m_i}^{n_i} \gamma_{\pi(j)}\, e^j, a^i\rangle| = |\sum_{j\in K_i} \gamma_{\pi(j)} a_j^i + \sum_{j\in J_i} \gamma_{\pi(j)}\, a_j^i|$$

$$\geq |k \sum_{j\in J_i} a_j^i| - \sum_{j\in J_i} |\gamma_{\pi(j)} - k|\ |a_j^i| - \sum_{j\in K_i} |\gamma_{\pi(j)}\, a_j^i|$$

$$\geq 2\varepsilon|k| - \frac{\varepsilon}{2}|k| - \frac{\varepsilon}{2}|k| = \varepsilon|k|.$$

However, this shows that $\sum_{j\geq 1} \gamma_{\pi(i)} e^i$ does not converge in (δ, T^{**}), which is a contradiction. □

We can now offer the following characterization of an sy.S.b. In fact, we have something more to say in terms of

<u>THEOREM 8.2.2</u> (i) Let δ be an s.s.s. such that $\delta \subset c_0$ or $\delta \not\subset c$, then $\{x_n;f_n\}$ is an sy.S.b. (ii) If δ is an s.s.s. with $\delta \subset c$ and $\delta \not\subset c_0$, then $\{x_n;f_n\}$ is an sy.S.b. if and only if $T^{**} \subset S$.

<u>PROOF</u>. (i) Let $\delta \not\subset c$. If $\delta \not\subset \ell^\infty$, then by Theorem 7.2.2, $T^{**} \approx \sigma(\delta,\phi)$. Hence $\{f_n\}$ is a Hamel base for X^* and $T \approx \sigma(X,X^*)$. Therefore, the map $F_\sigma : X \to X$, $\sigma \in P$ with

$$F_\sigma(x) = \sum_{n\geq 1} f_{\sigma(n)}(x)x_n$$

is continuous and so $\{x_n;f_n\}$ is an sy.S.b. (cf. Proposition 7.3.5).

Next, let $\delta \subset \ell^\infty$ but $\delta \not\subset c$. Then by Proposition 8.2.1, $T^{**} \subset S$. Consider the space (δ,S). Since for α in δ and a compact subset A of ℓ^1

$$\sup_{\beta\in A} \left| \langle \alpha - \sum_{i=1}^{n} \alpha_i e^i, \beta\rangle \right| \leq \|\alpha\|_\infty \sup_{\beta\in A} \sum_{i>n} |\beta_i|,$$

$\{e^n;e^n\}$ is an S.b. for (δ,S). If A is any compact subset of ℓ^1 and $\sigma \in P$, then $B_{A,\sigma} = \{\{\varepsilon_n\alpha_{\sigma(n)}\} : |\varepsilon_n| = 1, n \geq 1$ and $\alpha \in A\}$ is also compact in ℓ^1 and

$$(*) \qquad \sup_{\beta\in A} |\langle \alpha_\sigma, \beta\rangle| \leq \sup_{\gamma\in B_{A,\rho}} |\langle \alpha,\gamma\rangle|, \; \rho = \sigma^{-1}.$$

Consider the map $F_\sigma : \delta \to \delta$, $F_\sigma(\alpha) = \alpha_\sigma$; then by (*), F_σ is continuous on (δ,S). Therefore, using Proposition 7.3.5, we find that $\{e^n;e^n\}$ is an sy.S.b. for (δ,S) and so $\{x_n;f_n\}$ is an sy.S.b. for (X,T).

Let now $\delta \subset c_0$. If we could prove that $\{x_n;f_n\}$ is a u-S.b., then Theorem 7.3.3 would yield the symmetric character of $\{x_n;f_n\}$.

On the other hand, let $\{x_n;f_n\}$ not be a u-S.b. for (X,T). Hence there exist x in X, p in D_T, π in P and $\{n_j\}$ in I with

$$(*) \quad p(x - \sum_{i=1}^{n_j} f_{\pi(i)}(x)x_{\pi(i)}) \geq 3, \ \forall\, j \geq 1.$$

At the same time there exists q_0 in $\mathbb{N}$ with

$$(**) \quad p(x - \sum_{i=1}^{n} f_i(x)x_i) \leq 1, \ \forall\, n \geq q_0.$$

By induction, we can show the existence of $\{j_i\}$, $\{m_i\}$, $\{p_i\}$ and $\{q_i\}$ in I_∞, sequences $\{J_i\}$ and $\{K_i\}$ in Φ and a sequence $\{\Theta_i\}$, each Θ_i being a 1-1 map from K_i into $\mathbb{N}$ such that

(i) $\quad J_i = \pi[1,n_{j_i}] \supset [1,q_{i-1}]$;

(ii) $\quad m_i = \sup\{j : j \in J_i\} + 1$;

(iii) $\quad K_i = [q_{i-1},m_i] \smallsetminus J_i$;

(iv) $\quad |f_k(x)| \leq [\sum_{j=1}^{m_i} p(x_j)]^{-1}, \ \forall\, k \geq p_i$;

(v) $\quad p_i > m_i$;

(vi) $\quad \Theta_i : K_i \to [p_i,\infty),\ \Theta_i$ is 1-1;

and

(vii) $\quad q_i = \sup\{j : j \in \Theta_i[K_i]\} + 1$,

where $i = 1,2,\ldots$.

Indeed, to begin with $i = 1$, proceed as in the next paragraph for $r = 1$ with q_0 as obtained in (**). Suppose that we have already established (i) to (vii) for $i = 1,\ldots,r-1$ $(r > 1)$.

Since π maps $\mathbb{N}$ onto itself, we can find j_r satisfying (i), where n_{j_r} satisfies (*). Thus J_r, m_r and K_r are well defined satisfying respectively (i), (ii) and (iii). Since $f_k(x) \to 0$, there exists p_r in $\mathbb{N}$ for which (iv) and (v) are satisfied. Finally observe that $K_r \neq \emptyset$ and finite and this yields (vi) and (vii). Consequently the induction process is complete.

Define σ in P as follows:

$$\sigma(j) = \begin{cases} \Theta_i(j), & j \in K_i; \\ \Theta_i^{-1}(j), & j \in \Theta_i[K_i]; \\ j, & \text{otherwise.} \end{cases}$$

Now we have

$$\sum_{k=q_{i-1}+1}^{m_i} f_{\sigma(k)}(x)x_k = \sum_{k\in J_i} f_k(x)x_k + \sum_{k\in K_i} f_{\sigma(k)}(x)x_k - \sum_{k=1}^{q_{i-1}} f_k(x)x_k$$

$$= \left(\sum_{k=1}^{n_{j_i}} f_{\pi(k)}(x)x_{\pi(k)} - x\right) + \sum_{k\in K_i} f_{\sigma(k)}(x)x_k + \left(x - \sum_{k=1}^{q_{i-1}} f_k(x)x_k\right)$$

Hence, using (*) and (**), we have

$$(+) \quad p\left(\sum_{j=q_{i-1}+1}^{m_i} f_{\sigma(j)}(x)x_j\right) \geq 2 - \left(\sup_{j\in K_i} |f_{\sigma(j)}(x)|\right) \sum_{j\in K_i} p(x_j).$$

But from the properties of Θ_i, p_i and σ, we find from (iv) that

$$(++) \quad \sup_{j\in K_i} |f_{\sigma(j)}(x)| \leq \left(\sum_{j=1}^{m_i} p(x_j)\right)^{-1}.$$

Therefore, it follows from (+) and (++) that $\sum_{n\geq 1} f_{\sigma(n)}(x)x_n$ does not converge in (X,T) and so we arrive at a contradiction.

(ii) Let $T^{**} \subset S$. Proceeding as at the beginning of this proof (second paragraph), we conclude that $\{x_n;f_n\}$ is an sy.S.b. for (X,T). The converse follows from Theorem 6.2.1. □

Simple topological properties on (X,T) also ensure a better location of the size of the s.s.s. δ and this in turn yields another characterization of an sy.S.b. We first have (compare this result with Theorem 7.2.1)

<u>PROPOSITION 8.2.3</u> Let δ be an s.s.s. and (X,T) an ω-complete space. Then either $\delta \subset c_0$ or $\delta = c$ or $\delta = \ell^\infty$ or $\delta = \omega$.

<u>PROOF</u>. If $\delta \not\subset \ell^\infty$, then from Theorem 7.2.2, $T^{**} \approx \sigma(\delta,\phi)$. Hence for each α in ω, $\{\alpha^{(n)}\}$ is T^{**}-Cauchy; thus $\alpha \in \delta$ and so $\delta = \omega$.

Let now $\delta \subset \ell^\infty$ but $\delta \not\subset c$. By Proposition 8.2.1, $T^{**} \subset S$. If $\alpha \in \ell^\infty$, then it is easily seen that $\{\alpha^{(n)}\}$ is Cauchy in (δ,S) and so $\alpha \in \delta$; that is, $\delta = \ell^\infty$.

Finally, let $\delta \subset c$ but $\delta \not\subset c_0$. If $A \subset \delta^* = \{\{f(x_n)\} : f \in X^*\} \equiv (\delta,T^{**})^*$ is T^{**}-equicontinuous, we then proceed to show that A is bounded in $(\ell^1, \|\cdot\|_1)$; in particular, the method of proof followed hereafter also shows that δ^* is contained in ℓ^1.

Since A is equicontinuous, it is coordinatewise bounded and so

$$B_n \equiv \sup\{\sum_{i=1}^{n} |a_i| : a \in A\} < \infty, \ \forall\, n \geq 1.$$

As outlined in the proof of Theorem 6.2.1, there exists α in δ with $\alpha_n \to k \neq 0$ and $|\alpha_n - k| \leq |k|/8$ for $n \geq 1$.

Assume that A is not bounded in ℓ^1. Hence for n in $\mathbb{N}$ and $M > 0$, there exist a in A and J in Φ_n so that (cf. the proof of Theorem 6.2.1)

$$\left|\sum_{j\in J} a_j\right| \geq \frac{1}{4} \sum_{j\in J} |a_j| \geq M.$$

Inductively, it is now possible to construct sequences $\{a^i\}$ in A, $\{n_i\}$ in I_∞ and $\{J_i\}$ in Φ so that

$$\left|\sum_{j\in J_1} a^1_j\right| \geq \frac{1}{4} \sum_{j\in J_1} |a^1_j| \geq 1, \quad J_1 \subset [1,n_1];$$

$$\left|\sum_{j\in J_i} a^i_j\right| \geq \frac{1}{4} \sum_{j\in J_i} |a^i_j| \geq 2^i B_{n_{i-1}}, J_i \subset [2n_{i-1}+1, n_i], i \geq 2.$$

Put $K_1 = [1,n_1] \setminus J_1$ and $K_i = [2n_{i-1}+1, n_i] \setminus J_i$, $i \geq 2$.

Note that $B_{n_i} > 1$ for $i \geq 1$; also $\{\sum_{i=1}^{n} \alpha_i e^i\}$, being convergent in (δ, T^{**}), is bounded. Since (δ, T^{**}) is ω-complete, $\beta \in \delta$ where

$$\beta = \sum_{j=1}^{n_1} \alpha_j e^j + \sum_{i\geq 2} 2^{-i} B^{-1}_{n_{i-1}} \left(\sum_{j=n_{i-1}+1}^{n_i} \alpha_j e^j \right).$$

Define σ in P with $\sigma([1,2n_i]) \subset [1,2n_i]$; $\sigma[K_i] \subset [n_i+1, 2n_i]$, $i \geq 1$ and

$$\sigma(j) = j, \text{ if } j \in \cup \{J_i : i \geq 1\}.$$

If $i \geq 2$, then

$$|\langle \sum_{j\in J_i} \beta_{\sigma(j)} e^j, a^i\rangle| = |\sum_{j\in J_i} a^i_j \beta_j|$$

$$= 2^{-i} B^{-1}_{n_{i-1}} \quad |\sum_{j\in J_i} a^i_j \alpha_j|$$

$$\geq 2^{-i} B^{-1}_{n_{i-1}} \{|k \sum_{j\in J_i} a^i_j| - \sum_{j\in J_i} |a^i_j||\alpha_j - k|\}$$

$$\geq \frac{|k|}{2} \ ;$$

also, for $i \geq 2$ (cf. the construction of β over $[n_i+1, n_{i+1}]$)

$$|\langle \sum_{j\in K_i} \beta_{\sigma(j)} e^j, a^i\rangle| \leq B_{n_i} \sup \{|\beta_{\sigma(j)}| : j \in K_i\},$$

$$\leq 2^{-i-1} \ ||\alpha||_\infty$$

$$\leq 2^{-i} \frac{9}{2^4} |k| \leq 9.2^{-6}|k|.$$

Hence

$$|\langle \sum_{j=2n_{i-1}+1}^{n_i} \beta_{\sigma(j)} e^j, a^i\rangle| \geq \frac{23}{64} |k|.$$

But $\beta_\sigma \in \delta$ and so the preceding inequality leads to a contradiction. Hence A is bounded in $(\ell^1, \|\cdot\|_1)$; and consequently $T^{**} \subset \beta(\ell^\infty, \ell^1)|\delta$.

Since for each b in c_o, $\{\sum_{i=1}^{n} b_i e^i\}$ is $\beta(\ell^\infty, \ell^1)|\delta$-Cauchy in δ, $b \in \delta$. In other words, $c_o \subset \delta$. But $\delta \not\subset c_o$ and so $\delta = c$.

Finally the choice of δ being contained in c_o is a consequence of the cases considered above. □

The next result is the major result of this section, giving a neat characterization of an sy.S.b. in terms of the symmetric character of $\delta \neq c$ and a mild restriction on (X,T). Indeed, we have

<u>THEOREM 8.2.4</u> Let (X,T) be ω-complete. Then $\{x_n;f_n\}$ is an sy.S.b. if and only if δ is an s.s.s. with $\delta \neq c$.

<u>PROOF</u>. Let $\{x_n;f_n\}$ be an sy.S.b., then δ is clearly an sy.S.b. Also $\{x_n;f_n\}$ is an u-S.b. and so it is a b.m-S.b. By Propositions 6.3.1 and 8.2.3, $\delta \neq c$.

Conversely, on account of Proposition 8.2.3, either $\delta \subset c_0$ or $\delta = \ell^\infty$ or $\delta = \omega$ and the result follows from Theorem 8.2.2. □

<u>PROPOSITION 8.2.5</u> Let (λ,T) be an ω-complete AK-space with λ being symmetric. Then $\lambda = c$ or λ is normal.

<u>PROOF</u>. If $\lambda \neq c$, then $\{e^n;e^n\}$ is an sy.S.b. for (λ,T). In particular, by Proposition 6.3.1, λ is normal. □

<u>Illustrations of the size of δ</u>

This subsection is essentially devoted to offering a few examples, illustrating the occurrence of the size of the s.s.s. δ associated with $\{x_n;f_n\}$.

<u>EXAMPLE 8.2.6</u> See Example 7.4.9, but $\delta = \phi \subsetneq c_0$.

<u>EXAMPLE 8.2.7</u> See Example 2.3.6, but $\delta = c_0$ and $T = \tau(c_0,\ell^1)$.

<u>EXAMPLE 8.2.8</u> This is the familiar space $(\omega,\sigma(\omega,\phi))$ with $\delta = \omega$.

Finally, let us pass on to an uncommon example dealing with the last case when $\delta = c$.

EXAMPLE 8.2.9 Let

$$\mathfrak{S} = \{A \subset \ell^1 : A \text{ is bounded in } \ell^1 \text{ with } \sup_{a\in A} |\sum_{i\geq n} a_i| \to 0\}$$

and T be the l.c. topology on c generated by the polar of A in $\mathfrak{S}$. For α in c and A in $\mathfrak{S}$,

$$\begin{aligned}\sup_{a\in A} |\langle\alpha^{(n)}-\alpha,a\rangle| &\leq \sup_{a\in A} |\sum_{i>n} (\alpha_i-k)a_i| + |k| \sup_{a\in A} |\sum_{i>n} a_i| \\ &\leq M \sup_{i>n} |\alpha_i-k| + |k| \sup_{a\in A} |\sum_{i>n} a_i| \\ &\to 0 \text{ as } n \to \infty.\end{aligned}$$

where $k = \lim \alpha_n$ and $M = \sup \{\|a\|_1 : a \in A\}$. Therefore, (c,T) is an AK-space; in other words, $\{e^n;e^n\}$ is an S.b. for (c,T).

Here δ = c and in order to complete the required discussion, it is enough to show that (c,T) is complete.

Let $\bar{T}$ be the l.c. topology on ℓ^∞ generated by the polars of the members of $\mathfrak{S}$. Then $\bar{T}|c = T$. Since $\mathcal{B}_T$ consists of $\tau(\ell^\infty,\ell^1)$-closed neighbourhoods and ℓ^∞ is $\tau(\ell^\infty,\ell^1)$ complete (cf. [129], p. 189 or [130], p. 40), the space $(\ell^\infty,\bar{T})$ is complete.

It now remains to prove that c is closed in $(\ell^\infty,\bar{T})$. So, let $\alpha \in \bar{c}$ (the $\bar{T}$-closure of c) and suppose that $\alpha \in \ell^\infty \setminus c$. We may find $\{m_i\}$ and $\{n_i\}$ in I with $n_i < m \leq n_{i+1}$ such that $\alpha_{n_i} \to \mu$; $\alpha_{m_i} \to \nu$ with $\mu \neq \nu$ and

$$|\alpha_{n_i} - \mu|; \alpha_{m_i} - \nu| \leq \frac{1}{4} |\mu-\nu|, \quad \forall i \geq 1.$$

Put

$$a^s = 2^{-s} \sum_{i=2^{s-1}+1}^{2^s} (e^{n_i}-e^{m_i}),$$

then $A = \{a^s : s \geq 1\} \in \mathfrak{S}$.

If $\beta \in c$, we can find N in $\mathbb{N}$ such that $|\beta_i - \beta_j| \leq |\mu-\nu|/4$ for all $i,j \geq N$. Hence for all sufficiently large s with $m_{2^{s-1}}, n_{2^{s-1}} > N$,

$$|\langle\alpha-\beta, a^s\rangle| \geq \frac{1}{2^s}\{\sum_{i=2^{s-1}+1}^{2^s}[|\mu-\nu| - |\alpha_{n_i} - \mu| - |\nu - \alpha_{m_i}| - |\beta_{m_i} - \beta_{n_i}|]\}$$

$$\geq \frac{1}{8}|\mu-\nu|.$$

Therefore

$$\sup_{s\geq 1} |\langle\alpha-\beta,\ a^s\rangle| \geq \frac{1}{8}|\mu-\nu|, \quad \forall\ \beta \in c$$

and so $\alpha \notin \bar{c}$, a contradiction. Hence $\bar{c} = c$ in $(\ell^\infty, \bar{T})$.

8.3 DISSECTION OF SUFFICIENT CONDITIONS

In view of the importance of symmetric bases, enough attention has been paid in this and in previous chapters for us to discover the conditions on δ, as well as on sums related to $\{x_n; f_n\}$, that ensure the symmetric character of the base. Normally these conditions, namely, δ is an s.s.s. and $R_p(x)$ is finite, are quite strong in character and hence an attempt will be made in this section to have a close look at these conditions and to produce weaker restrictions ensuring the symmetric character of the S.b. in question. Needless to say, we will endeavour to find the interrelationship, if any, among these restrictions as well.

To be precise in our discussion, let us once again recall the notation and terminology from section 8.1. For the sake of convenience, let us put the conditions on δ and $R_p(x)$ in terms of

(A) δ is an s.s.s.

and

(B) The function $R_p(x)$ of (7.4.2) is finite for every x in X and p in D_T.

Let us further introduce the following conditions:

(C) The family $\{P_{\pi,n}(x) \equiv \sum_{i=1}^{n} f_i(x)x_{\pi(i)} : \pi \in P,\ n \geqq 1\}$ is bounded in (X,T) for each x.

(D) The sequence $\{P_{\pi,n}(x) : n \geqq 1\}$ is bounded in (X,T) for each π and x.

(E) The family $\{F_{\pi,n}(x) \equiv \sum_{i=1}^{n} f_{\pi(i)}(x)x_i :\ \pi \in P,\ n \geqq 1\}$ is bounded in (X,T) for each x.

(F) The sequence $\{F_{\pi,n}(x) :\ n \geqq 1\}$ is bounded in (X,T) for each π and x.

Finally, let us write

(G) $\{x_n;f_n\}$ is an sy.S.b. for (X,T).

Conditions (A) through (G) are related as follows:

PROPOSITION 8.3.1 We have the following implication diagram:

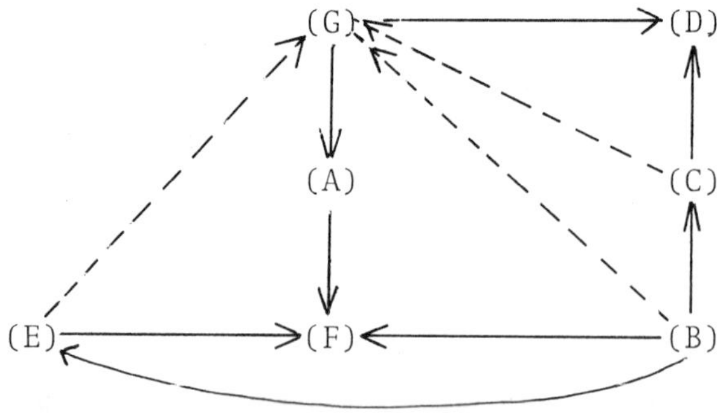

Here dotted line implications are true when (X,T) is ω-complete and barrelled.

PROOF. The implication (B) $\Rightarrow$ (G) is a consequence of Proposition 7.4.16 whereas (C), (E) $\Rightarrow$ (G) will be taken up in Proposition 8.3.7. The other implications are trivially true. □

Earlier, we have discussed some of the reverse implications involving (G) in the above diagram. However, it would also be interesting to investigate conditions which will yield the reverse implications in Proposition 8.3.1, for this will ultimately help in characterizing bases having symmetric character.

To begin with, let us first of all observe that $\{f_n;\Psi x_n\}$ is a $\sigma(X^*,X)$-S.b. for X^*, $\Psi: X \to X^{**}$ being the usual canonical injection with $\langle f,\Psi x\rangle = \langle x,f\rangle; x \in X,\ f \in X^*$. In what follows, we say that one of the conditions listed in (A) to (G) is true for X^* if the same holds for $(X^*,\sigma(X^*,X))$, thereby replacing X,T, $\{x_n;f_n\}$ and δ by $X^*,\sigma(X^*,X)$, $\{f_n;\Psi x_n\}$ and $\mu = \{\{f(x_n)\}: f \in X^*\}$ respectively in that condition appropriately. With this background, we have the following simple

LEMMA 8.3.2 We have the following:

(i) The condition (C) holds for X^* if and only if (E) is satisfied for $X \equiv (X,T)$.

(ii) The condition (C) holds for X if and only if (E) is satisfied for X^*.

(iii) The condition (D) holds for X^* if and only if (F) is satisfied for X.

(iv) The condition (D) holds for X if and only if (F) is satisfied for X^*.

(v) The condition (B) is satisfied for X if and only if it is satisfied for X^*.

PROOF. This is a direct consequence of the fact that weakly and T-bounded subsets of X are the same. □

Recalling a *dyadic complex number* β as a complex number of the form $(p+iq)/2^k$ where p,q,k are integers, we have

PROPOSITION 8.3.3 Let (E) be satisfied with $\delta \subset c_0$; then (B) holds.

PROOF. Fix x in X, p in D_T and σ in P arbitrarily and let

$$M_p(x) = \sup_{n\geqq 1} \sup_{\pi\in P} p(\sum_{i=1}^{n} f_{\pi(i)}(x)x_i). \tag{8.3.4}$$

By (E), M_p is a seminorm on X for each p in D_T.

Let $J \in \Phi$, $m = \max\{\sigma(i): i \in J\}$ and $L = [1,m]\smallsetminus\sigma[J]$. There exists $n_0 \equiv n_0(m)$ such that

$$|f_i(x)| \leqq (\sum_{j=1}^{m} p(x_j)+1)^{-1}, \ \forall\ i \geqq n_0.$$

Choose ρ in P with $\rho \circ \sigma(j) = j$ for j in J and $\rho(i) \geqq n_0$ for i in L. Since

$$\sum_{j\in J} f_j(x)x_{\sigma(j)} = \sum_{k=1}^{m} f_{\rho(k)}(x)x_k - \sum_{k\in L} f_{\rho(k)}(x)x_k,$$

we get

$$p(\sum_{j\in J} f_j(x)x_{\sigma(j)}) \leqq 1+M_p(x), \ \forall\ J \in \Phi.$$

Fix n in $\mathbb{N}$ and choose arbitrary complex numbers $\alpha_1,\ldots,\alpha_n$ with $|\alpha_i| \leqq 1$, $1 \leqq i \leqq n$. Since dyadic complex numbers are dense in the closed unit disc, one finds a set of dyadic complex numbers $\beta_1,\ldots,\beta_n$ with $|\beta_i| \leqq 1$ so that

$$p(\sum_{j=1}^{n} (\alpha_j-\beta_j)f_j(x)x_{\sigma(j)}) \leqq 1.$$

Since $|\beta_j| \leqq 1$, we can write $\beta_j = (p_j+iq_j)/2^{k_j}$, where p_j,q_j are integers and $k_j \in \mathbb{N}_0$, $1 \leqq j \leqq n$. Let $r = \sup\{k_j : 1 \leqq j \leqq n\}$;

then we can write

$$\sum_{j=1}^{n} \beta_j f_j(x) x_{\sigma(j)} = \sum_{k=0}^{r} 2^{-k} \{ \sum_{j \in A_k} - \sum_{j \in B_k} + i \sum_{j \in C_k} - i \sum_{j \in D_k} \} f_j(x) x_{\sigma(j)},$$

where A_k, B_k, C_k, D_k are suitable subsets of $[1,n] = \{i \in \mathbb{N} : 1 \leqq j \leqq n\}$. Hence

$$p(\sum_{j=1}^{n} \beta_j f_j(x) x_{\sigma(j)}) \leqq 4(1+M_p(x)) \sum_{k \geqq 0} 2^{-k}.$$

Therefore, for each n, σ and α, $|\alpha_i| \leqq 1$,

$$p(\sum_{j=1}^{n} \alpha_j f_j(x) x_{\sigma(j)}) \leqq 9 + 8M_p(x),$$

and we have completed the proof. □

<u>EXERCISE 8.3.5</u> Let (C) be satisfied for X with $\delta \subset c$ and $\delta \not\subset c_o$. Show that (B) is satisfied for X. [Hint: For some x in X and some $\alpha > 0$, $|f_n(x)| \geqq \alpha$ eventually in n. Conclude that $\delta^* = \mu \subset c_o$. Now apply Lemma 8.3.2 and Proposition 8.3.3.]

<u>THEOREM 8.3.6</u> Let $\delta \subset \ell^\infty$ and $\delta \not\subset c$. Then the following conditions are equivalent:

(i) The condition (B) is true for X.

(ii) The condition (C) is true for X.

(iii) The condition (E) is true for X.

(iv) $\delta^* \subset \ell^1$.

<u>PROOF</u>. Our line of proof is explained below

$$\begin{array}{ccc} (i) & \longleftarrow & (ii) \\ \downarrow & & \uparrow \\ (iii) & \longrightarrow & (iv) \end{array}$$

Assume the truth of (iv). Then for x in X and f in X^*,

$$\sup\left\{\sum_{i=1}^{n}|f_i(x)f(x_{\sigma(i)})| : n \geqq 1,\ \sigma \in P\right\}$$

$$\leqq \|\{f_n(x)\}\|_\infty \sum_{n \geqq 1} |f(x_n)|,$$

and so (iv) implies (ii). It is clear that (i) implies (ii) as well as (iii).

(iii) $\Rightarrow$ (iv). If (iv) is not true, then $a \equiv \{f_n(x)\} \in \ell^1$ for some f in X^*. Also $b \equiv \{f_n(x)\} \in \delta \setminus c$ with b in ℓ^∞ for some x in X. There exist numbers α and β, $\alpha \neq \beta$ and $\{m_i\}$, $\{n_i\}$ in I_∞ with $\{m_i\} \subset \mathbb{N} \setminus \{n_i\}$ such that $b_{n_i} \to \alpha$ and $b_{m_i} \to \beta$.

Consider an arbitrary number $M > 0$; then we can find a finite subset J of $\mathbb{N}$ so that

$$\left|\sum_{j\in J} a_j\right| \geqq \frac{1}{4}\sum_{j\in J}|a_j| \geqq \frac{6M}{|\alpha-\beta|}.$$

Put $m = \sup\{j : j \in J\}$ and $K = [1,m]\setminus J$. Select ρ, σ in P so that $\rho(j) = \sigma(j)$ for j in K and $|b_{\sigma(j)} - \alpha|$, $|b_{\rho(j)} - \beta| \leqq |\alpha - \beta|/16$ for j in J. Then

$$\left|\sum_{j=1}^{m}(b_{\sigma(j)} - b_{\rho(j)})a_j\right|$$

$$\geqq \left|\sum_{j\in J}(\alpha-\beta)a_j\right| - \sum_{j\in J}|(b_{\sigma(j)}-\alpha)a_j| - \sum_{j\in J}|(b_{\rho(j)}-\beta)a_j|$$

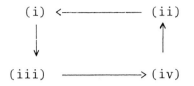

$$\geq \frac{|\alpha-\beta|}{4} \{ \sum_{j\in J} |a_j| - \frac{1}{4} \sum_{j\in J} |a_j| - \frac{1}{4} \sum_{j\in J} |a_j|\}$$

$$\geq 3M.$$

Choosing $M = \sup \{|f(F_{\pi,n}(x))| : n \geq 1, \pi \in P\}$, the preceding inequality yields a contradiction, giving $\delta^* \subset \ell^1$.

(ii) $\Rightarrow$ (i). It is enough to show that $\delta^* \subset c_o$. To begin with, let us observe that $\delta^* \subset \ell^\infty$. In fact, for x in X and f in X*, there exists $M \equiv M(f,x) > 0$ so that

(*) $|f_n(x)f(x_{\sigma(n)})| \leq M;\ \forall\ n \geq 1,\ \sigma \in P.$

If $\delta^* \not\subset \ell^\infty$, then for some f in X*, there exists $\{n_k\}$ in I_∞ so that $|f(x_{n_k})| \geq M2^k$, $k \geq 1$. Also, there exist x in X, $\alpha \neq 0$ and $\{m_k\}$ in I_∞ with $|f_{m_k}(x)| \geq |\alpha|$, $k \geq 1$. Let $\sigma \in P$ with $\sigma(m_k) = n_k$, $k \geq 1$.

Then

$$|f(x_{\sigma(m_k)})f_{m_k}(x)| \geq 2^k M|\alpha|$$

and this contradicts (*).

Next, let $\delta^* \not\subset c$. Then for some f, $a \equiv \{f(x_n)\} \notin c$. We can find γ, $\eta(\gamma\neq\eta)$ and $\{p_i\}$, $\{q_i\}$ in I_∞ with $\{p_i\} \subset N \setminus \{q_i\}$ such that $a_{p_i} \to \gamma$ and $a_{q_i} \to \eta$. We may suppose that $N\setminus(\{p_i\} \cup \{q_i\}) \in I_\infty$. Further, let b, α,β and $\{m_i\}$, $\{n_i\}$ be as in (iii) $\Rightarrow$ (iv) and we may assume again that $N \setminus(\{m_i\} \cup \{n_i\}) \in I_\infty$. It is now possible to construct members π and σ in P with the following properties:

$$\pi(m_i) = q_i,\ \pi(n_i) = p_i;\ \sigma(m_i) = p_i,\ \sigma(n_i) = q_i,\ i \geq 1$$

and $\sigma(i) = \pi(i)$ elsewhere. Then

$$\sup_{n\geq 1} \left| \sum_{i=1}^{n} b_i(a_{\sigma(i)} - a_{\pi(i)}) \right| = \infty,$$

and this contradicts (ii). Thus $\delta^* \subset c$.

Now suppose that $\delta^* \not\subset c_o$. Then there exists a in δ^* such that $a \in c \setminus c_o$ and so $\sum_{n\geq 1} a_n b_n$ is not convergent. However, this is not true and so $\delta^* \subset c_o$.

By Lemma 8.3.2 (ii) and Proposition 8.3.3, (B) is satisfied for X* and hence by Lemma 8.3.2(V), (B) is satisfied for X. □

Finally, we prove two more results relating to the reverse implications in the diagram of Proposition 8.3.1.

PROPOSITION 8.3.7 Let $\{x_n\}$ be bounded and $\delta \subset \ell^\infty$. If (F) (resp. (D)) is satisfied, so is then (E) (resp. (C)).

PROOF. We will prove (F) ⇒ (E) and a similar argument will yield (C) from (D).

Let (E) not be satisfied. Then for some x in X and p in D_T,

$$(+) \qquad \sup_{\sigma\in P} \sup_{n\geq 1} p\left(\sum_{i=1}^{n} f_{\sigma(i)}(x)x_i \right) = \infty.$$

We claim the existence of $\{\sigma_i\}$ in P and that of $\{n_i\}$ in I such that the following relations are satisfied:

$$\text{(i)} \quad p\left(\sum_{j=1}^{n_i} f_{\sigma_i(j)}(x)x_j \right) \geq i, \ i = 1,2,\ldots$$

$$\text{(ii)} \ i \in \sigma_i[1,n_i], \ i = 1,2,\ldots;$$

$$\text{(iii)} \ \sigma_i(j) = \sigma_{i-1}(j), \ 1 \leq j \leq n_{i-1}, \ i = 2,3,\ldots .$$

Indeed, there exist m in $\mathbb{N}$ and σ in P so that (cf. (+))

$$(*)\qquad p\Big(\sum_{j=1}^{m} f_{\sigma(j)}(x)x_j\Big) \geqq 1 + \|\{f_i(x)\}\|_\infty \sup_i p(x_i),$$

where the right-hand side is finite by the hypothesis. If $\sigma^{-1}(1) \leq m$, we take $n_1 = m$ and $\sigma_1 = \sigma$, thus giving (i) and (ii). If $\sigma^{-1}(1) > m$, define ρ in P by setting $\rho(m+1) = 1$, $\rho\sigma^{-1}(1) = \sigma(m+1)$ and $\rho(i) = \sigma(i)$ elsewhere. Now, let $\sigma_1 = \rho$ and $n_1 = m+1$, and thus (ii) is satisfied. Also

$$p\Big(\sum_{j=1}^{n_1} f_{\sigma_1(j)}(x)x_j\Big) \geqq p\Big(\sum_{j=1}^{m} f_{\sigma(j)}(x)x_j\Big) - |f_1(x)|p(x_{m+1})$$

$$\geqq 1,$$

by (*) and we get (i) for i = 1.

Suppose that we have obtained $\sigma_1,\ldots,\sigma_{r-1}$ and $n_1,\ldots,n_{r-1}$ so that (i) to (iii) are satisfied. On account of (+), we can find ρ in P and k in $\mathbb{N}$ with $k > n_{r-1}$ so that

$$p\Big(\sum_{j=1}^{k} f_{\rho(j)}(x)x_j\Big) \geqq r + (4n_{r-1}+1)\,\|\{f_i(x)\}\|_\infty \sup_i p(x_i).$$

Set

$$R = \sigma_{r-1}^{-1} \circ \rho([1,n_{r-1}]) \setminus [1,n_{r-1}],\ S = \rho([1,n_{r-1}]) \setminus$$

$$\sigma_{r-1}([1,n_{r-1}]) \quad\text{and}\quad T = (R \cap [1,k]) \cup [1,n_{r-1}].$$

Then #(R) = #(S). Hence there exists a 1-1 map θ from R onto S. Define π in P by

$$\pi(i) = \begin{cases} \sigma_{r-1}(i), & 1 \leqq i \leqq n_{r-1}; \\ \theta(i), & i \in R; \\ \rho(i), & \text{otherwise.} \end{cases}$$

Now

$$p\left(\sum_{j=1}^{k} f_{\pi(j)}(x)x_j\right) = p\left(\sum_{j\in T} f_{\pi(j)}(x)x_j + \sum_{j\in[1,k]\setminus T} f_{\rho(j)}(x)x_j\right)$$

$$\geq p\left(\sum_{j=1}^{k} f_{\rho(j)}(x)x_j\right) - p\left(\sum_{j\in T} f_{\rho(j)}(x)x_j\right) - p\left(\sum_{j\in T} f_{\pi(j)}(x)x_j\right)$$

$$\geq r + \{(4n_{r-1}+1) - 2\#(T)\} \, \|\{f_i(x)\}\|_\infty \sup_i p(x_i)$$

$$(++) \qquad \geq r + \|\{f_i(x)\}\|_\infty \sup_i p(x_i),$$

since $\#(T) \leq 2n_{r-1}$. If $\pi^{-1}(r) \leq k$, then let $\sigma_r = \pi$ and $n_r = k$. Then (i) follows from (++) with $i = r$ and (ii), (iii) are obviously satisfied for $i = r$. If $\pi^{-1}(r) > k$, put $n_r = k+1$ and define μ in P with $\mu(k+1) = r$, $\mu \circ \pi^{-1}(r) = \pi(k+1)$ and $\mu(i) = \pi(i)$, otherwise. Let now $\sigma_r = \mu$. Then $\sigma_r(n_r) = r$ and so (ii) is satisfied for $i = r$. If $1 \leq j \leq n_{r-1}$, then $\pi(j) = \mu(j) = \sigma_r(j)$ and so $\sigma_r(j) = \sigma_{r-1}(j)$, yielding thereby (iii) for $i = r$. Next

$$p\left(\sum_{j=1}^{n_r} f_{\sigma_r(j)}(x)x_j\right) \geq p\left(\sum_{j=1}^{k} f_{\pi(j)}(x)x_j\right) - |f_{\sigma_r(k+1)}(x)| \times p(x_{k+1})$$

$$\geq r,$$

by (++). This proves (i) for $i = r$ and completes the required induction process.

Finally put

$$\nu(i) = \begin{cases} \sigma_1(i), & i \leq n_1; \\ \sigma_j(i), & n_{j-1} < i \leq n_j,\ j \geq 2. \end{cases}$$

By (iii), ν is 1-1 on $\mathbb{N}$ and using (ii), we find that ν is onto $\mathbb{N}$. Thus $\nu \in P$ and observe that

$$p\left(\sum_{j=1}^{n_i} f_{\nu(j)}(x)x_j\right) \geq i,$$

by using (iii) successively. But this contradicts (F). □

PROPOSITION 8.3.8 Let (X,T) be ω-complete and barrelled. If (C) or (E) is satisfied, then so is (G).

PROOF. Let (E) be satisfied. Put

$$N_p(x) = \sup_{n\geq 1} \sup_{\sigma\in P} p(F_{\sigma,n}(x)), \quad x \in X.$$

Then each N_p is a lower semicontinuous seminorm on (X,T) and so for each p in D_T, there exists q in D_T with $N_p(x) \leq q(x)$ for all x in X. Clearly $p(x) \leq N_p(x)$ for each x in X. Thus T is also generated by $\{N_p : p \in D_T\}$. Further, for x in X, σ in P and p in D_T,

$$N_p\left(\sum_{i=m}^{n} f_{\sigma(i)}(x)x_i\right) = N_p\left(\sum_{i=m}^{n} f_i(x)x_i\right),$$

and it follows that δ is symmetric. Also, the map F_σ defined in Proposition 7.3.5 is continuous and consequently by the same proposition, $\{x_n;f_n\}$ is an sy.S.b. The other part of the result is a consequence of Proposition 7.4.16. □

Two counter-examples

In this subsection, we present two examples exhibiting respectively the importance of barrelledness and ω-completeness in Proposition 8.3.8. In other words, none of these two conditions can be dropped to infer the condition (G), that is, the symmetric character of $\{x_n;f_n\}$.

EXAMPLE 8.3.9 This is the space (c,T) which is, by virtue of Example 8.2.9, an ω-complete space having $\{e^n;e^n\}$ as its S.b. By Theorem 8.2.4, $\{e^n;e^n\}$ is not an sy.S.b. On the other hand, for x in c, σ in P and a in $A \in \mathcal{S}$,

$$\left|<\sum_{i=1}^{n} <x,e^i>e^{\sigma(i)}, a>\right| \leq \|x\|_\infty \sum_{n \geq 1} |a_n|;$$

thus

$$\sup_{\sigma \in P} \sup_{n \geq 1} \sup_{a \in A} \left|<\sum_{i=1}^{n} x_i e^{\sigma(i)}, a>\right| < \infty .$$

Therefore (C) is satisfied for (c,T). This space is, however, not barrelled.

EXAMPLE 8.3.10 Let O be the subset of ω considered in Example 7.3.8 and let $\ell_0 \equiv \ell_0^1 = \{y = \alpha x : \alpha \in O,\ x \in \ell^1\}$. Since ℓ^1 is normal and $\tau(\ell^1,\ell^\infty)$ is generated by $\|\cdot\|_1$, the space $(\ell_0,\tau(\ell_0,\ell^\infty))$ is barrelled (cf. [129], p. 199). Thus $\tau(\ell_0,\ell^\infty) \approx \beta(\ell_0,\ell^\infty)$. But $\beta(\ell_0,\ell^\infty) = \beta(\ell^1,\ell^\infty)|\ \ell_0$ (cf. [129], Propositions 4.2.20 and 4.2.21). Hence ℓ_0 is a barrelled subspace of $(\ell^1, \|\cdot\|_1)$ and $\{e^n;e^n\}$ is clearly an S.b. for ℓ_0. Further, this space cannot be ω-complete, for then $\ell_0 = \ell^1$. Next, observe that (C) is satisfied for $(\ell_0, \|\cdot\|_1)$. Finally, we show that $\delta = \ell_0$ is not an s.s.s. and this will prove that (G) is not satisfied. So, choose y in ℓ_0, $y = \alpha x$ such that $x_n = 1/2^n$ and $\alpha_n = 1$ for $n = 2^m + m$, $m \geq 1$ and x_i's, α_i's are zero elsewhere. Let σ be the member of P described in Example 7.3.8. Then $y_{\sigma(n)} = \beta_n z_n$, where

$$\beta_n = \begin{cases} 1, n \text{ is odd}; \\ 0, n \text{ is even}, \end{cases} \quad \text{and } z_n = \frac{1}{2^{2^{\frac{n+1}{2}}} + \frac{n+1}{2}}, \ n = 1,3,\ldots .$$

Since $\beta \notin O$, $y_\sigma \notin \ell_0$.

8.4 THE SIZE OF δ AND ITS IMPLICATIONS

In the last section, several conditions were offered with the sole aim of relating them to the symmetric character of either $\{x_n;f_n\}$ or δ. As we have seen earlier, we could relate some of these conditions by restricting the size of δ. Consequently, it apparently follows that, if δ is not restricted in the way it has been, the resulting conclusions might not be true and we are going to demonstrate this by some examples. Nevertheless, these examples will also exhibit the independent character, to some extent, of various conditions laid down in Proposition 8.3.1.

EXAMPLE 8.4.1 Let $\delta = \phi$ and endow it with $\sigma(\phi,\omega)$. In this case (A), (D), (F) and (G) are evidently true for ϕ. However, (E) is not satisfied; for otherwise, given $\{i_k\}$ in I and x in ϕ, we can find K > 0 so that

$$(*) \qquad \sup_{\sigma\in P}\ \sup_{n\geq 1}\ \left|\sum_{i=1}^{n} i_k x_{\sigma(k)}\right| \leq K.$$

Suppose that ℓ is the length of x and $|x_{\ell_o}| = \max(|x_1|,\ldots,|x_\ell|)$. Choose $N > \ell$ so large that

$$i_N|x_{\ell_o}| > 2K + \sum_{i=1}^{\ell} i_k|x_k|.$$

Define σ in P such that $\sigma(N) = \ell_o$, $\sigma(\ell_o) = N$ and $\sigma(i) = i$ elsewhere. Then from (*)

$$K \geq \left|\sum_{k=1}^{N} i_k x_{\sigma(k)}\right| \geq i_N|x_{\sigma(N)}| - \sum_{k=1}^{\ell} i_k|x_{\sigma(k)}|$$

$$> 2K,$$

since $x_{\sigma(\ell_o)} = x_N = 0$. The absurdity arrived at disproves (*). Similarly, (B) and (C) are not true for ϕ.

EXAMPLE 8.4.2 Let $\phi \subsetneqq \delta \subset c_o$ and equip δ with the sup norm topology. If δ is symmetric, then (G) holds. Indeed, let $x \in \delta$; then for each σ in P, $\sum_{n \geqq 1} x_n e^{\sigma(n)}$ converges to y (say) in c_o, thus $y_{\sigma(i)} = x_i$, $i \geq 1$. Therefore $y_\sigma \in \delta$ and this shows that $y \in \delta$. Hence, for symmetric δ, (A) <==> (G). Let now (G) be satisfied for δ; in particular δ is symmetric. Here $\delta^* = \delta^\times = \delta^\beta$. Also, $\delta^* \subset \ell^\infty$ (cf. the proof of Theorem 7.2.2). Following the proof of Proposition 7.4.14, we find that δ satisfies (B). Further, using Propositions 8.3.3 and 8.3.7, we conclude that (B) <==> (E) <==> (F) and (C) <==> (D). Summing up, the following implication diagram is true for δ:

$$\begin{array}{ccccc}
(B) & \Longleftrightarrow & (E) & \Longleftrightarrow & (F) \\
\Uparrow & & & & \Downarrow \\
(A) & & & & (C) \\
\Updownarrow & & & & \Updownarrow \\
(G) & & & & (D)
\end{array}$$

EXAMPLE 8.4.3 Consider $\delta = cs$ along with the topology $\sigma(\delta,k)$. By the definition of cs, there exist x in cs and σ in P so that $\{\sum_{i=1}^{n} x_{\sigma(i)} : n \geq 1\}$ is unbounded. But

$$\left|\left\langle \sum_{i=1}^{n} \langle x, e^{\sigma(i)}\rangle e^i, e \right\rangle\right| = \left|\sum_{i=1}^{n} x_{\sigma(i)}\right|,$$

and so (F) and hence (E) are not satisfied. But for x in cs, y in k and σ in P,

$$\left|\left\langle \sum_{i=1}^{n} \langle x, e^i\rangle e^{\sigma(i)}, y \right\rangle\right| = \left|\sum_{i=1}^{n} x_i y_{\sigma(i)}\right|.$$

Thus (D) is satisfied for δ and consequently (C) is also satisfied (cf. Proposition 8.3.7).

EXAMPLE 8.4.4 Let $\delta \subset c$ and $\delta \not\subset c_o$. Endow δ with the topology inherited from T of Example 8.2.9. By Exercise 8.3.5 and Proposition 8.3.7, (B) <==> (C) <==> (D) and (E) <==> (F). Thus the following implication diagram is valid:

$$\begin{array}{ccccc} & & (D) & & \\ & & \Updownarrow & & \\ (G) & \Longleftrightarrow & (C) & \Longleftrightarrow & (B) \\ \Downarrow & & \Downarrow & & \\ (A) & \Longrightarrow & (F) & \Longleftrightarrow & (E) \end{array}$$

NOTE. In the above example, (A) does not imply (G) (cf. Example 8.3.9). The situation here is opposite to that which we explained in Example 8.4.2. In the following example, we exhibit the truth of (A) whereas (C) is not satisfied.

EXAMPLE 8.4.5 Consider $\delta = k$ equipped with the topology $\sigma(k,k^{\beta})$. Then (A) is true. Now for e in k, y in k^{β} and σ in P,

$$|<\sum_{i=1}^{n} <e,e^{i}>e^{\sigma(i)}, y>| = |\sum_{i=1}^{n} y_{\sigma(i)}|,$$

and it easily follows that (C) is not satisfied (observe that $y \in cs$).

EXAMPLE 8.4.6 Let $\delta \subset \ell^{\infty}$ but $\delta \not\subset c$ and endow δ with $\sigma(\delta,\ell^{1})$. Then by Theorems 8.2.2, 8.3.6 and Proposition 8.3.7, we have the following implication diagram for δ:

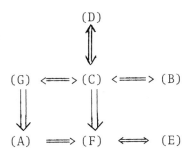

$$\begin{array}{ccccccc} (C) & \Longleftrightarrow & (D) & \Longleftrightarrow & (E) & \Longleftrightarrow & (F) \\ \Updownarrow & & & & & & \\ (B) & \Longleftarrow & (A) & \Longleftrightarrow & (G) & & \end{array}$$

Finally we have

<u>EXAMPLE 8.4.7</u> Consider the space $(\delta, \sigma(\delta, \phi))$, where $\delta \not\subset \ell^{\infty}$. Here $\{e^n; e^n\}$ is an S.b. for this space. Following the arguments of the proof of Theorem 8.2.2 (i), we find that (A) <===> (G) in this case. Here neither (C) nor (E) is satisfied (cf. the working of Example 7.4.19).

9 Shrinking and γ-complete bases

9.1 PURPOSE AND REQUIREMENT

The importance of shrinking and γ-complete bases has already been realized in terms of their applications, especially in characterizing reflexivity of the space in question (cf. [130]). Whereas the introduction of a shrinking base is a consequence of a natural question related to an S.b. $\{x_n;f_n\}$ in an l.c. TVS (X,T), the justification of the concept of a γ-complete base is derived from the dual character of the former. As such, we may expect a good relationship between these two notions of an S.b. Earlier, in [130], we did not pay enough attention to this aspect of the duality relationship between these types of S.b. and, as promised there, we will now make a comparatively detailed study of this aspect. Nevertheless, we will also take a more analytical approach to the study of shrinking bases. Above all, characterizations of shrinking and γ-complete bases will also be discussed.

Throughout, we write (X,T) to mean an arbitrary l.c. TVS; further, whenever we write $\{x_n;f_n\}$, it will mean an arbitrary S.b. for (X,T). The context will make it clear whether a given l.c. TVS (X,T) contains an S.b. $\{x_n;f_n\}$ or not. For convenience, let us recall the following definitions.

DEFINITION 9.1.1 An l.c. TVS (X,T) is said to be a *non-ℓ^1-space* [resp. *non-c_0-space*] if it has no subspace topologically isomorphic to $(\ell^1, \|\cdot\|_1)$ (resp. $(c_0, \|\cdot\|_\infty)$).

Finally, we recall the canonical embedding $\Psi: X \to X^{**}$, $\Psi(x)(f) = f(x)$. We will write Ψx for $\Psi(x)$. For a subspace

H of X^*, equipped with the topology $\beta_H \equiv \beta(X^*,X)|H$, we will often denote by J the canonical embedding from X into $H^* \equiv (H,\beta_H)^*$ defined by $Jx(h) = \langle h,x\rangle$; $h \in H$, $x \in X$. Thus, for each x in X, $Jx = \Psi x|H$ and unless we wish to attach some significance to the functional Jx on H, this will often be written as Ψx rather than $\Psi x|H$.

9.2 $\mathfrak{S}$-UNIFORM BASES

Let $\mathfrak{S}$ stand for an arbitrary collection of bounded subsets of (X,T) with the property that $\mathfrak{S}$ covers X and suppose that S is the l.c. topology on X^* generated by the polars of members of $\mathfrak{S}$.

DEFINITION 9.2.1 $\{x_n;f_n\}$ is called *$\mathfrak{S}$-uniform* (an $\mathfrak{S}$-S.b.) if

$$\lim_{n\to\infty} \sup \{|f(x-S_n(x))| : x \in B\} = 0,$$

for each f in X^* and B in $\mathfrak{S}$. Further, if $\mathfrak{S}$ is the family of all bounded subsets of (X,T), then an $\mathfrak{S}$-uniform base is called a *weakly uniform* S.b. (w.u.S.b.).

In this section, we will essentially confine ourselves to two cases: (a) $\mathfrak{S}$ is the family of all $\sigma(X,X^*)$-bounded subsets on X, and (b) $\mathfrak{S}$ is the family of all balanced convex and $\sigma(X,X^*)$-compact subsets of X.

However, first we have ([123])

THEOREM 9.2.2 $\{x_n;f_n\}$ is an $\mathfrak{S}$-S.b. for (X,T) if and only if $\{f_n;\Psi x_n\}$ is an S.b. for (X^*,S).

PROOF. Let $\{x_n;f_n\}$ be an $\mathfrak{S}$-S.b. Pick up $\varepsilon > 0$, f in X^* and B in $\mathfrak{S}$ arbitrarily. Then there exists N in $\mathbb{N}$ so that $f-S_n^*(f) \in \varepsilon B^0$ for all $n \geq N$. Thus $\{f_n;\Psi x_n\}$ is an S.b. for (X^*,S).

Conversely, in a similar manner,

$$|f(x) - \sum_{i=1}^{n} f(x_i)f_i(x)| \leqq \varepsilon; \forall n \geqq N, x \in B$$

and hence $\{x_n;f_n\}$ is an $\mathfrak{S}$-S.b. □

The next theorem which improves an earlier result of [123], is useful in characterizing shrinking bases.

PROPOSITION 9.2.3 Let $\{x_n;f_n\}$ be $\mathfrak{S}$-uniform. Then

(i) Each sequence $\{y_n\} \in \mathfrak{S}$ for which $f_i(y_n) \to 0$ as $n \to \infty$ for $i \geqq 1$ implies that $y_n \to 0$ in $\sigma(X,X^*)$.

Conversely, let (X,T) be a W-space and $\mathfrak{S}$ satisfy the additional property:

(ii) Each bounded sequence of $A_B = \{x-S_n(x):x \in B, n \geqq 1\}$ belongs to $\mathfrak{S}$ for each B in $\mathfrak{S}$.

Then (i) and (ii) imply that $\{x_n;f_n\}$ is $\mathfrak{S}$-S.b.

PROOF. For the first part, let $f \in X^*$ and $\varepsilon > 0$ be given. There exists $N \equiv N(f,\varepsilon,\{y_k\})$ such that

$$|f(y_k-S_n(y_k))| \leqq \frac{\varepsilon}{2}; \ \forall\ n \geqq N,\ k \geqq 1.$$

Also

$$|f_i(y_k)| \leqq \frac{\varepsilon}{2aN} : \forall\ k \geqq k_0,\ 1 \leqq i \leqq N,$$

where $a = \max\{|f(x_i)|: 1 \leqq i \leqq N\} + 1$. Therefore

$$|f(y_k)| \leqq \frac{\varepsilon}{2} + \sum_{i=1}^{N} |f_i(y_k)f(x_i)| \leqq \varepsilon,\ \forall\ k \geqq k_0.$$

Conversely, let $\{x_n;f_n\}$ be not an $\mathfrak{S}$-S.b. Hence there exist $\varepsilon > 0$, f in X^*, $\{n_k\}$ in I and $\{y_k\}$ in B for some B in $\mathfrak{S}$ so that

(*) $|f(y_k - s_{n_k}(y_k))| > \varepsilon, \forall k \geq 1.$

Observe that A_B is bounded. In fact, since (X,T) is a W-space and $\{S_n^*(g)\}$ is $\sigma(X^*,X)$-bounded, $\{S_n^*(g)\}$ is $\beta(X^*,X)$-bounded for each g in X^*. Thus the inequality

$$\sup_{x\in B} |g(x-S_n(x))| \leq \sup_{x\in B} |g(x)| + \sup_{x\in B} |\langle x, S_n^*(g)\rangle|$$

shows that A_B is bounded. Hence the sequence $\{u_k\} \in \mathfrak{S}$, where $u_k = y_k - S_{n_k}(y_k)$. Observe that $f_i(u_k) \to 0$ for each $i \geq 1$ and so $g(u_k) \to 0$ for each g in X^*, a contradiction to (*). □

Shrinking bases

As one might expect, a shrinking base is a special case of an $\mathfrak{S}$-S.b.; more precisely, we have

PROPOSITION 9.2.4 $\{x_n;f_n\}$ is shrinking if and only if it is a w.u.S.b.

PROOF. Cf. Theorem 9.2.2. □

THEOREM 9.2.5 Let $\{x_n;f_n\}$ be a w.u.S.b. Then for each bounded subset B of X and f in X^*,

$$\lim_{n\to\infty} \sup \{|f(y)| : y \in B \cap \mathcal{R}_n\} = 0, \tag{9.2.6}$$

where $\mathcal{R}_n = \{R_n(x) : x \in X\} = \overline{sp}\{x_{n+1}, x_{n+2}, \ldots\}$. Conversely, if (X,T) is a W-space and $\{x_n;f_n\}$ satisfies (9.2.6) for each bounded subset B of X, then $\{x_n;f_n\}$ is a w.u.S.b.

PROOF. Let $\{x_n;f_n\}$ be a w.u.S.b. If $x \in B \cap \mathcal{R}_n$, then

$f_i(x) = 0$, $1 \leqq i \leqq n$, giving $|f(x)| = |f(x-S_n(x))|$.

(9.2.6) is now a consequence of Definition 9.2.1.

Conversely, let (9.2.6) be true. For any bounded subset B of X, $A_B = \{x-S_n(x) : x \in B, n \geqq 1\}$ is bounded (cf. the proof of Proposition 9.2.3). Thus

$$\lim_{n\to\infty} \sup \{|f(y)| : y \in A_B \cap \mathcal{R}_n\} = 0,$$

for every bounded subset B of X and so $\{x_n;f_n\}$ is a w.u.S.b. □

NOTE. According to [226], an S.b. $\{x_n;f_n\}$ in an l.c. TVS (X,T) is called *stretching* if it satisfies (9.2.6).

COROLLARY 9.2.7 If (X,T) is a W-space, then $\{x_n;f_n\}$ is shrinking if and only if it is a w.u.S.b. if and only if it is stretching.

Almost shrinking bases

Shrinking bases were introduced because of certain natural reasons, especially to demand the S.b. character of $\{f_n;\Psi x_n\}$ for X* equipped with its natural topology $\beta(X^*,X)$. However, it appears it is too much to ask about the S.b. character of $\{f_n;\Psi x_n\}$ for $(X^*,\beta(X^*,X))$, for this space, in general, may not be even separable, for example, consider $X = \ell^1$, then $(\ell^\infty,\beta(\ell^\infty,\ell^1))$ is not separable. On the other hand, if an l.c. TVS (X,T) is equipped with an S.b. $\{x_n;f_n\}$, then $(X^*,\sigma(X^*,X))$ is separable and so is $(X^*,\tau(X^*,X))$; indeed, if A is countable and $\sigma(X^*,X)$-dense in X*, then finite linear combinations of A with rational coefficients is $\tau(X^*,X)$-dense in X*. This discussion motivates the introduction (cf. [98]) of

DEFINITION 9.2.8 As S.b. $\{x_n;f_n\}$ is said to be *almost shrinking* (a.s.S.b.) if $\{f_n;\Psi x_n\}$ is an S.b. for $(X^*,\tau(X^*,X))$.

The foregoing definition is further justified by the following two examples.

EXAMPLE 9.2.9 Consider the space $(\ell^1, \tau(\ell^1, \ell^\infty))$ with its S.b. $\{e^n; e^n\}$. Then $\{e^n; e^n\}$ is also an S.b. for $(\ell^\infty, \tau(\ell^\infty, \ell^1))$.

EXAMPLE 9.2.10 The space $(\ell^1, \sigma(\ell^1, k))$ has an S.b. $\{e^n; e^n\}$; but $\{e^n; e^n\}$ is not an S.b. for $(k, \tau(k, \ell^1))$ (cf. [129], p. 123 or [130], p. 90).

If $\mathfrak{S}_c$ is the family of all balanced, convex and $\sigma(X, X^*)$-compact subsets of X, then we have

PROPOSITION 9.2.11 $\{x_n; f_n\}$ is an a.s.S.b. if and only if it is an $\mathfrak{S}_c$-S.b.

PROOF. Obvious from Theorem 9.2.2. □

The following results extend similar results in [98].

THEOREM 9.2.12 Let $(X, \tau(X, X^*))$ be quasi-complete. Then $\{x_n; f_n\}$ is an a.s.S.b. if and only if for any $y_k \to y$ in $\sigma(X, X^*)$ and $\{n_k\}$ in I, $S_{n_k}(y_k) \to y$ in $\sigma(X, X^*)$.

PROOF. Assume that $\{x_n; f_n\}$ is an a.s.S.b. If $K = \{y_k\} \cup \{y\}$, then by Krein's Theorem (cf. [143], p. 325), K^{oo} is $\sigma(X, X^*)$-compact. Hence for $\varepsilon > 0$ and f in X^*, there exists k_o such that

$$\sup_{j \geq 1} \left| \left(f - \sum_{i=1}^{n_k} \Psi x_i(f) f_i\right)(y_j)\right| \leq \varepsilon, \ \forall\ k \geq k_o.$$

In particular,

$$|f(y_k) - f(S_{n_k}(y_k))| \leq \varepsilon\ , \ \forall\ k \geq k_o$$

and this proves the necessary part.

For the converse, we make use of the theorems of Eberlein and Šmulian. Let W be a $\sigma(X,X^*)$-compact subset of X; we first show that

$$K \equiv K(W) = \bigcup_{n\geq 1} S_n[W]$$

is $\sigma(X,X^*)$-relatively compact. Since sequential compactness implies countable compactness, appealing to Eberlein's Theorem (cf. [89], p. 189) it is enough to show that K is $\sigma(X,X^*)$-relatively sequentially compact. Consider, therefore, $\{y_k\}$ in K. Then $y_k = S_{n_k}(u_k)$; $u_k \in W$, $k \geq 1$. Next $(X^*,\sigma(X^*,X))$ is separable and so by the Hahn-Banach theorem, $(X,\sigma(X,X^*))$ admits a weaker metrizable l.c. topology. Thus from the extended form of Šmulian's Theorem (cf. [89], 191; [143], 311) we may assume that $\{u_k\}$ has a subsequence which we denote by itself such that $u_k \to u$ in $(X,\sigma(X,X^*))$.

If $\{n_k\}$ is bounded, then $\{y_k\} \subset \cup \{S_n[W] : n = 1,\ldots,N\}$ for some N in $\mathbb{N}$ and since the last set is relatively compact (being finite dimensional), $\{y_k\}$ has convergent subsequence. On the other hand, assuming $n_k \to \infty$, then by the hypothesis, $S_{n_k}(u_k) \to u$. Thus, in any case, K is $\sigma(X,X^*)$-relatively sequentially compact.

Replacing W by a balanced, convex and $\sigma(X,X^*)$-compact subset of X and noting that $K^{oo} = (\bigcup_{n\geq 1} S_n[W])^{oo}$ is $\sigma(X,X^*)$-compact, we easily conclude that $\{S_n^*\}$ is equicontinuous on $(X^*,\tau(X^*,X))$. Further, $[f_n]^\sigma = [f_n]^\tau = X^*$, where $\sigma \equiv \sigma(X^*,X)$ and $\tau \equiv \tau(X^*,X)$. Therefore, by Theorem 2.1.3, $\{f_n;\Psi x_n\}$ is an S.b. for $(X^*,\tau(X^*,X))$. □

<u>EXERCISE 9.2.13</u> Show that if $\{x_n;f_n\}$ is an a.s.S.b. for (X,T) with $(X,\tau(X,X^*))$ being quasi-complete, then $\{f_n;\Psi x_n\}$ is e-Schauder (Definition 2.1.7) for $(X^*,\tau(X^*,X))$.

To obtain other characterizations of an a.s.S.b., we depend heavily on the characterization of compact subsets of X* under different polar topologies. Because of the repeated use of certain symbols in this subsection, let us abbreviate, the topologies $\sigma(X,X^*)$, $\sigma(X^*,X)$, $\tau(X,X^*)$ and $\tau(X^*,X)$ as σ, σ_*, τ and τ_* respectively. Let us also recall the symbols G_c^σ; G_n^σ and σ_c; σ_n from chapter 1 (T-limited subsets etc., replacing T by σ).

The following is a simple consequence of Grothendieck's completion theorem.

<u>LEMMA 9.2.14</u> If (X,T) is a Mazur space, then (X^*,σ_n) is complete; in addition, if (X,τ) is also quasi-complete, then (X^*,τ_*) is complete.

<u>LEMMA 9.2.15</u> Let (X,T) be a Mazur space such that (X,τ) is quasi-complete. Then for $K \subset X^*$, the following statements are equivalent

(i) K is σ-limited.

(ii) K is σ_n-relatively compact.

If (X,T) also satisfies the condition that X^* is σ_*-separable or alternatively (X,T) admits a weaker metrizable l.c. topology, then (i) and (ii) are equivalent to

(iii) K is τ_*-relatively compact.

<u>PROOF</u>. (i) and (ii) are equivalent on account of Proposition 1.2.13 and Lemma 9.2.14.

(iii) $\Rightarrow$ (ii). By Krein's Theorem, $\sigma_n \subset \tau_*$.

(ii) $\Rightarrow$ (iii). By Proposition 1.2.13, K is σ_c-precompact. By Šmulian's Theorem ([143], p. 311), each balanced, convex and σ-compact subset A of X belongs to G^σ, and so K is τ_*-precompact. Hence (iii) holds by Lemma 9.2.14. □

Proposition 1.2.11 and Lemma 9.2.15 immediately yield

<u>LEMMA 9.2.16</u> Let (X,T) be a Mazur space such that (X,τ) is quasi-complete and X^* is σ_*-separable. Then σ^+ coincides with the l.c. topology on X, generated by the polars of all

τ_*-compact subsets of X^*

THEOREM 9.2.17 Let (X,T) be a Mazur and a τ-quasi complete space having an S.b. $\{x_n;f_n\}$. Then the following statements are equivalent:

(i) $\{x_n;f_n\}$ is an a.s.S.b.

(ii) $\{f_n;\Psi x_n\}$ is an e-Schauder base for (X^*,τ_*).

(iii) $\{x_n;f_n\}$ is e-Schauder for (X,σ^+).

PROOF. (i) $\Rightarrow$ (ii) follows from Exercise 9.2.13.

(ii) $\Rightarrow$ (iii). It is enough to show that $\{S_n\}$ is σ^+-equicontinuous. In other words, by Lemma 9.2.16, if K is τ_*-compact subset of X^*, we have to find a τ_*-compact subset J of X^* such that

$$(*)\qquad \sup_{f\in K} |\langle S_n(x),f\rangle \leq \sup_{g\in J} |\langle x,g\rangle|,\ \forall x \in X.$$

Since $\{S_n^*\}$ is τ_*-equicontinuous, for every given τ_*-continuous seminorm p there exists a similar seminorm q so that $p(S_n^*(f)) \leq q(f)$ for all f in X^* and $n \geq 1$. Let $r = \max(p,q)$, $\varepsilon > 0$. Then there exist $g_1,\ldots,g_m$ in K so that $K \subset \{g_1,\ldots,g_m\} + \{g \in X^* : r(g) < \varepsilon/3\}$. There exists N such that

$$p(S_n^*(g_j)-g_j) < \varepsilon/3;\ \forall n \geq N,\ 1 \leq j \leq m.$$

If $f \in K$, we can find j_o with $r(f-g_{j_o}) < \varepsilon/3$. It now easily follows that $p(S_n^*(f)-f) < \varepsilon$, for all $n \geq N$ and uniformly in $f \in K$; that is, $S_n^*(f) \to f$ uniformly on K. Hence, if

$$H = \bigcup_{m\geq 1} S_m^*[K],$$

we conclude that $S_n^*[H]$ is τ_*-precompact for $n \geq 1$; also

$S_n^*(f) \to f$ uniformly on H relative to the topology τ_*. Therefore, by Lemma 1.2.14 (see also Proposition 12.2.1), H is τ_*-precompact and so from Lemma 9.2.14, $J = \bar{H}$ is τ_*-compact. Finally, for x in X,

$$\sup_{f\in K} |\langle S_n(x), f\rangle| = \sup_{f\in K} |\langle x, S_n^*(f)\rangle| \leq \sup_{g\in J} |\langle x, g\rangle|,$$

thereby proving (*).

(iii) ⇒ (i). If $y_k \to y$ in σ, then $S_n(y_k) \to S_n(y)$ in σ^+ uniformly in n by σ^+-equicontinuity of $\{S_n\}$. As $S_n(y) \to y$ in σ^+, it follows that $S_{n_k}(y_k) \to y$ in σ^+. Now (i) follows from Theorem 9.2.12. □

PROPOSITION 9.2.18 Every s-S.b. $\{x_n;f_n\}$ for an S-space (X,T) is an a.s.S.b.

PROOF. If $f \in X^*$, then

$$f = \sum_{n\geq 1} f(x_n)f_n,$$

where the series is subseries convergent in $(X^*,\sigma(X^*,X))$ by the S-character of X; and hence it is subseries convergent in $(X^*,\tau(X^*,X))$ by Theorem 1.5.4. □

REMARK. The condition that $(X^*,\sigma(X^*,X))$ is ω-complete in Proposition 9.2.18 cannot be dropped, for we have

EXAMPLE 9.2.19 By Proposition 6.3.1, $\{e^n;e^n\}$ is an s-S.b. for $(\ell^1,\sigma(\ell^1,k))$. However, $(k,\sigma(k,\ell^1))$ is not ω-complete and observe that $\{e^n;e^n\}$ is not an S.b. for $(k,\tau(k,\ell^1))$.

Next, we pass on to an example of a space having no a.s.S.b. and so by Proposition 9.2.18, this space is also devoid of any s-S.b. In fact, following [98] (cf. also [130], p. 90) we have

EXAMPLE 9.2.20 Let $C = C[0,1]$ be the usual Banach space of continuous functions on $[0,1]$. As observed in [130], p. 90, $(C^*,\tau(C^*,C))$ has no t.b., although C possesses an S.b. (for instance, the Schauder system of C). By Proposition 9.2.18, it also follows that no S.b. of C can be a u-S.b., a fact also observed in [154], p. 24.

Finally, we ask a natural relevant question, namely, when an a.s.S.b. is shrinking. Right now we have the following answer [98] and further discussion on this topic is postponed to the next chapter (Theorem 10.5.8).

THEOREM 9.2.21 An a.s.S.b. for a Banach space $(X, \|\cdot\|)$ is shrinking if and only if X^* is $\|\cdot\|^*$-separable.

For an arbitrary l.c. TVS (X,T), denote by G^σ_ρ the family of all balanced, convex, bounded and σ-metrizable subsets of X and by σ_ρ the l.c. topology on X^* generated by the polars of members of G^σ_ρ. Taking the metric completion of A in G^σ_ρ and observing that A is σ-precompact, we conclude that $\sigma_\rho \subset \sigma_c$.

LEMMA 9.2.22 If (X,T) is an arbitrary metrizable l.c. TVS, then (X^*,σ_ρ) is complete.

PROOF. This follows by Grothendieck's completion theorem and a result in [143], p. 385, 5(2). □

LEMMA 9.2.23 Let (X,T) be an arbitrary Fréchet space such that X^* is σ_*-separable. Then a subset K of X^* is (i) τ_*-relatively compact if and only if (ii) K is γ-relatively compact for any l.c. topology γ on X^* such that (X^*,γ) is separable and γ is $\langle X,X^*\rangle$ polar.

PROOF. (ii) ⇒ (i). Here X^* is also τ_*-separable and τ_* is $\langle X,X^*\rangle$ polar.

(i) ⇒ (ii). By Lemma 9.2.15, K is σ_n-relatively compact and so from Lemma 9.2.22 and Proposition 1.2.13, K is σ_ρ-

relatively compact (since $\sigma_\rho \subset \sigma_c$). It now suffices to prove that $\gamma \subset \sigma_\rho$. Let $B = \{g_n\}$ with $\bar{B}^\gamma = X^*$ and A be any balanced, convex and bounded subset of X so that A^o is a γ-neighbourhood of zero in X*. Then $\sigma(X,X^*)|A = \sigma(X,B)|A$ (cf. [80], p. 252; [143], p. 259) and so A is σ-metrizable. Thus $A \in G_\rho^\sigma$ and so $\gamma \subset \sigma_\rho$. □

<u>LEMMA 9.2.24</u> Let $(X, \|\cdot\|)$ be an arbitrary Banach space such that X* is $\|\cdot\|^*$-separable. Then τ_* and $\|\cdot\|^*$ define the same compact sets and convergent sequences.

<u>PROOF</u>. The part involving compact sets directly follows from Lemma 9.2.23, since $(X, \|\cdot\|)$ is separable and so is (X^*,σ_*) (cf. [217], p. 187 and [143], p. 259). The other part follows from the first part and the fact that $(X^*, \|\cdot\|^*)$ is metrizable. □

<u>PROOF OF THEOREM 9.2.21</u> Cf. Lemma 9.2.24. □

<u>Shrinking and unconditional bases</u>

Shrinking, and several other types of, bases have close relationship; however, we consider here the relationship of the former with that of a u-S.b. Let us first quote the following result of Weill [232], which has already been explained in [130], p. 129.

<u>THEOREM 9.2.25</u> Let (X,T) be an ω-complete barrelled space having a u-S.b. $\{x_n;f_n\}$. Then the following statements are equivalent:

(i) $\{x_n;f_n\}$is shrinking

(ii) X* is $\beta(X^*,X)$-separable

(iii) (X,T) is a non-ℓ^1-space.

The implication (iii) ⇒ (i) is further strengthened in [95]; indeed, we have

PROPOSITION 9.2.26 Let $\{x_n;f_n\}$ be a u-S.b. for an ω-complete barrelled space (X,T) satisfying the condition:

(iv) (X,T) has no complemented subspace Y topologically isomorphic to ℓ^1.

Then $\{x_n;f_n\}$ is shrinking.

PROOF. This is a modification of the proof of Theorem 9.2.25, (iii) $\Rightarrow$ (i) given in [130], p. 129 and hence we present here only the relevant features of the proof.

Suppose $\{x_n;f_n\}$ is not shrinking. Then there exist g in X^*, a bounded sequence $\{z_i\}$ and $\{n_i\}$ in I such that

$$\sum_{k=n_{i-1}+1}^{n_i} g(x_k)f_k(z_i) = 1, \ \forall\ i \geqq 1 \quad (n_o = 0)$$

If $y_i = S_{n_i}(z_i) - S_{n_{i-1}}(z_i)$, then by the barrelledness of (X,T), $\{y_i\}$ is bounded and $g(y_i) = 1$, $i \geqq 1$. Consequently, $\{y_i\}$ is an u-S.b. for $Y = \overline{sp}\{y_i\} = [y_i]$ (cf. Proposition 6.1.4) with $(Y,T|Y) \simeq (\ell^1, \|\cdot\|_1)$ (cf. [130], p. 129) and

$$Y = \{y \equiv \sum_{n\geqq 1} \alpha_n y_n : \alpha \in \ell^1\}.$$

The topology T is also given by $\{P_p : p \in D_T\}$ (cf. (6.1.3)). There exist p and then q in D_T so that $|g(u)| \leqq P_p(u) \leqq q(u)$ for every u in X. Therefore, for $x \in X$

$$\sum_{n\geqq 1} |f_n(x)g(x_n)| \leqq q(x)$$

and hence

$$\sum_{j\geqq 1} |g(\sum_{i=n_{j-1}+1}^{n_j} f_i(x)x_i)| \leqq q(x).$$

Consequently the map $P : X \to Y$ is well defined where

$$Px = \sum_{j\geqq 1} g(\sum_{i=n_{j-1}+1}^{n_j} f_i(x)x_i)y_i.$$

Also, for every r in D_T, $r(Px) \leq k_r\, g(x)$.

Let $y = Px$, $x \in X$. Then

$$P^2x = Py = \sum_{j\geq 1} g\Big(\sum_{i=n_{j-1}+1}^{n_j} f_i(y)x_i\Big)y_j,$$

where

$$f_i(y) = \sum_{m\geq 1} g\Big(\sum_{r=n_{m-1}+1}^{n_m} f_r(x)x_r\Big)f_i(y_m).$$

Writing

$$y_m = \sum_{j=n_{m-1}+1}^{n_m} \beta_j x_j;\ m \geq 1,$$

we obtain

$$f_i(y) = g\Big(\sum_{r=n_{j-1}+1}^{n_j} f_r(x)x_r\Big)\ \beta_i,\ n_{j-1}+1 \leq i \leq n_j.$$

Therefore, using $g(y_m) = 1$, we have

$$g\Big(\sum_{i=n_{j-1}+1}^{n_j} f_i(y)x_i\Big) = g\Big(\sum_{r=n_{j-1}+1}^{n_j} f_r(x)x_r\Big),$$

and so $P^2x = Px$. This contradicts (iv) and hence the result follows. □

The following result is now immediate.

THEOREM 9.2.27 Let (X,T) be an ω-complete barrelled space having a u-S.b. $\{x_n;f_n\}$. Then (i), (ii) and (iii) of Theorem 9.2.25 and (iv) of Proposition 9.2.26 are equivalent.

REMARK. The condition (iv) in Proposition 9.2.26 is indispensible, for consider

EXAMPLE 9.2.28 This is the space $(\ell^1, \tau(\ell^1, \ell^\infty))$ equipped with its u-S.b. $\{e^n; e^n\}$.

NOTE. Further characterizations of shrinking bases are given in Theorems 10.5.5 and 10.5.8.

9.3 DIFFERENT TYPES OF COMPLETE BASES

Whereas the reason of introducing γ-complete bases is already spelled out in the beginning of this chapter, the concept of β-complete and complete bases is an obvious extension of the former. In fact, after [92] and [94], we have

DEFINITION 9.3.1 An S.b. $\{x_n; f_n\}$ for an l.c. TVS (X,T) is called *complete* (resp. *β-complete*) provided $\{\sum_{i=1}^{n} \alpha_i x_i\}$ is convergent in (X,T), whenever for α in ω, the sequence $\{\sum_{i=1}^{n} \alpha_i x_i\}$ is T-Cauchy (resp. $\sigma(X,X^*)$-Cauchy) in X.

Obviously, γ-complete S.b. ==>β-complete S.b. ==> complete S.b. The reverse implications are not always true as justified by the next two examples.

EXAMPLE 9.3.2 Consider the S.b. $\{e^n; e^n\}$ for the Banach space $(cs, \|\cdot\|_{cs})$. Let $\alpha \in \omega$ be such that $\{\sum_{i=1}^{n} \alpha_i e^i\}$ is $\sigma(cs,bv)$-Cauchy. Since $k \subset bv$, the sequence $\{\sum_{i=1}^{n} \alpha_i \beta_i\}$ is Cauchy for each β in k. Hence $\{\sum_{i=1}^{n} \alpha_i\}$ is Cauchy; that is, $\alpha \in cs$. On the other hand, if $\alpha_i = (-1)^i$; $i \geq 1$, then $\|\sum_{i=1}^{n} \alpha_i e^i\|_{cs} \leq 1$ for $n \geq 1$. But $\alpha \notin cs$. Therefore, $\{e^n; e^n\}$ is β-complete but not γ-complete.

EXAMPLE 9.3.3 The S.b. $\{e^n; e^n\}$ for $(c_0, \|\cdot\|_\infty)$ is complete but not β-complete, the former being clear by the Banach character of c_0. Here $\{\sum_{i=1}^{n} \alpha_i e^i\}$ is $\sigma(c_0,\ell^1)$-Cauchy with $\alpha_i = 1$, $i \geq 1$. However, $\{e^{(n)}\}$ does not converge in c_0.

Another example of an S.b. which is β-complete but not γ-complete is contained in

EXAMPLE 9.3.4 Consider the Banach space $(bv_0, \|\cdot\|_{bv_0})$. Then $\{e^n;e^n\}$ is an S.b. for this space; indeed for x in bv_0,

$$\|x^{(n)}-x\|_{bv_0} = \sum_{i\geq n+1} |x_{i+1}-x_i| + |x_{n+1}| \to 0 \text{ as } n \to \infty.$$

It is clear that $e^n \in (bv_0)^* = bs$. Choose α in ω so that $\{\sum_{i=1}^{n} \alpha_i e^i\}$ is $\sigma(bv_0,bs)$-Cauchy. Then $\sum_{i\geq 1} \alpha_i\beta_i$ converges for each β in bs; that is, $\alpha \in (bs)^\beta = bv_0$. Hence $\{e^n;e^n\}$ is β-complete. Observe that $\|e^{(n)}\|_{bv_0} \leq 1$ for all $n \geq 1$. But $e \notin bv_0$ and so this S.b. cannot be γ-complete.

EXERCISE 9.3.5 Show that $\{e^n;e^n\}$ is not a complete S.b. for $(\phi, \|\cdot\|_\infty)$. (Hint: For any α in $c_0 \setminus \phi$, $\{\sum_{i=1}^{n} \alpha_i e^i\}$ is Cauchy without being convergent.)

There is an interesting application of the presence of an a.s.S.b. which is also a β-complete S.b. More precisely, we have (cf. also [98], p. 78)

PROPOSITION 9.3.6 Let (X,T) be a Mazur $\tau(X,X^*)$-quasi complete space having a β-complete a.s.S.b. $\{x_n;f_n\}$. Then $(X,\sigma(X,X^*))$ is ω-complete.

The proof depends upon the following lemma [109].

LEMMA 9.3.7 Suppose $\{H_i : i \in \Lambda\}$ is an equicontinuous net of linear maps from an l.c. TVS (X,S) to another l.c. TVS (Y,T) and let $\{x_\alpha : \alpha \in \Delta\}$ be an S-Cauchy net in X such that for each i in Λ, $\{H_i(x_\alpha) : \alpha \in \Lambda\}$ is convergent in (Y,T). (A) Let $\{H_i(x_\alpha) : i \in \Lambda\}$ be T-Cauchy in Y for each α in Δ, then (i) $\{\lim_\alpha H_i(x_\alpha) : i \in \Lambda\}$ is T-Cauchy in Y and (ii) $y = T\text{-}\lim_i \lim_\alpha H_i(x_\alpha)$, provided that $y = \sigma(Y,Y^*)$ - $\lim_i \lim_\alpha H_i(x_\alpha)$. (B) If $X = Y$ and $x_\alpha = T\text{-}\lim_i H_i(x_\alpha)$ for each α in Δ and $y = \sigma(X,X^*)\text{-}\lim_i \lim_\alpha H_i(x_\alpha)$, then $y = T\text{-}\lim_\alpha x_\alpha$.

PROOF. The proof is omitted as it is analogous to that of its particular case given in [130], p. 37.

PROOF OF PROPOSITION 9.3.6 By Theorem 9.2.17, $\{x_n;f_n\}$ is e-Schauder for (X,σ^+) and so $\{S_n\}$ is σ^+-σ^+ equicontinuous. In the above lemma, replace Y by X, T by σ^+, S by σ^+ and $\{H_i : i \in \Lambda\}$ by $\{S_n : n \geq 1\}$. Let now $\{y_p\}$ be a σ^+-Cauchy net in X and $\alpha_i = \lim_p f_i(y_p)$, $i \geq 1$; then $S_n(y_p) \to \sum_{i=1}^{n} \alpha_i x_i$ in (X,σ^+) for each $n \geq 1$. Hence from the lemma (part (A)), $\{\sum_{i=1}^{n} \alpha_i x_i\}$ is σ-Cauchy. Therefore for some y in X,

$$y = \sigma^+ - \lim_{n\to\infty} \lim_p S_n(y_p).$$

By part (B) of the lemma, $y_p \to y$ in σ^+ and so (X,σ^+) is complete. □

EXERCISE 9.3.8 Give an example of a space showing that the β-completeness of $\{x_n;f_n\}$ in Proposition 9.3.6 cannot be dropped. (Hint: Consider $(c_0, \|\cdot\|_\infty)$.)

EXERCISE 9.3.9 If an S.b. $\{x_n;f_n\}$ in an l.c. TVS (X,T) is β-complete and $\sigma(X,X^*)$-u-S.b., prove that it is a b.m-S.b. for (X,T).

Next, we have (cf. [94])

PROPOSITION 9.3.10 Let $\{x_n;f_n\}$ be an e-Schauder base for an l.c. TVS (X,T). Then the following statements are equivalent:

(i) (X,T) is complete

(ii) (X,T) is ω-complete.

(iii) $\{x_n;f_n\}$ is complete.

PROOF. It is enough to show that (iii) $\Longrightarrow$ (i). In Lemma 9.3.7, replace Y by X, S by T and $\{H_i : i \in \Lambda\}$ by the equicontinuous sequence $\{S_n : n \geq 1\}$ from (X,T) into itself. Now proceed as in the proof of Proposition 9.3.6. □

The $\sigma\gamma$-topology

Corresponding to an arbitrary S.b. $\{x_n;f_n\}$ for an l.c. TVS (X,T), recall the topology $\sigma\gamma(X,X^*)$ on X generated by the seminorms p_f, $f \in X^*$, where $p_f(x) = \sup\{|\langle S_n(x),f\rangle| : n \geq 1\}$. The topology $\sigma\gamma$ is the same topology as $\tilde{\sigma}$ introduced in [130], p. 31. We recall ([130], p. 32; cf. also [92], p. 384).

PROPOSITION 9.3.11 Let $\{x_n;f_n\}$ be an S.b. for an l.c. TVS. Then $\sigma\gamma$ is the smallest polar topology on X, finer than σ, for which $\{x_n;f_n\}$ is an e-Schauder base.

We deduce (cf. [92])

PROPOSITION 9.3.12 If $\{x_n;f_n\}$ is an S.b. for an l.c. TVS (X,T), then $(X,\sigma\gamma)$ is complete (resp. ω-complete, quasi-complete) if and only if $\{x_n;f_n\}$ is a β-complete S.b. for (X,T).

PROOF. Let $(X,\sigma\gamma)$ be ω-complete. By Propositions 9.3.10 and 9.3.11, $\{x_n;f_n\}$ is a complete S.b. for $(X,\sigma\gamma)$. Next observe that if $\alpha \in \omega$, then $\{\sum_{i=1}^{n} \alpha_i x_i\}$ is $\sigma(X,X^*)$-

Cauchy if and only if it is $\sigma\gamma$-Cauchy. This shows that $\{x_n;f_n\}$ is β-complete for (X,T). Conversely, $(X,\sigma\gamma)$ is complete by Proposition 9.3.10. □

γ-complete and unconditional bases

To begin with, let us recall the following result from [232] already proved in [130], p. 133.

THEOREM 9.3.13 Let $\{x_n;f_n\}$ be a u-S.b. for a barrelled space (X,T). Then the following statements are equivalent:

(i) $\{x_n;f_n\}$ is γ-complete.

(ii) $(x,\sigma(X,X^*))$ is ω-complete

(iii) (X,T) is a ω-complete non-c_0-space.

The implication (iii) ==> (i) can be further strengthened in the form of (cf. [95]).

PROPOSITION 9.3.14 Let $\{x_n;f_n\}$ be a u-S.b. for a barrelled space (X,T) satisfying the condition

(iv) (X,T) is ω-complete having no complemented subspace Y with $(Y,T|Y) \simeq (c_0, \|\cdot\|_\infty)$.

Then $\{x_n;f_n\}$ is γ-complete.

PROOF. The proof of (iv) ==> (i) is contained in [130], p. 133. □

We thus have

THEOREM 9.3.15 Let $\{x_n;f_n\}$ be a u-S.b. for a barrelled space (X,T). Then the statements (i) - (iii) of Theorem 9.3.13 and (iv) of Proposition 9.3.14 are all equivalent.

EXERCISE 9.3.16 Prove that none of the statements mentioned in the preceding theorem is valid for $(c_0,\tau(c_0,\ell^1))$ although the hypothesis is satisfied.

9.4 APPLICATIONS

As remarked on previous occasions, the essential use of shrinking and γ-complete bases is amply reflected in the following classical theorem of James [87], namely,

THEOREM 9.4.1 A Banach space $(X, \|\cdot\|)$ with an S.b. $\{x_n;f_n\}$ is reflexive if and only if $\{x_n;f_n\}$ is both shrinking and γ-complete.

Since its appearance, the above result has undergone several extensions of itself, either in the direction of considering wider classes of locally convex spaces or weakening the hypothesis of Schauder bases or both. We have partially touched upon this discussion in [130]; in particular, let us recall the following result (cf. [130], p. 134 or [25])

THEOREM 9.4.2 An l.c. TVS (X,T) having an S.b. $\{x_n;f_n\}$ is semireflexive if and only if the S.b. is shrinking and γ-complete.

An extension of the above result is given in (cf. [92])

THEOREM 9.4.3 Let $\{x_n;f_n\}$ be an S.b. for an l.c. TVS (X,T). Then the following statements are equivalent:

(i) (X,T) is semireflexive.

(ii) $\{x_n;f_n\}$ is shrinking and γ-complete.

(iii) $\{x_n;f_n\}$ is shrinking and β-complete.

PROOF. By virtue of Theorem 9.4.2, we have to show only that (iii) ==> (i). So, if $x^{**} \in X^{**}$, then using the $\beta(X^*,X)$-S.b. character of $\{f_n;\Psi x_n\}$ for X^*, we find that $\{\sum_{i=1}^{n} \langle f_i,x^{**}\rangle x_i\}$ is $\sigma(X,X^*)$-Cauchy in X and so $\Psi x = x^{**}$ for some x in X. □

M-bases and semireflexivity

In [90], it is shown that in characterizing semireflexivity, we demand too many characteristics for the space; at the outset we want the space to possess an S.b. In what follows, we prove a result (cf. [90]) which characterizes semireflexivity for a wider class of spaces than those mentioned in Theorem 9.4.2. Let us first recall ([130], p. 108)

DEFINITION 9.4.4 Let (X,T) be an l.c. TVS. A biorthogonal sequence (b.s.) $\{x_n;f_n\}$ for $\langle X,X^*\rangle$ is called an *M-base* if $[x_n] = X$ and $\{f_n\}$ is total on X.

Corresponding to a b.s. $\{x_n;f_n\}$ for $\langle X,X^*\rangle$ where $X \equiv (X,T)$ is an arbitrary l.c. TVS, it will be further convenient to recall a few notations and simple results from [130]. Accordingly, let us write $G \equiv [x_n]$, the T-closure of $sp\{x_n\}$; $\beta \equiv \beta(X,X^*)$; $H \equiv [f_n]$, the β-closure of $sp\{f_n\}$; $\beta_H \equiv \beta|H$; $H^* \equiv (H,\beta_H)^*$; and $J : X \to H^*$ is the map of Section 9.1. With $\{x_n;f_n\}$ and X as mentioned above, we recall

DEFINITION 9.4.5 A b.s. $\{x_n;f_n\}$ for $\langle X,X^*\rangle$ is called *quasiregular* (q.r.b.s.) provided $\{S_n(x)\}$ is $\sigma(X,X^*)$-bounded for each x in X, where $S_n(x) = \sum_{i=1}^{n} f_i(x)x_i$, $n \geq 1$.

From [130], p. 115 and 116, we mention the following two results:

PROPOSITION 9.4.6 Let $\{x_n;f_n\}$ be a q.r.b.s. for $\langle X,X^*\rangle$, X being a W-space. Then $\{f_n;Jx_n\}$ is an S.b. for (H,β_H) and $\{x_n;f_n\}$ is an S.b. for $(E,\beta(X,X^*)|E)$, $E \equiv$ the $\beta(X,X^*)$-closure of $sp\{x_n\}$.

THEOREM 9.4.7 Let $\{x_n;f_n\}$ be a q.r.b.s. for $\langle X,X^*\rangle$, X being a barrelled space. Then the following statements are valid:

(i) For every α in ω such that $\{\sum_{i=1}^{n} \alpha_i f_i\}$ is $\sigma(X^*,X)$-bounded, there exists f in X* with $f(x_i) = \alpha_i$, $i \geq 1$.

(ii) For each ϕ in X**, the sequence $\{\sum_{i=1}^{n} \phi(f_i)x_i\}$ is $\sigma(X,X^*)$-bounded.

(iii) For every α in ω such that $\{\sum_{i=1}^{n} \alpha_i x_i\}$ is $\sigma(X,X^*)$-bounded, there exists ϕ in X** with $\phi(f_i) = \alpha_i$, $i \geq 1$.

Following [90], we have

DEFINITION 9.4.8 An M-base $\{x_n;f_n\}$ for an l.c. TVS (X,T) is called (i) *shrinking* if H = X* and (ii) γ-*complete* if whenever $\{y_\alpha : \alpha \in \Lambda\}$ is a bounded net in (X,T) such that $\{f_i(y_\alpha): \alpha \in \Lambda\}$ converges in $\mathbb{K}$ for every $i \geq 1$, then there is x in X with

$$\lim_{\alpha} f_i(y_\alpha) = f_i(x), \quad \forall i \geq 1.$$

NOTE. Let $\{x_n;f_n\}$ be an S.b. for an l.c. TVS. If $\{x_n;f_n\}$ is shrinking as an S.b., then it is also shrinking as an M-base; if $\{x_n;f_n\}$ is γ-complete as an M-base, then it is also γ-complete as an S.b. Conversely we have

PROPOSITION 9.4.9 Let $\{x_n;f_n\}$ be a simple S.b. for an l.c. TVS (X,T). (i) If $\{x_n;f_n\}$ is γ-complete as an S.b., then it is γ-complete as an M-base. (ii) If $\{x_n;f_n\}$ is shrinking as an M-base, it is shrinking as an S.b.

PROOF. At the outset let us observe that if A is bounded in (X,T), by Proposition 1.2.4 there exists a bounded set B in (X,T) such that

$$(*) \quad \sup_{x \in A} |\langle x, S_n^*(f)\rangle| \leq \sup_{y \in B} |\langle y, f\rangle|; \ \forall\, n \geq 1,\ f \in X^*$$

(i) Now let $\{y_\alpha\}$ be as in Definition 9.4.8 (ii) with $\beta = \lim_\alpha f_i(y_\alpha)$. Then $S_n(y_\alpha) \to \sum_{i=1}^{n} \beta_i x_i$, for $n \geq 1$. Since $\{y_\alpha\}$ is bounded, for each f in X* there exists $k_f > 0$ so that $|\langle y_\alpha, S_n^*(f)\rangle| \leq k_f$ for all $n \geq 1$ and $\alpha \in \Lambda$ (cf. (*)). It follows that $\{\sum_{i=1}^{n} \beta_i x_i\}$ is bounded and so for some x in X, $\beta_i = f_i(x)$ for all $i \geq 1$.

(ii) By (*), $\{S_n^*\}$ is equicontinuous on $(X^*, \beta(X^*,X))$. Since $H = [f_n] = X^*$, the required result follows by an application of Theorem 2.1.3. □

The next result [90] extends Theorem 9.4.2.

<u>THEOREM 9.4.10</u> Let $\{x_n;f_n\}$ be an M-base for an l.c. TVS (X,T). Then (X,T) is semireflexive if and only if $\{x_n;f_n\}$ is both shrinking and γ-complete.

<u>PROOF</u>. Let (X,T) be semireflexive. Since $\{f_n\}$ is total over X, $[f_n]^\sigma = X^*$, where $\sigma \equiv \sigma(X^*,X)$. But $\sigma(X^*,X) = \sigma(X^*,X^{**})$ and so $H \equiv [f_n]^\beta = X^*$, $\beta \equiv \beta(X^*,X)$. Thus $\{x_n;f_n\}$ is shrinking. Next, let $\{y_\alpha\}$ be as in Definition 9.4.8(ii) with $f_i(y_\alpha) \to \beta_i$, $i \geq 1$. By the semireflexivity of (X,T), $\{y_\alpha\}$ has a $\sigma(X,X^*)$-adherent point x in X. Thus $\beta_i = f_i(x)$ for $i \geq 1$.

Conversely, let $\{x_n;f_n\}$ be both shrinking and γ-complete. It is enough to show that each $\sigma(X,X^*)$-bounded Cauchy net $\{y_\alpha : \alpha \in \Lambda\}$ in X converges in $(X,\sigma(X,X^*))$. We may suppose $\beta_i = \lim f_i(y_\alpha)$, since $\{f_i(y_\alpha) : \alpha \in \Lambda\}$ is Cauchy. Therefore, for some x in X, $\beta_i = f_i(x)$ for $i \geq 1$. If $f \in X^*$ and $\varepsilon > 0$, then there exists g in $sp\{f_i\}$ so that $\sup\{|(f-g)(y)| : y \in \{y_\alpha\} \cup \{x\}\} \leq \varepsilon/3$ and we easily find that $|f(y_\alpha - x)| \leq \varepsilon$ eventually in α. □

<u>EXERCISE 9.4.11</u> Assuming the truth of Theorem 9.4.10, derive Theorem 9.4.2 (Hint: observe that the S.b. is simple; make use of the note after Definition 9.4.8 and Proposition

9.4.9.)

The next two results on the characterization of semireflexivity in terms of M-bases are taken from [236]. However, we first recall the following existence theorem [90] (cf. also [130], p. 110) on M-bases which suits our purpose.

THEOREM 9.4.12 Let (X,T) be a separable l.c. TVS and Y a $\beta \equiv \beta(X^*,X)$-closed separable subspace of X^* with Y being total over X. Then there exist $\{y_n\}$ in X and $\{g_n\}$ in X^* so that $\{y_n;g_n\}$ is an M-base for $\langle X,X^*\rangle$; that is, for (X,T) and $Y = [g_n]^{\beta}$.

THEOREM 9.4.13 Let an l.c. TVS (X,T) have an M-base $\{x_n;f_n\}$. If each M-base for (X,T) is γ-complete, then (X,T) is semireflexive.

PROOF. We invoke both the notation and proof of the converse part of Theorem 9.4.10. Then $\beta_i \equiv \lim_{\alpha} f_i(y_\alpha) = f_i(x)$, $i \geq 1$. Now, let $f_o \in X^*$. Then $Y = [f_n : n \geq 0]^{\beta}$ satisfies the condition of Theorem 9.4.12 (for separability, consider the linear combinations of $\{f_o,f_1,\ldots\}$ with rational coefficients) and so this theorem yields an M-base $\{u_n;g_n : n \geq 1\}$ for (X,T) with $Y = [g_n]^{\beta}$. Since $\{u_n;g_n\}$ is γ-complete, there exists y in X so that $g_i(y) = \lim_{\alpha} g_i(y_\alpha)$, $i \geq 1$. Now $f_k \in [g_n]^{\beta}$, $k \geq 0$. Thus taking the bounded set $A = \{y_\alpha\} \cup \{y\}$ and proceeding as in the converse part of Theorem 9.4.10, we conclude that $f_k(y) = \lim_{\alpha} f_k(y_\alpha)$, $k = 0,1,\ldots$. Consequently $x = y$; in particular, $f_o(x) = \lim_{\alpha} f_o(y_\alpha)$ showing that each $\sigma(X,X^*)$-bounded Cauchy net in X converges in $(X,\sigma(X,X^*))$. □

THEOREM 9.4.14 Let an l.c. TVS (X,T) have an M-base $\{x_n;f_n\}$. If each M-base for (X,T) is shrinking, then

(X,T) is semireflexive.

PROOF. Suppose X is not semireflexive and so there exists x^{**} in X^{**} such that $\Psi x \neq x^{**}$ for any x in X. Let $Y = \ker(x^{**})$. Then Y is a maximal $\beta(X^*,X)$-closed subspace of X^*. It is clear that $\bar{Y}^{\sigma} = X^*$, $\sigma \equiv \sigma(X^*,X)$ (use Hahn-Banach Theorem and Theorem 3 of [233], p. 39). Hence Y is total over X. Since $[f_n : n \geq 1]^{\beta} = X^*$, X^* is $\beta(X^*,X)$-separable (cf. the preceding proof). As $X^* = Y \oplus Z$ with Z being of dimension 1, there exists a continuous projection from X^* onto Y. Hence Y is a separable subspace of $(X^*,\beta(X^*,X))$. By Theorem 9.4.12, there is an M-base $\{y_n;g_n\}$ for (X,T) and as such $Y = [g_n]^{\beta} = X^*$, a contradiction.

9.5 λ-BASES FOR DUAL SPACES

We have already emphasized in chapter 4 the importance of the presence of various notions of 'λ-bases' in an arbitrary l.c. TVS. Given one or the other type of λ-base in an l.c. TVS (X,T), we would like to explore in this chapter a similar character of $\{f_n;\Psi x_n\}$ for $(X^*,\beta(X^*,X))$ and we follow [35] for the rest of this section.

PROPOSITION 9.5.1 Let $\{x_n;f_n\}$ be an S.b. for an l.c. TVS (X,T) and λ a normal simple s.s. If any of the following conditions:

(i) (X,T) is an S-space and $\{x_n;f_n\}$ is a shrinking semi λ-base, or

(ii) $(X^*,\beta(X^*,X))$ is ω-complete and $\{x_n;f_n\}$ is a fully λ-base where λ satisfies (4.5.4) is true, then $\{f_n;\Psi x_n\}$ is a semi λ-base for $(X^*,\beta(X^*,X))$.

PROOF. Case (i) By Proposition 4.6.1, for each bounded subset B of (X,T), $\phi_f[B]$ is $\eta(\lambda,\lambda^{\times})$-bounded, where $\phi_f : X \to \lambda$ with $\phi_f(x) = \{f_n(x)f(x_n)\}$ for all x in X and f in X^*. Since λ is simple, there exists α in λ such that

$\sup\{|f_n(x)| : x \in B\}\ |f(x_n)| \leq |\alpha_n|$ for all $n \geq 1$. Thus $\{\Psi(x_n)(f)p_B(f_n)\} \in \lambda$ for every f in X* and each bounded subset B of (X,T).

Case (ii) Here $\{p(x_n)p_B(f_n)\} \in \lambda$ by Proposition 4.5.17 for each p in D_T and each bounded subset B of (X,T). By (4.5.4), we conclude the convergence of $\sum_{n\geq 1} |f(x_n)|p_B(f_n)$ for each f in X* and every bounded subset B of (X,T). Thus $\{f_n;\Psi x_n\}$ is an S.b. for $(X^*,\beta(X^*,X))$ with $\{f(x_n)p_B(f_n)\} \in \lambda$. □

<u>THEOREM 9.5.2</u> Let $\{x_n;f_n\}$ be an S.b. for a reflexive space (X,T) and λ an arbitrary normal simple s.s. Then $\{x_n;f_n\}$ is a semi (resp. fully) λ-base for (X,T) if and only if $\{f_n;\Psi x_n\}$ is a semi (resp. fully) λ-base for $(X^*,\beta(X^*,X))$.

<u>PROOF</u>. For the semi λ-base character, it is enough to prove the 'if' part only. Hence if $\{f_n;\Psi x_n\}$ is a semi λ-base for $(X^*,\beta(X^*,X))$, then $\{x_n;f_n\}$ is also a semi λ-base for $(X,\sigma(X,X^*))$. By using Proposition 4.6.3, $\{x_n;f_n\}$ is even a fully λ-base.

Now, let $\{x_n;f_n\}$ be a fully λ-base for (X,T). Hence $\{f_n;\Psi x_n\}$ is a fully λ-base for (X,T). Hence $\{f_n;\Psi x_n\}$ is a semi λ-base for $(X^*,\beta(X^*,X))$ (cf. Proposition 4.5.17). Now $(X^*,\beta(X^*,X))$ is barrelled and $(X^*,\sigma(X^*,X^{**})$ is ω-complete and so $\{f_n;\Psi x_n\}$ is a fully λ-base for $(X^*,\beta(X^*,X))$ by Proposition 4.6.3. The other part is already contained in the earlier arguments. □

<u>EXERCISE 9.5.3</u> Let λ be a normal simple s.s. satisfying (4.5.4). Suppose $\{x_n;f_n\}$ is a fully λ-base for an ω-complete l.c. TVS (X,T) such that $(X^*,\beta(X^*,X))$ is also ω-complete. Prove that (X,T) is semireflexive. (Hint: Observe that $\{x_n;f_n\}$ is shrinking and prove that it is γ-complete by using its fully λ-base character and ω-completeness of the

space; then apply Theorem 9.4.2).

Particular cases of λ-bases

In this subsection, we take up the dual character of λ-bases when $\lambda = \ell^p$, $1 \leq p \leq \infty$ and refer to chapter 5 for notation and terminology. We essentially follow [99] for the rest of this subsection.

THEOREM 9.5.4 Let $\{x_n;f_n\}$ be an S.b. for an ω-complete l.c. TVS (X,T). Then the following statements are equivalent:

(i) $\{f_n;\Psi x_n\}$ is a fully ℓ^1-base for $(X^*,\beta(X^*,X))$.

(ii) $\{f_n;\Psi x_n\}$ is a semi ℓ^1-base for $(X^*,\beta(X^*,X))$.

(iii) $\{x_n;f_n\}$ is an ∞-absolute base.

PROOF. (i) ==> (ii). Obvious.

(iii) ==> (i). In view of Proposition 5.6.5, it is enough to show that $\beta(X^*,X)$ is also generated by $\{Q_B : B$ is bounded in $(X,T)\}$, where

$$Q_B(f) = \sum_{n\geq 1} |f(x_n)|p_B(f_n). \tag{9.5.5}$$

For a given bounded subset B of (X,T), let $\mu_n = p_B(f_n)$ and $C = \{\sum_{n\geq 1} \alpha_n\mu_n x_n : \alpha \in c_0, |\alpha_n| \leq 1\}$. Following the proof of Proposition 5.6.5, we easily see that C is bounded in (X,T). For f in X^* and $\varepsilon > 0$, choose N so that $\sum_{i>N} \mu_i|f(x_i)| < \varepsilon/2$. On the other hand

$$p_C(f) \geq \left|\sum_{i=1}^{N} \alpha_i\mu_i\, f(x_i)\right| - \frac{\varepsilon}{2},$$

for each α in c_0 with $|\alpha_n| \leq 1$, $n \geq 1$. For $1 \leq i \leq N$, let $\alpha_i = |f(x_i)|/f(x_i)$ if $f(x_i) \neq 0$, otherwise $\alpha_i = 0$ and $\alpha_i = 0$ for $i > N$. Then

$$p_C(f) \geq \sum_{i=1}^{N} \mu_i |f(x_i)| - \frac{\varepsilon}{2}$$

and so $Q_B(f) \leq p_C(f)$. Hence (i) follows.

(ii) ==> (iii). Consider α in ω such that $B = \{\alpha_n x_n\}$ is bounded in (X,T). For f in X*, let $\nu(f) = \sup \{|\alpha_n f(x_n)| : n \geq 1\}$. By the hypothesis, $\sum_{n\geq 1} |f(x_n)| p_B(f_n) < \infty$ for each f in X* and so $\sum_{n\geq 1} |\alpha_n f(x_n)| < \infty$ for each f in X*. Thus the set

$$E = \{ \sum_{i=1}^{n} \theta_i \alpha_i x_i : \theta \in \ell^\infty, \ |\theta_i| \leq 1; \ n \geq 1\}$$

is bounded in (X,T). Hence for β in c_o,

$$\sum_{i=m+1}^{n} \beta_i \alpha_i x_i \in (\sup_{m+1\leq i\leq n} |\beta_i|)E$$

and this yields the convergence of $\sum_{n\geq 1} \beta_n \alpha_n x_n$. □

PROPOSITION 9.5.6 Let $\{x_n;f_n\}$ be a semi ℓ^1-base for a barrelled space (X,T). Then $\{f_n;\Psi x_n\}$ is an ∞-absolute base for $(X^*,\sigma(X^*,X))$ and an ∞-absolute base for $[f_n]$, the $\beta(X^*,X)$-closure of $sp\{x_n\}$.

PROOF. The first part follows from Theorem 9.5.4, (ii) ==> (iii) by applying the same to the space $(X^*,\sigma(X^*,X))$. In fact, in this case, $\{x_n;f_n\}$ is a semi ℓ^1-base for $(X,\beta(X,X^*))$.

For the second part, let us observe that $\{x_n;f_n\}$ is simple and so $\{f_n;\Psi x_n\}$ is S.basic in $(X^*,\beta(X^*,X))$ (cf. the proof of Proposition 9.4.9 (ii)).

Let $\{\alpha_n f_n\}$ be $\beta(X^*,X)$-bounded for α in ω; then for some p in D_T, $\sup |\alpha_n f_n(x)| \leq p(x)$ for all x in X. There exists q in D_T such that $Q_{p,e}(x) \leq q(x)$ (cf. (4.2.4) and

Proposition 4.4.5(a)). Hence

$$(*) \qquad \sum_{n\geq 1} |\alpha_n f_n(x)| \leqq q(x), \quad \forall x \in X.$$

Now, if $\beta \in c_o$, then $\sum_{n\geq 1} \alpha_n \beta_n f_n$ converges in $(X^*, \sigma(X^*,X))$ on account of (*). Also

$$\left| \sum_{i=M+1}^{N} \alpha_i \beta_i f_i(x) \right| \leqq 2 \sup_{m+1\leqq i\leqq N} |\beta_i| \, q(x)$$

and so $\sum_{n\geq 1} \alpha_n \beta_n f_n$ is $\beta(X^*,X)$-Cauchy. Therefore this series converges in $(X^*, \beta(X^*,X))$. □

The next result not only helps in deriving the p-Köthe character of $\{f_n; \Psi x_n\}$ but also represents an interesting application of the presence of ℓ^p-bases in an l.c. TVS.

PROPOSITION 9.5.7 If an ω-complete l.c TVS (X,T) contains an ℓ^p-base $\{x_n;f_n\}$, $1 < p < \infty$, then it is semi-reflexive.

PROOF. At the outset, let us recall the definition of ν_p, $\nu \in D_T$ (cf. (5.4.4)). Let now $y_\alpha \to y$. For $\varepsilon > 0$, there exists N in $\mathbb{N}$ such that

$$(\nu_p(y)) \leqq \left(\sum_{i=1}^{N} |f_i(y)|^p \, (\nu(x_i))^p \right)^{1/p} + \frac{\varepsilon}{2}$$

$$\leqq (\nu_p(y_\alpha)) + \varepsilon$$

eventually in α and so ν_p is lower semicontinuous. Hence each T-bounded subset of X is T*-bounded, T* being the topology generated by $\{\nu_p : \nu \in D_T\}$. This fact easily leads to the γ-bounded character of $\{x_n;f_n\}$.

If $\{x_n;f_n\}$ were not shrinking, then by the simple character of this base, we would obtain an f in X* and a bounded bl.s.$\{y_n\}$ such that (see the proof of Proposition 9.2.26) $f(y_n) = 1$, $n \geq 1$. Choose α in ℓ^p such that $\alpha \notin cs$. Now for ν in D_T, $\sup \{\nu_p(y_n) : n \geq 1\} < \infty$ and so

$$\sum_{n\geq 1} |\alpha_n|^p \nu_p^p(y_n) < \infty.$$

Therefore, with $y_n = \sum_{i=m_{n-1}+1}^{m_n} f_i(y_n)x_i$ ($m_0 = 0$) we get

$$\sum_{n\geq 1} \sum_{i=m_{n-1}+1}^{m_n} |f_i(\alpha_n y_n)|^p(\nu(x_i))^p < \infty,$$

and hence using the ℓ^p-base character of $\{x_n;f_n\}$, we conclude the convergence of $\sum_{n\geq 1} \alpha_n y_n$. Since $f(y_n) = 1$, $\sum_{n\geq 1} \alpha_n$ converges and this gives a contradiction.

It only remains now to apply Theorem 9.4.2 to conclude the semireflexivity of (X,T). □

EXERCISE 9.5.8 By considering spaces ℓ^1 and c_0, show that the above result is not true when $p = 1$ or $p = \infty$.

The next result is due to Schock [208].

THEOREM 9.5.9 Let $\{x_n;f_n\}$ be an ℓ^p-base for a Fréchet space (X,T), where $1 < p < \infty$. Then $\{f_n;\Psi x_n\}$ is a q-Köthe base for $(X^*,\beta(X^*,X))$, $p^{-1}+q^{-1} = 1$.

PROOF. At the outset, let us observe that $\{x_n;f_n\}$ is shrinking (cf. Proposition 9.5.7 and Theorem 9.4.2).

We now invoke both the notation and proof of Theorem 5.4.2. First of all, let us observe that $\beta(X^*,X)$ is generated by seminorms given by p_L^*, $p_L^*(f) = \sup \{|f(x)|: x \in L\}$.

By Hölder's inequality, for f in X^*, we have

$$p_L^*(f) \leq [\sum_{n\geq 1} \left|\frac{f(x_n)}{\sigma_n}\right|^q]^{1/q} \sup_{x\in L} \{\sum_{n\geq 1} |f_n(x)\sigma_n|^p\}^{1/p}.$$

On the other hand, define y_f^n in X_L by

$$y_f^n = \sum_{i=1}^n \left|\frac{f(x_i)}{\sigma_i}\right|^{q-1} . \operatorname{sgn} f(x_i) . \frac{x_i}{\sigma_i} .$$

Then

$$\| y_f^n \|_L = \{\sum_{i=1}^n \left|\frac{f(x_i)}{\sigma_i}\right|^q]^{1/p}.$$

Let $z_f^n = y_f^n / \| y_f^n \|_L$; then $z_f^n \in L$ and

$$p_L^*(f) \geq |f(z_f^n)| = \{\sum_{i=1}^n \left|\frac{f(x_i)}{\sigma_i}\right|^q]^{1/q}.$$

Since $p_L^*(f_n) = 1/\sigma_n$, we find that

$$p_L^*(f) = \{\sum_{n\geq 1} (|\Psi x_n(f)| p_L^*(f_n))^q \}^{1/q}$$

and this shows the q-Köthe character of the S.b. $\{f_n;\Psi x_n\}$ for $(X^*,\beta(X^*,X))$. □

Finally, we pass on to

PROPOSITION 9.5.10 Every Fréchet space (X,T) equipped with an ℓ^p-base and a different ℓ^q-base with $1 \leq q < p \leq \infty$ is nuclear.

PROOF. In view of Theorem 6.4.13, it suffices to confine our attention to the case when $1 \leq q < p \leq 2$.

Let $q = 1$. By Proposition 5.4.1, the ℓ^p-base is p-Köthe and so it is a u-S.b. Hence it is also 1-Köthe by Proposition 6.4.2(i) and Theorem 6.4.4. Therefore (X,T) is nuclear by Theorem 5.5.5.

Finally, we suppose that $1 < q < p \leq 2$. By Theorem 9.5.9, $(X^*,\beta(X^*,X))$ has an r-Köthe and a different s-Köthe base, where $r^{-1}+p^{-1} = 1$ and $s^{-1}+q^{-1} = 1$. Then $2 \leq r < s < \infty$. By Theorem 6.5.10, $(X^*,\beta(X^*,X))$ is nuclear and so is (X,T). □

Finally, let us recall the following result of [176]:

PROPOSITION 9.5.11 Let $\{x_n;f_n\}$ be an ℓ^1-base for a Fréchet space (X,T) such that $\{f_n;\Psi x_n\}$ is an ℓ^1-base for $(X^*,\beta(X^*,X))$. Then (X,T) is nuclear.

PROOF. By Theorem 9.5.4, $\{x_n;f_n\}$ is ∞-absolute and hence ∞-Köthe (Proposition 5.6.4). In view of Proposition 4.4.5, $\{x_n;f_n\}$ is also 1-Köthe and now make use of Proposition 5.6.11. □

9.6 DUALITY OF BASES

In this section, we study the duality relationship between shrinking and γ-complete bases. Throughout this section we write (X,T) for an arbitrary l.c. TVS having an S.b. $\{x_n;f_n\}$. For brevity, we write β^* for $\beta(X^*,X)$ and H for the β^*-closure of $sp\{f_n\}$, H being equipped with the topology $\beta^*|H$.

Consider the following statements:

(I) $\{x_n;f_n\}$ is shrinking for (X,T).

(II) $\{f_n;\Psi x_n\}$ is γ-complete for (X^*,β^*).

(III) $\{x_n;f_n\}$ is γ-complete for (X,T).

(IV) $\{f_n;\Psi x_n\}$ is shrinking for $(H,\beta^*|H)$.

The above statements were essentially considered by Singer [211] who showed that if (X,T) is a Banach space, then (I) $\Longleftrightarrow$ (II) and (III) $\Longleftrightarrow$ (IV). Thus in a Banach space, shrinking and γ-complete bases exhibit by and large a dual relationship to each other. A natural question is whether such a duality relationship occurs when we confine ourselves to an arbitrary l.c. TVS and we answer this question subsequently with the help of a few counter examples and propositions. Observe that (II) always implies (I) in any l.c. TVS.

Counter-examples

We follow [46] and we first give an example where (I) does not imply (II).

EXAMPLE 9.6.1 Consider the S.b. $\{e^n;e^n\}$ for $(\ell^1, \tau(\ell^1,c_o))$ (cf. Example 5.3.5). Since $\beta(c_o, \ell^1)$ is the sup norm topology of c_o, (I) follows. By Example 9.3.3, (II) is not satisfied.

To justify that (III) =≠=> (IV) always, consider

EXAMPLE 9.6.2 The S.b. $\{e^n;e^n\}$ for $(\ell^\infty, \tau(\ell^\infty,\ell^1))$ is clearly γ-complete (cf. Proposition 2.3.5 (V)). Here $H = \ell^1$ and as $\{e^n;e^n\}$ is not an S.b. for $(\ell^\infty, \beta(\ell^\infty,\ell^1))$ [here $e^{(n)} \not\to e$ in $\beta(\ell^\infty,\ell^1)$], (IV) is not satisfied.

Next, to justify that (IV) does not necessarily imply (III), let us have

EXAMPLE 9.6.3 Consider the S.b. $\{e^n;e^n\}$ for $(\phi,\sigma(\phi,\phi))$. Since $\{e^{(n)}\}$ is bounded but not covnergent in $(\phi,\sigma(\phi,\phi))$, the S.b. in question is not γ-complete. On the other hand, $H = \phi$. Observe that $\beta(\phi,\phi) = \beta(\phi,\omega) = \eta(\phi,\omega)$ (cf. [234], p. 263 and [129], p. 106) and so $(H,\beta(X^*,X)|H)^* = (\phi,\beta(\phi,\phi))^* = \omega$. Thus $\{e^n;e^n\}$ is shrinking for $(H,\beta(X^*,X)|H)$ and (IV) follows.

Results concerning duality

We always have (II) ==> (I). Concerning (I) ==> (II), we prove the following result (cf. [123]) which extends an earlier proposition in [46], p. 272. We recall the notation from section 9.2 and pass on to a more general

PROPOSITION 9.6.4 Let $\{x_n;f_n\}$ be an S.b. for an l.c. TVS (X,T). If $\mathcal{S}$ satisfies the additional property that every S-bounded sequence in X^* is equicontinuous on (X,T) and $\{x_n;f_n\}$ is $\mathcal{S}$-uniform, then $\{f_n;\Psi x_n\}$ is a γ-complete S.b. for (X^*,S).

PROOF. In view of Theorem 9.2.2, it suffices to establish the γ-bounded character of $\{f_n;\Psi x_n\}$. Consider any α in ω so that $\{\sum_{i=1}^{n} \alpha_i f_i\}$ is S-bounded in X^*. Then the set

$$A = \{\sum_{i=k}^{k+p} \alpha_i f_i : k,p \geqq 1\}$$

is equicontinuous, and consequently for $\varepsilon > 0$ and x in X, we find an $N \equiv N(\varepsilon,x,A)$ in $\mathbb{N}$ so that

$$|<\sum_{i\geqq N} f_i(x)x_i, \sum_{j=k}^{k+p} \alpha_j f_j>| \leqq \varepsilon \ ; \quad \forall\ k,p \geqq 1.$$

$$\Rightarrow |<x, \sum_{j=N}^{N+p} \alpha_j f_j>| \leqq \varepsilon , \quad \forall\ p \geqq 1.$$

Therefore $\{\sum_{i=1}^{n} \alpha_i f_i\}$ is $\sigma(X^*,X)$-Cauchy. Also, by the Alaoglu-Bourbski theorem, A is $\sigma(X^*,X)$-relatively compact. Thus the $\sigma(X^*,X)$-Cauchy sequence $\{\sum_{i=1}^{n} \alpha_i f_i\}$ has a $\sigma(X^*,X)$-adherent point f in X^* and it follows that $\sum_{i=1}^{n} \alpha_i f_i \to f$ in S. □

REMARK. In general, the additional restirction on $\mathcal{S}$ may not be relaxed in the preceding result; for instance, one may confine to Example 9.6.1.

Let us recall the notation H, β_H, H^* and J mentioned after Definition 9.4.4 corresponding to an arbitrary S.b. $\{x_n;f_n\}$ for an l.c. TVS (X,T).

The next result [92] includes a partial sharpening of Proposition 9.4.6.

<u>PROPOSITION 9.6.5</u> Let $\{x_n;f_n\}$ be a simple S.b. for an l.c. TVS (X,T). Then $\{f_n;Jx_n\}$ is an S.b. for (H,β_H) and J is a topological isomorphism from $(X,\beta^*(X,X^*))$ into $(H^*,\beta(H^*,H))$.

<u>PROOF</u>. By the hypothesis and Proposition 1.2.4, $\{S_n^*\}$ is β-β equicontinuous, $\beta \equiv \beta(X^*,X)$. It is now easy to show (cf. Theorem 2.1.3) that $\{f_n;Jx_n\}$ is an S.b. for (H,β_H).

The map J is clearly linear and 1-1. For the other part, let A be $\sigma(H,H^*)$-bounded subset of H. Then A is β_H-bounded and so it is $\beta(X^*,X)$-bounded. Hence J is continuous.

Conversely, let $Jy_\alpha \to 0$ in $\beta(H^*,H)$. Consider any $\beta(X^*,X)$-bounded subset M of X^*. By the β-β equicontinuity of $\{S_n^*\}$, the set

$$S^*(M) = \bigcup_{n\geq 1} S_n^*[M]$$

is $\beta(X^*,X)$-bounded and hence it is β_H-bounded. Therefore, for each $\varepsilon > 0$

$$Jy_\alpha \in \varepsilon(S^*(M))^{\bullet}, \text{ eventually in } \alpha$$

$$\Rightarrow \sum_{i=1}^{n} f_i(y_\alpha)x_i \in \varepsilon M^\circ, \quad \forall\, n \geq 1 \text{ and eventually in } \alpha.$$

Here $\bullet$ and $\circ$ denote polars of sets with respect to the dual systems $\langle H,H^*\rangle$ and $\langle X,X^*\rangle$. Since εM° is T-closed and $\{x_n;f_n\}$ is an S.b., $y_\alpha \in \varepsilon M^\circ$ eventually in α. □

Returning to the case (III) ==> (IV), we offer the following proposition which improves an earlier result of [46].

PROPOSITION 9.6.6 Let $\{x_n;f_n\}$ be a simple base for an infrabarrelled space (X,T). If $\{x_n;f_n\}$ is also γ-complete, then $\{f_n;Jx_n\}$ is a shrinking S.b. for (H,β_H).

PROOF. Here $T \approx \beta^*(X,X^*)$. In view of Proposition 9.6.5, we have only to show that $J[X] = H^*$. Let $F \in H^*$. Since for each f in X^*, $\{\sum_{i=1}^{n} f(x_i)f_i\}$ is β_H-bounded,

$$\sup_{n\geq 1} |<\sum_{i=1}^{n} F(f_i)x_i, f>| < \infty.$$

Hence there exists x in X with $F(f_i) = f_i(x)$ for $i \geq 1$. Thus $Jx(f_i) = F(f_i)$ and so $Jx = F$. □

NOTE. Variations of the foregoing and the following results may be traced in [46] and [92] (cf. [92] for further extensions).

Concerning (IV) ==> (III), we have (cf. [46]) the following simple

PROPOSITION 9.6.7 Let $\{x_n;f_n\}$ be an S.b. for an l.c. TVS (X,T) which is $\sigma(X,H)$-ω-complete. If $\{f_n\}$ is an S.b. for (H,β_H), then $\{x_n;f_n\}$ is γ-complete.

PROOF. For α in ω, let $A = \{\sum_{i=1}^{n} \alpha_i x_i : n \geq 1\}$ be bounded in (X,T). Fix f in H and $\varepsilon > 0$. Then we find $N \equiv N(\varepsilon,A,f)$ so that

$$|<f - \sum_{i=1}^{N} f(x_i)f_i, \sum_{j=1}^{m} \alpha_j x_j>| \leq \frac{\varepsilon}{2}, \ \forall\ m \geq 1.$$

Hence we easily find that $|<\sum_{j=P+1}^{Q} \alpha_j x_j, f>| \leq \varepsilon$ for all $Q > P \geq N$. Therefore $\{\sum_{i=1}^{n} \alpha_i x_i\}$ converges in (X,T). □

EXERCISE 9.6.8 Construct examples exhibiting the importance of conditions laid down on (X,T) in the foregoing Propositions 9.6.6 and 9.6.7.

NOTE. For results relating the duality relationship between shrinking and γ-complete M-bases, one is referred to [90].

The final result of this subsection is closer to the implication (IV) ==> (III) and is implicit in the work of [92], p. 386-388.

PROPOSITION 9.6.9 Let $\{x_n; f_n\}$ be an S.b. for an ω-complete l.c. TVS (X,T) such that $\{f_n; Jx_n\}$ is a shrinking S.b. for (H, β_H). Then $H^* = J[X]$ and $\{x_n; f_n\}$ is γ-complete.

PROOF. As (X,T) is a W-space, the base $\{x_n; f_n\}$ is simple and $\beta^*(X,X^*) = \beta(X,X^*)$.

If $F \in X^*$, then $\{\sum_{i=1}^{n} F(f_i)Jx_i\}$ is Cauchy in $(J[X], \beta(H^*,H))$ and so $\{\sum_{i=1}^{n} F(f_i)x_i\}$ is Cauchy in $(X, \beta(X,X^*))$ by Proposition 9.6.5. Hence for some x in X, $F(f_i) = f_i(x)$, for each $i \geq 1$ and so $Jx = F$. Thus $(X, \beta(X,X^*) \simeq (H^*, \beta(H^*,H))$ under J.

Let $B \equiv \{\sum_{i=1}^{n} \alpha_i x_i\}$ be bounded in (X,T). Then

$$(*) \quad |(\sum_{i=1}^{n} \alpha_i Jx_i)(f)| \leq p_B(f) \equiv \sup\{|f(x)| : x \in B\};$$
$$\forall\ f \in X^*,\ n \geq 1.$$

Since by (*)

$$|(\sum_{i=m+1}^{n} \alpha_i Jx_i)(f)| \leq 2p_B(f) : \forall\ f \in X^*,\ n > m \geq 1,$$

we find that

$$G = \{f \in X^* : \sum_{n\geq 1} \alpha_n f(x_n) \text{ converges}\}$$

is a $\beta(X^*,X)$-closed subspace of X^*. Hence $H \subset G$ and so for f in H

$$F(f) = \sum_{n\geq 1} \alpha_n f(x_n)$$

is well defined. By (*), $|F(f)| \leq p_B(f)$; thus $F \in H^*$. Consequently, for some x in X, $Jx = F$. But

$$Jx = \beta(H^*,H) - \lim_{n\to\infty} \sum_{i=1}^{n} \alpha_i Jx_i$$

and so $\sum_{i=1}^{n} \alpha_i x_i \to x$ in $\beta(X,X^*)$. Hence $x = \sum_{n\geq 1} \alpha_n x_n$. □

10 Bounded and regular bases

10.1 PRELIMINARY DISCUSSION

In this chapter, we examine certain natural topological properties of the ingredients $\{x_n\}$ and $\{f_n\}$ forming an S.b. $\{x_n;f_n\}$ for a TVS (X,T). A question which appears to be of immediate attention for us in this direction, is to study the boundedness and regularity of $\{x_n\}$ as well as the equicontinuity of $\{f_n\}$. Nevertheless, questions like the boundedness and regularity of $\{f_n\}$ should also be examined. Besides, we would also like to know if there is a duality relationship between these properties of $\{x_n\}$ and $\{f_n\}$. Above all, we would endeavour to find applications of this kind of study on $\{x_n\}$ and $\{f_n\}$, more specifically, to discover the impact of the normalized character of $\{x_n\}$ and the equicontinuity of $\{f_n\}$ on the structure of (X,T) and vice-versa.

Two general results

The following two elementary lemmas (cf. [104]) on boundedness and regularity of an arbitrary sequence will be found useful.

LEMMA 10.1.1 Let $\{x_n\}$ be a nonzero sequence in an l.c. TVS (X,T). Consider the following statements:

(i) $\{x_n\}$ is regular in (X,T).

(ii) Whenever $\sum_{n\geq 1} \alpha_n x_n$ converges in (X,T), then $\alpha \in c_0$.

(iii) Whenever $\sum_{n\geq 1} \alpha_n x_n$ converges in (X,T), then $\alpha \in \ell^\infty$.

Then (i) ==> (ii) ==> (iii). If (X,T) is also a Fréchet space, then (i) <==> (ii) <==> (iii).

PROOF. (i) ==> (ii). There exists p in D_T with $p(x_n) \geqq 1$, $n \geqq 1$. Now $|\alpha_n| \leqq p(\alpha_n x_n) \to 0$ and so $\alpha \in c_o$.
(ii) ==> (iii) is trivial.

(iii) ==> (i). Let (i) be not true. We may write $D_T = \{p_1 \leqq p_2 \leqq \ldots \leqq p_n \leqq \ldots\}$. Now there exists a subsequence $\{x_{n_k}\}$ so that $p_k(x_{n_k}) \leqq 1/2^{2k}$.

Let $\alpha_n = 2^k$, $n = n_k$, $k \geqq 1$ and zero otherwise. Then $\alpha \notin \ell^\infty$. On the other hand, for each $j \geqq 1$,

$$\sum_{n\geqq 1} p_j(\alpha_n x_n) \leqq \sum_{k\geqq j} p_k(\alpha_{n_k} x_{n_k}) + \sum_{k=1}^{j-1} p_j(\alpha_{n_k} x_{n_k}) < \infty,$$

and this contradicts (iii). □

LEMMA 10.1.2 Let $\{x_n\}$ be a sequence in an l.c TVS (X,T). Consider the following statements:

(i) $\{x_n\}$ is bounded in (X,T).

(ii) Whenever $\alpha \in \ell^1$, then $\sum_{n\geqq 1} \alpha_n x_n$ converges in (X,T).

Then (ii) ==> (i). If (X,T) is also ω-complete, then (i) ==> (ii).

PROOF. (i) ==> (ii). Trivial.

(ii) ==> (i). If (i) were not true, there would exist u in $\mathcal{B}_T$ such that $x_{n_k} \notin 2^k u$, $k \geqq 1$. Let $\alpha_n = 1/2^k$ for $n = n_k$, $k \geqq 1$ and zero otherwise. Then $\alpha \in \ell^1$ but $\alpha_{n_k} x_{n_k} \not\to 0$. This violates (ii). □

10.2 REGULARITY OF $\{x_n\}$, EQUICONTINUITY OF $\{f_n\}$

In this section, we take up a discussion on the regularity of $\{x_n\}$ and its possible impact on the boundedness, convergence and equicontinuity of $\{f_n\}$, where $\{x_n;f_n\}$ is an S.b. for a TVS (X,T). Most of the results of this and of a few subsequent sections are motivated by the following well-known result from the Schauder basis theory in Banach spaces (e.g. [215], p. 20) which says that an S.b. $\{x_n;f_n\}$ in a Banach space $(X, \|\cdot\|)$ satisfies the condition: $\inf \|x_n\| > 0$ if and only if $\sup \|f_n\| < \infty$. Let us begin with a simple (cf. [23])

PROPOSITION 10.2.1 Let (X,T) be a TVS. If $\{x_n;f_n\}$ is a $\sigma(X,X^*)$-S.b. (resp. T-S.b.) for X such that $\{x_n\}$ is $\sigma(X,X^*)$-regular (resp. T-regular), then $f_n \to 0$ in $\sigma(X^*,X)$.

PROOF. If $\{x_n;f_n\}$ is a T-S.b., then $f_n(x)x_n \to 0$ in T for each x in X. The required result clearly follows from this limit and the regularity of $\{x_n\}$. □

COROLLARY 10.2.2 Under the hypothesis of the foregoing proposition, $\{f_n\}$ is $\sigma(X^*,X)$-bounded, and if (X,T) is also a W-space, then $\{f_n\}$ is $\beta(X^*,X)$-bounded.

COROLLARY 10.2.3 Let $\{x_n;f_n\}$ be a regular S.b. for an ω-barrelled space (X,T), then $\{f_n\}$ is equicontinuous.

NOTE. Weaker versions of Corollary 10.2.2 are also found in [188], p. 41 and [231], p. 1046.

REMARK. Absence of the restrictions on $\{x_n;f_n\}$ and (X,T) in Corollary 10.2.2 does not necessarily prevent $\{f_n\}$ from being $\beta(X^*,X)$-bounded, for instance, we may consider

EXAMPLE 10.2.4 Confining ourselves to the S.b. $\{e^n;e^n\}$ for the space $(\phi,\sigma(\phi,\ell^1))$, observe that $\beta(\ell^1,\phi)$ is given by $\|\cdot\|_1$ (e.g. [234], p. 122). This space is not a W-space [consider $\{ne^n\}$ which is $\sigma(\ell^1,\phi)$-bounded but is not $\beta(\ell^1,\phi)$-bounded]. We also note that $\{e^n;e^n\}$ is not

$\sigma(\phi,\ell^1)$-regular; on the other hand, $\{e^n\}$ is $\beta(\ell^1,\phi)$-bounded.

EXERCISE 10.2.5 Show by an example that the regularity of $\{x_n\}$ in Proposition 10.2.1 is not always essential to get the desired conclusion.

EXERCISE 10.2.6 Let $\{x_n;f_n\}$ be a regular S.b. for $(X,\tau(X,X^*))$ such that $(X^*,\tau(X^*,X))$ is ω-complete. Prove that $\{f_n\}$ is equicontinuous (cf. [96], p. 92) [Hint: Conclude the ω-barrelledness of the space; cf. Proposition 1.2.9 or [234], p. 167].

EXERCISE 10.2.7 Let a TVS (X,T) contain a set of second category and a regular S.b. $\{x_n;f_n\}$. Prove that $\{f_n\}$ is equicontinuous. [Hint: $\{S_n\}$ is equicontinuous.] This result strengthens a theorem in [106].

Let us now examine the converse of the foregoing results. We begin with a simple

PROPOSITION 10.2.8 Let $\{x_n;f_n\}$ be an equicontinuous S.b. for a TVS (X,T). Then $\{x_n;f_n\}$ is regular.

PROOF. By the hypothesis, we can determine u in $\mathcal{B}_T$ such that $u \subset f_n^{-1}[\{\alpha : |\alpha| < 1/2\}]$ for all $n \geq 1$. Here $x_n \notin u$, $n \geq 1$. □

Equicontinuity in the above result is indispensable, for we have

EXAMPLE 10.2.9 This is the space of Example 10.2.4. It suffices to show that $\{e^n\}$ is not $\sigma(\phi,\ell^1)$-equicontinuous. On the other hand, let there be β in ℓ^1 with

$$|\alpha_n| = |\langle\alpha,e^n\rangle| \leq |\sum_{i\geq 1} \alpha_i\beta_i|; \quad \forall\alpha \in \phi,\ n \geq 1.$$

In particular, $|\beta_n| \geq 1$ for all $n \geq 1$ which is, however, not true.

PROPOSITION 10.2.10 Let $\{x_n;f_n\}$ be an S.b. for a σ-barrelled (resp. σ-infrabarrelled) space (X,T) such that $\{f_n\}$ is $\sigma(X^*,X)$-bounded (resp. $\beta(X^*,X)$-bounded). Then $\{x_n;f_n\}$ is regular.

PROOF. Straightforward.

EXERCISE 10.2.11 Prove that Propositions 10.2.8 and 10.2.10 are valid even when $\{x_n,f_n\}$ is an arbitrary b.s.

The following result says that the regularity of $\{x_n\}$ can be avoided to infer the equicontinuity of $\{f_n\}$; indeed, we have (cf. [96])

THEOREM 10.2.12 Let $\{x_n;f_n\}$ be an S.b. for a Mackey space $(X,\tau(X,X^*))$. Then $\{f_n\}$ is equicontinuous if and only if (i) $\delta \subset c_o$ and (ii) $\mu \supset \ell^1$.

PROOF. Let $\{f_n\}$ be equicontinuous. Then $p(x) = \sup\{|f_n(x)|: n \geqq 1\}$ defines a continuous norm on X. If $\alpha \in \delta$, then $p(\alpha_n x_n) \to 0$. But $p(x_n) = 1$ for all $n \geqq 1$, giving $\alpha \in c_o$. Next, let $\alpha \in \ell^1$ with $||\alpha||_1 \leqq 1$. Suppose $B = \overline{\Gamma\{f_n\}}^{\sigma}$, $\sigma \equiv \sigma(X^*,X)$. By the Alaoglu-Bourbaki theorem, B is $\sigma(X^*,X)$-compact. Since $\sum_{i=1}^{n} \alpha_i f_i \in B$ for all $n \geqq 1$, the sequence $\{\sum_{i=1}^{n} \alpha_i f_i\}$ has a $\sigma(X^*,X)$-adherent point f in B. In particular, for each $j \geqq 1$ and $\varepsilon > 0$, $\{\sum_{i=1}^{n} \alpha_i f_i\}$ belongs frequently to $f + \varepsilon\{x_j\}^o$ and so $\alpha_j = f(x_j)$, $j \geqq 1$. This shows that $\ell^1 \subset \mu$.

For the converse, define $F:X \to c_o$, by $F(x) = \{f_n(x)\} \in \delta \subset c_o$. For α in ℓ^1, $\alpha = \{f(x_n)\}$ with $f \in X^*$. Hence $\langle F(x),\alpha\rangle = f(x)$, showing that F is $\sigma(X,X^*)$-$\sigma(c_o,\ell^1)$ continuous. By Proposition 1.2.15, there exists a $\tau(X,X^*)$-continuous seminorm q on X such that $\|F(x)\|_\infty \leqq q(x)$ for all x in X, giving $|f_n(x)| \leqq q(x)$ for all $n \geqq 1$, $x \in X$. □

As an application of regular bases, let us quote the following result (cf. [104]; [130], p. 125) without proof (its proof is, in fact, based on Lemma 10.1.1).

<u>PROPOSITION 10.2.13</u> Let $\{x_n;f_n\}$ be a regular S.b. for an l.c. TVS (X,T) and $\alpha \in \omega$ with $\alpha_n \neq 0$, $n \geq 1$. Suppose $y_n = \sum_{i=1}^{n} \alpha_i x_i$. If $\{y_n/\alpha_{n+1}\}$ is bounded, then $\{y_n;g_n\}$ is an S.b. for (X,T), where $g_n = (f_n/\alpha_n)-(f_{n+1}/\alpha_{n+1})$.

<u>Regularization of</u> $\{x_n\}$

If $\{x_n;f_n\}$ is an S.b. for an l.c. TVS (X,T), then in some way equicontinuity of $\{f_n\}$ implies the regularity of $\{x_n\}$ and conversely. In fact, one of the advantages of studying regular bases is to infer the equicontinuity of the s.a.c.f. However, there are bases $\{x_n;f_n\}$ which are not regular and this raises the question whether the corresponding s.a.c.f. is equicontinuous or not. In such cases, it is sometimes possible to find scalars $\alpha_n \neq 0$, $n \geq 1$ so that $\{\alpha_n x_n\}$ is regular and that is what we often call the process of regularization of $\{x_n\}$ or $\{x_n;f_n\}$. Thus from a given S.b. which is possibly not regular, we get a dilated regular S.b. The next question, namely, under what circumstances can we regularize an arbitrary S.b., may be answered in terms of the following result for which we essentially follow [96].

<u>PROPOSITION 10.2.14</u> Let $\{x_n;f_n\}$ be an S.b. for an ω-barrelled space (X,T). Then the following statements are equivalent:

(i) There exists α in ω; $\alpha_n \neq 0$, $n \geq 1$ such that $\{\alpha_n x_n\}$ is regular in (X,T).

(ii) There exists β in ω; $\beta_n \neq 0$, $n \geq 1$ such that $\{\beta_n x_n\}$ is equicontinuous on (X,T).

(iii) (X,T) admits a continuous norm p.

PROOF. (i) ==> (ii). Put $\beta_n = 1/\alpha_n$ and use Corollary 10.2.3.

(ii) ==> (iii). Put $p(x) = \sup\{|\beta_n f_n(x)| : n \geq 1\}$, $x \in X$.

(iii) ==> (i). Put $\alpha_n = 1/p(x_n)$. Since $p \leq q$ for some q in D_T, $q(\alpha_n x_n) \geq 1$, $n \geq 1$. □

The next result [96] considerably strengthens Proposition 10.2.14 at the expense of the structure of (X,T) and improves Twierdzenie 5 of [19].

THEOREM 10.2.15 For an S.b. $\{x_n;f_n\}$ in a Fréchet space (X,T), the following statements are equivalent:

(i) There exists α in ω with $\alpha_n \neq 0$, $n \geq 1$ such that $\{\alpha_n x_n\}$ is regular.

(ii) There exists a continuous norm p on (X,T).

(iii) There exists no complemented subspace Y of (X,T) such that $(Y,T|Y) \simeq (\omega,\sigma(\omega,\phi))$.

(iv) There is no subsequence $\{x_{n_j}\}$ of $\{x_n\}$ such that $\sum_{j\geq 1} \alpha_j x_{n_j}$ converges for all α in ω.

PROOF. (i) <==> (ii). Cf. Proposition 10.2.14.

(ii) ==> (iii). Use the fact that $(\omega,\sigma(\omega,\phi))$ admits no continuous norm, which also follows from Proposition 10.2.14, (iii) ==> (i)

(iii) ==> (iv). If (iv) were not true, then for some $\{x_{n_j}\}$, we may determine a subspace Y of X with $Y = \{y = \sum_{j\geq 1} \alpha_j x_{n_j} : \alpha \in \omega\}$. If $F(x) = \lim_m F_m(x)$ where for x in X

$$F_m(x) = \sum_{j=1}^{m} f_{n_j}(x) x_{n_j},$$

then by Banach-Steinhaus theorem it immediately follows that F is a continuous projection from X onto Y. It is easy to see that $Y = [x_{n_j}]$. The map $G:Y \to \omega$ is clearly an algebraic isomorphism onto ω. It is easily seen to be continuous and so $(Y,T|Y) \simeq (\omega,\sigma(\omega,\phi))$, a contradiction.

(iv) ==> (ii). Assume that (ii) is not true. Without loss of generality, we may assume that $D_T = \{p_1 \leq p_2 \leq \cdots \leq p_n \leq \cdots\}$ where $p_n(x) = \sup\{p_n(S_m(x)) : m \geq 1\}$ (e.g. [130], p. 27).

Let $X_n = \ker(p_n)$; then $\dim X_n \nless \infty$. For otherwise, letting $\dim X_n = m$, we can find continuous extensions $g_1,\ldots,g_m$ of the s.a.c.f. corresponding to a finite S.b. for X_n. Put $q_n(x) = \sup\{|g_i(x)| : 1 \leq i \leq m\}$, then $p_n + q_n$ is a continuous norm on X, thereby contradicting our assumption.

Next, suppose $N_n = \{m \in \mathbb{N} : f_m(x) \neq 0$ for some x in $X_n\}$. Then N_n is an infinite set, otherwise $\dim X_n < \infty$. It is easily seen that $N_n = \{m \in \mathbb{N} : p_n(x_m) = 0\}$. We can choose $\{n_j\}$ in I so that if $k \leq j$, then $p_k(x_{n_j}) \leq p_j(x_{n_j}) = 0$. Hence $\{\sum_{j=1}^{m} \alpha_j x_{n_j}\}$ is Cauchy in (X,T) for each α in ω. This contradicts (iv). □

10.3 BOUNDEDNESS OF $\{x_n\}$, REGULARITY OF $\{f_n\}$

In this section, we examine the second natural question on the topological properties of an S.b. $\{x_n;f_n\}$, namely, the boundedness of $\{x_n\}$ and its possible impact on the regularity of $\{f_n\}$. Throughout this section, unless the contrary is mentioned, we write (X,T) for an arbitrary l.c. TVS containing an S.b. $\{x_n;f_n\}$. Let us begin with a simple (cf. [106])

PROPOSITION 10.3.1 Let $\{x_n;f_n\}$ be a b.s. for $\langle X,X^*\rangle$. If $\{x_n\}$ is T-bounded, then $\{f_n\}$ is $\beta(X^*,X)$-regular.

PROOF. If $A = \{2x_n\}$, then $f_n \notin A^o$, $n \geq 1$. □

REMARK. The boundedness in the above result cannot be dropped, for we have

EXAMPLE 10.3.2 The S.b. $\{n^2e^n; n^{-2}e^n\}$ for the space $(\phi, \sigma(\phi, \ell^1))$ is not bounded. Since $\beta(\ell^1, \phi)$ is given by $\| \cdot \|_1$, it readily follows that $\{n^{-2}e^n\}$ is not $\beta(\ell^1, \phi)$-regular.

EXERCISE 10.3.3 Construct an S.b. $\{x_n; f_n\}$ for $(c_o, \tau(c_o, \ell^1))$ so that neither $\{x_n\}$ is bounded nor $\{f_n\}$ is $\beta(\ell^1, c_o)$-regular.

To obtain a converse of Proposition 10.3.1, let us introduce a notion of an S.b. weaker than a simple base.

DEFINITION 10.3.4 An S.b. $\{x_n; f_n\}$ is called *semisimple* provided for each f in X^* and bounded subset A of (X,T)

$$\sup_{n \geq 1} \sup_{x \in A} |f_n(x) f(x_n)| < \infty. \tag{10.3.5}$$

PROPOSITION 10.3.6 Let $\{x_n; f_n\}$ be a b.s. for $\langle X, X^* \rangle$, satisfying (10.3.5). If $\{f_n\}$ is $\beta(X^*, X)$-regular, then $\{x_n\}$ is bounded in (X,T).

PROOF. There exists a bounded subset A of (X,T) containing a sequence $\{y_n\}$ such that $|f_n(y_n)| > 1$, $n \geq 1$. Hence for f in X^*, $|f(x_n)| \leq \sup |f_n(y_n) f(x_n)| < \infty$ by (10.3.5). □

REMARK. Without the restriction (10.3.5), the conclusion of the preceding result may not be true, for instance, one may consider

EXAMPLE 10.3.7 This is the space of Example 2.3.10.

Invoking both the notation and solution of this example, $\sup\{|\langle x,e^i\rangle\langle\alpha,e^i\rangle| : x \in A\} = n$ for $i = 2n, 2n-1$; $n \geq 1$ and so $\{e^n ; e^n\}$ is not semisimple. Here $\{e^n\}$ when considered in λ is $\beta(\lambda,\phi)$-regular, for $\sup\{|\langle e^{(2m)},e^n\rangle| : m \geq 1\} = 1$ for all $n \geq 1$. The sequence $\{e^n\}$ when considered in ϕ is not $\sigma(\phi,\lambda)$-bounded, for $\sup\{|\langle e^n,\alpha\rangle| : n \geq 1\} = \infty$.

PROPOSITION 10.3.8 If $\{x_n,f_n\}$ is a semisimple S.b., then $\{f_n\}$ is $\beta(X^*,X)$-regular if and only if $\mu \subset \ell^\infty$.

PROOF. For the 'if' part, use Proposition 10.3.1, whereas the other part follows with the help of Proposition 10.3.6. □

Next, we have (cf. [96])

PROPOSITION 10.3.9 If $\ell^1 \subset \delta$, then $\{x_n;f_n\}$ is a bounded S.b. for (X,T). Conversely, if (X,T) is ω-complete and $\{x_n;f_n\}$ is bounded, then $\ell^1 \subset \delta$.

PROOF. We have $\mu \subset \delta^\beta \subset \ell^\infty$ and this shows that $\{x_n\}$ is $\sigma(X,X^*)$-bounded.

Conversely, $\sum_{n\geq 1} \alpha_n x_n$ is absolutely convergent in (X,T) for each α in ℓ^1 and so for some y_α in X, $y_\alpha = \sum_{n\geq 1} \alpha_n x_n$.

Hence $\alpha = \{f_n(y_\alpha)\} \in \delta$. □

NOTE. The above result also follows by a direct application of Lemma 10.1.2.

The ω-completeness in the preceding result cannot be overlooked, for we have

EXAMPLE 10.3.10 Consider either of the spaces $(k,\sigma(k,\ell^1))$ or $(\phi, \|\cdot\|_\infty)$, each having the S.b. $\{e^n;e^n\}$. In both cases, the S.b. is bounded and $\ell^1 \not\subset \delta$. None of the spaces is ω-complete.

EXERCISE 10.3.11 (i) Let $\{x_n;f_n\}$ be a simple S.b. Show that $\{x_n\}$ is bounded if and only if $\{f_n\}$ is a regular S.b. for $[f_n]^\beta$, $\beta \equiv \beta(X^*,X)$. [Hint: Use Propositions 10.3.1, 10.3.6 and 9.4.6 (proof of second part).]

(ii) If $\{x_n;f_n\}$ is semisimple and regular, prove that $\{f_n\}$ is $\beta(X^*,X)$-bounded. [Hint: For each bounded subset A of (X,T), the set $\{f_n(x)x_n: x \in A, n \geq 1\}$ is bounded in (X,T).]

Formation of bounded bases

Boundedness of an S.b. is a useful property. However, there are bases which are not bounded, for example $\{e^n;e^n\}$ for $(\phi,\sigma(\phi,\omega))$. On the other hand, there is a class of spaces having bases which yield new bases that turn out to be bounded. In this direction, we have

PROPOSITION 10.3.12 Let $\{x_n;f_n\}$ be an S.b. for a metrizable l.c. TVS (X,T). Then there exists $\{\alpha_n\}$, $\alpha_n > 0$ such that $\{\alpha_n x_n;\alpha_n^{-1} f_n\}$ is a bounded S.b. for (X,T).

PROOF. Let d denote an invariant metric generating T. For each p in D_T, there exists $k_p > 0$ such that $p(x) \leq k_p d(x,0)$. Put $\alpha_n = 1/d(x_n,0)$. □

NOTE. In the preceding result, it is not necessary to take $\{x_n\}$ to be an S.b. In fact, it is enough to consider any nonzero sequence $\{x_n\}$.

Following [104], we have

PROPOSITION 10.3.13 Let $\{x_n;f_n\}$ be bounded and let (X,T) be ω-complete. Suppose that for some α in ω, $\alpha_n \neq 0$ for $n \geq 1$, the sequence $\{y_n\}$ is an S.b. for (X,T), where $y_n = \sum_{i=1}^{n} \alpha_i x_i$. Then $\{y_n/\alpha_{n+1}\}$ is bounded.

PROOF. At the outset, let us observe that the s.a.c.f. corresponding to $\{y_n\}$ is given by $\{g_n\}$ (cf. Proposition 10.2.13).

If the required result is not true, then there exist p in D_T and $\{n_k\}$ in I_∞ such that

$$p\left(\frac{y_{n_k}}{\alpha_{n_k} + 1}\right) \geq 2^{2k}, \ \forall\ k \geq 1.$$

Let $y = \sum_{n\geq 1} (x_{n_k+1}/2^k)$. Then $f_m(y) = 2^{-k}$ for $m = n_k+1$, $k \geq 1$ and zero otherwise. Therefore

$$(*) \qquad p\left(\frac{f_{n_k+1}(y)y_{n_k}}{\alpha_{n_k+1}}\right) \geq 2^k, \ \forall\ k \geq 1.$$

But

$$\sum_{i=1}^{n} g_i(y)y_i = \sum_{i=1}^{n} f_i(y)x_i - \frac{1}{\alpha_{n+1}} f_{n+1}(y)y_n,$$

and so $f_{n+1}(y)y_n/\alpha_{n+1} \to 0$, contradicting (*). □

Using Proposition 10.3.12, we obtain (cf. [96])

PROPOSITION 10.3.14 Let $\{x_n;f_n\}$ be simple and (X,T) be σ-infrabarrelled such that $(X^*,\beta(X^*,X))$ is a metrizable space. Then (X,T) admits a continuous norm.

PROOF. The space (H,β_H) is a metrizable space (cf. Proposition 9.6.6 for notation) having an S.b. $\{f_n,\Psi x_n\}$ (cf. proof of Proposition 9.4.9 (ii)). We can determine $\alpha_n \neq 0$, $n \geq 1$ such that $\{\alpha_n f_n\}$ is $\beta(X^*,X)$-bounded and so for some p in D_T, $\|x\| \equiv \sup\{|\alpha_n f_n(x)| : n \geq 1\} \leq p(x)$ for all x in X. □

REMARK. The metrizability restriction in Proposition 10.3.14 is indispensable.

EXAMPLE 10.3.15 Recall the familiar space $(\omega,\sigma(\omega,\phi))$ along with its S.b. $\{e^n;e^n\}$. It is clear that if $\beta \in \phi$ with length L and A is any $\sigma(\omega,\phi)$-bounded subset of ω, then

$$\sup_{\alpha \in A} |<\alpha, \sum_{i=1}^{n} <e^i,\beta>e^i>| \leqq \sup_{\alpha \in A} \sum_{i=1}^{L} |\alpha_i\beta_i| < \infty ,$$

and so $\{e^n;e^n\}$ is simple. The space $(\phi,\beta(\phi,\omega))$ is not metrizable, for otherwise, $(\phi,\eta(\phi,\omega))$ is a Baire space (here $\eta(\phi,\omega) = \beta(\phi,\omega)$) and hence a contradiction (cf. [89], p. 88). It is also a well known fact that $(\omega,\sigma(\omega,\phi))$ cannot admit a continuous norm.

10.4 NORMALIZED BASES

Normalized bases (nz-S.b. or nz-bases) enjoy a number of interesting properties. In particular, these bases are useful in characterizing new bases from the given ones. We will also find their applications in characterizing reflexivity of spaces. In the meantime, let us pass on to

THEOREM 10.4.1 Let $\{x_n;f_n\}$ be an nz-S.b. for an ω-complete l.c. TVS (X,T). Define $\{y_n\}$ and $\{g_n\}$ as in Proposition 10.2.13. Then $\{y_n;g_n\}$ is an S.b. for (X,T) if and only if $\{y_n/\alpha_{n+1}\}$ is bounded in (X,T).

PROOF. Combine Propositions 10.2.13 and 10.3.13. □

Earlier results on bounded and regular bases may be combined to give characterizations of normalized bases. A dual characterization of nz-bases is contained in (cf. [96])

THEOREM 10.4.2 Let $\{x_n;f_n\}$ be a simple S.b. for a σ-infrabarrelled space (X,T). Then $\{x_n;f_n\}$ is an nz-S.b. if and only if $\{f_n;\Psi x_n\}$ is an nz-S.b. for (H,β_H).

PROOF. At the outset, let us mention that the simple character of $\{x_n;f_n\}$ forces $\{f_n;\Psi x_n\}$ to be an S.b. for (H,β_H) (cf. Proposition 10.3.14).

If $\{x_n;f_n\}$ is an nz-S.b., then $\{f_n;\Psi x_n\}$ is so for (H,β_H) by Exercise 10.3.11.

The converse follows with the help of Proposition 10.2.10 and Exercise 10.3.11. □

EXERCISE 10.4.3 Let $\{x_n;f_n\}$ be an nz S.b. which is also a semi ℓ^1-base for an ω-complete l.c. TVS (X,T). Prove that $\delta = \ell^1$.

Normalized bases in normed spaces

Let $\{x_n;f_n\}$ be an S.b. in a normed space $(X, \|\cdot\|)$. Clearly $\{x_n;f_n\}$ is an nz-S.b. if and only if $0 < \inf \|x_n\| \leq \sup \|x_n\| < \infty$; and if $(X, \|\cdot\|)$ is also complete, then $\{x_n;f_n\}$ is an nz-S.b. if and only if $0 \leq \inf \|f_n\| \leq \sup \|f_n\| < \infty$ (cf. Theorem 10.4.2) [here $\{x_n;f_n\}$ is clearly simple because of the W-ness of (X,T)]. Given an S.b. $\{x_n;f_n\}$ in a normed space $(X, \|\cdot\|)$ it is always possible to find an S.b. $\{y_n\}$ with $\|y_n\| = 1$; on the other hand, if (X,T) is also complete, then it is possible (cf. [215], p. 22) to find a norm $\|\cdot\|^*$ equivalent to $\|\cdot\|$ such that $\|x_n\|^* = 1$, for $n \geq 1$. It is also well known that in every separable Hilbert space, one can find an S.b. $\{x_n;f_n\}$ such that $\|x_n\| = \|f_n\| = 1$ for all $n \geq 1$; in fact, every orthonormal base satisfies this condition.

DEFINITION 10.4.4 An S.b. $\{x_n;f_n\}$ for a normed space $(X, \|\cdot\|)$ is called *fully normalized* or an f.nz-S.b. if $\|x_n\| = \|f_n\| = 1$ for $n \geq 1$.

THEOREM 10.4.5 Let $\{x_n;f_n\}$ be an S.b. for a Hilbert space $(X, \|\cdot\|)$. Then $\{x_n;f_n\}$ is an f.nz-S.b. if and only if $\{x_n\}$ can be identified with $\{f_n\}$ for all $n \geq 1$, i.e. $\{x_n\}$ is orthonormal.

PROOF. Let us denote by $\langle .,. \rangle$ the inner product generating the norm $\|\cdot\|$.

The 'if' part is obvious.

For the 'only if' part, it is enough to show that $\{x_n\}$ is orthogonal, for then $\{x_n\}$ is orthonormal and $f_n(x) = \langle x, x_n \rangle$; $x \in X$, $n \geq 1$.

Let $\langle x_m, x_n \rangle \neq 0$ for some m,n with $m \neq n$. Set $Y = sp\{x_m, x_n\}$. Since dim Y = 2, we can find an orthonormal base $\{u, x_n\}$ in Y. Then $x_m = \langle x_m, u \rangle\ u + \langle x_m, x_n \rangle\ x_n$ with $0 < |\langle x_m, u \rangle| < 1$ and so $1 = |f_n(x_m)| < |f_m(u)|$. This contradicts $\|f_m\| = 1$. □

NOTE. The 'only if' of the above theorem was first established by Karlin [139]; however, the proof outlined above is taken from [26].

10.5 NORMAL BASES

We have had occasion to discover the importance of normalized bases. When we work with bases which are possibly not normalized, we try to find the process which may make these bases normalized. This underlying process is often referred to as the normalization of bases. To ask for normal bases is virtually to seek the normalization of such bases. More formally, recalling the definition of a normal sequence from chapter 1, let us introduce

DEFINITION 10.5.1 An S.b. $\{x_n; f_n\}$ for an l.c. TVS (X,T) is called *normal* (n-S.b.) if $\{x_n\}$ is normal.

Emphasizing more the normal character of $\{x_n; f_n\}$, we have

DEFINITION 10.5.2 An S.b. $\{x_n; f_n\}$ for an l.c. TVS (X,T) is called *completely normal* (c.n -S.b.) if every block sequence (bl.s.) $\{y_n\}$ of $\{x_n\}$ is normal.

On the other hand, an extreme case is contained in

DEFINITION 10.5.3 An S.b. $\{x_n;f_n\}$ for an l.c. TVS (X,T) is called *completely abnormal* (c.ab-S.b.) if no bl.s. of $\{x_n\}$ is normal.

REMARKS. Clearly every c.n-S.b. is an n-S.b. Every S.b. in a normed space is c.n-S.b. In fact, this happens because of the homogeneous property of the norm giving the topology of the space. For, if x is a nonzero element of a normed space $(X, \|\cdot\|)$, then $\|y\| = 1$ where $y = x/\|x\|$. It seems that the complete normality property of an S.b. for a normed space is an inherent feature of such a space and this may not be true in non-normable spaces as justified in the next subsection. However, in a non-normable space, this may not be the case, for example, the S.b. $\{e^n;e^n\}$ for $(\omega,\sigma(\omega,\phi))$ or $(\phi,\sigma(\phi,\omega))$ is not normal and hence it is a c.ab-S.b.

EXERCISE 10.5.4 Let $\{x_n;f_n\}$ be a c.ab-S.b. for an l.c. TVS (X,T). Show that $\{x_n;f_n\}$ is also a c.ab-S.b. for $(X,\sigma(X,X^*))$. [Hint: If a sequence $\{y_n\}$ is not T-regular, then it cannot be $\sigma(X,X^*)$-regular. Why?]

Applications

We reproduce from [96] the following two results which appear to be interesting applications of the bases discussed in the earlier paragraphs.

THEOREM 10.5.5 Let $\{x_n;f_n\}$ be a simple S.b. for an l.c. TVS (X,T). Then $\{x_n;f_n\}$ is a c.ab-S.b. for $(X,\sigma(X,X^*))$ if and only if it is shrinking.

PROOF. Let $\{x_n;f_n\}$ be shrinking but not a $\sigma(X,X^*)$-c.ab-S.b. Hence there exists a $\sigma(X,X^*)$- normal bl.s. $\{y_n'\}$ of $\{x_n\}$. Thus, for some α in ω, the sequence $\{y_n\}$ is $\sigma(X,X^*)$-regular, where $y_n = \alpha_n y_n'$. Hence there exist g in X^* and $\varepsilon > 0$ such that $|g(y_n)| > \varepsilon$ for infinitely many n. Thus

$\sup \{|<y_n - S_N(y_n), g>| : n \geq 1\} \leq \varepsilon$ for each $N \geq 1$; in particular, $S_N^*(g) - g \notin \frac{\varepsilon}{2} \{y_n\}^o$, $N \geq 1$. This gives, however, a contradiction.

Conversely, let $\{x_n; f_n\}$ be not shrinking. Hence following a known lemma on function spaces (cf. [140], p. 71), we can find g in X*, a bounded subset A of (X,T), an $\varepsilon > 0$ and $\{n_i\}$ in I such that

$$\sup_{x \in A} |<x, (S^*_{n_i} - S^*_{n_{i-1}})(g)>| > \frac{\varepsilon}{2} \; ; \quad \forall i \geq 1,$$

where $n_o = 0$. It is possible to select $\{y_i\}$ in A such that $|<S_{n_i}(y_i) - S_{n_{i-1}}(y_i), g>| > \varepsilon/2$ for $i \geq 1$. Put

$$u_i = S_{n_i}(y_i) - S_{n_{i-1}}(y_i) = \sum_{j=n_{i-1}+1}^{n_i} f_j(y_i) x_j.$$

Then $\{u_n\}$ is a $\sigma(X,X^*)$-regular bl.s. Further, if $f \in X^*$, then $|f(u_i)| \leq 2 \sup \{|S_n^*(f)(x)| : x \in A, n \geq 1\} < \infty$. Therefore, $\{x_n; f_n\}$ is not a c.ab-S.b. □

<u>THEOREM 10.5.6</u> An S.b. $\{x_n; f_n\}$ for a Fréchet space (X,T) is c.n-S.b. if and only if (X,T) is a Banach space.

<u>PROOF</u>. In view of the remarks made after Definition 10.5.3, it is sufficient to prove the 'only if' part of this theorem. On the other hand, let (X,T) not be a Banach space. We may, therefore, assume that $D_T = \{p_1 < p_2 < \ldots < p_n < \ldots\}$, where each p_n is a norm (cf. Proposition 10.2.14), $p_n(x) = \sup \{p_n(S_m(x)) : m \geq 1\}$ (use the barrelled character of (X,T) or see [130], p. 27); and if $m < n$, then

$$(*) \quad \sup_{x \in X} \frac{p_n(x)}{p_m(x)} = \infty.$$

Define $Y_n = sp\{x_1, \ldots, x_n\} = [x_i : 1 \leq i \leq n]$ and $Z_n = [x_i : i \geq n+1]$. Since all p_n's are equivalent on Y_n,

$\sup\{p_i(x)/p_j(x): x \in Y_n\} < \infty$, for any pair (i,j) of indices. Next, we claim that

$$(+) \qquad \sup_{x\in Z_n} \frac{p_i(x)}{p_j(x)} = \infty; \ \forall\ n \geqq 1,$$

provided $j < i$. In fact, if (+) were not true, then for some $n \geqq 1$ and $i > j$, $p(x) \leqq 3Kp_j(x)$ for all x in X, contradicting (*), the constant K being given by

$$K = \max\left\{\sup_{x\in Y_n} \frac{p_i(x)}{p_j(x)}\ ,\quad \sup_{x\in Z_n} \frac{p_i(x)}{p_j(x)}\right\}.$$

Thus, writing $Z_o = X$, for $i > j$, one has

$$(**) \qquad \sup_{x\in Z_m} \frac{p_i(x)}{p_j(x)} = \infty; \ m = 0,1,\ldots\ .$$

Let G denote the bijective map from $\mathbb{N}$ onto $I = \{(i,j,k) : i,j,k \in \mathbb{N},\ i > j\}$ and write $G(n) = (i_n, j_n, k_n)$, $n \geqq 1$.

Proceeding inductively and using (**), we can select (i_n, j_n, k_n) with $i_n > j_n$, $\{y_n\}$ and $\{m_{n-1}\}$, $\{k_n\}$ in I such that

$$\text{(i)} \quad y_n \in Z_{m_{n-1}}\ ; \quad \text{(ii)} \quad p_{j_n}(y_n) = 1,$$

$$\text{(iii)} \ p_{i_n}(y_n) > k_n; \ \text{(iv)} \quad p_{i_n}(y_n - S_{m_n}(y_n)) < \tfrac{1}{2},$$

where $n = 1,2,\ldots$ and $m_o = 0$ (in the first stage, let $m_o = 0$, choose $n = 1$ to get (i_1, j_1, k_1) with $i_1 > j_1$. Use (**) to obtain a nonzero y_1 in Z_o-this is possible since p_i's are norms - satisfying (ii) and (iii) and finally choose $m_1 > m_o$ to ascertain (iv)).

Next, put $u_n = S_{m_n}(y_n)$, $n \geqq 1$. Since $f_i(y_n) = 0$ for $1 \leqq i \leqq m_{n-1}$,

$$u_n = \sum_{i=m_{n-1}+1}^{m_n} f_i(y_n)x_i, \ n \geq 1.$$

Also, by (iii) and (iv), $p_{i_n}(u_n) > 0$ and this shows that $\{u_n\}$ is a bl.s. of $\{x_n\}$. By the hypothesis, we can find α in ω so that for each $i \geq 1$, $\sup p_i(\alpha_n u_n) < \infty$ and for some j, $\inf p_j(\alpha_n u_n) > 0$. Thus

$$\sup_{n\geq 1} \frac{p_i(\alpha_n u_n)}{p_j(\alpha_n u_n)} < \infty \quad , \ \forall i > j.$$

On the other hand, let us first observe that $p_{i_n}(u_n)/p_{j_n}(u_n) > k_n - 1/2$ and it easily follows for the above choice of j and i, $i > j$ that

$$\sup_{n\geq 1} \frac{p_i(u_n)}{p_j(u_n)} = \infty.$$

This contradicts the preceding inequality and hence (X,T) is normed. □

COROLLARY 10.5.7 A non-normable Fréchet space, in particular, any infinite dimensional nuclear Fréchet space cannot have a c.n-S.b.

REMARK. As mentioned earlier, each S.b. of a normed space is a c.n-S.b.; on the other hand, no S.b. of a non-normable Fréchet space (e.g. $(\omega,\sigma(\omega,\phi))$) can be a c.n-S.b. Concerning non-metrizable spaces, the S.b. $\{e^n;e^n\}$ for $(\phi,\sigma(\phi,\omega))$ is not a c.n-S.b. Besides, the following theorem [98] suggests that one can construct examples of non-metrizable spaces having c.n-S.b., for example the S.b. $\{e^n;e^n\}$ of $(\ell^1,\tau(\ell^1,c_0))$.

THEOREM 10.5.8 Let $\{x_n;f_n\}$ be an a.s.S.b. for a Banach space $(X, \|\cdot\|)$. Then $\{f_n;\psi x_n\}$ is an S.b. for $(X^*,\beta(X^*,X))$

if and only if it is a c.n-S.b. for $(X^*,\tau(X^*,X))$.

PROOF. Let $\{f_n;\Psi x_n\}$ be an S.b. for $(X^*,\beta(X^*,X))$ We again write $\|\cdot\|$ for the norm on X^*, giving the topology $\beta(X^*,X)$. Consider any bl.s. $\{g_n\}$ of $\{f_n\}$, say,

$$g_n = \sum_{i=m_{n-1}+1}^{m_n} \alpha_i f_i, \; 0 = m_o < m_1 < \ldots .$$

We may choose β in ω so that $\|\beta_n g_n\| = 1$, $n \geq 1$. Thus $\{\beta_n g_n\}$ is $\tau(X^*,X)$-bounded.

Put $X_n^* = sp\{f_{m_{n-1}+1},\ldots,f_{m_n}\}$. One can define a continuous linear functional ϕ_n on the Banach space X_n^* so that $\phi_n(\beta_n g_n) = 1 = \|\phi_n\|$. Let us write $c_i = \phi_n(f_i)$, $1+m_{n-1} \leq i \leq m_n$. Then, for f in X_n^*

$$\phi_n(f) = \sum_{i=m_{n-1}+1}^{m_n} c_i f(x_i), \; f = \sum_{i=m_{n-1}+1}^{m_n} f(x_i) f_i .$$

Define θ_n on X^* by

$$\theta_n(f) = \sum_{i=m_{n-1}+1}^{m_n} c_i f(x_i)$$

Suppose that

$$y_n = \sum_{i=m_{n-1}+1}^{m_n} c_i x_i$$

It is clear that $y_n \neq 0$ for all $n \geq 1$ by the successive choice of nonzero ϕ_n. Since θ_n is continuous on X^* and $\theta_n(f) = f(y_n)$ for f in X^*, $\|\theta_n\| = \|y_n\|$.

Next, let $K = \sup \|S_n\| < \infty$. Then we have $\sup \|S_n^*\| = K$. For f in X^*

$$|\theta_n(f)| = |\phi_n(\sum_{i=m_{n-1}+1}^{m_n} f(x_i)f_i)| \leq \|\langle S^*_{m_n} - S^*_{m_{n-1}})(f)\| .$$

and so $\|y_n\| \leq 2K$ for all $n \geq 1$. Therefore, $\{y_n\}$ is a bounded bl.s. of $\{x_n\}$. Hence $y_n \to 0$ in $\sigma(X,X^*)$ (Proposition 9.2.3 (i)) and so $\{y_n\}$ is $\tau(X^*,X)$-equicontinuous. Consequently, the set $V = \{f \in X^*: |f(y_n)| \leq 1/2, n \geq 1\}$ is a zero neighbourhood of $(X^*,\tau(X^*,X))$ and it follows that $\beta_n g_n \notin V$, $n \geq 1$. Therefore $\{g_n\}$ is normal in $(X^*,\tau(X^*,X))$.

Conversely, let $\{x_n;f_n\}$ be not shrinking. Then there exist $\{n_i\}$ in I , $\varepsilon > 0$ and g in X^* such that if

$$g_i = \sum_{j=n_{i-1}+1}^{n_i} g(x_j)f_j, \ n_0 = 0$$

then $\|g_i\| > \varepsilon$ for all $i \geq 1$ (see the proof of Theorem 10.5.5.) By the hypothesis, there exist α in ω and v in $\mathcal{B}_\tau$, $\tau \equiv \tau(X^*,X)$ so that $\{\alpha_n g_n\}$ is bounded in $(X^*,\tau(X^*,X))$ and $\alpha_n g_n \notin v$ for all $n \geq 1$. By the Banach-Steinhaus theorem, $\sup \|\alpha_n g_n\| < \infty$ and so $K \equiv \sup |\alpha_n| < \infty$ (use the fact that $\|g_n\| > \varepsilon$). Hence $g_n \notin K^{-1}v$ for all $n \geq 1$. However, $g_n \to 0$ in $(X^*,\tau(X^*,X))$ and hence a contradiction. □

REMARK. It appears that the norm structure of a Banach space plays a key role in the preceding theorem, especially in the first part. In fact, this theorem is, in general, not true for Fréchet space, for recall the familiar

EXAMPLE 10.5.9 The S.b. $\{e^n;e^n\}$ for the Fréchet space $(\omega,\sigma(\omega,\phi))$ is clearly an a.s.S.b. Since $\beta(\phi,\omega) = \tau(\phi,\omega)$, $\{e^n;e^n\}$ is shrinking. Using the fact that $\eta(\phi,\omega) = \tau(\phi,\omega)$, we conclude that $\{e^n;e^n\}$ is not a c.n-S.b. for $(\phi,\tau(\phi,\omega))$.

Semi-Montel spaces

We continue further with the applications of completely abnormal bases in characterizing semi-Montel spaces. Before we quote the main result, let us give the following (cf. [96])

LEMMA 10.5.10 Let $\{x_n;f_n\}$ be a simple S.b. for an l.c. TVS (X,T) for which $\sigma(X,X^*)$ and T-convergent sequences are not the same. Then there exists a normalized bl.s. $\{u_n\}$ of $\{x_n\}$.

PROOF. We may choose $\{y_n\}$ in X and p in D_T such that $y_n \to 0$ in $\sigma(X, X^*)$ and $p(y_n) \geq 1$, $n \geq 1$. Then $S_m(y_n) \to 0$, for each $m \geq 1$ and so we may find $\{m_j\}$ and $\{n_j\}$ in I_∞ with $m_1 = 1$ such that $p(y_{m_j} - S_{n_j}(y_{m_j}))$, $p(S_{n_j}(y_{m_{j+1}})) < 1/3$ for $j \geq 1$. Put $u_j = S_{n_j}(y_{m_j}) - S_{n_{j-1}}(y_{m_j})$, where $j \geq 1$ and $n_0 = 0$. It is readily verified that $p(u_j) \geq 1-2/3$. Since $\{y_m\}$ is bounded and $\{x_n;f_n\}$ is simple, $|f(u_j)| \leq 2 \sup \{|S_n^*(f)(y_m)| : m, n \geq 1\} < \infty$, for each f in X* and $j \geq 1$. □

Now we have (cf. [96])

THEOREM 10.5.11 An ω-complete l.c. TVS (X,T) with an S.b. $\{x_n;f_n\}$ is semi-Montel if and only if $\{x_n;f_n\}$ is a c.ab-S.b.

PROOF. First, let (X,T) be semi-Montel. Suppose that $\{y_n\}$ is an arbitrary bl.s. of $\{x_n\}$ and let it be normal. In particular, for some α in ω, $\alpha_n \neq 0$, the sequence $\{\alpha_n y_n\}$ is bounded and regular and so it has a cluster point y in $[y_n]$. Since $\{y_n\}$ is an S.b. for $[y_n]$ (cf. [130], p. 139), we may let $\{g_n\}$ to be the s.a.c.f. corresponding to $\{y_n\}$. For each $\varepsilon > 0$ and g_n, we can find $m > n$ so that $|g_n(y-\alpha_m y_m)| \leq \varepsilon$. Thus $g_n(y) = 0$ for $n \geq 1$, giving thereby

$y = 0$. Hence $\{\alpha_n y_n\}$ cannot be regular.

Conversely, let $\{x_n; f_n\}$ be a c.ab-S.b. Since (X,T) is a W-space, $\{x_n; f_n\}$ is simple also. Hence by Lemma 10.5.10, $\sigma(X,X^*)$ and T- convergent sequences in X are the same. Therefore from Proposition 1.2.15, Theorem 3.3.5 and a known result on the characterization of compact subsets in s.s. (cf. [129], p. 60-61), we easily conclude that $\sigma(X,X^*)$ and $\tau(X,X^*)$-compact subsets of X are the same.

By Exercise 10.5.4 and Theorem 10.5.5, $\{x_n; f_n\}$ is shrinking. Also, $\{x_n; f_n\}$ is γ-complete; indeed, choose α in ω such that $\{\sum_{i=1}^{n} \alpha_i x_i\}$ is bounded but not convergent. Hence for some $\varepsilon > 0$, p in D_T and $\{n_k\}$ in I_∞, $p(y_j) \geqq \varepsilon$, $j \geqq 1$, where

$$y_j = \sum_{i=n_{j-1}+1}^{n_j} \alpha_i x_i, \quad n_o = 0.$$

This shows the existence of a normal bl.s. of $\{x_n\}$, contrary to the hypothesis. Therefore (X,T) is semireflexive by Theorem 3.4.2. Using the result of the preceding paragraph it follows that (X,T) is semi-Montel. □

The above result may be applied to give another characterization of semi-Montel spaces having ∞-absolute bases. Indeed, one has (cf. [99])

<u>THEOREM 10.5.12</u> Let an ω-complete l.c. TVS (X,T) contain an ∞-absolute base $\{x_n; f_n\}$. Then the following statements are equivalent:

(i) If $\{\alpha_n x_n\}$ is bounded for α in ω, then $\alpha \in \delta$.

(ii) (X,T) is semi-Montel.

<u>PROOF</u>. (iii) ==> (ii). Obvious.

(ii) ==> (i). Let $\{\alpha_n x_n\}$ be bounded for α in ω.

By the ∞-absolute base character of $\{x_n; f_n\}$, $\{\alpha_n f(x_n)\} \in c_o^\beta = \ell^1$ for every f in X*. Thus $\{\sum_{i=1}^{n} \alpha_i x_i\}$ is bounded in (X,T) and so (i) follows by Theorem 9.4.2.

(i) ==> (iii). By virtue of Theorem 10.5.11, it suffices to prove that no bl.s. $\{y_n\}$ of $\{x_n\}$ is normal. On the contrary, assume that there exists a bl.s. $\{y'_n\}$ which is normal; that is, for some α in ω, $\{\alpha_n y'_n\}$ is regular and bounded. If $y_n = \alpha_n y'_n$, we may write

$$y_n = \sum_{i=m_{n-1}+1}^{m_n} a_i x_i, \; m_o = 0.$$

Since $\{x_n; f_n\}$ is simple and $\{y_n\}$ is bounded, $\{a_n x_n\}$ is weakly bounded. By (i), $y_n \to 0$ in (X,T), contradicting the regularity of $\{y_n\}$. □

11 Advances in normalized bases

11.1 INTRODUCTORY REMARKS

This chapter is in continuation of the study carried out in chapter 10 regarding normalized bases (nz-S.b.) and their applications. In fact, we discuss here further advances in normalized bases by retaining one of the underlying properties of these bases and replacing the other by a different one so that the resulting bases ultimately turn out to be of types P, P*, semishrinking and semi γ-complete. P and P* bases in Banach spaces were introduced in [211] whereas the other two types of bases in such spaces were studied in [175]. Later on, the study of P and P* bases in an arbitrary l.c. TVS was carried out in [46], [93], [104] and [184], while the other two concepts of bases were further investigated in [93] in the general setting of locally convex spaces. Our purpose in this chapter is to focus our attention on these different types of bases and to attempt to isolate their impact on the structure of the underlying locally convex spaces.

11.2 BIORTHOGONAL SEQUENCES OF TYPES P AND P*

It was pointed out in [93] that the definitions of P and P* bases could be extended to biorthogonal sequences to cover up a more meaningful situation and as such we pass on to

DEFINITION 11.2.1 Let (X,T) be an l.c. TVS and $\{x_n;f_n\}$ a biorthogonal sequence (b.s.) for the dual system $<X,X^*>$. Then $\{x_n;f_n\}$ is said to be (i) *of type* P if $\{x_n\}$ is

T-regular and $\{\sum_{i=1}^{n} x_i\}$ is T-bounded; and (ii) *of type* P* if $\{x_n\}$ is T-bounded and $\{\sum_{i=1}^{n} f_i\}$ $\beta(X^*,X)$-bounded.

We will also need

DEFINITION 11.2.2 Let $\{x_n;f_n\}$ be a b.s. for $<X,X^*>$, $X \equiv (X,T)$ being an l.c. TVS. For $\{n_i\}$ in I, $\{x_{n_i}\}$ is called a *type* P (resp. P*) *subsequence* of $\{x_n\}$, provided $\{x_{n_i};f_{n_i}\}$ is of type P (resp. P*) b.s. for $<X,X^*>$.

In this chapter, we take up a study of the possible duality relationship between biorthogonal sequences (b.s.) of types P and P* and then pass on to investigate the impact of types P and P* subsequences on the construction of new bases from the old ones. Finally, we touch upon a few applications depending upon these subsequences.

Lastly we recall a few abbreviations and notation which we will use without further reference. Throughout we write $\{x_n;f_n\}$ to mean a b.s. for $<X,X^*>$, $X \equiv (X,T)$ being an arbitrary l.c. TVS. Further, let $\beta \equiv \beta(X^*,X)$ and $H = [f_n] \equiv \overline{sp\{f_n\}}^{\beta}$, it being understood that H is endowed with its natural l.c. topology $\beta_H \equiv \beta|H$. Also, suppose $H^* = (H,\beta_H)^*$ and Ψ is the usual canonical mapping from X into $X^{**} \equiv (X^*,\beta)^*$, $\Psi x(f) = <x,f>$ with x in X and f in X^*.

Duality relationship

As before, we write $X \equiv (X,T)$ for an l.c. TVS. A b.s. $\{x_n;f_n\}$ for $<X,X^*>$ will interchangeably be written as a b.s. for (X,T); and since $\{f_n;\Psi x_n\}$ is also a b.s. for $<H,H^*>$, the same will be written as a b.s. for (H,β_H). In this section, we essentially confine ourselves to finding the duality relationship between P and P* biorthogonal sequences and these results generalize similar results on Schauder bases proved in [93] and [211]. We first need a result from locally convex spaces in the form of

LEMMA 11.2.3 Corresponding to an l.c. TVS (X,T), let Y be a subspace of X^* equipped with the l.c. topology $\beta_Y = \beta(X^*,X)|Y$ and suppose $Y^* = (Y,\beta_Y)^*$. Then the natural embedding $J : X \to Y^*$, $Jx(y) = \langle x,y\rangle$ is $\beta(X,X^*) - \beta(Y^*,Y)$ continuous.

PROOF. The proof of this result is implicit in Proposition 9.6.5 and the result follows by observing the simple fact that each $\sigma(Y,Y^*)$-bounded subset of Y is $\sigma(X^*,X)$-bounded. □

PROPOSITION 11.2.4 If $\{x_n;f_n\}$ is of type P^* for (X,T), then $\{f_n;\Psi x_n\}$ is of type P for (H,β_H). Now let, (X,T) be a W-space and $\{x_n;f_n\}$ a q.r.b.s for (X,T), then the following results are true: (i) if $\{f_n;\Psi x_n\}$ is of type P for (H,β_H), then $\{x_n;f_n\}$ is of type P^* and (ii) if $\{x_n;f_n\}$ is of type P for (X,T), then $\{f_n;\Psi x_n\}$ is of type P^* for (H,β_H).

PROOF. The first part is immediate from the definitions involved and Proposition 10.3.1.

(i) By the hypothesis, there exists a bounded subset A of (X,T) and a sequence $\{y_n\} \subset A$ so that $|f_n(y_n)| > 1$, $n \geq 1$. By the W-character of (X,T) and the quasi-regularity of $\{x_n;f_n\}$, for each f in X^*, the inequality (10.3.5) is clearly satisfied. Let $f \in X^*$. Since $|f(x_n)| < |f(f_n(y_n)x_n)|$ for $n \geq 1$; using (10.3.5), we find that $\{x_n\}$ is T-bounded. This completes the proof of (i).

(ii) Here the regularity of $\{x_n\}$ forces $\{f_n\}$ to be bounded in (H,β_H) (cf. Corollary 10.2.2). The $\beta(H^*,H)$-boundedness of $\{\sum_{i=1}^{n} \Psi x_i\}$ follows immediately from Lemma 11.2.3, since $\{\sum_{i=1}^{n} x_i\}$ is $\beta(X,X^*)$-bounded. □

The converse of Proposition 11.2.4 (ii) is contained in

PROPOSITION 11.2.5 Let (X,T) be σ-infrabarrelled and a W-space and let $\{x_n;f_n\}$ be a q.r.b.s. for (X,T). If $\{f_n;\Psi x_n\}$ is of type p* for (H,β_H), then $\{x_n;f_n\}$ is of type P for (X,T).

PROOF. The T-regularity of $\{x_n\}$ is a simple consequence of the $\beta(X^*,X)$-boundedness of $A \equiv \{f_n\}$.

To prove the T-boundedness of $\{\sum_{i=1}^{n} x_i\}$, let us first of all observe that $\{S_n^*\}$ is β-β equicontinuous (cf. Proposition 1.2.4). Hence for any $\sigma(X^*,X)$-bounded subset A of X*, the set

$$S^*[A] = \bigcup_{n\geq 1} S_n^*[A]$$

is β-bounded in X*. Note that $S^*[A] \subset H$ and we claim that $S^*[A]$ is $\sigma(H,H^*)$-bounded. Indeed, if $g \in H^*$, there exists $\hat{g}$ in X** so that $\hat{g}(h) = g(h)$ for all h in H. Also $\sigma(X^*,X^{**})$ and β are compatible, and so $g(S^*[A]) < \infty$, justifying the required claim. Therefore, from the hypothesis, for some $\lambda > 0$

$$\sum_{i=1}^{n} \Psi x_i \in \lambda(S^*[A])^{\bullet};\ \forall\, n \geq 1,$$

where $\bullet$ denotes the polar in H* relative to the dual system $\langle H,H^*\rangle$. Thus for any f in A,

$$|\langle \sum_{j=1}^{m} f(x_j)f_j,\ \sum_{i=1}^{n} \Psi x_i\rangle| \leq \lambda \quad ;\ \forall\, m,n \geq 1$$

$$\Longrightarrow |f(\sum_{j=1}^{m} f_j(\sum_{i=1}^{n} x_i)x_j)| \leq \lambda \quad ;\ \forall\, m,n \geq 1.$$

Consequently $\sum_{j=1}^{m} x_j \in \lambda A^\circ$ for all $m \geq 1$, $\circ$ being the polar in X with respect to $\langle X,X^*\rangle$. This proves the $\beta(X,X^*)$-

boundedness of $\{\sum_{i=1}^{n} x_i\}$ in X and so $\{x_n;f_n\}$ is of type P for (X,T). □

NOTE. The foregoing results envelop the following theorem of Singer [211] who, as remarked earlier, initiated the study of P and P* bases.

THEOREM 11.2.6 Let $\{x_n;f_n\}$ be an S.b. for a Banach space X. Then (i) $\{x_n;f_n\}$ is an S.b. of type P if and only if $\{f_n;\Psi x_n\}$ is an S.b. of type P* for (H,β_H); and (ii) $\{x_n;f_n\}$ is an S.b. of type P* if and only if $\{f_n;\Psi x_n\}$ is an S.b. of type P for (H,β_H).

REMARK. Most of the preceding results on the duality relationship between b.s. of types P and P* are proved under the hypothesis of W-ness of (X,T); however, in the last result, namely Proposition 11.2.5, an additional restriction of σ-infrabarrelledness on (X,T) is made. Indeed, this restriction is indispensable as shown in

EXAMPLE 11.2.7 Consider the b.s. $\{e^n;e^n\}$ for $(c_o,\sigma(c_o,\ell^1))$. Here $X = c_o$, $T = \sigma(c_o,\ell^1)$, $X^* = H = \ell^1$, $H^* = (\ell^1,\beta(\ell^1,c_o))^* = \ell^\infty$ (as $\beta(\ell^1,c_o)$ is the usual norm topology of ℓ^1) and $\Psi e^n = e^n$, for each $n \geqq 1$. Clearly, $\{e^n\}$ is $\beta(\ell^1,c_o)$-bounded and $\{\sum_{i=1}^{n} e^i\}$ is $\beta(\ell^\infty,\ell^1)$-bounded. Therefore $\{e^n,\Psi e^n\}$ is of type P* for $(\ell^1,\beta(\ell^1,c_o))$. However, $\{e^n;e^n\}$ is not of type P (cf. Proposition 2.3.5(x)). Finally, the space in question is not σ-infrabarrelled; indeed, $\{e^n: n \geqq 1\}$ is $\beta(\ell^1,c_o)$-bounded but it is not $\sigma(c_o,\ell^1)$-equicontinuous (cf. also [129], p. 120).

A characterization

The duality relationship between bases of types P and P* helps in characterizing one type of a base in terms of another. At the outset, we have the following characterization of bases of types P (cf. [104]).

<u>THEOREM 11.2.8</u> Let $\{x_n;f_n\}$ be an S.b. for an l.c. TVS (X,T) and consider the following statements.

(a) $\{x_n;f_n\}$ is of type P for (X,T).

(b) $\{x_n;f_n\}$ is normalized and the sequence $\{y_n\}$ is an S.b. for (X,T), where $y_n = \sum_{i=1}^{n} x_i$.

(c) The sequence $\{x_n\}$ is T-regular and for each p in D_T, there exists a constant $M \equiv M(p) > 0$ such that for every decreasing sequence α in c_o, the series $\sum_{n\geqq 1} \alpha_n x_n$ converges in (X,T) and

$$p(\sum_{n\geqq 1} \alpha_n x_n) \leqq M|\alpha_1|. \tag{11.2.9}$$

Then (i) (a) ==> (b) and if (X,T) is ω-complete, (b) ==> (a), and (ii) (c) ==> (a) and if (X,T) is ω-complete, (a) ==> (c).

<u>PROOF</u>. (i) Here (a) ==> (b) is a consequence of Proposition 10.2.13, whereas the implication (b) ==> (a) results from Proposition 10.3.13.

(ii) Putting $\alpha = e^{(n)}$, $n \geqq 1$ successively in (11.2.9), we easily obtain (a) from (c).

(a) ==> (c) Let $\alpha_1 \geqq \alpha_2 \geqq \cdots \geqq \alpha_n \geqq \cdots$, $\alpha_n \to 0$ as $n \to \infty$. Observe that for $n > m$,

$$\sum_{i=m}^{n} \alpha_i x_i = \sum_{i=m}^{n-1} (\alpha_i - \alpha_{i+1}) y_i + \alpha_n y_n,$$

where $y_n = \sum_{i=1}^{n} x_i$. If $M = \sup p(y_n)$, we find that

$$\sum_{i=m}^{n} p((\alpha_i - \alpha_{i+1}) y_i) \leqq M(|\alpha_m| + |\alpha_{n+1}|)$$

and so $\{\sum_{i=1}^{n} \alpha_i x_i\}$, is Cauchy in (X,T) and hence converges. The last two relations with $m = 1$ again yield (11.2.9). □

REMARK. In the above theorem, the hypothesis that $\{y_n\}$ is an S.b. is essnetial for the truth of (b) ==> (a), for consider

EXAMPLE 11.2.10 The space $(\ell^1, \|\cdot\|_1)$ has an nz - S.b. $\{e^n;e^n\}$, but clearly $\{\sum_{i=1}^{n} e^i\}$ is unbounded and so $\{e^n;e^n\}$ is not an S.b. of type P. On the other hand, $\{e^{(n)}\}$ is not an S.b. for $(\ell^1, \|\cdot\|_1)$, for if S_n denotes the partial sum corresponding to $\{e^{(n)}\}$, then for α in ℓ^1,

$$(*) \qquad \|\alpha - S_n(\alpha)\|_1 = \sum_{i \geq n+2} |\alpha_i| + (n+1)\,|\alpha_{n+1}|$$

must tend to zero as $n \to \infty$ However, for $\alpha_i = n^{-2}$, $i = n^4$, $n \geq 1$ and zero otherwise, we easily see that the extreme right term for $\|\alpha - S_{n^4-1}(\alpha)\|_1$ in (*) is n^2 and so $\|\alpha - S_n(\alpha)\|_1 \not\to 0$; (cf. also Proposition 10.3.13 for alternative proof).

EXERCISE 11.2.11 Using results of this section, prove Theorem 11.2.6.

11.3 SUBSEQUENCES OF TYPES P AND P*

In the course of study of bases of types P and P*, besides Theorem 11.2.6, Singer [211] also observed the following

THEOREM 11.3.1 A Banach space has an S.b. of type P if and only if it has an S.b. of type P*.

The preceding result was first extended in [184]; however, the next two theorems together envelop these results proved in [184] and [211] (cf. [93]).

THEOREM 11.3.2 Let $\{x_n;f_n\}$ be a semisimple base for an l.c. TVS (X,T) such that for $\{n_j\}$ in I, $\{x_{n_j}\}$ is a type P subsequence of $\{x_n\}$. If

$$y_k = \begin{cases} x_k; k \neq n_j, \ j \geq 1 \\ \sum_{i=1}^{j} x_{n_i}; k = n_j, \ j \geq 1 \end{cases}$$

$$g_k = \begin{cases} f_k; \ k \neq n_j, \ j \geq 1, \\ f_{n_j} - f_{n_{j+1}}; \ k = n_j, \ j \geq 1. \end{cases}$$

Then $\{y_n;g_n\}$ is an S.b. for (X,T) such that $\{y_{n_j}\}$ is a type P* subsequence of $\{y_n\}$.

PROOF. $\{y_n;g_n\}$ is clearly a b.s. for (X,T) or $\langle X,X^*\rangle$. For x in X and $n_j \leq m < n_{j+1}$,

$$\sum_{k=1}^{m} f_k(x)x_k - \sum_{k=1}^{m} g_k(x)y_k = \sum_{i=1}^{j} f_{n_i}(x)x_{n_i} - \sum_{i=1}^{j} g_{n_i}(x)y_{n_i}$$

$$(*) \qquad = f_{n_{j+1}}(x)y_{n_j}.$$

Since for some $p_o, p_o(x_{n_{j+1}}) \geq 1$, $f_{n_{j+1}}(x) \to 0$ as $j \to \infty$. Also $\{y_{n_j}\}$ is bounded; thus $\{y_n;g_n\}$ is an S.b. for (X,T).

Since $\{x_n;f_n\}$ is semisimple (Definition 10.3.4), the set $\{f_{n_j}(x)x_{n_j} : x \in A, \ j \geq 1\}$ is T-bounded for any given bounded subset A of (X,T). Hence using the fact that $p_0(x_{n_j}) \geq 1$, $j \geq 1$, we easily conclude that $\{f_{n_j}\}$ is $\beta(X^*,X)$-bounded and so is $\{\sum_{i=1}^{j} g_{n_i}\}$, since $\sum_{i=1}^{j} g_{n_i} = f_{n_1} - f_{n_{j+1}}$. Clearly, $\{y_{n_j}\}$ is T-bounded and so it is of type P*. □

The preceding proof yields

<u>COROLLARY 11.3.3</u> If a W-space (X,T) has an S.b. of type P, then (X,T) has an S.b. of type P*.

<u>THEOREM 11.3.4</u> Let $\{x_n;f_n\}$ be an S.b. for an l.c. TVS (X,T) and $\{n_j\}$ a member of I. (i) If $\{x_{n_j}\}$ is a bounded subsequence of $\{x_n\}$ and

$$y_k = \begin{cases} x_k; \ k \neq n_j, \ j \geq 1 \\ \\ x_{n_j} - x_{n_{j-1}}; \ k = n_j, \ j \geq 1, \ x_{n_0} = 0. \end{cases}$$

$$g_k = \begin{cases} f_k, \ k \neq n_j, \ j \geq 1, \\ \\ \sum_{i \geq j} f_{n_i}, \ k = n_j, \ j \geq 1, \end{cases}$$

where it is assumed that $\sum_{j \geq 1} f_{n_j}$ is convergent in $(X^*,\sigma(X^*,X))$; then $\{y_n;g_n\}$ is an S.b. for (X,T). (ii) If $\{x_{n_j}\}$ is a type P* subsequence of $\{x_n\}$ and $\{\sum_{i=1}^{k} f_{n_i}\}$ is equicontinuous on (X,T), then $\{y_n;g_n\}$ is an S.b. for (X,T) such that $\{y_{n_j}\}$ is a type P subsequence of $\{y_n\}$.

<u>PROOF</u>. (i) It is evident that $\{y_n;g_n\}$ is a b.s. for (X,T). For x in X and $n_j \leq m < n_{j+1}$,

$$\sum_{i=1}^{m} g_i(x)y_i - \sum_{i=1}^{m} f_i(x)x_i = \sum_{i=1}^{j} g_{n_i}(x)y_{n_i} - \sum_{i=1}^{j} f_{n_i}(x)x_{n_i}$$

$$= g_{n_{j+1}}(x)x_{n_j}.$$

Since $g_{n_{j+1}}(x) \to 0$ and $\{x_{n_j}\}$ is bounded, $\{y_n;g_n\}$ is an S.b. for (X,T).

(ii) By the Alaoglu-Bourbaki theorem, the sequence $\{\sum_{i=1}^{k} f_{n_i}\}$ has $\sigma(X^*,X)$-cluster point f in X*. Clearly, $f(x_{n_i}) = 1$, $i \geq 1$ and $f(x_j) = 0$ for $j \neq n_i$, $i \geq 1$ and so $f(x) = \sum_{i\geq 1} f_{n_i}(x)$, for each x in X. Thus $\{y_n;g_n\}$ is an S.b. for (X,T) and

$$g_{n_j}(x) = f(x) - \sum_{i=1}^{j-1} f_{n_i}(x), \quad \forall x \in X,\ j \geq 1.$$

This yields the equicontinuity of $\{g_{n_i}\}$ and hence $\{y_{n_i}\}$ is regular (cf. Proposition 10.2.8). Since $\{\sum_{i=1}^{j} y_{n_i}\}$ is clearly bounded, we have completed the proof. □

The foregoing proof clearly yields

<u>COROLLARY 11.3.5</u> If a barrelled space has an S.b. of type P*, then it also has an S.b. of type P.

The method of proofs of the last two theorems is helpful in deriving (cf. [93])

<u>THEOREM 11.3.6</u> Let an ω-complete space (X,T) possess an S.b. $\{x_n;f_n\}$.

(i) If each S. basic sequence in (X,T) is γ-complete, then (X,T) is semireflexive.

(ii) If each S. basic sequence in (X,T) is shrinking, then (X,T) is semireflexive.

PROOF. (i) Suppose the required result is not true. By Theorems 9.4.2 and 10.5.5, there exists a bl. s. $\{z_n\}$ of $\{x_n\}$ with

$$z_n = \sum_{i=m_{n-1}+1}^{m_n} \alpha_i x_i \qquad (m_o = 0)$$

such that $\{z_n\}$ is $\sigma(X,X^*)$-normal. Hence there exist f in X^* and β in $\omega(\beta_n \neq 0,\ n \geq 1)$ so that $\{\beta_n z_n\}$ is bounded in (X,T) and $|f(\beta_n z_n)| \geq 1$ for $n \geq 1$. Choose $\{\varepsilon_n\}$ so that $f(\varepsilon_n \beta_n z_n) = 1$, $n \geq 1$. Clearly $0 < |\varepsilon_n| \leq 1$, $n \geq 1$. Put $\gamma_i = \varepsilon_n \beta_n \alpha_i$, $1+m_{n-1} \leq i \leq m_n$, $n \geq 1$ and

$$(*) \qquad y_n = \sum_{i=m_{n-1}+1}^{m_n} \gamma_i x_i .$$

Then $\{y_n\}$ is a bounded bl.s. of $\{x_n\}$ with $f(y_n) = 1$, $n \geq 1$. Further, $\{y_n\}$ is S.basic (e.g. [130], p. 139) and let $\{g_n\}$ be the s.a.c.f. corresponding to $\{y_n\}$. For y in $Y \equiv [y_n]$,

$$f(y) = \sum_{n\geq 1} g_n(y) f(y_n) = \sum_{n\geq 1} g_n(y).$$

If $u_n = y_n - y_{n-1}\,(y_o = 0)$, it follows by Theorem 11.3.4 (i) and the hypothesis that $\{u_n\}$ is a γ-complete S.b. for $(Y,T|Y)$. Since $\{\sum_{i=1}^{n} u_i\}$ is bounded, the series $\sum_{n\geq 1} u_n$ converges in $(Y,T|Y)$ and since $f_i(\sum_{n\geq 1} u_n) = 0$ for each $i \geq 1$, it follows that $\sum_{n\geq 1} u_n = 0$. Consequently $f(\sum_{n\geq 1} u_n) = 0$. But this contradicts the fact that $f(\sum_{n\geq 1} u_n) = 1$. The contradiction arrived at disposes of (i).

(ii) If the required result is not true, $\{x_n;f_n\}$ would not be γ-complete by Theorem 9.4.2. Hence for some α in ω, the sequence $\{\sum_{i=1}^{n} \alpha_i x_i\}$ is bounded without being convergent in (X,T). There now exist p in D_T and $\{m_n\}$ in I with $m_o = 0$ so that if

$$y_n = \sum_{i=m_{n-1}+1}^{m_n} \alpha_i x_i,$$

then $\{y_n\} \equiv \{y_n;g_n\}$ is an S.b. for $Y \equiv [y_n]$ with $p(y_n) \geq 1$, $n \geq 1$, $\{g_n\}$ being the s.a.c.f. corresponding to $\{y_n\}$. Since $(Y,T|Y)$ is ω-complete, it is a W-space and therefore $\{y_n;g_n\}$ is a semisimple S.b. of type P for $(Y,T|Y)$. Following the proof of Theorem 11.3.2, we find the existence of an S.b. $\{u_n;h_n\}$ of type P* for $(Y,T|Y)$ such that $\sum_{n\geq 1} h_n$ converges, say, to h in $(Y^*,\sigma(Y^*,Y))$; indeed, $u_n = \sum_{i=1}^{n} y_i$ and $h_n = g_n - g_{n+1}$. But $\{u_n;h_n\}$ is shrinking also and so

$$h = \sum_{n\geq 1} \langle u_n,h\rangle h_n, \qquad [\text{in } \beta(Y^*,Y)].$$

From above, $\langle u_n,h\rangle = 1$ for $n \geq 1$ and it follows that $h_n \to 0$ in $\beta(Y^*,Y)$. But $\{u_n\}$ is $\sigma(Y,Y^*)$-bounded and so $h_n \in \{2u_m\}^o$ eventually in n. This is, however, not true. □

Relationship with γ-complete and shrinking bases

It is evident from the proof of Theorem 11.3.6 that S.basic sequences of types P and P* have an important role to play in deriving the reflexivity of the space having an S.b. On the other hand, looking at conclusions (i) and (ii) of the foregoing theorem and the fact that semireflexivity is characterized in terms of γ-complete and shrinking bases, one is obviously tempted to ask if there exist any relationships between γ-complete, shrinking and P, P* bases. The next four results answer this question.

Following [184], we have

PROPOSITION 11.3.7 An S.b. $\{x_n;f_n\}$ of type P for an l.c. TVS (X,T) is not a γ-complete S.b.

PROOF. If the base is γ-complete, then $\sum_{n\geq 1} x_n$ converges and so $\{x_n\}$ is not regular. □

PROPOSITION 11.3.8 An equicontinuous S.b. $\{x_n;f_n\}$ of type P* for an l.c. TVS (X,T) cannot be shrinking.

PROOF. Let $\{x_n;f_n\}$ be shrinking. Using Theorem 10.5.5 and the fact that $\{x_n\}$ is weakly bounded, we conclude that $\{x_n\}$ is not weakly regular. However, by the Alaoglu-Bourbaki theorem the sequence $\{f_n\}$ has a $\sigma(X^*,X)$-cluster point such that $f(x_n) = 1$ for $n \geq 1$, showing the weak regularity of $\{x_n\}$. □

PROPOSITION 11.3.9 Let an ω-complete space (X,T) contain an S.b. $\{x_n;f_n\}$ which is not γ-complete. Then (X,T) possesses an S.basic sequence of type P.

PROOF. Proceed as in the proof of Theorem 11.3.6 (ii) and observe that the sequence $\{y_n\}$ constructed therein does not belong to some neighourhood of zero in (X,T). □

PROPOSITION 11.3.10 Let an ω-complete l.c. TVS (X,T) contain an S.b. $\{x_n;f_n\}$ which is not shrinking. Then (X,T) possesses an S.basic sequence of type P*.

PROOF. Following the proof of Theorem 11.3.6 (i), we find the existence of an f in X* and a bounded S.basic sequence $\{y_n\} \equiv \{y_n;g_n\}$ with $f(y_n) = 1$, $n \geq 1$. For y in $Y \equiv [y_n]$, $\sum_{n\geq 1} g_n(y) = f(y)$ and as $(Y,T|Y)$ is also a W-space, we find that $\{\sum_{i=1}^{n} g_i\}$ is $\beta(Y^*,Y)$-bounded. □

EXERCISE 11.3.11 Let an l.c. TVS (X,T) contain an S.b. $\{x_n;f_n\}$ of type P such that $\mu = \delta^\times$. Prove that $(X,\sigma(X,X^*))$ is not ω-complete.

[Hint: Use Propositions 3.3.6 and 2.3.5 (v) (cf. [46]).]

11.4 SEMISHRINKING AND SEMI γ-COMPLETE BASES

Following [93], let us introduce the following two types of bases closely related to normalized bases in the form of

DEFINITION 11.4.1 An S.b. $\{x_n;f_n\}$ for an l.c. TVS (X,T) is called (i) *semishrinking* (*s-shrinking*) if $\{x_n\}$ is T-regular and $x_n \to 0$ in $\sigma(X,X^*)$; and (ii) *semi γ-complete* (*s. γ-complete*) if $\{x_n\}$ is T-bounded and whenever $\{\sum_{i=1}^{n} \alpha_i x_i\}$ is T-bounded for α in ω, then $\alpha \in c_0$.

EXERCISE 11.4.2 Prove that (i) an s-shrinking base is normalized, (ii) a normalized shrinking base is s-shrinking and (iii) a normalized γ-complete base is γ-complete. [Hint: (ii) $f(x_n)f_n \to 0$ in $\beta(X^*,X)$ for each f in X^* and note that $\{x_m\}$ is bounded; (iii) use Lemma 10.1.1.]

EXERCISE 11.4.3 Construct an example to show that an s. γ-complete base is not necessarily normalized.

The converses of Exercises 11.4.2 (ii) and (iii) are not true in general. For, we have ([33], [185]; cf. also [175])

EXAMPLE 11.4.4 Consider the s.s. d equipped with the norm $\|\cdot\|_d$, where

$$d = \{\alpha \in \omega: \alpha_i \in \mathbb{R},\ \|\alpha\|_d \equiv \sup_{\sigma\in P} \sum_{i\geq 1} \frac{|\alpha_{\sigma(i)}|}{i} < \infty\}.$$

The s.s. d is a special case of the more general Lorentz (sequence) space discussed in length in [129], p. 323 and 334. However, for the sake of completeness, let us hurriedly touch upon a few properties of the s.s. $(d, \|\cdot\|_d)$. In fact, it is a normal s.s.s. which is also an AK-BK space.

The normality and symmetric character of d readily follow and by considering a sequence $\{\sigma_n\}$ from P with $\sigma_n(1) = n$, $n \geq 1$ and proceeding in the usual manner, one can easily show that $(d, \|\cdot\|_d)$ is a Banach space.

Further, $d \subset c_o$; for otherwise, we can find α in d, $\varepsilon > 0$ and $\{n_i\}$ in I with $|\alpha_{n_i}| > \varepsilon$, for each $i \geq 1$ and this yields $\|\alpha\|_d = \infty$, by choosing ρ_k in P with $\rho_k(i) = n_i$ $(1 \leq i \leq k)$ for each $k \geq 1$.

Recalling the definition of the reduced form $\hat{\alpha}$ of an arbitrary α in c_o, we find (cf. [129], p. 94) that

$$\|\alpha\|_d = \sum_{n\geq 1} \frac{\hat{\alpha}_n}{n}, \quad \alpha \in d.$$

Next, we prove that $(d, \|\cdot\|_d)$ is an AK-space and this will establish the S.b. character of $\{e^n;e^n\}$ for this space. So, let $\alpha \in d$ and suppose that

$$\widehat{\alpha - \alpha^{(n)}} = \{\hat{x}_{i,n}\}.$$

Then $\{\hat{\alpha}_i : i \geq 1\} = \{\hat{x}_{i,n}\} \cup \{|\alpha_i| : 1 \leq i \leq n, |\alpha_i| \neq 0\}$. Define $m_n = \max\{i : \{\hat{\alpha}_1,\ldots,\hat{\alpha}_i\} \subset \{|\alpha_1|,\ldots,|\alpha_n|\}\}$. Then $0 \leq m_n \leq n$ and $m_n \to \infty$ with n. Since $\{\hat{\alpha}_1,\ldots,\hat{\alpha}_{m_n}\} \subset \{|\alpha_1|,\ldots,|\alpha_n|\}$, $\{\hat{x}_{i,n}\} \subset \{\hat{\alpha}_{m_n+1},\hat{\alpha}_{m_n+2},\ldots\}$. Hence $\hat{x}_{i,n} \leq \hat{\alpha}_{m_n+i}$, $i \geq 1$. Therefore

$$\|\alpha-\alpha^{(n)}\|_d = \sum_{i\geq 1} \frac{\hat{x}_{i,n}}{i} \leq \sum_{i\geq 1} \frac{\hat{\alpha}_{m_n+i}}{i}$$

$$\leq N\,\hat{\alpha}_{m_n} + \sum_{i\geq N+1} \frac{\hat{\alpha}_i}{i},$$

for any N in $\mathbb{N}$. By the above inequality and the fact that $\hat{\alpha}_{m_n} \to 0$ as $n \to \infty$ together with the convergence of $\sum_{i \geq 1} (\hat{\alpha}_i / i)$, we easily conclude that $\|\alpha^{(n)} - \alpha\|_d \to 0$ as $n \to \infty$.

It is obvious that $\|e^n\|_d = 1$, $n \geq 1$ and $\langle e^n, f\rangle \to 0$ as $n \to \infty$ for each f in d*. Indeed, if this were not true, then for some f in $d^* = d^\times$, $\varepsilon > 0$ and $\{n_k\}$ in I, we have $|f_{n_k}| = \langle e^{n_k}, f\rangle| > \varepsilon$, $k \geq 1$. Since $\{1/n\} \in d$, the sequence α also belongs to d, where

$$\alpha_i = \begin{cases} k^{-1}; & i = n_k, \ k \geq 1 \\ 0 \ , & \text{otherwise.} \end{cases}$$

But then, $\sum_{i \geq 1} |f_i \alpha_i|$ diverges, contradicting $f \in d^*$ and thus $\{e^n; e^n\}$ is s-shrinking.

To prove that $\{e^n; e^n\}$ is not shrinking, assume that $\{e^n; e^n\}$ is shrinking. Put $h_n = \sum_{i=1}^{n} (1/i)$ and $\beta^n = e^{(n)}/h_n$. Then $\|\beta^n\|_d = 1$, $n \geq 1$ and for f in d* with $f_n = 1/n$, i.e. $f(\alpha) = \sum_{i \geq 1} (\alpha_i / i)$ for α in d, it follows that $\langle \beta^n, f\rangle = 1$, $n \geq 1$. Thus $\{\beta^n\}$ is a bounded sequence in $(d, \|\cdot\|_d)$, which does not tend to zero in $\sigma(d, d^*)$. On the other hand, $\langle \beta^n, e^i\rangle = h_n^{-1}$ for each $i \geq 1$ and $n \geq 1$ and so for each $i \geq 1$, $\langle \beta^n, e^1\rangle \to 0$ as $n \to \infty$. Hence we arrive at a contradiction by virtue of Proposition 9.2.3.

Concerning the negation of the converse of Exercise 11.4.2 (iii), we have

<u>EXAMPLE 11.4.5</u> This is the space $(W, \|\cdot\|)$, where

$$W = \{\alpha \in \omega : \alpha_i\text{'s reals}, \ A_k(\alpha) \equiv \sup \left| \sum_{i=1}^{k+1} \alpha_{j_i} \right| \to 0 \text{ as } k \to \infty\},$$

the supremum being taken over k+1 integers j_i's in $\mathbb{N}$ with $k = j_1 < j_2 < \ldots < j_{k+1}$, and $\| \alpha \| = \sup \{A_k(\alpha) : 1 \leq k < \infty\}$. Clearly $(W, \| \cdot \|)$ is a Banach space with $\ell^1 \subset W$. For $\alpha \in W$, observe that

$$|\alpha_1| \leq A_1(\alpha) + \frac{1}{3} A_2(\alpha); \quad |\alpha_2| \leq A_2(\alpha) + \frac{1}{2} A_3(\alpha); \text{ and}$$

$$|\alpha_i| \leq 3 A_i(\alpha) + A_{i+1}(\alpha) + A_{2i+1}(\alpha), \quad \forall\, i \geq 3.$$

These inequalities immediately yield $W \subset c_0$ and $(W, \| \cdot \|)$ is a K-space.

For proving the AK-ness of $(W, \| \cdot \|)$, consider $\alpha \in W$ and write

$$\beta^n = \alpha - \alpha^{(n)}, \quad \forall\, n \geq 1.$$

For $\varepsilon > 0$, choose $N \equiv N(\varepsilon, \alpha) \geq 3$ in $\mathbb{N}$ such that $A_i(\alpha) < \varepsilon/5$, for each $i \geq N$. If we fix $n \geq 2N$ arbitrarily and write $k_n = [n/2]$, then we find

$$A_1(\beta^n) < \varepsilon; \; A_i(\beta^n) < \frac{7}{5} \varepsilon, \; 2 \leq i \leq 2\, k_n;$$

$$A_i(\beta^n) = A_i(\alpha) < \frac{\varepsilon}{5}, \; \forall\, i \geq 2\, k_n + 1.$$

Thus $\| \beta^n \| < (7/5)\varepsilon$, for each $n \geq 2N$, and this shows that $\{e^n; e^n\}$ is an S.b. for $(W, \| \cdot \|)$. One can easily verify that $\{e^n; e^n\}$ is s. γ-complete and $\| e^n \| = 1$, for each $n \geq 1$. However, it is not γ-complete; for instance consider the sequence $\{ \sum_{i=1}^{n} \alpha_i e^i \}$ in W, where $\alpha_i = 1/i$, $i \geq 1$. Then $\| \sum_{i=1}^{n} \alpha_i e^i \| \leq 2$, for each $n \geq 1$; but $\alpha \notin W$ (indeed, $A_n(\alpha) \geq 1/2$, for each $n \geq 1$).

Duality relationship

We have seen that there are certain types of bases which

enjoy the duality relationship among themselves (cf. this chapter and chapter 9). Such results have been found useful in characterizing reflexivity of the underlying spaces. Our aim in this subsection is similar and we prove a few results reflecting, in partiuclar, the duality relationship between bases which are s-shrinking and s-γ-complete.

First of all, let us prove

LEMMA 11.4.6 Let $\{x_n;f_n\}$ be a b.s. for $\langle X,X^*\rangle$, $X \equiv (X,T)$ being an l.c. TVS.

(i) If $x_n \to 0$ in $\sigma(X,X^*)$, then the equicontinuity of $\{\sum_{i=1}^{n} \alpha_i f_i\}$ for α in ω yields that α is in c_o.

(ii) If $\{x_n;f_n\}$ is simple and $\beta(X^*,X)$-boundedness of $\{\sum_{i=1}^{n} \alpha_i f_i\}$ implies that $\alpha \in c_o$, then $x_n \to 0$ in $\sigma(X,X^*)$.

PROOF. By the hypothesis, there exists a balanced, convex, $\sigma(X^*,X)$-closed and equicontinuous subset $M \equiv M(\alpha)$ of X^* such that

$$\left|\left(\sum_{i=1}^{n} \alpha_i f_i\right)(x)\right| \leq \sup_{f\in M} |f(x)| \quad ; \forall\, x \in X,\ n \geq 1.$$

Let $\{\beta_1,\ldots,\beta_n\}$ be any finite set of scalars. Then

$$\left|\sum_{i=1}^{n} \alpha_i\beta_i\right| = \left|\left(\sum_{i=1}^{n} \alpha_i f_i\right)\left(\sum_{j=1}^{n} \beta_j x_j\right)\right|$$

$$\leq \sup_{f\in M} \left|\sum_{i=1}^{n} \beta_i f(x_i)\right|$$

Hence by Helly's criterion ([140], H(c), p. 152), there exists f in X^* with $f(x_i) = \alpha_i$, $i \geq 1$ and so $\alpha \in c_o$.

(ii) Since for each f in X^*, $\{\sum_{i=1}^{n} f(x_i)f_i\}$ is $\beta(X^*,X)$-bounded, the result follows. □

COROLLARY 11.4.7 Let $\{x_n;f_n\}$ be a T-regular simple S.b. for a σ-infrabarrelled space (X,T). Then $\{x_n;f_n\}$ is s-shrinking if and only if the $\beta(X^*,X)$-boundedness of $\{\sum_{i=1}^{n} \alpha_i f_i\}$ implies that $\alpha \in c_o$.

Indeed, the 'only if' part follows from Lemma 11.4.6 (i) and the converse is trivial.

EXERCISE 11.4.8 Let $\{x_n;f_n\}$ be a q.r.b.s. for $\langle X,X^*\rangle$, $X \equiv (X,T)$ being a barrelled space such that $[x_n]^T = X$ and $\{x_n\}$ is T-regular. Prove that $\{x_n;f_n\}$ is an s-shrinking S.b. for (X,T) if and only if the $\beta(X^*,X)$-boundedness of $\{\sum_{i=1}^{n} \alpha_i f_i\}$ implies that $\alpha \in c_o$.

PROPOSITION 11.4.9 Let $\{x_n;f_n\}$ be a T-regular simple S.b. for a σ-infrabarrelled space (X,T). Then $\{x_n;f_n\}$ is s-shrinking if and only if $\{f_n;Jx_n\}$ is s-γ-complete S.b. for (H,β_H), where H, β_H and J are defined in Section 9.1.

PROOF. At the outset, let us observe by virtue of Propsoition 9.6.5 that $\{f_n;Jx_n\}$ is an S.b. for (H,β_H). Let $\{x_n;f_n\}$ be s-shrinking. The regularity of $\{x_n\}$ forces $\{f_n\}$ to be equicontinuous (Corollary 10.2.3) and hence it is β_H-bounded. Now, let $\{\sum_{i=1}^{n} \alpha_i f_i\}$ be β_H-bounded and hence by the σ-infrabarrelled character of (X,T), $\{\sum_{i=1}^{n} \alpha_i f_i\}$ is equicontinuous. Therefore by Lemma 11.4.6 (i), $\alpha \in c_o$.

Conversely, let $\{f_n;Jx_n\}$ be s-γ-complete. If $\{\sum_{i=1}^{n} \alpha_i f_i\}$ is $\beta(X^*,X)$-bounded, then it is β_H-bounded and so $\alpha \in c_o$. Now apply Corollary 11.4.7 to conclude the s-shrinking character of $\{x_n;f_n\}$. □

NOTE. The 'only' if part of the foregoing result is given in [93], Proposition 4.3 (ii).

PROPOSITION 11.4.10 Let $\{x_n;f_n\}$ be a simple S.b. for an infrabarrelled space (X,T) such that (H,β_H) is σ-infrabarrelled. If $\{f_n;Jx_n\}$ is s-shrinking, then $\{x_n;f_n\}$ is s-γ-complete.

PROOF. Since $\{f_n\}$ is $\beta(X^*,X)$-regular, $\{x_n\}$ is T-bounded (cf. Proposition 10.3.6). Let $\{\sum_{i=1}^{n} \alpha_i x_i\}$ be T-bounded and so it is $\beta^*(X,X^*)$-bounded (use the infrabarrelled property of (X,T)). By Proposition 9.6.5, $\{\sum_{i=1}^{n} \alpha_i Jx_i\}$ is $\beta(H^*,H)$-bounded. Further, $\{f_n;Jx_n\}$ is simple for (H,β_H). Indeed, let $F \in H^*$. Since for f in X^*, $\{\sum_{i=1}^{n} f(x_i)f_i\}$ is $\beta(X^*,X)$-bounded, we have

$$\sup_{n\geqq 1} \left| \sum_{i=1}^{n} f(x_i)F(f_i) \right| < \infty$$

$$\Longrightarrow \sup_{n\geqq 1} \left| f\left(\sum_{i=1}^{n} F(f_i)x_i \right) \right| < \infty .$$

This shows that $\{\sum_{i=1}^{n} F(f_i)x_i|$ is T-bounded and therefore $\beta^*(X,X^*)$-bounded. Hence using Proposition 9.6.5 again, we conclude that $\{f_n;Jx_n\}$ is a simple S.b. for (H,β_H). The required result now follows by an application of Corollary 11.4.7. □

Finally we have (cf. [93])

PROPOSITION 11.4.11 Let $\{x_n;f_n\}$ be a simple S.b. for an l.c. TVS (X,T). If $\{x_n;f_n\}$ is s-γ-complete, then $\{f_n;Jx_n\}$ is an s-shrinking S.b. for (H,β_H).

PROOF. Since $f_m \notin (1/2)\{x_n\}^\circ$, $m \geq 1$, $\{f_m\}$ is β_H-regular. Take any F in H^*. Then as in the proof of the preceding proposition, $\{\sum_{i=1}^{n} F(f_i)x_i\}$ is T-bounded and so $F(f_n) \to 0$. □

Block perturbations

The ultimate aim of this subsection is to derive conditions under which s-shrinking and s-γ-complete bases turn out to be shrinking and γ-complete respectively. In the ensuing process, we come across the concept of block perturbations of an S.b. initiated in [174], which ensures a new S.b. from the old one by perturbing a bl.s. of the given S.b. in a certain manner.

In what follows in this subsection, we assume throughout without further reference that $\{x_n\} \equiv \{x_n;f_n\}$ is a regular simple S.b. for an l.c. TVS (X,T) and $\{y_n\}$ is a bounded bl.s. of $\{x_n\}$ corresponding to the sequences $0 = m_0 < m_1 < \ldots < m_{n-1} < m_n < \ldots$ and $\{\alpha_n\}$ (cf. chapter 2). Then we have

DEFINITION 11.4.12 A *block-perturbation* (bl. ptb.) of $\{x_n\}$ (with respect to a bl.s. $\{y_n\}$) is a sequence $\{u_n\} \subset X$ of the form

$$u_n = x_n + \sum_{i\geqq 1} \delta_{np_i} y_i,$$

where $\{p_n\}$ is in I with $m_{i-1} + 1 \leqq p_i \leqq m_i$, and $\alpha_{p_i} = 0$, $i \geqq 1$.

NOTE. Thus, if $\{u_n\}$ is a bl.ptb. of $\{x_n\}$, then

$$u_i = \begin{cases} x_i : i \neq p_j \\ x_{p_j} + y_j ; i = p_j \end{cases}, \quad j = 1,2,\ldots .$$

We follow [93] for the rest of this subsection and begin with

PROPOSITION 11.4.13 Every bl. ptb. $\{u_n\}$ is a simple S.b. for (X,T).

PROOF. For each $i \geq 1$, there exists j with $m_{j-1} + 1 \leq i \leq m_j$ and now define g_i in X^* by $g_i = f_i - \alpha_i f_{p_j}$. Then $\{u_i;g_i\}$ is a b.s. for $\langle X,X^*\rangle$.

Given L in $\mathbb{N}$, there exists n in $\mathbb{N}$ so that for any x in X,

$$\sum_{i=1}^{L} g_i(x)u_i = \sum_{i=1}^{m_{n-1}} f_i(x)x_i + \sum_{i=m_{n-1}+1}^{L} g_i(x)u_i .$$

It then follows that

$$\sum_{i=1}^{L} g_i(x)u_i = \begin{cases} \sum_{i=1}^{L} f_i(x)x_i - f_{p_n}(x) \sum_{i=m_{n-1}+1}^{L} \alpha_i x_i ; & m_{n-1}+1 \leq L \leq p_n - 1 \\ \sum_{i=1}^{L} f_i(x)x_i + f_{p_n}(x) \sum_{i=L+1}^{m_n} \alpha_i x_i ; & p_n \leq L \leq m_n . \end{cases}$$

Hence

$$\sum_{i=1}^{L} g_i(x)u_i - \sum_{i=1}^{L} f_i(x)x_i = f_{p_n}(x)(\beta_L\, y_n - S_L(y_n)),$$

where $\beta_L = 0$ if $m_{n-1}+1 \leq L \leq p_n - 1$ and $\beta_L = 1$ if $p_n \leq L \leq m_n$. Since $\{y_n\}$ is bounded and $f(S_L(y_n)) = \langle y_n, S_L^*(f)\rangle$ for f in X^*, the sequence $\{\beta_L y_n - S_L(y_n) : L \geq 1\}$ is bounded by the simple character of $\{x_n;f_n\}$. Also $f_{p_n}(x) \to 0$ as $L \to \infty$.

Therefore $\{u_i;g_i\}$ is an S.b. for (X,T). Also, for f in X^*,

$$(*) \qquad \sum_{i=1}^{L} f(u_i)g_i - \sum_{i=1}^{L} f(x_i)f_i = (\beta_L f(y_n) - \langle y_n, S_L^*(f)\rangle) f_{p_n} .$$

By Exercise 10.3.11 (ii), $\{f_{p_n}\}$ is $\beta(X^*,X)$-bounded and therefore using (*), we conclude the simple character of $\{u_n;g_n\}$. □

EXERCISE 11.4.14 If $\{x_n;f_n\}$ is also equicontinuous, then prove that every bl. ptb. $\{u_n\}$ is an equicontinuous S.b. [Hint: Conclude the boundedness of $\{\alpha_n\}$ and use the definition of $\{g_n\}$].

Now we have one of the main results of this subsection.

PROPOSITION 11.4.15 An s-shrinking, regular simple S.b. $\{x_n\}$ of which every bl. ptb. $\{u_n\}$ is s-shrinking, is shrinking.

PROOF. Let $\{x_n;f_n\}$ be not shrinking. Using Theorem 10.5.5 and invoking the proof and notation of Theorem 11.3.6(i), there exist a bounded bl. s. $\{y_n\}$ (cf. (*) of the proof of Theorem 11.3.6) and f in X* so that $f(y_n) = 1$, $n \geqq 1$.

Put $n_j = m_{2j}; p_j = m_{2j-1}$, $j \geqq 1$ and $n_o = 0$. Let

$$z_j = \sum_{i=n_{j-1}+1}^{n_j} \alpha_i x_i,$$

where $\alpha_i = 0$, $n_{j-1} + 1 \leqq i \leqq m_{2j-1}$ and $\alpha_i = \gamma_i$, $m_{2j-1}+1 \leqq i \leqq n_j = m_{2j}$. Then $z_j = y_{2j}$. Let

$$u_i = \begin{cases} x_i, & i \neq p_j \\ & ; \ j \geqq 1 \\ x_{p_i}+z_j, & i = p_j \end{cases}$$

By assumption, $x_i \to 0$ and $u_i \to 0$ in $\sigma(X,X^*)$. Hence $z_j \to 0$ in $\sigma(X,X^*)$ and this contradicts the fact that $f(z_j) = 1$, $j \geqq 1$. □

Next, we pass on to the other main result of this subsection, namely,

PROPOSITION 11.4.16 Let (X,T) be an infrabarrelled space containing a normalized (nz.) simple S.b. $\{x_n;f_n\}$

and suppose that every nz. simple S.b. of (X,T) is s-γ-complete. Then $\{f_n;Jx_n\}$ is a shrinking S.b. for (H,β_H), and if (X,T) is also ω-complete, then $\{x_n;f_n\}$ is γ-complete.

PROOF. Here $\{f_n;Jx_n\}$ is a simple S.b. for (H,β_H) (see Proposition 9.6.5 and the proof of Proposition 11.4.10). By Proposition 11.4.11, $\{f_n;Jx_n\}$ is s-shrinking and so $\{f_n\}$ is β_H-regular.

Let $\{h_k\}$ be a $\beta(X^*,X)$-bounded bl.s. corresponding to $\{f_n\}$,

$$h_k = \sum_{i=m_{k-1}+1}^{m_k} \alpha_i f_i, \quad m_o = 0 < m_1 < \ldots < m_k < \ldots$$

with $m_{k-1}+1 \leqq p_k \leqq m_k$ and $\alpha_{p_k} = 0$, for $k \geqq 1$. Denote by $\{g_n\}$ a bl. ptb. of $\{f_n\}$ with respect to $\{h_n\}$ and let $\{u_n\}$ be the corresponding b.s. in H^* (cf. the proof of Proposition 11.4.13). Then $\{g_n;u_n\}$ is a simple S.b. for (H,β_H) by Proposition 11.4.13. Hence, again by Proposition 9.6.5 $\{u_n\}$ is an S.b. for $Z = [u_n]^\beta$, $\beta \equiv \beta(H^*,H)$. Since $sp\{u_n\} = sp\{Jx_n\}$ and $\beta^*(X,X^*) \approx T$, $\{y_n\} \equiv \{J^{-1}u_n\}$ is an S.b. for (X,T) by Proposition 9.6.5 (second part) such that for each x in (X,T), $x = \sum_{n\geq 1} g_n(x)y_n$. By Exercise 10.3.11 (ii), $\{f_n\}$ is equicontinuous. By construction, $\{g_n - f_n\}$ is $\beta(X^*,X)$-bounded, thus yielding the equicontinuity of $\{g_n\}$. Consequently, $\{y_n\}$ is T-regular (cf. Proposition 10.2.8). The boundedness of $\{x_n\}$ enforces β_H-equicontinuity of $\{Jx_n\}$ on H; therefore by Exercise 11.4.14, $\{u_n\}$ is β_H-equicontinuous. Thus $\{y_n\}$ is an nz. S.b.

In order to prove the simple character of $\{y_n;g_n\}$, we proceed as in Proposition 11.4.13 to conclude that for each f in X^*,

$$\sum_{i=1}^{L} f(y_i)g_i - \sum_{i=1}^{L} f(x_i)f_i = f(x_{p_n})(\beta_L h_n - S_L^*(h_n)),$$

where β_L stands as before. Since $\{S_L^*\}$ is $\beta(X^*,X)$-$\beta(X^*,X)$ equicontinuous (Proposition 1.2.4) and $\{h_n\}$ is $\beta(X^*,X)$-bounded, $\{\beta_L h_n - S_L(h_n) : L \geq 1\}$ is $\beta(X^*,X)$-bounded; also $\{f(x_{p_n})\}$ is bounded. Therefore $\{y_n;g_n\}$ is an nz. simple S.b.

Hence by the hypothesis, $\{y_n;g_n\}$ is s-γ-complete. Consequently, by Proposition 11.4.10, $\{g_n;u_n\}$ is s-shrinking for (H,β_H). Now make use of Proposition 11.4.15 to infer the shrinking character of $\{f_n;Jx_n\}$. The final part is a consequence of Proposition 9.6.9. □

11.5 REFLEXIVITY

The results proved in this chapter, as well as those proved in previous ones, finally result in a very strong characterization of reflexive spaces containing normalized Schauder bases in the form of (cf. [93])

THEOREM 11.5.1 Let a complete barrelled space (X,T) contain an nz. S.b. Then the following statements are equivalent:

(i) (X,T) is reflexive.

(ii) Every nz. S.b. for (X,T) is shrinking.

(iii) Every nz. S.b. for (X,T) is γ-complete.

(iv) Every nz. S.b. for (X,T) is s-shrinking.

(v) Every nz. S.b. for (X,T) is s-γ-complete.

The proof requires an intermediary

PROPOSITION 11.5.2 Let $\{x_n;f_n\}$ be an equicontinuous bounded and simple S.b. for an l.c. TVS(X,T) such that $\{x_n;f_n\}$ is not s-γ-complete. Then (X,T) has an nz. simple S.b. $\{y_n;g_n\}$ which is not s-shrinking such that $\{g_n\}$ may be extended by a single element g_0 to an S.b. $\{g_n;\ n \geq 1,\ g_0\}$ for (H,β_H).

PROOF. Hereafter, we call an equicontinuous bounded S.b. an *equi-normalized* S.b. (e-nz.S.b.).

By the hypothesis, there exists α in ω such that $\alpha \notin c_o$ and $\{\sum_{i=1}^{n} \alpha_i x_i\}$ is bounded. There exist $\varepsilon > 0$ and $\{p_j\}$ in I_∞ with $|\alpha_{p_j}| > \varepsilon$, $j \geq 1$ and such that for each $j \geq 1$ and some i in $p_{j-1}+1 \leq i \leq p_j-1$, $\alpha_i \neq 0$, where $p_o = 0$. Let $\gamma_{p_n} = \alpha_{p_n}$ $(n \geq 1)$ and $\gamma_n = 1$, otherwise. Also, let $m_n = p_n$ and

$$u_n = \sum_{i=m_{n-1}+1}^{m_n} \beta_i(\gamma_i x_i),$$

where $\beta_{p_n} = 0$ and $\beta_i = \alpha_i$, $m_{n-1}+1 \leq i \leq m_n-1$. Then $\{u_n\}$ is a bounded bl.s. of the S.b. $\{\gamma_n x_n\}$. If $z_n = x_n$ $(n \neq p_j)$ and $z_{p_j} = \alpha_{p_j} x_{p_j} + u_j$, then $\{z_n\}$ is a bl.ptb. of the S.b. $\{\gamma_n x_n\}$. Therefore by Proposition 11.4.3, and Exercise 11.4.14, $\{z_n;h_n\}$ is a simple e-nz. S.b. for (X,T), $\{h_n\}$ being the s.a.c.f. corresponding to $\{z_n\}$.

Observe that

$$\sum_{i=1}^{n} z_{p_i} = \sum_{i=1}^{p_n} \alpha_i x_i$$

and so $\{z_{p_n}\}$ is a type P subsequence of $\{z_n\}$. So, if

$$y_{p_n} = \sum_{i=1}^{n} z_{p_i} \text{ and } y_i = z_i;\ i \neq p_n,$$

then $\{y_n\}$ is an e-nz. S.b. for (X,T) by Theorem 11.3.2 such that $\{y_{p_n}\}$ is a type P* subsequence of $\{y_n\}$. Here the equicontinuity of the s.a.c.f. $\{g_n\}$ corresponding to $\{y_n\}$ is a consequence of its construction by the theorem referred to above. Since $h_{p_1}(y_{p_n}) = 1$ for all $n \geq 1$, $\{y_n;g_n\}$ is not s-shrinking.

Following (*) of the proof of Theorem 11.3.2 and the fact that $\{h_{p_{k+1}}\}$ is equicontinuous and $\{y_{p_k}\}$ is bounded, we easily conclude that $\{y_n;g_n\}$ is a simple S.b. and hence (cf. Proposition 9.6.5) $\{g_n;Jy_n\}$ is an S.b. for $G = [g_n]^{\beta}$, $\beta \equiv \beta(X^*,X)$. Let $g_o = h_{p_1}$. Then $sp\{g_i : i \geq 0\} = sp\{h_i : i \geq 1\} = sp\{f_i : i \geq 1\}$. If $h_{p_1} \in G$, then, as $h_{p_1}(y_{p_n}) = 1$, $n \geq 1$, we have

$$(*) \qquad h_{p_1} = \sum_{i \geq 1} g_{p_i} \text{ in } \beta(X^*,X).$$

But $\{g_n\}$ is $\beta(X^*,X)$-regular (Proposition 10.3.1) and so (*) is not true. Hence $g_o \notin G$ and so $H = sp\{g_o\} \oplus G$. □

<u>PROOF OF THEOREM 11.5.1</u> (iii) <==> (v) by Exercise 11.4.2 (iii) and Proposition 11.4.16. Using Exercise 10.3.11 (ii), Proposition 11.4.13 and Exercise 11.4.14, every bl. ptb. of a given nz. s-shrinking S.b. is an nz. S.b. and so the same is s-shrinking. Thus by Proposition 11.4.15 and Exercise 11.4.2(ii), (ii) <==> (iv). By Proposition 11.5.2 (iv) ==> (iii) and so (ii) ==> (iii) (cf. Exercise 11.4.2 (ii)). Therefore by Theorem 9.4.2 (i) <==> (ii). In order to complete proof, we have to show that (iii) ==> (iv).

(iii) ==> (iv). Let $\{x_n;f_n\}$ be an arbitrary nz. S.b. for (X,T). Then $\{x_n;f_n\}$ is γ-complete and so $J[X] = H^*$ (see the proof of Proposition 9.6.6). This shows that (H,β_H) is barrelled.

In order to conclude (iv), it suffices to show, in view of Proposition 11.4.9, that $\{f_n;Jx_n\}$ is s-γ-complete for (H,β_H). Observe that $\{f_n;Jx_n\}$ is already an S.b. for (H,β_H) by Propsoition 9.6.5; and as (H,β_H) is barrelled, it is also an e.-nz. simple S.b. (cf. also Exercise 10.3.11(ii)).

Assume that $\{f_n; Jx_n\}$ is not s-γ-complete. Then by Proposition 11.5.2, (H, β_H) has an nz. simple S.b. $\{g_n; u_n\}$ which is not s-shrinking, and has a type P* subsequence $\{g_{p_n}\}$; also $\{u_n\}$ may be extended by a single element u_o so that $\{u_n : n \geqq 0\}$ is an S.b. for $(H, \beta(H^*, H))$. Let $y_n = J^{-1}u_n$, $n = 0, 1, 2, \ldots$. Then $\{y_n : n \geqq 0\}$ is an S.b. for (X, T) (cf. Propositions 9.6.5 and 9.6.9). By Theorem 10.4.2, $\{u_n : n \geqq 1\}$, indeed, its extension $\{u_n : n \geqq 0\}$, is an nz. S.b. for $(H^*, \beta(H^*, H))$. Thus $\{y_n : n \geqq 0\}$ is an nz. S.b. for (X, T) and hence it is s-γ-complete (cf. Exercise 11.4.2(iii)). However the boundedness of $\{\sum_{i=1}^{n} u_{p_i}\}$ in $(H^*, \beta(H^*, H))$ yields that $\{\sum_{i=1}^{n} y_{p_i}\}$ is bounded and this contradicts the s-γ-complete character of $\{y_n : n \geqq 0\}$. □

12 e–Schauder bases

12.1 INTRODUCTION AND RELATED CONCEPTS

Spaces having e-Schauder bases (abbreviated hereafter as e-S.b.) enjoy a number of interesting properties, some of which have already been mentioned in earlier chapters and others are going to be discussed in this chapter. Often, we make use of a notion weaker than that of an e-S.b., namely, that of an uniformly equicontinuous S.b. (u.e-S.b.), to derive certain structural properties of the space containing this base.

It is clear that each S.b. of a barrelled space is e-S.b., although the converse is not true in general (cf. chapter 2). On the other hand, there are other types of bases which apparently look different from e-S.b., but are essentially the same as the latter ones. Before we proceed further, let us first dispose of them here. In fact, following [202], p. 34, we have

PROPOSITION 12.1.1 Every monotone S.b. (m-S.b.) in an l.c. TVS is an e-S.b. Conversely, an l.c. TVS (X,T) having an e-S.b. $\{x_n;f_n\}$ can be given an l.c. topology $\bar{T}$ such that $T \approx \bar{T}$ and $\{x_n;f_n\}$ is an m-S.b. for $(X,\bar{T})$.

PROOF. The first part easily follows; in fact, if $x \in X$ and $p \in D_T$, then for $n, k \geq 1$,

$$p\left(\sum_{i=1}^{n} f_i(x)x_i\right) \leq p\left(\sum_{i=1}^{n+k} f_i(x)x_i\right) \to p(x) \text{ as } k \to \infty.$$

For the other part, for x in X and p in D_T, let

$$\bar{p}(x) = \sup_{n \geq 1} p\left(\sum_{i=1}^{n} f_i(x)x_i\right).$$

Then for each p in D_T, there exists q in D_T so that $\bar{p}(x) \leq q(x)$ for all x in X. If $\bar{T}$ is the l.c. topology on X generated by $\bar{D} = \{\bar{p} : p \in D_T\}$, then $\bar{T} \subset T$. On the other hand, for each p in D_T there exists q in D_T such that for all x in X and $n \geq 1$,

$$\sup_{1 \leq m \leq n} p\left(\sum_{i=1}^{m} f_i(x)x_i\right) \leq q\left(\sum_{i=1}^{n} f_i(x)x_i\right) \leq \bar{q}(x);$$

thus $p(x) \leq \bar{q}(x)$. Therefore $T \approx \bar{T}$. Similarly

$$\bar{p}\left(\sum_{i=1}^{m} f_i(x)x_i\right) = \sup_{1 \leq i \leq m} p\left(\sum_{j=1}^{i} f_j(x)x_j\right) \leq \bar{p}\left(\sum_{i=1}^{m+1} f_i(x)x_i\right),$$

for x in X and $m \geq 1$. □

Following [105], we may pass on to

DEFINITION 12.1.2 An S.b. $\{x_n; f_n\}$ for an l.c. TVS (X,T) is called an *extended monotone* S.b. (e.m-S.b.) if for each p in D_T, there exists q in D_T such that for all x in X and $n \geq 1$, one has

$$p\left(\sum_{i=1}^{n} f_i(x)x_i\right) \leq q\left(\sum_{i=1}^{n+1} f_i(x)x_i\right).$$

It is clear that the notions of an e-S.b. and an e.m-S.b. are the same. Thus, in view of Proposition 12.1.1, the two notions of monotone bases are equivalent.

12.2 SPACES WITH e-S.b.

In this section, we explore certain properties of spaces containing e-S.b., which we have not explored fully in earlier chapters. Let us begin with (cf. [161])

PROPOSITION 12.2.1 Let an l.c. TVS (X,T) contain an e-S.b. $\{x_n;f_n\}$. Then a subset A of X is T-precompact if and only if $S_n(x) \to x$ uniformly for x in A and for each $i \geq 1$, $f_i[A]$ is bounded.

PROOF. The proof of the 'only if' part is easy and so omitted; however, a slightly more general result is contained in [130], p. 56.

Conversely let $S_n(x) \to x$ uniformly on A and let $f_i[A]$ be bounded for each $i \geq 1$. For each u in $\mathcal{B}_T$, we may determine v in $\mathcal{B}_T$ and N in $\mathbb{N}$ such that $v + v \subset u$ and

$$\sum_{i \geq N+1} f_i(x)x_i \in v, \quad \forall x \in A.$$

Since $\alpha \to \alpha x_i$ is continuous for each $i \geq 1$, the set

$$B = \{ \sum_{i=1}^{N} f_i(x)x_i \ : \ x \in A\}$$

is precompact. Hence for $y_1,\ldots,y_p$ in X

$$B \subset \bigcup_{i=1}^{p} (y_i+v)$$

and so A is precompact, as $A \subset B+v$. □

An immediate application of the foregoing result is (cf. [46])

PROPOSITION 12.2.2 Every l.c. TVS (X,T) with a γ-complete e-S.b. $\{x_n;f_n\}$ is ω-complete.

PROOF. Assume $\{y_k\}$ to be a T-Cauchy sequence in X; then the scalars $\alpha_k (k \geq 1)$ are well defined, where $\alpha_k = \lim_{i\to\infty} f_k(y_i)$.

Since $\{y_k\}$ is bounded, the e-S.b. character of $\{x_i;f_i\}$ forces the set $B = \{ \sum_{i=1}^{n} f_i(y_k)x_i \ : \ k,\ n \geq 1\}$ to be bounded in (X,T).

Hence for f in X^*, we find M_f such that

$$\left|\sum_{i=1}^{n} f_i(y_k) f(x_i)\right| \leqq M_f; \ \forall\ k,\ n \geqq 1.$$

Fix N in $\mathbb{N}$ and choose $k_o \equiv k_o(N)$ such that

$$|f_i(y_{k_o})-\alpha_i|\ |f(x_i)| \leqq \frac{1}{N}\ ,\ i = 1,\ldots,N.$$

Hence

$$\left|f\left(\sum_{i=1}^{N} \alpha_i x_i\right)\right| \leqq \left|\sum_{i=1}^{N} (f_i(y_{k_o})-\alpha_i) f(x_i)\right| + \left|\sum_{i=1}^{N} f_i(y_{k_o}) f(x_i)\right|$$

$$\leqq 1 + M_f,$$

and this shows that $\{\sum_{i=1}^{n} \alpha_i x_i\}$ is bounded. Let

$$x = \sum_{n \geqq 1} \alpha_n x_n.$$

We will prove that $y_k \to x$. Since $\{y_k\}$ is precompact, for each $\varepsilon > 0$ and p in D_T, we find N in $\mathbb{N}$ such that

$$p\left(\sum_{i=1}^{N} f_i(y_k) x_i - y_k\right) \leqq \frac{\varepsilon}{3}\ ,\ \forall\ k \geqq 1$$

and

$$p\left(\sum_{i=1}^{N} \alpha_i x_i - x\right) \leqq \frac{\varepsilon}{3}.$$

Determine $k_o \equiv k_o(N)$ so that

$$p\left(\sum_{i=1}^{N} (f_i(y_k) - \alpha_i) x_i\right) \leqq \frac{\varepsilon}{3},\ \ \forall k \geqq k_o.$$

Therefore $p(x-y_k) \leqq \varepsilon$, $k \geqq k_o$. □

NOTE. The preceding result merely demonstrates an application of Proposition 12.2.1, although it is already enveloped in the more general Proposition 9.3.10.

REMARK. The restriction on $\{x_n;f_n\}$ in Proposition 12.2.2 is essential to infer the ω-completeness of (X,T). The e-S.b. character alone of $\{x_n;f_n\}$ is not enough to conclude the ω-completeness of (X,T), for example, consider the incomplete space $(\phi, \|\cdot\|)$ for which $\{e^n;e^n\}$ is an e-S.b. but not γ-complete. Similarly, the γ-complete character alone of $\{x_n;f_n\}$ is not sufficient to infer the ω-completeness of (X,T). Before we actually proceed to the main example which is a Banach space, we need a result essentially contained in [87], p. 526 (cf. also [88], p. 633, [235], p. 126) in the form of

LEMMA 12.2.3 Let $\{x_n;f_n\}$ be an m-S.b. for a Banach space $(X, \|\cdot\|)$. Then for each F in X**

$$(*) \qquad \left\| \sum_{i=1}^{n} F(f_i)x_i \right\| \leq \|F\| \; ; \quad \forall n \geq 1,$$

where we denote by the symbol $\|\cdot\|$ the norms on X* and X**.

PROOF. We write S_n^{**} for the operator conjugate to S_n^*, $S_n^{**} : X^{**} \to X^{**}$ with $S_n^{**}(F) = \sum_{i=1}^{n} F(f_i)\, \Psi x_i$, Ψ being the usual canonical embedding satisfying $\|\Psi x\| = \|x\|$ for all x in X. Now

$$\left\| \sum_{i=1}^{n} F(f_i)x_i \right\| = \left\| \Psi\left(\sum_{i=1}^{n} F(f_i)x_i\right) \right\| \leq \|S_n^{**}\| \; \|F\| = \|S_n\| \; \|F\| .$$

But

$$\|S_n(x)\| \leq \|S_{n+k}(x)\| \to \|x\| \quad \text{as } k \to \infty.$$

Hence $\|S_n\| \leq 1$ and this proves the result. □

NOTE. If we don't assume the m-S.b. character of $\{x_n; f_n\}$, then the right-hand side of the inequality in (*) is replaced by $K \|F\|$, where K is a constant independent of F and $n \geq 1$.

We now pass on to the required

EXAMPLE 12.2.4 Here we confine ourselves to the space $(J^*, \sigma(J^*, J^{**}))$, where J is the s.s. of James [87] (cf. also [88]) defined below. We will show that $\{e^n; e^n\}$ is a γ-complete non-e-S.b. for this space and that this space is not ω-complete. The proof runs in several stages : I (showing the γ-completeness of S.b.), II (establishing ω-incompleteness) and III (proving the non-e-S.b. character).

I. The s.s. J is defined by

$$J = \{\alpha \in c_0 \cap \mathbb{R}^{\mathbb{N}} : \|\alpha\| \equiv \|\alpha\|_J \equiv \sup_{1 \leq k_1 < \dots < k_{2n+1}, n \geq 1} [\sum_{i=1}^{n} (\alpha_{k_{2i-1}} - \alpha_{k_{2i}})^2 + \alpha_{k_{2n+1}}^2]^{1/2} < \infty\} .$$

It is quite easy to verify that $(J, \|\cdot\|)$ is a Banach space. Observe that, if $\alpha \in J$, then

$$\|\alpha - \alpha^{(n)}\| = \sup_{k_1 < \dots < k_{2m+1}, m \geq 1} [\sum_{i=1}^{m} (\alpha_{k_{2i-1}} - \alpha_{k_{2i}})^2 + \alpha_{k_{2m+1}}^2]^{1/2},$$

where $k_i \geq n$. If $\|\alpha - \alpha^{(n)}\| \not\to 0$, then for some $\varepsilon > 0$ and $\geq$ $\{n_j\}$ in I_∞, $\|\alpha - \alpha^{(n_j)}\| \geq 3\varepsilon/2$, $j \geq 1$. Hence, after rearranging $\{n_j\}$ if required, we can find $\{m_j\}$ and integers $k_1^j, \dots, k_{2m_j+1}^j$ in $\mathbb{N}$ so that $n_j \leq k_1^j < \dots < k_{2m_{j+1}}^j < n_{j+1}$ and

$$\sum_{i=1}^{m_j} (\alpha_{k^j_{2i-1}} - \alpha_{k^j_{2i}})^2 + \alpha^2_{k^j_{2m_j+1}} \geqq \frac{9\varepsilon^2}{4}, \; j \geqq 1.$$

Find i_0 such that $|\alpha_i| \leqq \varepsilon/2$ for $i \geqq i_0$ and let j_0 be the first index with $n_{j_0} \geqq i_0$. For $r \geqq 0$, let $p_r = m_{j_0} + m_{j_0+1} + \ldots + m_{j_0+r}$ and

$$k_i = k_u^{j_0+s}, \; i = 2 \sum_{t=0}^{s} m'_{j_0+t-1} + u,$$

where $1 \leqq u \leqq 2m_{j_0+s}$, $s = 0,1,\ldots,r$, $m'_{j_0-1} = 0$ and $m'_{j_0+t-1} = m_{j_0+t-1}$ $(1 \leqq t \leqq r)$. With this choice of k_i's

$$||\alpha||^2 \geqq \sum_{i=1}^{p_r} (\alpha_{k_{2i-1}} - \alpha_{k_{2i}})^2 + \alpha^2_{k_{2p_r+1}}$$

$$= [\sum_{1}^{m_{j_0}} + \ldots + \sum_{m_{j_0}+\ldots+m_{j_0+r-1}+1}^{m_{j_0}+\ldots+m_{j_0+r}}](\ldots) + \alpha^2_{k_{2p_r+1}}$$

$$= \sum_{i=1}^{m_{j_0}} (\alpha_{k^{j_o}_{2i-1}} - \alpha_{k^{j_o}_{2i}})^2 + \sum_{i=1}^{m_{j_0+1}} (\alpha_{k^{j_o+1}_{2i-1}} - \alpha_{k^{j_0+1}_{2i}})^2 +$$

$$+ \ldots + \sum_{i=1}^{m_{j_0+r}} (\alpha_{k^{j_o+r}_{2i-1}} - \alpha_{k^{j_0+r}_{2i}})^2 + \alpha^2_{k_{2p_r+1}}$$

$$\geqq \frac{9r\varepsilon^2}{4} - \frac{r\varepsilon^2}{4} + \alpha^2_{k_{2p_r+1}} \to \infty \quad \text{as } r \to \infty.$$

Therefore $||\alpha|| = \infty$, a contradiction. Hence $\alpha^{(n)} \to \alpha$ in $(J, \|\cdot\|)$. By Theorem 1.3.13, $\{e^n;e^n\}$ is an S.b. for

$(J, \|\cdot\|)$.

Next, we show that $\{e^n; e^n\}$ is a shrinking S.b. for J. If this were not true, we would get an f in J* and a bounded bl. s. $\{y^n\}$ of $\{e^n\}$ in J with

$$y^n = \sum_{i=m_{n-1}+1}^{m_n} \alpha_i e^i$$

such that $f(y^n) = 1$; $n \geqq 1$ (see the proof of Theorem 11.3.8). If we could show that

$$(*) \qquad y = \sum_{n \geqq 1} \frac{1}{n} y^n \in J,$$

we would then find that $f(y) = \infty$, a contradiction. Hence it remains to establish(*).

Define $\{\gamma_i\}$ by $\gamma_i = \alpha_i/n$, $m_{n-1}+1 \leqq i \leqq m_n$, $n \geqq 1$. It is readily seen that $|x_i| \leqq \|x\|$ for each $i \geqq 1$ and x in J and so $\|e^i\| \leqq 1$ for each $i \geqq 1$, where $e^i \in J^*$. Hence, for $m_{n-1} + 1 \leqq i \leqq m_n$,

$$(+) \qquad |\gamma_i| = \frac{1}{n} |\langle y^n, e^i\rangle| \leqq \frac{1}{n} \|y^n\| \leqq \frac{K}{n},$$

where $K = \sup\{\|y^n\| : n \geqq 1\}$. At the same time, for terms of the sum

$$(++) \qquad \sum_{i=1}^{p} (\gamma_{k_{2i-1}} - \gamma_{k_{2i}})^2$$

we have indices $k_1 < k_2 < \ldots < k_{2p}$, satisfying either $m_{n-1}+1 \leqq k_{2r-1} < k_{2r} < \ldots < k_{2s-1} < k_{2s} \leqq m_n$, or $m_{n-1}+1 \leqq k_{2i-1} \leqq m_n$, $m_{n'-1} \leqq k_{2i} \leqq m_{n'}$ with $n' > n$. In the former case, the sum of the terms of (*) is bounded above by $K^2 \sum_{n=1}^{\infty} \frac{1}{n^2}$; for, by the definition of the norm $\|\cdot\|$ on J and $\{\langle y^n, e^i\rangle : i \geqq 1\} \in J$, one has

$$\sum_{i=r}^{s} (\alpha_{k_{2i-1}} - \alpha_{k_{2i}})^2 \leqq \|y^n\|^2 \leqq K^2, \ \forall\, n \geqq 1;$$

whereas in the latter case, using (+) we get

$$(\gamma_{k_{2i-1}} - \gamma_{k_{2i}})^2 \leqq K^2(\frac{1}{n} + \frac{1}{n'})^2 \leqq \frac{4K^2}{n^2}.$$

Thus, for any integer p and any choice of $k_1 < \ldots < k_{2p+1}$,

$$\sum_{i=1}^{p} (\gamma_{k_{2i-1}} - \gamma_{k_{2i}})^2 + \gamma_{k_{2p+1}}^2 \leqq 6K^2 \sum_{n\geqq 1} \frac{1}{n^2}.$$

Hence

$$y = \sum_{i\geqq 1} \gamma_i e^i = \sum_{n\geqq 1} \frac{1}{n} \sum_{i=m_{n-1}+1}^{m_n} \alpha_i e^i = \sum_{n\geqq 1} \frac{1}{n} y^n$$

belongs to J.

Summing up the discussion and making use of Proposition 9.6.4, we find that $\{e^n, \Psi e^n\}$ is a γ-complete S.b. for $(J^*, \sigma(J^*,J^{**}))$, where Ψ: $J \to J^{**}$ is the usual canonical embedding.

II. Here also, we establish a few facts before we turn to the proof of the required conclusion. At the outset, let us observe that J is not semireflexive, for otherwise $B = \{\alpha \in J : \|\alpha\| \leqq 1\}$ would be $\sigma(J,J^*)$-compact. Since $\|e^{(n)}\| = 1$ for $n \geqq 1$, $\{e^{(n)}\}$ will have a subsequence converging weakly to y in B. But $y = e \notin J$, a contradiction.

Next, we shall prove that J^{**} is strongly separable, for, once this is established, it will follow by a known result, namely "an ω-complete l.c. TVS with separable strong dual is semireflexive" (cf. [205], p. 194, Exer. 18(c)), that $(J^*, \sigma(J^*,J^{**}))$ is not ω-complete.

To derive the desired conclusion, let us introduce an

s.s. λ by $\lambda = \{\alpha \in c : ||\alpha|| \equiv ||\alpha||_J < \infty\}$ and proceed to show that $(J^{**}, \|\cdot\|)$ and $(\lambda, \|\cdot\|)$ are isometrically isomorphic to each other under the map $\Phi: J^{**} \to \lambda$, $\Phi(F) = \{F(e^n)\}$. In fact, let $F \in J^{**}$. By Lemma 12.2.3, $\|\{F(e^n)\}\|_J \leqq \|F\| < \infty$. Further, $\{F(e^n)\}$ converges, for otherwise, there exist $\varepsilon > 0$ and sequences $\{m_k\}$, $\{n_k\}$ with $m_1 < n_1 < m_2 < n_2 < \ldots$ such that $|F(e^{n_k}) - F(e^{m_k})| > \varepsilon$, $k \geqq 1$. However, this shows that $\|\{F(e^n)\}\| \equiv \|\{F(e^n)\}\|_J = \infty$. Thus Φ is well defined and $\|\Phi(F)\| \leqq \|F\|$. On the other hand, if $\alpha \in \lambda$, then for $n \geqq 1$,

$$\|\alpha^{(n)}\|_J^2 = \sup_{1 \leqq k_1 < \ldots < k_{2m+1}} \{\sum_{i=1}^{m} (\alpha_{k_{2i-1}}^{(n)} - \alpha_{k_{2i}}^{(n)})^2 + (\alpha_{k_{2m+1}}^{(n)})^2\},$$

where $\alpha_{k_i}^{(n)} = 0$ if $k_i > n$. Thus $\|\alpha^{(n)}\|_J \leqq \|\alpha\|_J$ for $n \geqq 1$. Thus $\{\sum_{i=1}^{n} \alpha_i e^i\}$ is bounded in J. Hence by Theorem 9.4.8, there exists F in J^{**} such $F(e^i) = \alpha_i$. Now for f in J^* and $n \geqq 1$,

$$|\sum_{i=1}^{n} F(e^i) f(e^i)| \leqq \|f\| \, \|\sum_{i=1}^{n} F(e^i) e^i\|_J \leqq \|f\| \, \|\alpha\|_J.$$

Since $\{e^n; \Psi e^n\}$ is an S.b. for J^*,

$$|F(f)| \leqq \|f\| \, \|\alpha\|_J = \|f\| \, \|\Phi(F)\|$$

$$\Longrightarrow \|F\| \leqq \|\Phi(F)\|.$$

Hence $\|\Phi(F)\| = \|F\|$.

The foregoing discussion suggests that $\lambda = J \oplus sp\{e\}$ and so J^{**} is separable. If $(J^*, \sigma(J^*, J^{**}))$ is ω-complete, then J is semireflexive, a contradiction.

III. <u>This part is a consequence of Corollary 12.3.2</u>

12.3 WEAKLY e-SCHAUDER BASES

There is an interesting characterization of the existence of weakly e-Schauder bases (w.e-S.b.) in an arbitrary l.c. TVS contained in [24]. Although this result is a consequence of a more general theorem in the sequence space theory (cf. [129], p. 65), we will reproduce the former for the sake of completeness.

<u>THEOREM 12.3.1</u> Let (X,T) be an l.c. TVS containing an arbitrary S.b. $\{x_n;f_n\}$ with respect to T or $\sigma(X,X^*)$. Then $\{x_n;f_n\}$ is w.e-S.b. for X if and only if $\{f_n\}$ is a Hamel base for X^*.

<u>PROOF</u>. If $\{f_n\}$ is a Hamel base for X^*, then for each f in X^*, $f = \sum_{i=1}^{m} f(x_i)f_i$. Since $|f(S_n(x))| = |f(x)|$ for each x in X and $n \geqq m$, $\{S_n\}$ is weakly equicontinuous.

Conversely, let $f \in X^*$. Then $\{f \circ S_n\}$ is a weakly equicontinuous sequence in X^* and so it is finite dimensional and $\sigma(X^*,X)$-bounded (cf. [140], p. 161). Hence for integers $k_1 \leqq \ldots \leqq k_p$, there exist a linearly independent set $\{g_{k_i} : 1 \leqq i \leqq p\}$ with $g_{k_i} = f \circ S_{k_i}$ $(1 \leqq i \leqq p)$ such that $\{f \circ S_n\} \subset sp\{g_{k_i} : 1 \leqq i \leqq p\}$. This shows that $f(x_i) = 0$, $i > k_p$ and so $f = \sum_{i=1}^{k_p} f(x_i)f_i$. □

<u>COROLLARY 12.3.2</u> No infinite dimensional Banach space X admits a w.e-S.b.

For otherwise, X^* would be of countable dimension, a contradiction.

<u>NOTE</u>. For another proof of Corollary 12.3.2, see Exercise 12.3.5.

After [161], we have

PROPOSITION 12.3.3 If $\{f_n\}$ is a $\sigma(X^*,X)$-S.b. for X^*, $(X, \|\cdot\|)$ being a Banach space, then $\{f_n\}$ cannot be an e-S.b. with respect to $\sigma(X^*,X)$.

PROOF. There is a sequence $\{x_n\}$ in X with $f_n(x_m) = \delta_{mn}$ such that for each f in X^* and x in X,

$$f(x) = \sum_{n\geq 1} f_n(x)\, f(x_n).$$

Hence by the weak basis theorem (cf. [129], p. 88), $\{x_n;f_n\}$ is an S.b. for $(X, \|\cdot\|)$. Set

$$x_o = \sum_{n\geq 1} (1/2^n \|x_n\|)x_n.$$

Let $u = \{f \in X^* : |f(x_o)| \leq 1\}$ and v be an arbitrary $\sigma(X^*,X)$-neighbourhood, say, $v = \{f \in X^* : |f(y_i)| \leq \varepsilon, 1 \leq i \leq m\}$. Suppose $Y = [y_1,\ldots,y_m]$, then $Y \subsetneq X$. Hence for some f in X^*, $f \neq 0$ and $f(y) = 0$ for all y in Y. Let n_o be the smallest integer such that $f(x_{n_o}) \neq 0$. Now for any α, $\alpha f \in v$. If $\{S_n^*\}$ were $\sigma(X^*,X)$ - $\sigma(X^*,X)$ equicontinuous, then there would exist v such that $S_n^*(v) \subset u$ for all $n \geq 1$. Choosing v as above, one gets $S_n^*(\alpha f) \in u$ for all $n \geq 1$ and all scalars α. But for $\alpha \neq 0$, $|S_{n_o}^*(\alpha f)(x_o)| = |\alpha|$ $|f(x_{n_o})f_{n_o}(x_o)|$ and this can be made greater than 1 by choosing $|\alpha|$ arbitrarily large. Therefore, for this choice of α, $S_n^*(\alpha f) \notin u$. □

As an application of Theorem 12.3.1, one has (cf. [24])

PROPOSITION 12.3.4 Let (X,T) be an ω-complete l.c. TVS containing an S.b. Then $\{x_n;f_n\}$ is a w.e-S.b. if and only if $(X,T) \simeq (\omega,\sigma(\omega,\phi))$.

PROOF. The 'if' part is obvious.

Conversely, let $\{x_n; f_n\}$ be a w.e-S.b. Then, using Theorem 12.3.1, the topology $\sigma(X,X^*)$ is metrizable. Hence $(X,\sigma(X,X^*))$ is bornological and so $\sigma(X,X^*) \approx \tau(X,X^*)$. Thus (X,T) is a Fréchet space with $T \approx \sigma(X,X^*)$. Let $\alpha \in \omega$. For f in X^*, $f(x_n) = 0$ for all but a finite number of integers n. Consequently $\{\sum_{i=1}^{n} \alpha_i x_i\}$ is $\sigma(X,X^*)$-Cauchy. Hence for some x in X, $\alpha = \{f_n(x)\}$. Therefore $\delta = \omega$; also it is clear that $\mu = \phi$. Now apply Theorem 3.3.5. □

EXERCISE 12.3.5 Prove Corollary 12.3.2 by using Proposition 12.3.4. [Hint: $\sigma(\omega,\phi)$ is non-normable.]

REMARK. The ω-completeness which appears to be essential in concluding the desired result in the 'only if' part of Proposition 12.3.4, is, indeed, unavoidable in general, as shown by

EXAMPLE 12.3.6 This is the ω-incomplete space $(\phi,\sigma(\phi,\phi))$ of Example 9.6.3 for which $\{e^n; e^n\}$ is an e-S.b. by Theorem 12.3.1.

12.4 UNIFORMLY EQUICONTINUOUS BASES

The basic purpose of this section is to exploit the presence of an uniformly equicontinuous S.b. (u.e-S.b.) and to produce a few more applications in addition to those already dealt with in chapter 5. In what follows, we will also include the converse of Propositions 5.5.4 and 5.6.8 as promised there.

Let us recall the definition of CSBS (section 2.1) and begin with (cf. [1]; [16])

PROPOSITION 12.4.1 Let $\{x_n\} \equiv \{x_n; f_n\}$ be a CSBS in an l.c. TVS (X,T). Then there exists a sequence $\{g_n\}$ in X^*

with $g_n(x_m) = \delta_{mn}$ such that for each p in D_T, the function p_∞ defined by

$$p_\infty(x) = \sup_{n \geq 1} |g_n(x)| \, p(x_n)$$

is a seminorm on X. In addition, if $\{x_n;f_n\}$ is also an u.e-S.b. for $Y \equiv [x_n]$, then for each p in D_T, there exists q in D_T with $p_\infty \leq q$. In particular, if $\{x_n;f_n\}$ is a u. e-S.b. for (X,T), then for each p in D_T there exists q in D_T so that $p_\infty(x) \equiv \sup\{|f_n(x)|p(x_n) : n \geq 1\} \leq q(x)$ for x in X.

PROOF. Define $g_n = f_n \circ P$, where P is a continuous projection from X onto Y. Clearly $g_n \in X^*$ and $g_n(x_m) = \delta_{mn}$. Let $p \in D_T$. For each x, $\{f_n(Px)x_n\}$ is bounded in Y and so in X. Thus p_∞ is well defined on X and is, indeed, a seminorm on X. For the second part, we find r in D_T so that $p_\infty(y) \leq r(y)$ for y in Y. Since $g_n(x) = g_n(Px)$ for x in X, $p_\infty(x) = p_\infty(Px) \leq r(Px)$. Put $q = r \circ P$. The last part is obviously true. □

Use of Kolmogorov's diameters

The next few results of this subsection make full use of Kolmogorov's diameters $\delta_n(v,u)$ of members of $\mathcal{B} \equiv \mathcal{B}_T$ related to an l.c. TVS (X,T).

Corresponding to a u.e-S.b. $\{x_n;f_n\}$ for an l.c. TVS (X,T), let us recall from (5.6.3) the seminorm p_∞ for p in D_T. Let us write u_p^∞ for u_{p_∞}

We begin with (cf. [220])

LEMMA 12.4.2 Let $\{x_n;f_n\}$ be a u.e-S.b. for an l.c TVS (X,T). Let ther exist for each p in D_T,q in D_T with $p_\infty \leq q$ such that $\delta_n(u_q,u_p^\infty) \to 0$. Then

$$\lim_{n\to\infty} \frac{p(x_n)}{q(x_n)} = 0, \quad (0/0 = 0). \qquad (12.4.3)$$

PROOF. Define $R : X \to \ell^\infty$ by $Rx = \{f_n(x)p(x_n)\}$. Then

$$u_q \subset u_p^\infty, \quad R[u_p^\infty] \subset u_\infty, \tag{12.4.4}$$

u_∞ being the closed unit ball in ℓ^∞. By (12.4.4) and Proposition 1.4.4 (cf. (iv) and (v))

$$\delta_n(R[u_q], u_\infty) \leqq \delta_n(u_q, u_p^\infty).$$

Hence by Proposition 1.4.8, $R[u_q]$ is relatively compact in $(\ell^\infty, \|\cdot\|_\infty)$

$$y_n = \begin{cases} x_n/q(x_n), & q(x_n) \neq 0; \\ 0 & , \text{ otherwise.} \end{cases}$$

Put $\theta^n = Ry_n$. Then $\|\theta^n\|_\infty = p(x_n)/q(x_n)$ and $\theta^n \in R[u_q]$. Let for some $\varepsilon > 0$ and $\{n_k\}$ in I_∞, $\|\theta^{n_k}\|_\infty > \varepsilon$. Since $\{\theta^{n_k}\}$ has a convergent subsequence, we may assume that $\{\theta^{n_k}\}$ is Cauchy, satisfying $\|\theta^{n_k}\|_\infty > \varepsilon$. If $\theta^{n_k} \to \theta$, then $\theta = 0$. On the other hand, $\|\theta\| > \varepsilon$, a contradiction. Therefore $\|\theta^n\|_\infty \to 0$. □

The following (cf. [220]) is a partial converse of Proposition 5.4.9.

PROPOSITION 12.4.5 If (X,T) is a Schwartz space containing a u.e-S.b. $\{x_n; f_n\}$, then for each p in D_T there exists q in D_T such that

$$\lim_{n\to\infty} \frac{p(x_n)}{q(x_n)} = 0. \qquad (0/0 = 0)$$

PROOF. By Theorem 1.4.12, for each p in D_T there exists v in $\mathcal{B}$ with $v < u_p^\infty$ such that $\delta_n(v, u_p^\infty) \to 0$. We can find q in D_T with $u_q \subset v < u_p^\infty$ and hence $\delta_n(u_q, u_p^\infty) \to 0$. Now for some $\alpha > 0$, $\alpha p_\infty \leqq q$ and so $\delta_n(u_{q_1}, u_p^\infty) \to 0$ where $q_1 = q/\alpha$.

The result now follows from Lemma 12.4.2. □

The corollary of the next theorem (the 'only if' part), namely, every base of a Fréchet nuclear space is an ℓ^1-base, generally known as the *Dynin-Mityagin theorem*, first appeared in a weaker form in [49] and subsequently was re-proved in [164] by a different method. This corollary, which finds an altogether different proof in [178], p. 173, was slightly extended in [194]. The following theorem was stated in [99] with the proof of its 'only if' part referred to [164]; however, its proof was subsequently completed in [111] with the help of Proposition 6 of [164]. Following [111] with some improvements, we state and prove the desired

THEOREM 12.4.6 Let $\{x_n;f_n\}$ be a u-e-S.b. for an l.c. TVS (X,T). Then (X,T) is nuclear if and only if for each $r > 0$ and p in D_T, there exists q in D_T with $p \leqq q$ and

$$\sum_{n\geqq 1} [p(x_n)/q(x_n)]^r < \infty, \qquad (0/0 = 0). \tag{12.4.7}$$

PROOF. Let $p \in D_T$, $J = \{i \in \mathbb{N} : p(x_i) \neq 0\}$, $J^\sim = \mathbb{N}\setminus J$ and $X_o = \{x \in X : f_i(x) = 0 \text{ for each } i \text{ in } J^\sim\}$. If J is finite, then (12.4.7) trivially follows. Assume, therefore, that J is infinite and let $J = \{m_1, m_2, \ldots\}$. By Proposition 12.4.5, Theorem 1.4.10 and the fact that X_o is also nuclear relative to $T|X_o$, we get q in D_T with

$$\text{(i)}\quad p_\infty \leqq q, \quad \text{(ii)}\quad p \leqq q, \quad \text{(iii)}\quad p(x_n)/q(x_n) \to 0$$

and

$$\text{(iv)}\quad \sum_{n\geqq 1} n\,\delta_n(u_q^o, u_p^{\infty,o}) < \infty,$$

where

$$u_q^o = X_o \cap u_q;\ u_q = \{i \in X : q(x) \leq 1\},$$

$$u_p^{\infty,0} = X_0 \cap u_p^\infty; \; u_p^\infty = \{x \in X : p_\infty(x) \leq 1\}.$$

Define

$$A_q = \{x \in X_0 : \sum_{i \geq 1} |f_i(x)| q(x_i) \leq 1\},$$

then

$$\text{(v)} \quad A_q \subset u_q^0 \subset u_p^{\infty,0} \implies \delta_n(A_q, u_p^{\infty,0}) \leq \delta_n(u_q^0, u_p^{\infty,0}).$$

Introduce the linear operator $R : X_0 \to \ell^\infty$, $Rx = \{f_i(x)p(x_i)\}$. Then $R[X_0]$ is a normed subspace of ℓ^∞ with unit ball $R[u_p^{\infty,0}]$. Here

$$R[u_p^{\infty,0}] = \{\{f_i(x)p(x_i)\} : \sup_{i \geq 1} |f_i(x)| p(x_i) \leq 1\};$$

$$R[A_q] = \{\{f_i(x)p(x_i)\} : \sum_{i \geq 1} |f_i(x)| q(x_i) \leq 1\}.$$

By (iii), there exists a permutation π of J so that for $i \geq j$,

$$\text{(vi)} \quad \beta_i \leq \beta_j, \; \beta_i \equiv p(x_{\pi(m_i)})/q(x_{\pi(m_i)}).$$

We next show that

$$(*) \quad n^{-1}\beta_n(L_n \cap R[u_p^{\infty,0}]) \subset R[A_q],$$

where

$$L_n = \text{sp}\{e^{\pi(m_i)} : 1 \leq i \leq n\}.$$

Indeed, let

$$z \in n^{-1}\beta_n(L_n \cap R[u_p^{\infty,0}])$$

$$\Rightarrow \; z = n^{-1}\beta_n y$$

with

$$(+) \quad y = \sum_{i=1}^{n} y_{\pi(m_i)} e^{\pi(m_i)}, \quad y = \{f_i(x)p(x_i)\}$$

for some x in $u_p^{\infty,o}$.

Thus

$$y_j = 0, \; j \in N \setminus \{\pi(m_1), \ldots, \pi(m_n)\}$$

and

$$y_{\pi(m_j)} = f_{\pi(m_j)}(x)p(x_{\pi(m_j)}), \quad 1 \leqq j \leqq n.$$

With x as in (+), define v in X_o by

$$v = n^{-1}\beta_n \sum_{i=1}^{n} f_{\pi(m_i)}(x)x_{\pi(m_i)}.$$

Clearly z = Rv. Thus to establish (*), it is enough to prove that $v \in A_q$; in fact,

$$\sum_{i \geqq 1} |f_i(v)|q(x_i) = n^{-1}\beta_n \sum_{i=1}^{n} |f_{\pi(m_i)}(x)|q(x_{\pi(m_i)})$$

$$= n^{-1}\beta_n \sum_{i=1}^{n} \frac{|f_{\pi(m_i)}(x)|p(x_{\pi(m_i)})}{p(x_{\pi(m_i)})/q(x_{\pi(m_i)})}$$

$$\leqq n^{-1}\beta_n \sum_{i=1}^{n} \beta_n^{-1} = 1,$$

by (vi) and (+). Hence we have finished with (*) and consequently by Propositions 1.4.4 and 1.4.7,

$$n^{-1}\beta_n \leqq \delta_n(R[A_q], R[u_p^{\infty,o}])$$

$$\leqq \delta_n(A_q, u_p^{\infty,o})$$

$$\leqq \delta_n(u_q^o, u_p^{\infty,o}),$$

by (v). Thus, using (iv), we find that $\sum_{n\geq 1} \beta_n < \infty$; that is,

$$\sum_{n\geq 1} p(x_{\pi(m_n)})/q(x_{\pi(m_n)}) < \infty$$

$$\Rightarrow \sum_{n\geq 1} p(x_n)/q(x_n) < \infty,$$

showing thereby the truth of (12.4.7) for r = 1. It now remains to make use of Lemma 5.5.1 to complete the 'only if' part of the theorem.

The converse of Theorem 12.4.6 is already proved in Proposition 5.5.4. □

NOTE. Theorem 12.4.6 in its present form is given in [131].

The following proposition strengthens a similar result given in [129], p. 288.

PROPOSITION 12.4.8 For any s.s. λ, the space $(\lambda,\eta(\lambda,\mu))$ is nuclear if and only for any $s > 0$ and α in μ, there exists β in μ such that $\alpha/\beta \in \ell^s$, where μ is a normal subspace of $\lambda^\times$ and as before $\alpha/\beta = \{\alpha_n/\beta_n\}$ where 0/0 is to be interpreted as 0.

PROOF. Observe that the l.c. topology $\eta(\lambda,\mu)$ is generated by the family $\{p_\alpha : \alpha \in \mu\}$ of seminorms on λ, where

$$p_\alpha(u) = \sum_{n\geq 1} |\alpha_n u_n|.$$

Further, $\{e^n;e^n\}$ is obviously a u.e-S.b. for $(\lambda,\eta(\lambda,\mu))$. One now needs to apply Theorem 12.4.6 to reach the desired result. □

The next exercise follows similarly

EXERCISE 12.4.9 The Köthe space $\Lambda(P)$ is nuclear if and only if for every $s > 0$ and $\alpha \in P$, there exists β in P such that $\alpha/\beta \in \ell^s$.

NOTE. The result contained in Exercise 12.4.9 for $s = 1$ is generally known as the *Grothendieck-Pietsch-Köthe criterion* for the nuclearity of $\Lambda(p)$.

EXERCISE 12.4.10 Prove that an ω-complete nuclear space (X,T) with a u.e-S.b. $\{x_n;f_n\}$ is topologically isomorphic with $(\Lambda(P),T_p)$, $P = \{\{p(x_n)\} : p \in D_T\}$. [Make use of Proposition 4.5.10.]

The next result is given in [220].

PROPOSITION 12.4.11 Let (X,T) be an ω-complete nuclear space having a u.e-S.b. $\{x_n;f_n\}$. Then a subset B of X is T-bounded if and only if there exists y in X such that

$$|f_n(x)| \leq |f_n(y)|; \quad \forall x \in B,\ n \geq 1. \tag{12.4.12}$$

PROOF. Let T_∞ be the topology generated by $\{p_\infty : p \in D_T\}$. By the u.e-S.b. character of $\{x_n;f_n\}$, $T_\infty \subset T$. Further, for p in D_T and $r = 1$, there exists q in D_T so that the inequality (*) in Theorem 12.4.6 is satisfied; thus for x in X,

$$\begin{aligned} p(x) \leq \sum_{n\geq 1} |f_n(x)|p(x_n) &\leq q_\infty(x) \sum_{n\geq 1} p(x_n)/q(x_n) \\ &\equiv k_{p,q}\, q_\infty(x), \end{aligned} \tag{12.4.13}$$

and so $T \approx T_\infty$.

If (12.4.12) is satisfied, then for all x in B, $p(x) \leq k_{p,q}\, q_\infty(x)$ (cf. (12.4.13)).

Let B be bounded. Then for each $n \geq 1$, the scalar $\alpha_n = \sup\{|f_n(x)| : x \in B\}$ is well defined. Consider now p in D_T. There exists q in D_T satisfying (12.4.13). Since $|f_n(x)|q(x_n) \leq q_\infty(x)$ for all $n \geq 1$, $\alpha_n\, q(x_n) \leq k_q \equiv \sup\{q_\infty(x)$:

$x \in B\}$. Hence

$$\sum_{n\geq 1} \alpha_n \, p(x_n) \leqq k_q \, k_{p,q} < \infty,$$

and so $\alpha_n = f_n(y)$, $n \geqq 1$ for some y in X. □

NOTE. There is an alternative proof of the 'only if' part contained in [142], (5), p. 270.

Let us note that the proof of the foregoing result (cf. (12.4.13)) yields

PROPOSITION 12.4.14 Let $\{x_n;f_n\}$ be a u.e-S.b. for a nuclear space (X,T). Then $T \approx T_\infty$.

Also, we have

PROPOSITION 12.4.15 Let $\{x_n;f_n\}$ be a u.e-S.b. for a nuclear space (X,T). Then for each p, $1 \leqq p < \infty$, $\{x_n;f_n\}$ is a semi p-Köthe base.

PROOF. Since $p \geqq 1$, for each ν in D_T, ν_p given by (5.4.4) is well defined. Also, $\nu_\infty \leqq \nu_p$. On the other hand, for each ν in D_T, there exists μ in D_T such that

$$k^p_{\nu,\mu} \equiv \sum_{n\geq 1} [\nu(x_n)/\mu(x_n)]^p < \infty,$$

and so

$$\nu_p \leqq k_{\nu,\mu} \, \mu_\infty \quad .$$

Thus the topology T^p generated by $\{\nu_p : \nu \in D_T\}$ is equivalent to T_∞ and now apply Proposition 12.4.14. □

The foregoing results yield (cf. [220])

PROPOSITION 12.4.16 Let $\{x_n;f_n\}$ be a u.e-S.b. for an ω-complete nuclear space (X,T). Then $\{f_n;Jx_n\}$ is a u.e-S.b. for $(X^*,\beta(X^*,X))$.

PROOF. For each y in X, let $A_y = \{x \in X : |f_n(x)| \leq |f_n(y)|, n \geq 1\}$. Then $\{A_y : y \in X\}$ represents the family of all bounded subsets of X. Fix y in X. For each f in X^*, define $\alpha \equiv \alpha_f$ in ω by $\alpha_n\, f(x_n)\, f_n(y) = |f(x_n)f_n(y)|$ so that $|\alpha_n| = 1$, $n \geq 1$. Since $\sum_{n\geq 1} \alpha_n f_n(y)x_n$ is absolutely convergent, there exists x in A_y with $x = \sum_{n\geq 1} \alpha_n f_n(y)x_n$. Therefore, for f as above

$$|f(x)| = \sum_{n\geq 1} |f(x_n)f_n(y)|;$$

in particular, if $f \in A_y^\circ$, then

$$\sum_{n\geq 1} |f(x_n)\, f_n(y)| \leq 1$$

$$\Rightarrow A_y^\circ \subset B_y \equiv \{f : \sum_{n\geq 1} |f(x_n)f_n(y)| \leq 1\}.$$

It is obvious that $B_y \subset A_y^\circ$ and so $A_y^o = B_y$, $y \in X$.

Let $f \in X^*$ and consider any A_y° for a given y in X. Then $f \in \alpha A_y^\circ$ for some $\alpha > 0$ and so

$$\sup_{x\in A_y} |(f - \sum_{i=1}^{n} f(x_i)f_i)(x)| \leq \sum_{i>n} |f(x_i)f_i(y)| \to 0$$

as $n \to \infty$. Thus $\{x_n;f_n\}$ is shrinking.

Also, for f in X^* and y in X

$$|f(x_n)| \sup \{|f_n(x)| : x \in A_y\}$$

$$\leq \sum_{n\geq 1} |f(x_n)f_n(y)|, \ \forall\ n \geq 1$$

and this shows that $\{f_n;Jx_n\}$ is a u.e-S.b. for the space $(X^*,\beta(X^*,X))$. □

A FINAL REMARK. Uniformly equicontinuous Schauder bases have other important applications; in fact, these bases can be used to characterize bounded and continuous linear operators from locally convex spaces to themselves. A discussion on this aspect is deferred until [132].

PART III

13 Domination of sequences

13.1 INTRODUCTORY REMARKS

In Part II, a good deal of attention has been paid to exploiting the presence of an S.b. in an l.c. TVS. This was possible because we confined ourselves to describing the different types of S.b. and then investigated the impact of these bases on the structure of the underlying space.

We carry out a somewhat different study of Schauder bases in this Part. Indeed, Part III is devoted to the study of two other aspects of Schauder bases: (A) if there are two S.b. in a TVS, then in what manner can these two be related; and (B) if an S.b. $\{x_n\}$ is given a 'displacement' in a suitable manner, will the resulting sequence still be an S.b., or, if there is a sequence $\{y_n\}$ 'near' to an S.b. $\{x_n\}$ in a certain suitable sense, will $\{y_n\}$ also be an S.b.?

The two questions contained in (A) and (B) are born out of some natural reasonings. Whereas (A) will be studied in generality in this and the next chapters, (B) will be taken up in chapter 15.

13.2 DOMINATING AND EQUIVALENT SEQUENCES

Several authors in the past have considered the concept of domination of one sequence over the other, with slight variations of terminology; in particular, in 1925, Banach ([11], p. 1638; [12], p. 11) considered domination between a pair of sequences. Later, Arsove ([5]; [7]; [8]) used this concept in the study of Schauder bases and constructed examples of dominating sequences from the theory of spaces of analytic functions. Subsequently, Singer [214] pointed out the importance of such sequences in reformulating certain results on the properties of sequences in Banach spaces. More recently, quite a few results from [214] have been sharpened in [69] in the general setting of locally convex spaces and we follow the latter for most of the work on this section.

Throughout we write (X,T) and (Y,S) for an arbitrary TVS or l.c. TVS (the context will make this clear) over the same field $\mathbb{K}$ of scalars, containing respectively the sequences $\{x_n\}$ and $\{y_n\}$.

<u>DEFINITION 13.2.1</u> $\{x_n\}$ is said (i) to *dominate* $\{y_n\}$, to be abbreviated as $\{x_n\} > \{y_n\}$ if for α in ω

Convergence of $\sum_{n\geq 1} \alpha_n x_n \Rightarrow$ convergence of $\sum_{n\geq 1} \alpha_n y_n$;

(ii) to *strictly dominate* $\{y_n\}$, to be written as $\{x_n\} >> \{y_n\}$, provided there exists a continuous linear operator

$$R : [x_n] \to [y_n] \text{ with } Rx_n = y_n, \; n \geq 1;$$

(iii) to be *equivalent* or *similar* to $\{y_n\}$ (resp. *strictly equivalent* to $\{y_n\}$) to be written as $\{x_n\} \sim \{y_n\}$ (resp. $\{x_n\} \approx \{y_n\}$) providee $\{x_n\} > \{y_n\} > \{x_n\}$ (resp. $\{x_n\} >> \{y_n\} >> \{x_n\}$); and (iv) to be *fully equivalent* to $\{y_n\}$, to be denoted as $\{x_n\} \approxeq \{y_n\}$, if there exists a topological isomorphism R from (X,T) onto (Y,S) with $Rx_n = y_n$, $n \geq 1$.

It is clear that $\{x_n\} >> \{y_n\} \Rightarrow \{x_n\} > \{y_n\}$, and $\{x_n\} \approxeq \{y_n\} \Rightarrow \{x_n\} \approx \{y_n\} \Rightarrow \{x_n\} \sim \{y_n\}$. The converse implications are not true in general as exhibited in the next two examples.

<u>EXAMPLE 13.2.2</u> Let $\{x_n\}$ be any sequence in a TVS (X,T) such that x_1 and x_2 are linearly independent (1.i). Define $y_1 = x_1$; $y_2 = \alpha x_1$, $\alpha \neq 0, 1$; $y_n = x_n$, $n \geq 3$. Then

$\{y_n\} \sim \{x_n\}$; but $\{y_n\} \not\gg \{x_n\}$, for otherwise $Ry_1 = x_1$, $Ry_2 = x_1$ yields $\alpha x_1 = x_2$.

<u>EXAMPLE 13.2.3</u> Let $X = Y = c_o \equiv (c_o, \|\cdot\|_\infty)$, $x^n = e^n$ and $y^n = e^{2n-1}$, $n \geq 1$. If $R_1 x^n = y^n$, $n \geq 1$, then R_1 can be extended as a continuous linear map from $[x^n]$ to $[y^n]$. In a similar manner, by letting $R_2 y^n = x^n$, $R_2 : [y^n] \to [x^n]$ is a continuous linear map. Thus $\{x^n\} \approx \{y^n\}$.

On the other hand, $\{x^n\} \not\approxeq \{y^n\}$. For otherwise, $c_o \simeq c_o$, say, under R with $Re^n = e^{2n-1}$, $n \geq 1$. In particular, $\alpha \equiv R^{-1} e^{2n} \in c_o$; so that

$$e^{2n} = \sum_{i \geq 1} \alpha_i Re^i = \sum_{i \geq 1} \alpha_i e^{2i-1}$$

which is clearly impossible.

Concerning the converse implications, we have

<u>PROPOSITION 13.2.4</u> For sequences $\{x_n\}$ in (X,T) and $\{y_n\}$ in (Y,S), if $\{x_n\} \approx \{y_n\}$, then $[x_n] \simeq [y_n]$ under the map R.

<u>PROOF</u>. Let $R : [X_n] \to [y_n]$ and $P : [y_n] \to [x_n]$ be the continuous linear maps with $Rx_n = y_n$; $Py_n = x_n$, $n \geq 1$. Let $x \in [x_n]$ and $Rx = 0$. Then there exists a net $\{u_\delta\}$ in sp $\{x_n\}$ with $u_\delta \to x$. Since

$$u_\delta = \sum_{i \in F_\delta} \alpha_i^\delta x_i, \quad F_\delta = \text{a finite subset of } \mathbb{N}$$

we conclude that

$$\lim_\delta \sum_{i \in F_\delta} \alpha_i^\delta y_i = 0 \Rightarrow \lim_\delta \sum_{i \in F_\delta} \alpha_i^\delta Py_i = 0.$$

Hence $x = 0$ and so R is 1-1. Similarly, R is onto $[y_n]$. In a similar manner, P is 1-1 and onto $[x_n]$. Now observe that $R^{-1} = P$. □

NOTE. The map R in Proposition 13.2.4 with $Rx_n = y_n$, $n \geq 1$ is uniquely determined.

COROLLARY 13.2.5 Let $\{x_n\}$, $\{y_n\}$, (X,T) and (Y,S) be as above. Let $\{x_n\} \approx \{y_n\}$, $[x_n] = X$ and $[y_n] = Y$. Then $\{x_n\} \approxeq \{y_n\}$.

13.3 CHARACTERIZATIONS OF DOMINATING SEQUENCES

Throughout this section, let us write (X,T_1) and (Y,T_2) for two arbitrary l.c. TVS containing sequences $\{x_n\}$ and $\{y_n\}$ respectively. Let us write $D_1 \equiv D_{T_1} = \{p_\lambda : \lambda \in \Lambda_1\}$ and $D_2 \equiv D_{T_2} = \{q : \delta \in \Lambda_2\}$ and if (X,T_1) [resp. (Y,T_2)] is metrizable, then without loss of generality, we would let $D_1 = \{p_1 \leq p_2 \leq \cdots\}$ (resp. $D_2 = \{q_1 \leq q_2 \leq \cdots\}$). Then we have (cf. [7])

THEOREM 13.3.1 Let (X,T_1) be a Fréchet space and (Y,T_2) an ω-complete l.c. TVS. Then the following two statements are equivalent:

(i) $\{x_n\} \succ \{y_n\}$.

(ii) For each δ in Λ_2, there exist N, n_o in $\mathbb{N}$ and a positive constant K so that

$$q_\delta\left(\sum_{i=n_o}^{k} \alpha_i y_i\right) \leq K \sup_{n_o \leq j \leq k} p_N\left(\sum_{i=n_o}^{j} \alpha_i x_i\right), \tag{13.3.2}$$

for all finite sequences $\{\alpha_{n_o}, \ldots, \alpha_k\}$ of scalars.

In addition, if $\{x_n\}$ is an S.b. for (X,T_1), then (i) and (ii) are equivalent to

(iii) For each δ in Λ_2, there exist N, n_o in $\mathbb{N}$ and $C > 0$ so that

$$q_\delta\left(\sum_{i=n_o}^{k} \alpha_i y_i\right) \leq C p_N\left(\sum_{i=n_o}^{k} \alpha_i x_i\right), \tag{13.3.3}$$

for all finite sequences $\{\alpha_{n_o},\dots,\alpha_p\}$ of scalars.

<u>PROOF</u>. (ii) ==> (i) by the ω-completeness of (Y,T_2). (i) ==> (ii). If (ii) were not true, then for some δ in Λ_2 and each N, n_o, K in $\mathbb{N}$, we can find $\alpha_{n_o},\dots,\alpha_k$ such that

$$(*)\qquad q_\delta\Big(\sum_{i=n_o}^{k}\alpha_i y_i\Big) > K \sup_{n_o\leq j\leq k} p_N\Big(\sum_{i=n_o}^{j}\alpha_i x_i\Big).$$

Since the left-hand side in (*) is positive, we may scale the scalars in (*) so that

$$(**)\qquad q_\delta\Big(\sum_{i=n_o}^{k}\alpha_i y_i\Big) = 1.$$

Fix δ as in (*) and let N = 1, n_o = 1, K = 2^1. We can find $k_1 > 1$ and scalars $\alpha_1,\dots,\alpha_{k_1}$, such that (cf (*) and (**))

$$1 = q_\delta\Big(\sum_{i=1}^{k_1}\alpha_i y_i\Big) > 2 \sup_{1\leq j\leq k_1} p_1\Big(\sum_{i=1}^{j}\alpha_i x_i\Big).$$

Next, let N = 2, $n_o = 1 + k_1$ and K = 2^2. We can find $k_2 > k_1$ and scalars $\alpha_{k_1+1},\dots,\alpha_{k_2}$ so that (cf. (*) and (**))

$$1 = q_\delta\Big(\sum_{i=k_1+1}^{k_2}\alpha_i y_i\Big) > 2^2 \sup_{k_1+1\leq j\leq k_2} p_2\Big(\sum_{i=k_1+1}^{j}\alpha_i x_i\Big)$$

Thus we get an increasing sequence $\{k_n\}$ of integers ($k_{n-1}+1 < k_n$, $n \geq 1$, $k_o = 1$) and a sequence $\{\alpha_n\}$ of scalars so that, for n = 1,2,...,

$$1 = q_\delta\Big(\sum_{i=1+k_{n-1}}^{k_n}\alpha_i y_i\Big) > 2^n \sup_{1+k_{n-1}\leq j\leq k_n} p_n\Big(\sum_{i=1+k_{n-1}}^{j}\alpha_i x_i\Big).$$

By the monotonic character of $\{p_n\}$, the series $\sum_{i\geq 1}\alpha_i x_i$

is Cauchy and hence converges in (X,T_1); but $\sum_{i\geq 1} \alpha_i y_i$ diverges in (Y,T_2). This contradicts (i).

(iii) ==> (ii). Obvious (cf. Theorem 2.1.3).

(ii) ==> (iii). It suffices to observe here that in view of the barrelledness of (X,T_1), T_1 is also generated by $\{p_n^*\}$, where for x in X, $p_n^*(x) = \sup\{p_n(S_m(x)) : m \geq 1\}$. □

<u>NOTE.</u> The above result is enveloped in a similar result of [135] which, however, requires a minor modification in the light of the foregoing proof.

The next result is reproduced from [69].

<u>THEOREM 13.3.4</u> Let (Y,T_2) be complete with $\{x_n\} \subset X$ and $\{y_n\} \subset Y$.

<u>I</u>. Then the following statements are equivalent:

(i) $\{x_n\} >> \{y_n\}$.

(ii) For each δ in Λ_2, there correspond λ in Λ_1 and $C > 0$ such that

$$q_\delta\left(\sum_{i=1}^{m} \alpha_i y_i\right) \leq C p_\lambda\left(\sum_{i=1}^{m} \alpha_i x_i\right), \tag{13.3.5}$$

for all finite sequences $\{\alpha_1,\ldots,\alpha_m\}$ of scalars.

<u>II</u>. Assume that $P = sp\{x_n\}$ is bornological.

Then (i) and (ii) of I are equivalent to

(iii) For every $g \in [y_n]^*$, the system of equations

$$f(x_n) = g(y_n), \ n \geq 1 \tag{13.3.6}$$

has a unique solution f in $[x_n]^*$.

<u>III</u>. If $\{x_n\}$ is also minimal, the conditions (i), (ii) and (iii) are equivalent to

(iv) There exists n_0 in $\mathbb{N}$ so that $\{x_n : n \geq n_0\} \gg \{y_n : n \geq n_0\}$.

IV. If (X,T_1) is complete, (X,T_1), (Y,T_2) are metrizable and $\{x_n;f_n\}$, $\{y_n;g_n\}$ are b.s. with $\{f_n\} \subset [x_n]^*$, $\{g_n\} \subset [y_n]^*$, $\{g_n\}$ being total on $[y_n]$, then (i)-(iv) are equivalent to

(v) For each x in $[x_n]$, the system of equations

$$f_n(x) = g_n(y),\ n \geq 1 \tag{13.3.7}$$

has a unique solution y in $[y_n]$.

PROOF. I. (i) ==> (ii) is trivial.

(ii) ==> (i). Let $R_0: P \to [y_n]$ be defined by

$$(*) \qquad R_0\left(\sum_{i=1}^{m} \alpha_i x_i\right) = \sum_{i=1}^{n} \alpha_i y_i .$$

Then R_0 can be continuously extended as a linear map $R:[x_n] \to [y_n]$ with $Rx_n = y_n$, $n \geq 1$ and this proves (i).

II. We prove here (i) <==> (iii).

(i) ==> (iii). The truth of (i) yields a continuous linear map $R : [x_n] \to [y_n]$ with $Rx_n = y_n$, $n \geq 1$. Then $R^* : [y_n]^* \to [x_n]^*$ is a well defined linear map and so if g is in $[y_n]^*$, then $f^* = Rg$ is the unique member in $[x_n]^*$ satisfying (13.3.6).

(iii) ==> (i). Define R_0 as in (*), in I above. Let B be bounded in $(P,T_1|P)$. Consider g in $[y_n]^*$, then we have a unique f satisfying (13.3.6). Now for some $k_f > 0$, $\sup\{|f(b)| : b \in B\} \leq k_f$. If $b \in B$, then $b = \sum_{i=1}^{n} \alpha_i x_i$ and so

$$|g(R_0 b)| = |f(b)| \leq k_f .$$

Thus $R_o[B]$ is $\sigma([y_n], [y_n]^*)$-bounded and therefore it is bounded in $[y_n]$. Since P is bornological, R_o is continuous and so it can be continuously extended as a linear map $R:[x_n] \to [y_n]$ with $Rx_n = y_n$, $n \geq 1$. This proves (i).

III. Since (i) ==> (iv) is trivial, we prove (iv) ==> (i). Define $R_1, R_2:P \to [y_n]$ by

$$R_1x = \sum_{i=1}^{\min(n,n_o-1)} \alpha_i y_i, \quad R_2x = \begin{cases} \sum_{i=n_o}^{n} \alpha_i y_i; & n \geq n_o, \\ 0 & ; n \leq n_o-1, \end{cases}$$

for $x = \sum_{i=1}^{n} \alpha_i x_i$. Since $\{x_n\}$ is minimal, we can find $\{f_n\}$ in X^* with $f_n(x_m) = \delta_{mn}$, see for instance [130], p. 105. We may then define linear operators $J_1:P \to X_{n_o} \equiv \text{sp}\,\{x_1,\ldots,x_{n_o-1}\}$ and $J_2:P \to X^{n_o} \equiv \text{sp}\,\{x_{n_o}, x_{n_o+1},\ldots\}$ by

$$J_1x = \sum_{i=1}^{\min(n,n_o-1)} f_i(x)x_i;$$

$$J_2x = \begin{cases} \sum_{i=n_o}^{n} f_i(x)x_i, & n \geq n_o; \\ 0 & , n \leq n_o-1, \end{cases}$$

where

$$x = \sum_{i=1}^{n} f_i(x)x_i.$$

Observe that J_1 and J_2 are continuous. (The continuity of J_2 is a consequence of splitting members of P into members of X_{n_o} and X^{n_o}; since $\{x_1,\ldots,x_{n_o-1}\}$ is a Hamel base for X_{n_o}, a net $\{u_\delta\}$ in X_{n_o} converges to u if and only if

$\{f_i(u_\delta)\}$ converges to $f_i(u)$ for $1 \leq i \leq n_0-1$.).

Define linear mappins H_1 and H_2, $H_1 : X_{n_0} \to [y_n]$, $H_2 : X^{n_0} \to [y_n]$ by

$$H_1\left(\sum_{i=1}^{n_0-1} \alpha_i x_i\right) = \sum_{i=1}^{n_0-1} \alpha_i y_i; \; H_2\left(\sum_{i=n_0}^{n} \alpha_i x_i\right) = \sum_{i=n_0}^{n} \alpha_i y_i.$$

H_1 is trivially continuous, while H_2 is continuous by (i). Therefore, $R_1 = H_1 \circ J_1$ and $R_2 = H_2 \circ J_2$ are continuous, giving the continuity of the linear mapping $R_0 = R_1 + R_2$: $P \to [y_n]$. Extend R_0 continuously from $[x_n]$ into $[y_n]$. Thus (iv) ==> (i).

IV. (i) ==> (v). We have a continuous linear map $R:[X_n] \to [y_n]$ with $Rx_n = y_n$, $n \geq 1$. Since $f_i(x_j) = \delta_{ij} = g_i(y_j) = g_i(Rx_j)$,

$$f_i(x) = g_i(Rx), \; \forall \; x \in [x_n]; \; i \geq 1.$$

If $y = Rx$, then y is the unique solution satisfying (13.3.7) corresponding to each x in $[x_n]$, the uniqueness being the consequence of the totality of $\{g_n\}$.

(v) ==> (i). The truth of (13.3.7) yields a natural linear map $R:[x_n] \to [y_n]$, $Rx = y$ where y is the one given by (13.3.7). It is also clear that $Rx_n = y_n$, $n \geq 1$. For proving the continuity, we show that the graph of R is closed. So, let $y_n \to y$ in $[x_n]$ and $Ry_n \to y_0$. Then for $i \geq 1$,

$$g_i(Ry) = f_i(y) = \lim_{n\to\infty} f_i(y_n) = \lim_{n\to\infty} g_i(Ry_n) = g_i(y_0),$$

and so $y_0 = Ry$. □

REMARKS. Theorem 13.3.4 remains valid if (X,T_1) is taken as a metrizable l.c. TVS and (Y,T_2) is considered to be ω-complete. A few other variations of this result are also possible; for instance, in IV., (v) ==> (i), one may consider $[x_n]$ to be barrelled and $[y_n]$ to be a Pták space. Secondly, one may ask about the validity of the above result for sequences of subspaces as initiated in [135]. As pointed out earlier, in Part I, (i) ==> (ii) can be extended to sequences of subspaces as done in [135] by slightly modifying the proof given therein.

We know by an example presented earlier, that $\{x_n\} > \{y_n\}$ does not, in general, imply $\{x_n\} \gg \{y_n\}$. However, the following result yields the converse of this situation, namely,

PROPOSITION 13.3.8 Let (X,T_1) be a Fréchet space whereas (Y,T_2) is an ω-complete l.c. TVS. Assume further that $\{x_n\}$ is an S.b. Then $\{x_n\} > \{y_n\}$ if and only if $\{x_n\} \gg \{y_n\}$.

PROOF. This is a direct consequence of Theorems 13.3.1, (i) <==> (iii) and 13.3.4, I., (i) <==> (ii). □

13.4 EXISTENCE OF EQUIVALENT SEQUENCES

The following result [69] exhibits that any non-zero sequence in a complete l.c. TVS is equivalent to an S.b. of some other complete l.c. TVS and is, indeed, strictly dominated by the latter.

THEOREM 13.4.1 Let a complete l.c. TVS (X,T) contain a sequence $\{x_n\}$ with $x_n \neq 0$, $n \geqq 1$. Then there exists a complete AK-space (λ,T^*) equipped with its S.b. $\{e^n\}$ such that

(i) $\{x_n\} \sim \{e^n\}$

and

(ii) $\{e^n\} \gg \{x_n\}$.

PROOF. Let

$$\lambda = \{\alpha \in \omega : \sum_{n\geq 1} \alpha_i x_i \text{ converges in } (X,T)\}$$

and

$$p^*(\alpha) = \sup_{n\geq 1} p(\sum_{i=1}^{n} \alpha_i x_i), \quad \forall p \in D_T.$$

Let T* be the l.c. topology on the s.s. λ generated by $\{p^*: p \in D_T\}$. It is a routine exercise to verify that (λ, T^*) is a complete l.c. TVS.

For α in λ and p in D_T,

$$p^*(\alpha^{(n)} - \alpha) \leq \sup_{n+1\leq m\leq k} p(\sum_{i=m}^{k} \alpha_i x_i) \to 0 \text{ as } n \to \infty.$$

Hence $\alpha^{(n)} \to \alpha$ in (λ, T^*). Fix $i \geq 1$ and determine p in D_T so that $p(x_o) \neq 0$. Let $\alpha^\delta \to \alpha$ in (λ, T^*). Then

$$|\alpha_i^\delta - \alpha_i| = \frac{1}{p(x_i)} p[(\sum_{j=1}^{i} - \sum_{j=1}^{i-1})(\alpha_j^\delta - \alpha_j)x_j] \leq \frac{2}{p(x_i)} p^*(\alpha^\delta - \alpha)$$

and so $\alpha_i^\delta \to \alpha_i$. Therefore (λ, T^*) is an AK-space. In other words, $\{e^n\}$ is an S.b., the corresponding s.a.c.f. being given by the projections $\alpha \to \alpha_n$ or equivalently by $\{e^n\}$.

Coming to (i), by the definition of λ and its AK-character, $\{x_n\} > \{e^n\}$. On the other hand, let $\sum_{i\geq 1} \alpha_i e^i$ converge in (λ, T^*). Then for p in D_T and $\varepsilon > 0$, there exists N in $\mathbb{N}$ such that

$$p^*(\sum_{i\geq 0} \alpha_i e^i) \leq \varepsilon, \quad \forall\, n \geq N$$

$$\Rightarrow p(\sum_{i=n}^{n_o} \alpha_i x_i) \leq \varepsilon, \quad \forall\, m \geq n \geq N.$$

Therefore $\sum_{i\geq 1} \alpha_i x_i$ converges, giving $\{e^n\} \sim \{x_n\}$.

To prove (ii), define $R:sp\{e^n\} \to sp\{x_n\}$ by

$$R\left(\sum_{i=1}^{n} \alpha_i e^i\right) = \sum_{i=1}^{n} \alpha_i x_i .$$

Since for p in D_T

$$p\left(R\left(\sum_{i=1}^{n} \alpha_i e^i\right)\right) \leq \sup_{1\leq j\leq n} p\left(\sum_{i=1}^{j} \alpha_i x_i\right) = p^*\left(\sum_{i=1}^{n} \alpha_i e^i\right),$$

R is a continuous linear map from $sp\{e^n\}$ into $[x^n]$ and its natural continuous extension to $[e^n]$ yields (ii). □

<u>COROLLARY 13.4.2</u> Let a complete l.c. TVS (X,T) contain a sequence $\{x_n\}$, $x_n \neq 0$ for $n \geq 1$ such that $\{x_n\}$ is not a t.b. for $[x_n]$. Then there exists an AK-space (λ,T^*) containing $\{e^n\}$ for which $\{x_n\} \succ \{e^n\}$ but $\{x_n\} \not\gg \{e^n\}$.

In fact, invoking the notation and proof of the preceding theorem, $\{x_n\} \succ \{e^n\}$. If $\{x_n\} \gg \{e^n\}$, then $\{x_n\} \approx \{e^n\}$ and so $[x_n] \simeq [e^n]$ by Proposition 13.2.4. Hence $\{x_n\}$, is a t.b. for $[x_n]$, a contradiction.

<u>REMARK</u>. The foregoing theorem and its corollary essentially speak of the fact that domination and equivalence do not necessarily imply strict domination and strict equivalence as supported by Example 13.2.2.

As a consequence of Theorem 13.4.1, we have

<u>PROPOSITION 13.4.3</u> Let $\{x_n;f_n\}$ be a t.b. for a complete l.c. TVS (X,T). Then there exists a complete AK-space (λ,T^*) for which $\{e^n;e^n\}$ is an S.b. and there exists a 1-1 linear continuous map R from (λ,T^*) onto (X,T). For p in D_T and x in X, if $\bar{p}(x) = \sup\{p(S_n(x)): n \geq 1\}$ and $\bar{T}$ is the topology on X generated by $\bar{D} = \{\bar{p}:p \in D_T\}$, then each f_i is $\bar{T}$-continuous. In addition, if (X,T) is metrizable, then so is (λ,T^*) and $\{x_n;f_n\}$ is an S.b. for (X,T).

PROOF. Invoking both the notation and proof of Theorem 13.4.1, we find the existence of an s.s. (λ, T^*) and a continuous linear map $R: X = [e^n] \to [x_n] = X$ such that $Re^n = x_n$, $n \geq 1$. Since $\{x_n\}$ is a t.b. and $\lambda = [e^n]$, R is the map required in the first part of the result. Next, observe that $\lambda = \{\{f_i(x)\}: x \in X\}$ and $p^*(\{f_i(x)\}) = \bar{p}(x)$. If $y^\alpha \to y$ in $(X, \bar{T})$, then $\{f_i(y^\alpha)\} \to \{f_i(y)\}$ in (λ, T^*) and since (λ, T^*) is a K-space, $f_i(y^\alpha) \to f_i(y)$ for $i \geq 1$. Finally, if (X,T) is metrizable, then $\bar{T}$ is also metrizable. Since $(X, \bar{T})$ is complete and $T \subset \bar{T}$, $T \approx \bar{T}$. Thus $\{x_n; f_n\}$ is an S.b. for (X,T). □

13.5 RELATIONSHIP BETWEEN DOMINATING SEQUENCES

By choosing $\{x_n\}$ to be an S.b. in Example 13.2.2, we find that the equivalence of two sequences does not necessarily preserve the completeness, linear independence of any kind, the S.b. character, the minimality, etc. of the underlying sequences. Similarly, some of these properties are also not always preserved under strict equivalence (for example, completeness and S.b. character, see Example 13.2.3). On the other hand, the following result provides information concerning the preservation of some of the properties of sequences under strict domination.

PROPOSITION 13.5.1 Let (X, T_1) and (Y, T_2) be two l.c. TVS containing sequences $\{x_n\}$ and $\{y_n\}$ respectively such that $\{y_n\} \gg \{x_n\}$. Then the following statements are valid:

(i) If $x_n \neq 0$, $n \geq 1$, then $y_n \neq 0$, $n \geq 1$.

(ii) If $\{x_n\}$ is l.i., so is $\{y_n\}$.

(iii) If $\{x_n\}$ is ω-l.i., so is $\{y_n\}$.

(iv) If $\{x_n\}$ is minimal, so is $\{y_n\}$.

PROOF. Since $\{y_n\} \gg \{x_n\}$, there exists a continuous linear map $R: [y_n] \to [x_n]$ with $Ry_n = x_n$, $n \geq 1$. This immediately yields (i)-(iii).

To prove (iv), let us observe that we can find f_n in $[x_n]^*$ (in fact, $f_n \in X^*$) for $n \geq 1$ so that $f_n(x_m) = \delta_{mn}$. If R^* is the adjoint of R, put $g_i = R^*f_i$, $i \geq 1$. Then

$$g_n(y_m) = \langle R^*f_n, y_m\rangle = \langle f_n, x_m\rangle = \delta_{nm}$$

and so $\{y_n\}$ is minimal. □

Proposition 13.5.1 and its symmetrization immediately lead to

<u>THEOREM 13.5.2</u> Let (X,T_1) and (Y,T_2) be two l.c. TVS containing sequences $\{x_n\}$ and $\{y_n\}$ respectively. If $\{x_n\} \approx \{y_n\}$, then the following statements are valid:

(i) $x_n \neq 0$ for $n \geq 1$ if and only if $y_n \neq 0$ for $n \geq 1$.

(ii) $\{x_n\}$ is l.i. if and only if $\{y_n\}$ is l.i.

(iii) $\{x_n\}$ is ω-l.i. if and only if $\{y_n\}$ is ω-l.i.

(iv) $\{x_n\}$ is minimal if and only if $\{y_n\}$ is minimal.

<u>PROPOSITION 13.5.3</u> Let (X,T_1) and (Y,T_2) be two l.c. TVS containing sequences $\{x_n\}$ and $\{y_n\}$ respectively such that $[x_n]$ and $[y_n]$ are barrelled. If $\{x_n\} \approx \{y_n\}$, then $\{x_n\}$ is S-basic if and only if $\{y_n\}$ is S-basic.

<u>PROOF</u>. Denote by $R_1:[x_n] \to [y_n]$ and $R_2:[y_n] \to [x_n]$ the continuous linear maps with $R_1x_n = y_n$, $R_2y_n = x_n$, for $n \geq 1$ (the existence of R_1, R_2 is guaranteed by $\{x_n\} \approx \{y_n\}$). Let $\{x_n\}$ be S-basic. Choose q in D_2 and scalars $\alpha_1, \ldots, \alpha_m, \ldots, \alpha_n$ arbitrarily. Since R_1 is continuous, there exists p in D_1 such that

$$q\left(\sum_{i=1}^{m} \alpha_i y_i\right) = q\left(R\left(\sum_{i=1}^{m} \alpha_i x_i\right)\right) \leq p\left(\sum_{i=1}^{m} \alpha_i x_i\right).$$

By the converse part in Theorem 2.1.3, there exists r in D_1 so that (cf. (2.1.4)

$$q\left(\sum_{i=1}^{m} \alpha_i y_i\right) \leqq r\left(\sum_{i=1}^{n} \alpha_i x_i\right).$$

Also, R_2 is continuous and so there exists s in D_2 such that

$$q\left(\sum_{i=1}^{m} \alpha_i y_i\right) \leqq r\left(R_2\left(\sum_{i=1}^{n} \alpha_i y_i\right)\right) \leqq s\left(\sum_{i=1}^{n} \alpha_i y_i\right).$$

Therefore, making use of the first part of Theorem 2.1.3, we conclude that $\{y_n\}$ is S-basic in (Y,T_2). Similarly the converse follows. □

NOTE. If (X,T_1), (Y,T_2) are TVS and $[x_n]$, $[y_n]$ are spaces containing sets of the second category, then the preceding result also remains valid. Therefore, for all practical purposes, the following theorem holds.

THEOREM 13.5.4 Let (X,T_1), (Y,T_2) be two Fréchet spaces (resp. F-spaces). Let $x_n \in X$, $y_n \in Y$ for $n \geqq 1$. If $\{x_n\} \approx \{y_n\}$, then $\{x_n\}$ is S-basic in (X,T_1) if and only if $\{y_n\}$ is S-basic in (Y,T_2).

13.6 SIMILAR BASES AND ISOMORPHISMS

The basic aim of introducing dominance of one sequence over the other, has been to study equivalent or similar bases. However, hereafter, we prefer to use the term similar bases or sequences rather than equivalent bases or sequences and begin with the following result from [7].

THEOREM 13.6.1 Let (X,T_1) and (Y,T_2) be Fréchet spaces containing sequences $\{x_n\}$ and $\{y_n\}$ respectively. Then the following two statements are equivalent:

(i) $\{x_n\} \sim \{y_n\}$.

(ii) The condition (13.3.2) holds with δ replaced by $M(\geqq 1)$, and for each $N \geqq 1$, there exist M, n_o in $\mathbb{N}$ and a constant $J > 0$ so that

$$p_N\left(\sum_{i=n_o}^{k} \alpha_i x_i\right) \leq J \sup_{n_o \leq j \leq k} q_M\left(\sum_{i=n_o}^{j} \alpha_i y_i\right) \qquad (13.6.2)$$

for all finite sequences $\{\alpha_1,\ldots,\alpha_k\}$ of scalars.

In addition, if $\{x_n\}$ and $\{y_n\}$ are S.b., then (i) and (ii) are equivalent to

(iii) The condition (13.3.3) holds with δ replaced by $M(\geq 1)$, and for each $N \geq 1$, there exist M, n_o in $\mathbb{N}$ and a constant $D > 0$ so that

$$p_N\left(\sum_{i=n_o}^{k} \alpha_i x_i\right) \leq D q_M\left(\sum_{i=1}^{k} \alpha_i y_i\right) \qquad (13.6.3)$$

for all finite sequences $\{\alpha_1,\ldots,\alpha_k\}$ of scalars.

PROOF. Immediate from Theorem 13.3.1. □

Concerning strictly similar sequences, we have

THEOREM 13.6.4 Let (X,T_1) and (Y,T_2) be two complete l.c. TVS containing sequences $\{x_n\}$ and $\{y_n\}$ respectively.

I. Then the following statements are equivalent:

(i) $\{x_n\} \approx \{y_n\}$

(ii) The condition (13.3.5) holds, and for each λ in Λ_1, there corresponds δ in Λ_2 and $D > 0$ so that

$$p_\lambda\left(\sum_{i=1}^{m} \alpha_i x_i\right) \leq D q_\delta\left(\sum_{i=1}^{m} \alpha_i y_i\right) \qquad (13.6.5)$$

for all finite sequences $\{\alpha_1,\ldots,\alpha_m\}$ of scalars.

II. Let $sp\{x_n\}$ and $sp\{y_n\}$ be bornological.
Then (i) and (ii) of I are equivalent to

(iii) For each f in $[x_n]^*$ (resp. g in $[y_n]^*$) the system of equations (13.3.6) has a unique solution g in $[y_n]^*$

(resp. f in $[x_n]^*$).

<u>III</u>. If $\{x_n\}$ and $\{y_n\}$ are minimal, the conditions (i)-(iii) are equivalent to

(iv) There exists n_0 in $\mathbb{N}$ so that $\{x_n : n \geq n_0\} \approx \{y_n : n \geq n_0\}$.

<u>IV</u>. Let (X,T_1), (Y,T_2) be metrizable and $\{x_n;f_n\}$, $\{y_n;g_n\}$ be b.s. with $\{f_n\}$, $\{g_n\}$ being total respectively on $[x_n]$, $[y_n]$ ($\{f_n\} \subset [x_n]^*$, $\{g_n\} \subset [y_n]^*$), then (i) - (iv) are equivalent to

(v) For each x in $[x_n]$ (resp. y in $[y_n]$), the system of equations (13.3.7) has a unique solution y in $[y_n]$ (resp. x in $[x_n]$).

<u>PROOF</u>. Use theorem 13.3.4. □

Recalling that a TVS has t-*property* if closure of each absorbing set contains an open set, we have the following characterization of similarity of two bases in terms of isomorphisms (cf. [91]).

<u>THEOREM 13.6.6</u> Let (X,T_1) and (Y,T_2) be TVS with the t-property or simply barrelled spaces having S.b. $\{x_n;f_n\}$ and $\{y_n;g_n\}$ respectively. Then $\{x_n\} \sim \{y_n\}$ if and only if $(X,T_1) \simeq (Y,T_2)$ under R (say) with $Rx_n = y_n$, $n \geq 1$.

<u>PROOF</u>. It is enough to establish the necessary part. For $m \geq 1$, define the linear maps $R_m : X \to Y$ by

$$R_m x = \sum_{i=1}^{m} f_i(x) y_i, \quad x = \sum_{i \geq 1} f_i(x) x_i.$$

Also, define $R : X \to Y$ by

$$Rx = \sum_{i \geq 1} f_i(x) y_i,$$

the convergence of the infinite series being assured by the similarity of $\{x_n\}$ and $\{y_n\}$. Then $R_m x \to Rx$ for each x in X and so by the Banach-Steinhaus theorem, R is a continuous linear map. R is clearly 1-1 with $Rx_n = y_n$, $n \geq 1$. By the similarity of the two bases, R is onto. Hence $(X,T_1) \simeq (Y,T_2)$ under R. □

NOTE. Theorem 13.6.6, generally known as the *isomorphism theorem for similarity of Schauder bases*, was earlier proved in [7] (cf. also [8]) for Fréchet spaces, where full use of the Fréchet character of the spaces was made in deriving Theorem 13.6.6 from Theorem 13.6.1.

REMARK. The restriction of barrelledness of spaces, as we observed above, appears to be essential in order to prove the necessary part of Theorem 13.6.6; in fact, this restriction cannot be avoided as exhibited by the next two examples (cf. [91]).

EXAMPLE 13.6.7 Let (X,T_1) be an arbitrary infinite dimensional Banach space having an S.b. $\{x_n;f_n\}$. Put $Y = X$, $T_2 = \sigma(X,X^*)$, $y_n = x_n$, $g_n = f_n$, $n \geq 1$. Then $\{x_n\} \sim \{y_n\}$ and $(X,T_1) \not\simeq (Y,T_2)$.

EXAMPLE 13.6.8 Let $X = Y = \ell^1$; T_1 is the topology on X generated by $\|\cdot\|_1$ and T_2 is the topology on X generated by $\|\cdot\|_\infty$. Then $\{e^n;e^n\}$ is an S.b. for (X,T_1) and (Y,T_2), and the bases for these two spaces are similar. However, $(X,T_1) \not\simeq (Y,T_2)$ since (X,T_1) is complete and (Y,T_2) is incomplete.

Let (X,T_1) and (Y,T_2) be Fréchet spaces in the following exercises (cf. [7]).

EXERCISE 13.6.9 Let $\{x_n\}$ be an S.b. for (X,T_1) and $R:(X,T_1) \to (Y,T_2)$ a topological isomorphism from X into

Y with $Rx_n = y_n$, $n \geq 1$. Prove that $\{y_n\}$ is an S.b. for a closed subspace Y_0 of (Y,T_2) such that $\{x_n\} \sim \{y_n\}$. Further, if $\{y_n\}$ is an S.b. for some closed subspace Y_0 of (Y,T_2) such that $\{x_n\} \sim \{y_n\}$, prove that $(X,T_1) \simeq (Y_0,T_2)$, say, under R such that $Rx_n = y_n$; $n \geq 1$.

EXERCISE 13.6.10 Let $\{x_n\}$ and $\{y_n\}$ be S.b. for (X,T_1) and (Y,T_2) respectively. For k in $\mathbb{N}$, let $X_k = sp\{x_n : n \geq k\}$ and $Y_k = sp\{y_n : n \geq k\}$. If $X_k \simeq Y_k$, say, under R_k with $R_k x_n = y_n$ for $n \geq k$, prove that R_k can be extended to a topological isomorphism R from (X,T_1) onto (Y,T_2) with $Rx_n = y_n$ for $n \geq 1$. If $(X,T_1) \simeq (Y,T_2)$, say, under R with $Rx_n = y_n$, $n \geq 1$, show that $X_k \simeq Y_k$ under $R_k \equiv R|X_k$.

HINT. Use Theorem 13.6.6 in all these exercises.

13.7 APPLICATIONS

The results of the previous sections may be used to characterize sequences as bases and λ-unconditionally Cauchy series introduced in [128].

We begin with (cf. [69])

THEOREM 13.7.1 Let (X,T) be a Fréchet space containing a complete sequence $\{x_n\}$, $x_n \neq 0$, $n \geq 1$. Then $\{x_n\}$ is an S.b. for (X,T) if and only if

$$\{\beta_n\} \in \omega,\ \{x_n\} > \{\beta_n\} \Rightarrow \{x_n\} \gg \{\beta_n\} \tag{13.7.2}$$

PROOF. The necessary condition is contained in Proposition 13.3.8.

Conversely, let (13.7.2) be true. Let λ be the s.s. of Theorem 13.4.1. Since $\{e^n\}$ is an S.b. for (λ,T^*), for each g in $\lambda^* \equiv (\lambda,T^*)^*$ we have $\{x_n\} > \{g(e^n)\}$. Hence $\{x_n\} \gg \{g(e^n)\}$ for each g in λ^* and so there exists a continuous linear map $R_g : [x_n] \to [g(e^n)]$ such that $R_g(x_n) = g(e^n)$, $n \geq 1$. Since $[x_n] = X$, $R_g \in X^*$. The uniqueness of R_g is easily seen and so from Theorem 13.3.4 (iii) ==> (i) we get $\{x_n\} \gg \{e^n\}$. Therefore, by (ii) of Theorem 13.4.1, $\{x_n\} \approx \{e^n\}$. Observe that (λ, T^*) is also a Fréchet space (cf. the proof of Theorem 13.4.1). Hence by Proposition 13.5.3, $\{x_n\}$ is an S.b. for (X,T). □

Finally, let us recall (cf. [128])

DEFINITION 13.7.3 Corresponding to an s.s. λ, a series $\sum_{n\geq 1} x_n$ in an l.c. TVS (X,T) is said to be a *weakly* λ-*unconditionally Cauchy* (w.λ-u.C.) provided $\sum_{n\geq 1} \alpha_n x_n$ converges in (X,T) for each α in λ.

If $\lambda \equiv (\lambda, T_\lambda)$ is an AK-space, the definition of a w.λ-u.C. series can be reformulated in the form of

PROPOSITION 13.7.4 Let λ be an AK-space. Then a series $\sum_{n\geq 1} x_n$ in an l.c. TVS (X,T) is w.λ-u.C. if and only if $\{e^n\} > \{x_n\}$

PROOF. Straightforward. □

Next, observe that if λ is an AK-space, then $\lambda^\beta = \{\alpha \in \omega : \{e^n\} > \{\alpha_n\}\}$. Now we have the main result

PROPOSITION 13.7.5 Let $\lambda \equiv (\lambda, T_2)$ be an FK-AK-space and (X,T) an ω-complete l.c. TVS. Then a series $\sum_{n\geq 1} x_n$ is w.λ-u.C. if and only if $\{e^n\} > \{f(x_n)\}$ for each f in X^*.

PROOF. If $\sum_{n\geq 1} x_n$ is w.λ-u.C., then from Proposition 13.7.4, $\{e^n\} > \{x_n\}$ and so $\{e^n\} > \{f(x_n)\}$.

Conversely, let $\{e^n\} > \{f(x_n)\}$ for each f in X^*. By Theorem 13.7.1,

$$\{e^n\} \gg \{f(x_n)\}, \quad \forall f \in X^*.$$

Hence there exists a continuous linear map $R_f : \lambda \equiv [e^n] \to [f(x_n)]$ with $R_f e^n = f(x_n)$, $n \geq 1$. Clearly $R_f \in \lambda^*$ for each f in X* and such a R_f is uniquely defined for f in X*. Therefore, from Theorem 13.3.4, (iii)==>(i) (cf. especially remarks following its proof), $\{e^n\} \gg \{x_n\}$ and we have completed the proof.

NOTES. The preceding theorem is a more general version of a similar result proved in [128] (cf. Theorem 3.2) and, in particular, includes similar results to those of [162], p. 117 and [232], p. 471, when $\lambda = c_o$, and those of [15], p. 21 and [214], p. 131, when $\lambda = \ell^p$ $(1 < p < \infty)$. For a further detailed study of w.λ-u.C. series, the reader is referred to [129], chapter 3.

14 Further discussion about similar bases

14.1 INTRODUCTION

In this chapter, we essentially discuss two aspects of the basis theory: absolute similarity of bases and certain concrete examples of similar bases in the spaces of analytic functions. The present chapter is a continuation of the last one.

14.2 ABSOLUTELY SIMILAR BASES

We have seen in the last chapter that the concept of similar bases essentially depends upon the convergence of infinite series. Since there are basically three modes of convergence of an infinite series, namely, ordinary, unconditional or subseries or bounded multiplier and absolute, one may obviously talk of other types of similar bases depending upon absolute, unconditional convergence, etc. of the underlying series. There is a huge class of spaces having similar bases related by the absolute convergence of the infinite series (e.g. spaces of analytic functions); we, however, prefer to take up a detailed study of similar bases depending upon the absolute convergence of the resulting expansions. To be very general, let us therefore introduce

DEFINITION 14.2.1 Let (X,T_1) and (Y,T_2) be two l.c. TVS containing sequences $\{x_n\}$ and $\{y_n\}$ respectively. Then $\{x_n\}$ is *absolutely similar* to $\{y_n\}$ (to be written as $\{x_n\} \bar{\underset{-}{\sim}} \{y_n\}$) provided

$$\sum_{n\geq 1} p(\alpha_n x_n) < \infty, \ \forall p \in D_{T_1} \iff \sum_{n\geq 1} q(\alpha_n y_n) < \infty, \ \forall q \in D_{T_2}.$$

Unless otherwise specified, our notation for (X,T_1), (Y,T_2), D_1, D_2, $\{x_n\}$, $\{y_n\}$, etc. will remain the same as that given in section 13.3.

NOTE. Without some restrictions of one kind or the other, it is not possible to relate similar sequences with absolutely similar sequences. However, we have

PROPOSITION 14.2.2 Let (X,T_1) and (Y,T_2) be ω-complete. Suppose $\{x_n\}$ and $\{y_n\}$ are semi-ℓ^1-bases for (X,T_1) and (Y,T_2) respectively. Then $\{x_n\} \sim \{y_n\}$ if and only if $\{x_n\} \bar{\underline{\sim}} \{y_n\}$.

PROOF. In fact, let $\{x_n\} \sim \{y_n\}$. If $\sum_{n\geq 1} \alpha_n x_n$ converges absolutely, then $\sum_{n\geq 1} \alpha_n x_n$ converges. Hence $\sum_{n\geq 1} \alpha_n y_n$ converges and therefore this series converges absolutely by Proposition 4.4.2. Thus, proceeding in a similar manner, we find that $\{x_n\} \bar{\underline{\sim}} \{y_n\}$. The other part follows similarly.□

The following result is a variation of Proposition 14.2.2.

PROPOSITION 14.2.3 Let $\{x_n\}$ and $\{y_n\}$ be ℓ^1-bases for (X,T_1) and (Y,T_2) respectively. Then $\{x_n\} \sim \{y_n\}$ if and only if $\{x_n\} \bar{\underline{\sim}} \{y_n\}$.

NOTE. Propositions 14.2.2 and 14.2.3 envelop a similar result of [7] (cf. Theorem 9, p. 292) which is derived by a different method.

14.3 CHARACTERIZATIONS OF ABSOLUTE SIMILARITY

The ultimate aim of this section is to provide an isomorphism theorem for characterizing absolutely similar bases i.e. a result similar to Theorem 13.6.6. To achieve the objective, we need to discuss a few related results.

The following result appears to be quite natural for

providing sufficient conditions for absolutely similar sequences.

PROPOSITION 14.3.1 Two sequences $\{x_n\} \subset (X,T_1)$ and $\{y_n\} \subset (Y,T_2)$ satisfy the relationship that $\{x_n\} \bar{\underset{-}{\sim}} \{y_n\}$ provided that for every λ in Λ_1 there exists δ in Λ_2 and for γ in Λ_2 there exists μ in Λ_1 such that

$$p_\lambda(x_n) \leq q_\delta(y_n), \ n \geq 1 \tag{14.3.2}$$

and

$$q_\gamma(y_n) \leq p_\mu(x_n), \ n \geq 1. \tag{14.3.3}$$

PROOF. Straightforward. □

The converse of the above result is obtained at the expense of restricting the spaces (X,T_1), (Y,T_2) and we follow [7] for the corresponding discussion.

If (X,T_1) and (Y,T_2) are metrizable, the condition (14.3.3) can be rephrased as: for every j in $\mathbb{N}$ there exist i in $\mathbb{N}$ and $M > 0$ such that

$$q_j(y_n) \leq Mp_i(x_n), \ n \geq 1, \tag{14.3.4}$$

while (14.3.2) takes the form: for each i in $\mathbb{N}$, there exists $j \geq 1$ and $K > 0$ such that

$$p_i(x_n) \leq Kq_j(y_n), \ n \geq 1. \tag{14.3.5}$$

LEMMA 14.3.6 Let (X,T_1) and (Y,T_2) be metrizable. Let $x_n \in X$, $y_n \in Y$ for $n \geq 1$. Suppose that for each $i \geq 1$,

$$\sum_{n\geq 1} p_i(\alpha_n x_n) < \infty \implies \alpha_n y_n \to 0, \tag{14.3.7}$$

where $\alpha \in \omega$, then $\{x_n\}$ and $\{y_n\}$ satisfy (14.3.4).

PROOF. If (14.3.4) were not true, we would get some j in N and $\{n_i\}$ in I such that

$$q_j(y_{n_i}) > 2^i p_i(x_{n_i}), \quad \forall i \geq 1.$$

Let

$$\alpha_{n_i} = 1/q_j(y_n); \ \alpha_n = 0, \ n \neq n_i, \ i \geq 1.$$

Then for each $m \geq 1$

$$\sum_{n \geq n_m} p_m(\alpha_n x_n) = \sum_{i \geq m} p_m(\alpha_{n_i} x_{n_i}) \leq \sum_{i \geq 1} 1/2^i < \infty .$$

But $q_j(\alpha_{n_i} y_{n_i}) = 1$ for $i \geq 1$. This contradicts (14.3.7). □

Symmetry considerations in Lemma 14.3.6 then yield the following

PROPOSITION 14.3.8 Let (X,T_1) and (Y,T_2) be metrizable with $x_n \in X$ and $y_n \in Y$, $n \geq 1$. Then $\{x_n\} \underset{=}{\sim} \{y_n\}$ if and only if (14.3.4) and (14.2.5) hold.

It is natural to enquire if conditions (14.3.4) and (14.3.5) can, in some way, be related with the similarity of $\{x_n\}$ and $\{y_n\}$. In this direction, we have

PROPOSITION 14.3.9 Let (X,T_1), (Y,T_2) be metrizable and $\{x_n\}$, $\{y_n\}$ be ℓ^1-bases for (X,T_1) and (Y,T_2) respectively. Then $\{x_n\} \sim \{y_n\}$ if and only if (14.3.4) and (14.3.5) hold.

PROOF. It is immediate from Propositions 14.2.3 and 14.3.8. □

A variation of the above result is contained in

PROPOSITION 14.3.10 Let $\{x_n\}$ and $\{y_n\}$ be semi ℓ^1-bases for Fréchet spaces (X,T_1) and (Y,T_2) respectively. Then (14.3.4) and (14.3.5) are valid if and only if $\{x_n\} \sim \{y_n\}$.

PROOF. Follows from Propositions 14.2.2 and 14.3.8. □

In order to characterize absolute similarity of $\{x_n\}$ and $\{y_n\}$ in terms of isomorphisms, we try to find conditions equivalent to (14.3.2), (14.3.3) or (14.3.4), (14.3.5). To begin with, we prove

PROPOSITION 14.3.11 Let (X,T_1) contain a fully ℓ^1-base $\{x_n;f_n\}$ and $\{y_n\}$ be an arbitrary sequence in (Y,T_2). Then (14.3.3) and the following condition are equivalent: for each γ in Λ_2, there exists μ in Λ_1 such that

$$q(\sum_{i=1}^{k} \alpha_i y_i) \leq p_\mu(\sum_{i=1}^{k} \alpha_i x_i), \tag{14.3.12}$$

for all finite sequences of scalars $\{\alpha_1,\ldots,\alpha_k\}$.

PROOF. (14.3.10) readily yields (14.3.3).

Conversely, assume the truth of (14.3.3). Observe that T_1 is also generated by $\{Q_p : p \in D_1\}$(cf. the remark following the proof of Proposition 6.4.2). Write Q_μ for Q_{p_μ}, $\mu \in \Lambda_1$. Then, for given scalars $\alpha_1,\ldots,\alpha_k$, we have

$$q_\gamma(\sum_{i=1}^{k} \alpha_i y_i) \leq \sum_{i=1}^{k} p_\mu(\alpha_i x_i) = Q_\mu(\sum_{i=1}^{k} \alpha_i x_i).$$

But there exists η in Λ_1 so that $Q_\mu(x) \leq p_\eta(x)$ for all x in X. Hence (14.3.10) holds with μ replaced by η. □

Let us also consider the following set of conditions for $\{x_n\} \subset (X,T_1)$, $\{y_n\} \subset (Y,T_2)$ and $\alpha \in \omega$.

$$\sum_{n\geq 1} p_\lambda(\alpha_n x_n) < \infty, \ \forall\lambda \in \Lambda_1 \implies \sum_{n\geq 1} q_\delta(\alpha_n y_n) < \infty, \quad \forall\, \delta \in \Lambda_2. \tag{14.3.13}$$

$$\sum_{n\geq 1} p_\lambda(\alpha_n x_n) < \infty, \ \forall\lambda \in \Lambda_1 \implies \sum_{n\geq 1} \alpha_n y_n \text{ converges.} \tag{14.3.14}$$

$$\sum_{n\geq 1} p_\lambda(\alpha_n x_n) < \infty, \ \forall\lambda \in \Lambda_1 \implies \alpha_n y_n \to 0. \qquad (14.3.15)$$

Then we have

<u>PROPOSITION 14.3.16</u>: (14.3.3) ==> (14.3.13), (14.3.14) ==> (14.3.15) and if (Y,T_2) is ω-complete, then (14.3.13) ==> (14.3.14). Conversely, if (X,T_1) and (Y,T_2) are metrizable, then (14.3.15) ==> (14.3.4).

<u>PROOF</u>. See Lemma 14.3.6 for the last implication; other parts are obvious. □

<u>PROPOSITION 14.3.17</u> Let $\{x_n\}$ be an arbitrary sequence in a metrizable space (X,T_1) and R a continuous linear map from (X,T_1) into a Fréchet space (Y,T_2) with $Rx_n = y_n$ (say). Then the equivalent conditions (14.3.4), (14.3.13), (14.3.14) and (14.3.15) hold. Conversely, if $\{x_n\}$ is a fully ℓ^1-base for a metrizable space (X,T_1) and $\{y_n\}$ an arbitrary sequence in a Fréchet space for which any of the equivalent conditions (14.3.4), (14.3.13)-(14.3.15) holds, then there exists a continuous linear map $R:(X,T_1) \to (Y,T_2)$ with $Rx_n = y_n$, $n \geq 1$.

<u>PROOF</u>. The first statement follows from the continuity of R and Proposition 14.3.11; and the converse statement is a consequence of Proposition 14.3.11, for $R(x) = \sum_{i=1}^{\infty} f_i(x)y_i$. □

Suppose $\{x_n;f_n\}$ is an S.b. for (X,T_1) and R is a topological isomorphism from (X,T_1) onto (Y,T_2) with $Rx_n = y_n$. For y in Y, $y = R^{-1}x$ for a unique x in X. Thus $y = \sum_{n\geq 1} f_n(x)y_n$ and this expansion is unique. Let $\{g_n\}$ be the s.a.c.f. corresponding to $\{y_n\}$. Hence for each y in Y, there exists a unique x in X with $g_n(y) = f_n(x)$. Also, there exist p in D_1 and q in D_2 such that $|g_n(y)| = |f_n(x)| \leq p(x) = p(R^{-1}y) \leq q(y)$ for all y in Y. Hence $\{y_n\} \equiv \{y_n;g_n\}$ is an S.b. for (Y,T_2) such that $\{x_n\} \sim \{y_n\}$. Since (14.3.2) and (14.3.3) are satisfied in the present situation, the semi ℓ^1-character (resp. ℓ^1-character) of $\{x_n\}$ implies the corresponding one for $\{y_n\}$ as well. Thus by Proposition 14.2.2, if (X,T_1), (Y,T_2) are ω-complete, $\{x_n\}$ is a semi ℓ^1-base and R is as above, then $\{y_n\}$ is a semi ℓ^1-base for (Y,T_2) with $\{x_n\} \bar{\underset{-}{\sim}} \{y_n\}$. This discussion leads to the first part of the isomorphism theorem, namely,

THEOREM 14.3.18 Let (X,T_1) and (Y,T_2) be two ω-complete l.c. TVS such that $(X,T_1) \simeq (Y,T_2)$, say, under R. Suppose $\{x_n\}$ is a semi ℓ^1-base for (X,T_1). If $y_n = Rx_n$; $n \geq 1$, then $\{y_n\}$ is a semi ℓ^1-base for (Y,T_2) such that $\{x_n\} \bar{\underset{-}{\sim}} \{y_n\}$.

Conversely, we have

THEOREM 14.3.19 Let (X,T_1) and (Y,T_2) be two ω-complete barrelled spaces possessing semi ℓ^1-bases $\{x_n\}$ and $\{y_n\}$ respectively such that $\{x_n\} \bar{\underset{-}{\sim}} \{y_n\}$. Then $(X,T_1) \simeq (Y,T_2)$, say, under R with $Rx_n = y_n$, $n \geq 1$.

PROOF. Apply Proposition 14.2.2 and Theorem 13.6.6. □

NOTE. The result contained in Theorems 14.3.18 and 14.3.19 is generally referred to as the *isomorphism theorem for absolute similarity* of Schauder bases.

The following special case of the isomorphism theorem will be found useful in chapter 19. In fact, the previous considerations lead to the following (cf. also [39])

THEOREM 14.3.20 Let (X,T_1) and (Y,T_2) be two nuclear Fréchet spaces having S.b. $\{x_n;f_n\}$ and $\{y_n;g_n\}$ respectively with D_1 and D_2 as before. Then $\{x_n\} \sim \{y_n\}$ if and only if $\{x_n\} \bar{\underset{-}{\sim}} \{y_n\}$ if and only if (14.3.4) and (14.3.5) hold.

PROOF. In view of Proposition 14.2.15, $\{x_n\}$ and $\{y_n\}$ are both fully ℓ^1-bases. Now apply Propositions 14.2.2 and 14.3.10. □

FINAL REMARKS

Absolute similarity of sequences take a very simple form in normed spaces. In fact, if $x_n \in (X, \|\cdot\|_1)$ and $y_n \in (Y, \|\cdot\|_2)$, then from Proposition 14.3.8, $\{x_n\} \bar{\sim} \{y_n\}$ if and only if there exist M, N > 0 such that

$$M \|x_n\|_1 \leqq \|y_n\|_2 \leqq N \|x_n\|_1, \; n \geqq 1. \tag{14.3.21}$$

Let us now observe that the basis hypothesis in Proposition 14.2.2 is essential, otherwise the desired conclusion may not be true, for instance, consider (cf. [7])

EXAMPLE 14.3.22 Let $x^n = e^n$, $n \geqq 1$; $y^1 = e^1$, $y^n = e^n - e^{n-1}$, $n \geqq 2$ and $(X,T_1) = (Y,T_2) = (\ell^1, \|\cdot\|_1)$. Then $\{x^n\}$ is a semi ℓ^1-base for (X,T_1) and $\{y^n\}$ is not a t.b. for (Y,T_2). Here $\{x^n\} \not\sim \{y^n\}$; indeed, $\sum_{n\geqq 1} (1/n)x^n$ diverges and $\sum_{n\geqq 1} (1/n)y^n$ converges. Since $\|y^1\|_1 = \|x^1\|_1$ and $\|y^n\|_1 = 2\|x^n\|_1$, $n \geqq 2$, one gets

$$\frac{1}{2}\|x^n\|_1 < \|y^n\|_1 \leqq 2\|x^n\|_1, \; n \geqq 1$$

and so from (14.3.21), $\{x^n\} \bar{\sim} \{y^n\}$.

NOTE. The discussion involving absolutely similar bases can be further extended by introducing the concept of λ-similar bases corresponding to an s.s. λ equipped with its normal topology $\eta(\lambda,\lambda^\times)$. In fact, let us agree to call two S.b. $\{x_n;f_n\}$ and $\{y_n;g_n\}$ for l.c. TVS (X,T) and (Y,S), respectively, *λ-similar* (to be expressed as $\{x_n\} \overset{\lambda}{\sim} \{y_n\}$), provided for α in ω, $\{p(\alpha_n x_n)\} \in \lambda$ for each p in D_T <==> $\{q(\alpha_n y_n)\} \in \lambda$ for each q in D_S. It is easily seen that if $\{x_n;f_n\}$ and $\{y_n;g_n\}$ are λ-bases, then $\{x_n\} \sim \{y_n\}$ <==> $\{x_n\} \overset{\lambda}{\sim} \{y_n\}$. This generalizes Proposition 14.2.3. For further discussion, see [70].

14.4 CONSTRUCTION OF SIMILAR BASES

Because of their vast importance in several off-shoots of analysis, spaces of analytic functions occupy a privileged place among the family of all known classes of locally convex spaces. As such, we would like to discuss in this section the precise forms of bases in spaces of analytic functions, which are similar to the classical ones appearing in Taylor or Dirichlet series.

Functions of several complex variables

Similar bases have been constructed in different types of spaces of analytic functions of one or more complex variables represented by Taylor series, for example [3], [5], [6], [103], [118], [119], [126], etc. and we pick up [118] to reproduce one such instance.

Fix four positive real numbers $\rho_1, \rho_2; \sigma_1, \sigma_2$. Let X be the space of all entire functions of two variables given by Taylor series. Let $Y \equiv Y(\rho_1, \rho_2; \sigma_1, \sigma_2)$ be a subspace of X, consisting of those f from X such that to every $\varepsilon > 0$

$$\sup_{|z_i| \leq r_1, |z_2| \leq r_2} |f(z_1, z_2)| \equiv M(f; f_1, r_2) \qquad (14.4.1)$$

$$< \exp\{(\sigma_1+\varepsilon) r_1^{\rho_1} + (\sigma_2+\varepsilon) r_2^{\rho_2}\},$$

valid for all sufficiently large values of r_1 and r_2. Taking into consideration the Taylor expansion of each f in Y as in (*) of Example 2.3.3, after [54], we find that (14.4.1) is equivalent to: for each $\varepsilon > 0$

$$|a_{mn}| < \left[\frac{e\rho_1(\sigma_1+\varepsilon)}{m}\right]^{m/\rho_1} \left[\frac{e\rho_2(\sigma_2+\varepsilon)}{n}\right]^{n/\rho_2}, \qquad (14.4.2)$$

valid for all large values of m+n. Thus, on Y, one may define a natural l.c. topology T generated by the family

$\{\|\cdot\,;\,\sigma_1+\delta,\,\sigma_2+\delta\|:\,\delta>0\}$ of norms on Y, where for f in Y

$$\|f;\,\sigma_1+\delta,\,\sigma_2+\delta\| \tag{14.4.3}$$

$$= |a_{oo}| + \sum_{m+n\geqq 0} |a_{mn}| \left[\frac{m}{(\sigma_1+\delta)e\rho_1}\right]^{m/\rho_1}\left[\frac{n}{(\sigma_2+\delta)e\rho_2}\right]^{n/\rho_2},$$

the convergence of the series in (14.4.3) being guaranteed by (14.4.2). It is not difficult to see that (Y,T) is a Fréchet space (cf. [118], Theorem 2.1). Following Example 2.3.3, we can show that the infinite matrix (e_{mn}) is an S.b. for (Y,T) (cf. also [118], p. 373); in particular, let us observe that the convergence of $\sum\sum_{m+n\geqq 0} a_{mn}e_{mn}$ in (Y,T) is equivalent to the truth of (14.4.2). In fact, the convergence of the infinite double series in T implies the convergence of this series in the compact-open topology (e.g., [66], p. 107) and so (a_{mn}) satisfies (14.4.2). On the other hand, if (a_{mn}) satisfies (14.4.2), then the double infinite series in question is Cauchy and so it converges in (Y,T).

In order to characterize bases in (Y,T) similar to (e_{mn}), we prove two intermediary lemmas.

LEMMA 14.4.4 For any infinite matrix (α_{mn}) in Y, the following conditions are equivalent:

(A) For each $\delta > 0$, there exist $k,\ \varepsilon' > 0$ such that

$$\|\alpha_{mn};\,\sigma_1+\delta,\,\sigma_2+\delta\| < k\left[\frac{m}{(\sigma_1+\varepsilon')e\rho_1}\right]^{m/\rho_1}\left[\frac{n}{(\sigma_2+\varepsilon')e\rho_2}\right]^{n/\rho_2},$$

for all $m+n \geqq 0$.

(B) For all infinite matrices (a_{mn}) of complex numbers,

$$\text{conv. of } \sum\sum_{m+n\geqq 0} a_{mn}e_{mn} \implies \text{conv. of } \sum\sum_{m+n\geqq 0} a_{mn}\alpha_{mn}$$

in (Y,T).

(C) For all infinite matrices (a_{mn}) of complex numbers,

$$\text{conv. of } \sum_{m+n\geqq 0}\sum a_{mn}e_{mn} \implies a_{mn}\alpha_{mn} \to 0$$

in (Y,T).

<u>PROOF</u>. (A) ==> (B). Choose ε with $0 < \varepsilon < \varepsilon'$. Then we can find $N \equiv N(\varepsilon')$ such that (14.4.2) holds for all $m+n \geqq N$. Hence

$$|a_{mn}| \;\|\alpha_{mn};\, \sigma_1+\delta,\, \sigma_2+\delta\| \leqq k \left(\frac{\sigma_1+\varepsilon}{\sigma_1+\varepsilon'}\right)^{m/\rho_1} \left(\frac{\sigma_2+\varepsilon}{\sigma_2+\varepsilon'}\right)^{n/\rho_2},$$

for all $m+n \geqq N$. This shows that the series $\sum_{m+n\geqq 0}\sum a_{mn}\alpha_{mn}$ converges absolutely and this establishes (B).

(B) ==> (C). Trivial.

(C) ==> (A). Suppose that (A) is not true. Then for some δ and each integer $k \geqq 1$, we find increasing sequences $\{m_k\}$ and $\{n_k\}$ such that

$$(*)\quad \|\alpha_{m_k n_k};\, \sigma_1+\delta,\, \sigma_2+\delta\| \geqq k \left[\frac{m_k}{(\sigma_1+1/k)e\rho_1}\right]^{m_k/\rho_1} \times \left[\frac{n_k}{(\sigma_2+1/k)e\rho_2}\right]^{n_k/\rho_2}.$$

Define

$$a_{mn} = \begin{cases} \dfrac{1}{\|\alpha_{mn};\sigma_1+\delta,\sigma_2+\delta\|}; & m = m_k,\ n = n_k,\ k \geqq 1, \\ 0, & \text{otherwise.} \end{cases}$$

Making use of (*), we find that (a_{mn}) satisfies (14.4.2). Consequently, the infinite series $\sum_{m+n\geqq 0}\sum a_{mn}e_{mn}$ converges.

Hence, by (C), $(a_{mn}\alpha_{mn})$ must tend to zero. However, by the construction of (a_{mn}), this is not true. The contradiction arrived at proves (A). □

<u>LEMMA 14.4.5</u> For an infinite matrix (α_{mn}) in Y, the following conditions are equivalent:

(a) For each $\eta > 0$ there exist $\delta \equiv \delta(\eta)$ and $N \equiv N(\eta)$ in $\mathbb{N}$, so that for $m+n \geq N$,

$$\|\alpha_{mn};\sigma_1+\delta,\sigma_2+\delta\| \geq \left[\frac{m}{(\sigma_1+\eta)e\rho_1}\right]^{m/\rho_1}\left[\frac{n}{(\sigma_2+\eta)e\rho_2}\right]^{n/\rho_2},$$

(b) For all infinite matrices (a_{mn}) of complex numbers,

$$\text{conv. of } \sum_{m+n\geq 0}\sum a_{mn}\alpha_{mn} \implies \text{conv. of } \sum_{m+n\geq 0}\sum a_{mn}e_{mn}$$

in (Y,T).

(c) For all infinite matrices (a_{mn}) of complex numbers, convergence of $(a_{mn}\alpha_{mn})$ to zero implies convergence of $\sum_{m+n\geq 0}\sum a_{mn}e_{mn}$ in (Y,T).

<u>PROOF</u>. (c) ==> (b). Obvious.

(b) ==> (a). Let (a) be untrue. Then there exist $\eta > 0$ and increasing sequences $\{m_k\}$, $\{n_k\}$ in I such that for $k \geq 1$,

$$\|\alpha_{m_k n_k};\ \sigma_1+k^{-1},\sigma_2+k^{-1}\| \leq \left[\frac{m_k}{(\sigma_1+\eta)e\rho_1}\right]^{m_k/\rho_1} \times \left[\frac{n_k}{(\sigma_2+\eta)e\rho_2}\right]^{n_k/\rho_2}.$$

Let $0 < \gamma < \eta$ and define (a_{mn}) by

$$a_{mn} = \begin{cases} \left[\frac{(\sigma_1+\gamma)e\rho_1}{m}\right]^{m/\rho_1} \left[\frac{(\sigma_2+\eta)e\rho_2}{n}\right]^{n/\rho_2}; & m=m_k,\ n=n_k, \\ 0, & \text{otherwise.} \end{cases}$$

Fix $\delta > 0$ and find k in $\mathbb{N}$ so that $k^{-1} < \delta$. Since $\|\cdot;\ \sigma_1+\delta,\ \sigma_2+\delta\|$ decreases as δ increases, the series $\sum\sum_{m+n\geq 0} a_{mn}\alpha_{mn}$ converges in (Y,T). Hence, using (b), (a_{mn}) must satisfy (14.4.2). However, (a_{mn}) does not satisfy (14.4.2) for each $\varepsilon > 0$ and hence there is a contradiction.

(a) ==> (c). Let (c) be untrue. This means there exists (a_{mn}) such that $a_{mn}\alpha_{mn} \to 0$ in (Y,T) but (a_{mn}) does not satisfy (14.4.2). Hence we find $\lambda > 0$ and $\{m_k\}$, $\{n_k\}$ in I so that

$$|a_{m_k n_k}| \geq \left[\frac{(\sigma_1+\lambda)e\rho_1}{m_k}\right]^{m_k/\rho_1} \left[\frac{(\sigma_2+\lambda)e\rho_2}{n_k}\right]^{n_k/\rho_2}, \quad k \geq 1.$$

Choose $\eta > 0$ with $\eta < 2\lambda/3$. By the hypothesis, the inequality in (a) is satisfied for this choice of η and the corresponding $\delta > 0$ and the integer N. Now

$$\begin{aligned} \|a_{m_k n_k}\alpha_{m_k n_k}; \sigma_1+\delta, \sigma_2+\delta\| &\geq \left[\frac{\sigma_1+\lambda}{\sigma_1+\eta}\right]^{m_k/\rho_1} \left[\frac{\sigma_2+\lambda}{\sigma_2+\eta}\right]^{n_k/\rho_2} \\ &\geq 1, \end{aligned}$$

for all large k. This shows that $a_{mn}\alpha_{mn} \not\to 0$ in (Y,T), a contradiction. Hence (a) ==> (c). □

Combining the above two lemmas, we obtain

<u>THEOREM 14.4.6</u> An S.b. (α_{mn}) for a closed subspace Y_o of (Y,T) is similar to (e_{mn}) if and only if (α_{mn}) satisfies (A) and (a).

Functions represented by Dirichlet series

Next, we pass on to an example of an S.b. similar to the classical S.b. present in the space of analytic functions represented by Dirichlet series and we follow [114]. For similar examples, one may see [30], [113], [115], [116], [117] and [239].

Let $0 = \lambda_1 < \lambda_2 < \dots < \lambda_n < \dots$ be a fixed sequence of reals with $\lambda_n \to \infty$ as $n \to \infty$ and $(\log n)/\lambda_n \to 0$ ans $n \to \infty$. Consider the space X_A of all functions $f : \mathbb{C} \to \mathbb{C}$, where

$$f(s) = \sum_{n\geq 1} a_n \exp(s\lambda_n), \quad s = \sigma + it$$

satisfying the condition:

$$\limsup_{n\to\infty} \frac{\log |a_n|}{\lambda_n} \leq -A, \tag{14.4.7}$$

A being a fixed positive number. For convenience, we write X for X_A. If $f \in X$, let us define

$$\|f\|_\sigma = \sum_{n\geq 1} |a_n| \exp(\sigma\lambda_n), \quad \sigma < A$$

and let T be the l.c. topology on X generated by $\{\|\cdot\|_\sigma : \sigma < A\}$. It is easily verified that (X,T) is a Fréchet space, e.g. [114], p. 1068.

Define $\{e_n\}$ in X by $e_n(s) = \exp(s\lambda_n)$, $n \geq 1$. Then following Example 2.3.1, we find that $\{e_n\}$ is an S.b. for (X,T).

To construct bases similar to $\{e_n\}$, we establish two lemmas. Hereafter all scalar sequences are to be considered in $\mathbb{C}$.

LEMMA 14.4.8 For $\{\alpha_n\}$ in X, the following conditions are equivalent:

(D) $\displaystyle\limsup_{n\to\infty} \frac{\log \|\alpha_n\|_\sigma}{\lambda_n} < A, \ \forall\sigma < A.$

(E) For $a \in \omega$,

conv. of $\sum_{n\geq 1} a_n e_n$ ==> conv. of $\sum_{n\geq 1} a_n \alpha_n$ in (X,T).

(F) For $a \in \omega$

conv. of $\sum_{n\geq 1} a_n e_n$ ==> $a_n \alpha_n \to 0$ in (X,T).

<u>PROOF</u>. (D) ==> (E). Choose $\sigma < A$ arbitrarily. By (D), there exist $\varepsilon > 0$ and $N_1 \equiv N_1(\sigma,\varepsilon)$ in $\mathbb{N}$ so that $\|\alpha_n\|_\sigma < \exp\{(A-\varepsilon)\lambda_n\}$ for all $n \geq N_1$. Since the convergence of $\sum_{n\geq 1} a_n e_n$ is equivalent to the truth of (14.4.7), by letting $0 < \delta < \varepsilon$, one finds N_2 in $\mathbb{N}$ so that $|a_n| < \exp\{(\delta - A)\lambda_n\}$ for all $n \geq N_2$. Hence

$$|a_n| \; \|\alpha_n\|_\sigma < \exp\{(\delta-\varepsilon)\lambda_n\}, \; n \geq \max(N_1,N_2).$$

This shows that $\sum_{n\geq 1} |a_n| \; \|\alpha_n\|_\sigma$ converges for each $\sigma < A$ and so (E) follows.

(E) ==> (F). Obvious.

(F) ==> (D). Suppose that (D) is not satisfied. Hence there exist $\sigma_1 < A$ and $\{n_k\}$ in I such that

$$\frac{\log \|\alpha_{n_k}\|_{\sigma_1}}{\lambda_{n_k}} > A - \frac{1}{k}, \; k \geq 1.$$

Define a in ω by

$$a_n = \begin{cases} \exp\{-(A-k^{-1})\lambda_{n_k}\}; \; n = n_k, \; k \geq 1, \\ 0; \quad \text{otherwise.} \end{cases}$$

Then for $\sigma < A$

$$|a_{n_k}| \exp(\sigma\lambda_{n_k}) = \exp[-\{A-(\sigma+k^{-1})\}\lambda_{n_k}].$$

We can find $k \geqq 1$ such that $A > \sigma + k^{-1}$ and it follows that $\Sigma |a_n| \exp(\sigma\lambda_n)$ converges for all $\sigma < A$. Consequently $\sum_{n \geqq 1} a_n e_n$ converges in (X,T). However, $|a_{n_k}| \, \|\alpha_{n_k}\|_{\sigma_1} \geqq 1$ for $k \geqq 1$ and so $a_n \alpha_n \not\to 0$ in (X,T), a contradiction. □

LEMMA 14.4.9 For $\{\alpha_n\}$ in X, the following conditions are equivalent:

(d) $\lim_{\sigma \to A} \{\liminf_{n \to \infty} \dfrac{\log \|\alpha_n\|_\sigma}{\lambda_n}\} \geqq A.$

(e) For $a \in \omega$,

conv. of $\sum_{n \geqq 1} a_n \alpha_n$ ==> conv. of $\sum_{n \geqq 1} a_n e_n$ in (X,T).

(f) For $a \in \omega$,

$a_n \alpha_n \to 0$ ==> conv. of $\sum_{n \geqq 1} a_n e_n$ in (X,T).

PROOF. (f) ==> (e) is obvious.

(e) ==> (d). Let (d) not be satisfied. Since $\|\alpha_n\|_\sigma$ increases with σ, we have

$$\liminf_{n \to \infty} \frac{\log \|\alpha_n\|_\sigma}{\lambda_n} < A_1 < A_2 < A$$

for all $\sigma < A$. Take any sequence $\{\sigma_k\}$ with $\sigma_k < A$ and $\sigma_k \to A$. Then we find $\{n_k\}$ in I so that

$$\log \|\alpha_{n_k}\|_{\sigma_k} < A_1 \lambda_{n_k}, \; k \geqq 1.$$

Define a in ω by

$$a_n = \begin{cases} \exp(-A_2\lambda_n); \; n = n_k, \; k \geqq 1, \\ 0, \quad \text{otherwise.} \end{cases}$$

Let $\sigma < A$. Then for $k \geq k_0$, $\sigma \leq \sigma_k$. Hence

$$\sum_{n\geq 1} |a_n| \, \|\alpha_n\|_\sigma \leq \sum_{k=1}^{k_0-1} |a_{n_k}| \, \|\alpha_{n_k}\|_\sigma +$$

$$\sum_{k\geq k_0} \exp\{(A_1-A_2)\lambda_{n_k}\}$$

$$< \infty ,$$

which in turn yields the convergence of $\sum_{n\geq 1} a_n\alpha_n$ in (X,T). On the other hand, if $A_2 < \sigma < A$, then $\sum_{n\geq 1} a_n e_n$ is not Cauchy relative to $\|\cdot\|_\sigma$ and so this series is not convergent in (X,T). This contradicts (e).

(d) ==> (f). If (f) is not true, then there exists a in ω such that $a_n\alpha_n \to 0$ but $\sum_{n\geq 1} a_n e_n$ does not converge in (X,T). Hence (14.4.7) is not satisfied and so we can find $\varepsilon > 0$ and $\{n_k\}$ in I with

$$(*) \quad |a_{n_k}| > \exp\{(\varepsilon-A)\lambda_{n_k}\}, \; k \geq 1.$$

Choose $\eta > 0$ with $\eta < \varepsilon$. There exist $\sigma \equiv \sigma(\eta)$ and $N \equiv N(\sigma,\eta)$ so that (cf. (d))

$$(**) \quad \|\alpha_n\|_\sigma \geq \exp\{(A-\eta)\lambda_n\}, \; n \geq N.$$

(*) and (**) show that $|a_n| \, \|\alpha_n\|_\sigma \not\to 0$, a contradiction. □

<u>THEOREM 14.4.10</u> An S.b. $\{\alpha_n\}$ for a closed subsapce Y of (X,T) is similar to $\{e_n\}$ if and only if $\{\alpha_n\}$ satisfies (D) and (d).

<u>PROOF</u>. A direct consequence of Lemmas 14.4.8 and 14.4.9.□

15 Paley–Wiener stability theorems

15.1 MOTIVATION

We know that $\{\delta_n\}$ ($\delta_n(z) = z^n$) is an S.b. for X_r equipped with the compact-open topology T_r for each r, $0 < r \leqq \infty$, X_r being the space of all functions analytic in $D_r = \{z \in \mathbb{C} : |z| < r\}$. If X_r is restricted in one or the other manner, that is, if we consider a subspace Y_r of X_r and endow it with an appropriate natural l.c. topology (equipping Y_r with T_r often prevents this space from being complete), then it has been observed in a number of cases that $\{\delta_n\}$ ceases to be an S.b. for Y_r although this sequence happens to be complete in these spaces, for example, A_c (the disc algebra), H_p and many other Fréchet spaces appearing in the theory of distributions (cf. [62], [63]). In such a situation as this, we want to locate a sequence $\{g_n\}$ in Y_r, provided it exists at all, which is an S.b. for the space in question. In general, this appears to be a difficult problem and as such we look forward to finding two properties of a sequence $\{g_n\}$, namely, (i) completeness and (ii) ω-linear independence which may ultimately yield the S.b. character of $\{g_n\}$ for Y_r. One may achieve (i) or (ii) in different ways. However, to accomplish (i) (and similarly (ii)), the sequence $\{g_n\}$ is restricted suitably so as to come "nearer" to a known complete sequence, say, $\{\delta_n\}$ and as we will see in subsequent sections, this in turn assures (i) (and similarly (ii)).

On the other hand, if an l.c. TVS X has an S.b. $\{x_n\}$, one may ask whether any sequence $\{y_n\}$ is an S.b. for X, provided $\{y_n\}$ is chosen to be "sufficiently nearer" to $\{x_n\}$

in a suitable sense.

In both the problems, whenever we get the second sequence to be of the same nature as the first one via the process of "nearness", the former is referred to as a "stable" sequence with respect to the properties conserved. However, the delicacy lies in appropriately defining the term "nearness" between two sequences under study.

Motivated possibly by the first problem mentioned above, Paley and Wiener [173] brought in for the first time a concept of "nearness" of two sequences in their study of non-harmonic Fourier expansions of functions in $L^2[-\pi,\pi]$. Since then, this notion of "nearness" has assumed different dimensions which will be taken up in this and the next chapter, and the basic problem we are faced with can be rephrased analytically as follows. We start with an S.b. for an l.c. TVS, consider another sequence with appropriate conditions of its "nearness" to this base and show that this sequence is an S.b. for the underlying space.

Most of the main results of sections 15.2 to 15.4 are proved in the setting of complete locally convex spaces. In each of these results, it is required to have a scalar λ_p corresponding to each continuous seminorm p. It is trivially seen that these results envelop the corresponding ones on Banach or F-spaces in which case we would have only a single scalar λ relative to the underlying norm or the invariant metric.

15.2 THE MAIN THEOREM

As mentioned earlier, it was Paley and Wiener who introduced conditions on two sequences in $L^2[-\pi,\pi]$ by which one could infer the completeness or linear independence of one sequence from another. Their result ([173], p. 100) which is generally referred to as the Paley-Wiener (P-W) stability criterion or theorem on sequences (see Theorem 15.2.1(b), (f) and (h)) was neatly abstracted to an arbitrary Banach space by Boas ([20], p. 469; cf. also [213] which gives the proof of (III) and (IV) for Banach spaces). Its subsequent extension to F- and Fréchet spaces was carried out by Arsove [9]. For further historical accounts of the P-W criterion, one may see [9] and [215], p. 205. In this section, we go a step further and present a still generalized version of the P-W criterion in an arbitrary complete l.c. TVS. Indeed, we have (cf. [71])

<u>THEOREM 15.2.1</u> Let (X,T) be a complete l.c. TVS containing sequences $\{x_n\}$ and $\{y_n\}$. For p in D_T, let $0 < \lambda_p < 1$ and suppose that

$$\text{(P-W)} \quad p\left(\sum_{i=1}^{m} \alpha_i(x_i - y_i)\right) \leq \lambda_p \, p\left(\sum_{i=1}^{m} \alpha_i x_i\right)$$

valid for all finite sequences $\{\alpha_1, \ldots, \alpha_m\}$ of scalars. Then $\{x_n\} \approx \{y_n\}$.

(I) Hence the following statements are valid:

(a) $x_n \neq 0$ if and only if $y_n \neq 0$.

(b) $\{x_n\}$ is l.i. if and only if $\{y_n\}$ is l.i.

(c) $\{x_n\}$ is ω-l.i. if and only if $\{y_n\}$ is ω-l.i.

(d) $\{x_n\}$ is minimal if and only if $\{y_n\}$ is minimal.

(II) If $[x_n]$ and $[y_n]$ are barrelled subspaces, then

(e) $\{x_n\}$ is S-basic if and only if $\{y_n\}$ is S-basic.

(III) If $[x_n] = X$, then

(f) $\{x_n\} \underset{\approx}{\sim} \{y_n\}$; $[y_n] = X$.

(IV) If $\{x_n\}$ is an S.b. for (X,T), say, $\{x_n\} \equiv \{x_n; f_n\}$ then

(h) $\{y_n\}$ is an S.b. for X; and for each p in D_T, and x in X

(i) $p\left(\sum_{i \geq 1} f_i(x) x_i\right) \leq \dfrac{1}{1-\lambda_p} \, p\left(\sum_{i \geq 1} f_i(x) y_i\right)$.

To complete the proof we require

LEMMA 15.2.2 Let (X,T) be an ω-complete l.c. TVS and $S : X \to X$ a linear operator. Let one of the conditions be satisfied, namely, (i) for each p in D_T, there exists λ_p, $0 < \lambda_p < 1$ such that $p(Sx) \leq \lambda_p\, p(x)$ for all x in X, or (ii) there exist p_0 in D_T and δ, $0 < \delta < 1$ such that $p_0(Sx) \leq \delta p_0(x)$ for all x in X and for each p in D_T there exist $k_p > 0$ with $p(Sx) \leq k_p\, p_0(x)$ for all x in X. Then $R = I-S$ is a topological isomorphism from (X,T) onto itself, where I is the identity map of X.

PROOF. (i) We have $p(S^n x) \leq \lambda_p^n\, p(x)$; $x \in X$, $n = 0,1,\ldots$. Hence the operator $Q : X \to X$

$$Qx = \sum_{n \geq 0} S^n x$$

is well defined (since $\sum_{n\geq 0} S^n x$ is absolutely convergent and hence convergent), continuous and

$$(+) \quad p(Qx) \leq \frac{1}{1-\lambda_p}\, p(x), \quad \forall x \in X.$$

Next observe that $(I-S)Q = Q(I-S) = I$. This shows that I-S has the required properties.

(ii) Here $p(S^n x) \leq k_p\, \delta^{n-1}\, p_0(x)$ and now proceed as in (i). □

PROOF OF THEOREM 15.2.1 The (P-W) inequality results in

$$(1-\lambda_p)p\left(\sum_{i=1}^{m} \alpha_i x_i\right) \leq p\left(\sum_{i=1}^{m} \alpha_i y_i\right) \leq (1+\lambda_p)p\left(\sum_{i=1}^{m} \alpha_i x_i\right). \tag{15.2.3}$$

Consequently by Theorem 13.2.4, I, (ii) ==> (i) and its symmetrization, $\{x_n\} \approx \{y_n\}$.

(I). The statements (a) - (d) are now contained in Theorem 13.5.2.

(II). The statement (e) follows from Proposition 13.5.3.

(III). Since $\{x_n\} \approx \{y_n\}$, $[x_n] \simeq [y_n]$ by Proposition 13.2.4 under a uniquely determined topological isomorphism R with $Rx_n = y_n$, $n \geq 1$; R may be regarded from $X = [x_n]$ onto $[y_n]$ or into X. In order to prove that R is onto X, let us observe that R is defined on the whole of X and so if I is the identity map of X, by using (P-W), it is easily seen that

$$p((I-R)x) \leq \lambda_p \, p(x), \quad \forall x \in X.$$

Therefore by Lemma 15.2.2(i), R = I-(I-R) is a topological isomorphism from (X,T) onto itself.

(IV) Here (h) follows from (III), whereas (i) results from (+) (cf. the proof of Lemma 15.2.2) in replacing x by Rx and observe that $R^*Rx = x$ and $Rx_n = y_n$, $n \geq 1$. □

REMARK. One would obviously like to know if $\lambda_p = 1$ in(P-W) may prevent the theorem from being true. For instance if $\{x_n\}$ is an S.b. and $y_n = 0$ for $n \geq 1$. A non-trivial example in this direction is the following (cf. [162] where it is used for a similar, but different, purpose)

EXAMPLE 15.2.4 Consider the usual S.b. $\{x^n;f^n\}$ for $(c_0, \|\cdot\|_\infty)$, where $x^n = e^n$; $f^n = e^n$, $n \geq 1$. Let $y^n = (-1)^{n+1}x^1 + x^n$, $n \geq 2$ and $y^1 = x^1$. Then $x^1-y^1 = 0$ and $x^n-y^n = (-1)^n x^1$ for $n \geq 2$. For any finite set $\alpha_1,\ldots,\alpha_m$ of scalars $\|\sum_{i=1}^{m} \alpha_i(x^i-y^i)\| = \sup\{|\alpha_i| : 2 \leq i \leq m\}$ and $\|\sum_{i=1}^{m} \alpha_i x^i\| = \sup\{|\alpha_i| : 1 \leq i \leq m\}$. Since

$\sup\{|\alpha_i| : 2 \leq i \leq m\} \leq \sup\{|\alpha_i| : 1 \leq i \leq m\}$, (P-W) is satisfied with $\lambda = 1$. However, $\{y^n\}$ is not even S. basic, for otherwise one finds a g in ℓ^1 with $g(y^1) = 1$, $g(y^n) = 0$ for $n \geq 2$. Observe that $g_n \to 0$ and so $g(x^n) \to 0$. If

$n = 2m-1$, then $g(y^{2m-1}) = g(x^1) + g(x^{2m-1})$, $m \geq 2$ and so $0 = g(x^1) + \lim g(x^{2m-1})$ which is absurd.

15.3 VARIANTS OF THE PALEY-WIENER CRITERION

In this section, we throughout consider $\{x_n\} \equiv \{x_n; f_n\}$ to be an arbitrary S.b. for an l.c. TVS (X,T) and endeavour to find the S. basis character of another sequence $\{y_n\}$ in X such that $\{x_n\}$ and $\{y_n\}$ satisfy the (P-W) condition with some variation in their summands. To begin with, we require (cf. [9])

DEFINITION 15.3.1 A sequence $\{y_n\}$ is called *triangular* with respect to $\{x_n\}$ or merely $\{x_n\}$-*triangular* provided

$$y_n = x_n + \sum_{i \geq n+1} f_i(y_n)x_i, \quad n \geq 1. \tag{15.3.2}$$

Let $X_k = [x_i : i \geq k]$. It is readily verified that $X_k = \{x \in X : x = \sum_{i \geq k} f_i(x)x_i\}$, $X = X_1 \supset X_2 \supset \ldots$ and $y_k \in X_k$ $(k \geq 1)$, where $\{y_n\}$ is given by (15.3.2).

LEMMA 15.3.3 If $\{y_n\}$ is $\{x_n\}$-triangular, then $\{y_n\}$ is semibasic, i.e. $y_n \notin [y_i : i \geq n+1]$, $n \geq 1$.

PROOF Straightforward. □

LEMMA 15.3.4 If $\{y_n\}$ is $\{x_n\}$-triangular, then

(i) $\{y_n\}$ is ω-l.i.

(ii) There exists a unique sequence $\{g_n\} \subset [y_n]^*$ such that $g_m(y_n) = \delta_{mn}$.

PROOF. (i) Let

$$\sum_{i \geq 1} \alpha_i y_i = 0.$$

Since $y_i \in X_2$ $(i \geq 2)$, $\sum_{i \geq 2} \alpha_i y_i \in X_2$. Hence $\alpha_1 x_1 + u = 0$ for some u in X_2. Bur $x_1 \notin X_2$ and so $\alpha_1 = 0$. The required result now follows by induction.

(ii) Since $\{y_n\}$ is l.i. and semibasic, there exists $\{g_n\} \subset (sp\{y_n\})^*$ with $g_m(y_n) = \delta_{mn}$ (cf. [130], Proposition 9.1.5). Denote by g_n itself the unique continuous extension of g_n to $[y_n]$. The uniqueness of $\{g_n\} \subset [y_n]^*$ is obviously true. □

<u>PROPOSITION 15.3.5</u> Let $\{y_n\}$ be $\{x_n\}$-triangular. Then $\{y_n\}$ is an S.b. for (X,T) if and only if there exists $k \geq 1$ such that $\{y_i : i \geq k\}$ is an t.b. for $X_k \equiv (X_k, T|X_k)$.

<u>PROOF</u>. It is enough to assume that $\{y_i : i \geq k\}$ is a t.b. for X_k, $k > 1$ and then establish the S.b. character of $\{y_n\}$ for (X,T).

Let $y \in X$. Then we can find α_1 such that $y - \alpha_1 y_1 \in X_2$. We can now find α_2 with $y - \alpha_1 y_1 - \alpha_2 y_2 \in X_3$. Therefore, inductively we can determine $\alpha_1, \ldots, \alpha_{k-1}$ so that $y - \alpha_1 y_1 - \ldots - \alpha_{k-1} y_{k-1} \in X_k$. Hence by the hypothesis $\{y_n\}$ ω-spans X and so by Lemma 15.3.4(i), $\{y_n\}$ is a t.b. for (X,T). Now apply Lemma 15.3.4(ii) to infer the required result. □

<u>THEOREM 15.3.6</u> Let (X,T) be a complete l.c. TVS containing an S.b. $\{x_n; f_n\}$ and $\{y_n\}$ be an $\{x_n\}$-triangular sequence in X. Let for p in D_T, there exist k_p in $\mathbb{N}$ and λ_p with $0 < \lambda_p < 1$ such that

$$p\Big(\sum_{i=k_p}^{m} \alpha_i (y_i - x_i)\Big) \leq \lambda_p \, p\Big(\sum_{i=k_p}^{m} \alpha_i x_i\Big) \qquad (15.3.7)$$

valid for all finite sequences $\{\alpha_{k_p}, \ldots, \alpha_m\}$ of scalars. Then

(i) $\{y_n\}$ is an S.b. for (X,T)

and

(ii) $\{x_n\} \approx \{y_n\}$, say, under R with $Rx_n = y_n$, $n \geq 1$.

PROOF. (i) Suppose that $k_p \leq k \in \mathbb{N}$, for all p in D_T. Then for p in D_T, (15.3.7) results in

$$p\left(\sum_{i=k}^{m} \alpha_i(y_i - x_i)\right) \leq \lambda_p \, p\left(\sum_{i=k}^{m} \alpha_i x_i\right)$$

valid for all scalars $\alpha_k, \ldots, \alpha_m$. Since $\{x_i : i \geq k\}$ is an S.b. for X_k and $y_i \in X_k$, $i \geq k$, the sequence $\{y_i : i \geq k\}$ is an S.b. for X_k by Theorem 15.2.1, IV(h). The required result follows by the preceding proposition.

If $\{k_p : p \in D_T\}$ is not bounded we proceed as follows.

Define $P : X \to X$ by

$$Px = \sum_{n \geq 1} f_n(x)(y_n - x_n)$$

whose existence is guaranteed by the Cauchy character of infinite series in question. To prove the continuity of P, choose p in D_T arbitrarily. Given any x in X, we can write it as $x = u_p + z_p$, where $z_p \in X_{k_p}$ and

$$u_p = \sum_{i=1}^{k_p - 1} f_i(x) x_i .$$

Since $|f_i(x)| \leq r_p(x)$ for all x and $1 \leq i \leq k_p - 1$ for some $r_p \in D_T$, $p(Pz_p) \leq \lambda_p \, p(z_p)$ and $c_p = \sum_{i=1}^{k_p - 1} p(y_i - x_i) < \infty$, we get

$$p(Px) \leq (\lambda_p c_p + \lambda_p + c_p) \, q_p(x), \quad \forall x \in X$$

where $q_p = \max(p, r_p)$. Therefore P is a continuous linear map.

Now observe that for p in D_T

$$p(Px) \leqq \lambda_p\, p(x), \quad \forall x \in X_{k_p};$$

$$P^n x \in X_{n+1}; \quad \forall n \geqq 1,\ x \in X$$

and so for $n \geqq k_p$ and x in X,

$$p(P^n x) \leqq \lambda_p^{\,n-k_p}\, p(P^{k_p} x).$$

Therefore the map $U : X \to X$ given by

$$Ux = \sum_{n \geqq 0} (-P)^n x, \; x \in X$$

is well defined and continuous. Clearly U is the inverse of $P + I = R$ (say). Then R is a topological isomorphism of (X,T) onto itself. This proves (i).

(ii). This follows from the proof of (i). □

NOTE. The proof in the second case requires only the ω-completeness of (X,T).

Theorem 15.3.6 yields

PROPOSITION 15.3.8 Let $\{x_n; f_n\}$ be a fully ℓ^1-base for a complete l.c. TVS (X,T) and $\{y_n\}$ an $\{x_n\}$-triangular sequence in X. Let, for p in D_T,

$$\limsup_{n\to\infty} \frac{\sum_{i \geqq n+1} p(f_i(y_n)x_i)}{p(x_n)} < 1. \tag{15.3.9}$$

Then (i) $\{y_n\}$ is a fully ℓ^1-base for (X,T) and (ii) $\{x_n\} \overset{\approx}{\approx} \{y_n\}$ under R with $Rx_n = y_n$, $n \geqq 1$.

PROOF. By Proposition 4.5.5, T is also generated by $\{Q_p : p \in D_T\}$ and by (15.3.9), for each p in D_T, there exist k_p in $\mathbb{N}$ and λ_p, $0 < \lambda_p < 1$ so that $Q_p(y_n - x_n) \leqq \lambda_p\, Q_p(x_n)$, for all $n \geqq k_p$. Therefore, for all scalars

$\alpha_{k_p},\ldots,\alpha_m$,

$$Q_p(\sum_{i=k_p}^{m} \alpha_i(y_i-x_i)) \leq \lambda_p \sum_{i=k_p}^{m} |\alpha_i| p(x_i)$$

$$= \lambda_p Q_p(\sum_{i=k_p}^{m} \alpha_i x_i).$$

Hence by Theorem 15.3.6, (i) follows partially and (ii) holds. If $\{g_n\}$ is the s.a.c.f. associated with $\{y_n\}$, then for x in X and p in D_T,

$$p(x) \leq \sum_{n\geq 1} |g_n(x)| p(y_n) \leq \sum_{n\geq 1} |g_n(x)| q(x_n) = Q_q(R^{-1}x)$$

$$\leq s(x)$$

for some $s \in D_T$, where $p(Rx) \leq q(x)$, $x \in X$ and $g_n = f_n \circ R^{-1}$, $n \geq 1$. □

15.4 WEAKER STABILITY CRITERIA

Following the (P-W) criterion on the stability of two sequences, a number of mathematicians have proposed different stability criteria of two sequences in Banach spaces. We mention below the locally convex analogues of these conditions and name them according to the first English letters of the last names of the mathematicians who invented them for Banach or Hilbert spaces. To be precise, we have

DEFINITION 15.4.1 Let $\{x_n\}$ and $\{y_n\}$ be two sequences in an l.c. TVS (X,T). Consider the following statements:

(1) For each p in D_T, there exists λ_p, $0 < \lambda_p < 1$ so that

$$\text{(P-W)} \quad p(\sum_{i=1}^{m} \alpha_i(x_i-y_i)) \leq \lambda_p \; p(\sum_{i=1}^{m} \alpha_i x_i),$$

valid for all finite scalars $\alpha_1,\ldots,\alpha_m$.

(2) For each p in D_T and any positive real k, there exist $\lambda_{1,p}$, $\lambda_{2,p}$ $(0 < \lambda_{i,p} < \min(1, 2^{1-1/k}),\ i = 1,2)$ such that

$$\text{(P-H)}\quad p^k\Big(\sum_{i=1}^{m} \alpha_i(x_i - y_i)\Big) \leqq \lambda_{1,p}\, p^k\Big(\sum_{i=1}^{m} \alpha_i x_i\Big) + \lambda_{2,p}\, p^k\Big(\sum_{i=1}^{m} \alpha_i y_i\Big),$$

valid for all finite scalars $\alpha_1, \ldots, \alpha_m$.

(3) For each p in D_T, there exist η_p, μ_p and ν_p with $0 < \eta_p, \nu_p < 1$; $0 \leqq \mu_p$, $\mu_p^2 \leqq (1 - \eta_p)(1 - \nu_p)^{-1}$ so that

$$\text{(N)}\quad p^2\Big(\sum_{i=1}^{m} \alpha_i(x_i - y_i)\Big) \leqq \eta_p\, p^2\Big(\sum_{i=1}^{m} \alpha_i x_i\Big) + \nu_p\, p^2\Big(\sum_{i=1}^{m} \alpha_i y_i\Big)$$

$$+ 2\mu_p\, p\Big(\sum_{i=1}^{m} \alpha_i x_i\Big) p\Big(\sum_{i=1}^{m} \alpha_i y_i\Big),$$

valid for all finite scalars $\alpha_1, \ldots, \alpha_m$.

(4) For each p in D_T, there exists λ_p, $0 < \lambda_p < 1$ such that

$$\text{(R)}\quad p\Big(\sum_{i=1}^{m} \alpha_i(x_i - y_i)\Big) \leqq \lambda_p \Big[p\Big(\sum_{i=1}^{m} \alpha_i x_i\Big) + p\Big(\sum_{i=1}^{m} \alpha_i y_i\Big)\Big],$$

valid for all finite scalars $\alpha_1, \ldots, \alpha_m$.

Then $\{x_n\}$ and $\{y_n\}$ are said to be (P-W)-, (P-H)-, (N)- or (R)-*stable* according to whether they satisfy (P-W), (P-H), (N) or (R).

NOTE. The conditions (P-W), (P-H), (N) and (R) are respectively contained in the work of [173], [179], [79], [166] and [183].

The next result is essentially contained in [183] .

LEMMA 15.4.2 Let $\{x_n\}$, $\{y_n\}$, etc. be as in Definition 15.4.1. Then (P-W) ==> (P-H), (N), (R) ==> (R).

PROOF. Let (P-H) be satisfied. Put $\lambda_p = \{\max(\lambda_{1,p}, \lambda_{2,p}) . \max(1, 2^{1-k})\}^{1/k}$ and apply the inequality

$$a^k + b^k \leq c\,(a + b)^k, \quad a, b \geq 0$$

where

$$c = \begin{cases} 2^{1-k}, & 0 < k < 1 \\ 1, & k \geq 1 \end{cases}$$

If (N) is satisfied, let $\lambda_p^2 = \max(\eta_p, \mu_p, \nu_p)$. □

Hence, amongst all stability conditions enunciated so far, (R) happens to be the weakest one. It is, therefore, natural to raise

PROBLEM 15.4.3 Is Theorem 15.2.1 valid when (P-W) is replaced by (R)?

NOTE. For Banach spaces, Problem 15.4.3 is solved in [186]. However, we are not aware of the complete solution of this problem for general locally convex spaces. On the other hand, if $0 < \lambda_p < 1/3$ in (R), then (f) and (h) of Theorem 15.2.1 holds (if $0 < \lambda_p < 1$, then $\{x_n\} \approx \{y_n\}$ is always true). Indeed, let (R) be true, $0 < \lambda_p < 1$. Then

$$\left(\frac{1 - \lambda_p}{1 + \lambda_p}\right) p\left(\sum_{i=1}^{m} \alpha_i x_i\right) \leq p\left(\sum_{i=1}^{m} \alpha_i y_i\right) \leq \left(\frac{1 + \lambda_p}{1 - \lambda_p}\right) p\left(\sum_{i=1}^{m} \alpha_i x_i\right)$$

and so following the proof of Theorem 15.2.1, $\{x_n\} \approx \{y_n\}$.

Next, let $0 < \lambda_p < 1/3$ and put $\mu_p = 2\lambda_p/(1-\lambda_p)$. Then $\mu_p < 1$ and

$$p\left(\sum_{i=1}^{m} \alpha_i (x_i - y_i)\right) \leq \lambda_p \left(1 + \frac{1+\lambda_p}{1-\lambda_p}\right) p\left(\sum_{i=1}^{m} \alpha_i x_i\right)$$

$$= \mu_p\, p\left(\sum_{i=1}^{m} \alpha_i x_i\right),$$

and hence if $[x_n] = X$, then $\{x_n\} \approx \{y_n\}$ and if $\{x_n\}$ is an S.b. for (X,T) then so is $\{y_n\}$(cf. the proof of Theorem 15.2.1(f) and (h)).

<u>A further criterion</u>

It is clear that condition (R) is good enough to prove II(e) of Theorem 15.2.1. However, one may still go a step further to prove Theorem 15.2.1, II(e) under a condition (R*) given below (cf. [183]) which is weaker than (R).

<u>PROPOSITION 15.4.4</u> Let $\{x_n\}$ and $\{y_n\}$ be two sequences in an l.c. TVS(X,T) such that $[x_n]$ is a barrelled subspace of (X,T). For each p in D_T, let there be λ_p, $0 < \lambda_p < 1$ so that for any choice of finite scalars $\alpha_1,\ldots,\alpha_n$, there exist scalars $\beta_1,\ldots,\beta_n$ satisfying

$$(R^*) \quad p\left(\sum_{i=1}^{m} (\alpha_i y_i - \beta_i x_i)\right) \leqq \lambda_p\left[p\left(\sum_{i=1}^{m} \alpha_i y_i\right) + p\left(\sum_{i=1}^{m} \beta_i x_i\right)\right], \quad 1 \leqq m \leqq n.$$

Then if $\{x_n\}$ is S-basic in (X,T), then so is $\{y_n\}$.

<u>PROOF</u>. We have

$$p\left(\sum_{i=1}^{m} \alpha_i y_i\right) \leqq p\left(\sum_{i=1}^{m} (\alpha_i y_i - \beta_i x_i)\right) + p\left(\sum_{i=1}^{m} \beta_i x_i\right)$$

$$(*) \qquad \leqq \frac{1+\lambda_p}{1-\lambda_p}\, p\left(\sum_{i=1}^{m} \beta_i x_i\right), \quad 1 \leqq m \leqq n.$$

Similarly, we have for q in D_T

$$(**) \quad q\left(\sum_{i=1}^{m} \beta_i x_i\right) \leqq \frac{1+\lambda_q}{1-\lambda_q}\, q\left(\sum_{i=1}^{m} \alpha_i y_i\right), \quad 1 \leqq m \leqq n.$$

Since $\{x_n\}$ is an S.b. for the barrelled subspace $[x_n]$, there exists $q \in D_T$ (cf. Theorem 2.1.3) so that

$$(+)\quad p\Big(\sum_{i=1}^{m} \beta_i x_i\Big) \leqq q\Big(\sum_{i=1}^{n} \beta_i x_i\Big),\ 1 \leqq m \leqq n.$$

Therefore, by (*) and (+) we have

$$p\Big(\sum_{i=1}^{m} \alpha_i y_i\Big) \leqq \frac{1+\lambda_p}{1-\lambda_p}\, q\Big(\sum_{i=1}^{n} \beta_i x_i\Big),\ 1 \leqq m \leqq n$$

and hence using (**), one has

$$p\Big(\sum_{i=1}^{m} \alpha_i y_i\Big) \leqq \frac{1+\lambda_p}{1-\lambda_p}\,\frac{1+\lambda_q}{1-\lambda_q}\, q\Big(\sum_{i=1}^{n} \alpha_i y_i\Big).$$

The required result now follows by applying Theorem 2.1.3. □

15.5 COMPARISON BETWEEN COEFFICIENT FUNCTIONALS

In earlier sections, we have seen that if $\{x_n;f_n\}$ is an S.b. for some l.c. TVS X, then a sequence $\{y_n\}$ satisfying certain stability conditions, also turns out to be an S.b., say, $\{y_n;g_n\}$ for X. One may then ask a pertinent question, namely, if there exists any relationship between $\{f_n\}$ and $\{g_n\}$. In general, this may not be an answerable question; however, in specific situations, we may get a solution to this problem and this is what we are going to do in this section.

To begin with, let us observe the following lemma of which the proof is identical to that of a similar result of [9] proved for Fréchet spaces.

LEMMA 15.5.1 Let $\{x_n;f_n\}$ be an S.b. and $\{y_n;g_n\}$ an $\{x_n\}$-triangular S.b. for an l.c. TVS (X,T). Then for each x in X,

$$f_1(x) = g_1(x);\ f_n(x) = g_n(x) + \sum_{i=1}^{n-1} f_n(y_{n-i})g_{n-i}(x),\ n \geqq 2.$$

PROOF. Here $f_n(y_i) = 0$, $i > n$ and $f_i(y_i) = 1$, $i \geqq 1$. Now

$$x = \sum_{i\geq 1} g_i(x)y_i$$

$$\Longrightarrow f_j(x) = \sum_{i=1}^{j} g_i(x)f_j(y_i)$$

$$\Longrightarrow f_1(x) = g_1(x);\ f_j(x) = g_j(x) + \sum_{i=1}^{j-1} g_i(x)f_j(y_i)$$

$$= g_j(x) + \sum_{k=1}^{j-1} f_j(y_{j-k})g_{j-k}(x),$$

and this proves the result. □

The main result of this section is to estimate $|g_i(x)|$ in terms of the corresponding estimates for $|f_i(x)|$ and following [9], we have

THEOREM 15.5.2 Let $\{x_n;f_n\}$ be an S.b. for an l.c. TVS (X,T) and $\{y_n\}$ an $\{x_n\}$-triangular S.b. for (X,T). Let there exist for p in D_T and x in X reals $Q_p(x)$ and $H_p(y_n)$ such that

$$|f_n(x)|\ p(x_n) \leq Q_p(x);\ n \geq 1, \tag{15.5.3}$$

and

$$|f_i(y_n)|p(x_i) \leq H_p(y_n)p(x_n),\ i \geq n+1. \tag{15.5.4}$$

Then

$$|g_1(x)|p(x_1) \leq Q_p(x);\ |g_n(x)|p(x_n) \leq$$

$$Q_p(x) \prod_{i=1}^{n-1} (1 + H_p(y_i)),\ n \geq 2.$$

PROOF. By Lemma 15.5.1, (15.5.3) and (15.5.4), we have

(+) $|g_1(x)|p(x_1) \leqq Q_p(x);$

$$|g_n(x)|p(x_n) \leqq |f_n(x)|p(x_n) + \sum_{i=1}^{n-1} |g_{n-i}(x)||f_n(y_{n-i})|p(x_n)$$

$$\leqq Q_p(x) + \sum_{i=1}^{n-1} |g_{n-i}(x)|H_p(y_{n-i})p(x_{n-i}), n \geqq 2.$$

Following [167], we now proceed as follows. Define $\{b_n\}$ in ω by

(*) $b_1 p(x_1) = Q_p(x),$

(**) $b_n\, p(x_n) = Q_p(x) + \sum_{i=1}^{n-1} b_{n-i}\, H_p(y_{n-i})p(x_{n-i}),\ n \geqq 2.$

Then for $n \geqq 2$

$$b_n\, p(x_n) - b_{n-1}\, p(x_{n-1}) = b_{n-1}\, H_p(y_{n-1})\, p(x_{n-1})$$

$$\Longrightarrow b_n\, p(x_n) = [1 + H_p(y_{n-1})]\, b_{n-1}\, p(x_{n-1})$$

(++) $\Longrightarrow b_n\, p(x_n) = Q_p(x) \prod_{i=1}^{n-1} [1 + H_p(y_i)],$

by using (*). The required result now follows from (+) and combining (**) with (++). □

NOTE. Inequalities of the type (15.5.3) and (15.5.4) do occur, for instance, if we consider semi ℓ^1-bases, then we have (15.5.3) and this remark also carries over to (15.5.4).

COROLLARY 15.5.5 In addition to the hypothesis of Theorem 15.5.2, define M_p by

$$\limsup_{n \to \infty} H_p(y_n) < M_p,$$

then there exist constants K_p so that

$$|g_n(x)| \; p(x_n) < (1 + M_p)^n \; K_p, \; n \geq 1.$$

15.6 ILLUSTRATIVE EXAMPLES

In this final section, we use two previous results for specific spaces to yield some known results.

In order to produce the first example, let us introduce the set P of real sequences defined by

$$P = \{\alpha \in \omega: \sup (\alpha_{n-1}/\alpha_n) < 1, \; \alpha_1 > 0\}.$$

PROPOSITION 15.6.1 P is a Köthe set of infinite type.

PROOF. For showing that P is a Köthe set, let $\alpha, \beta \in P$ and what we need to prove is the existence of γ in P with $\alpha, \beta \leqq \gamma$. Let

$$a = \sup \left\{\frac{\alpha_{n-1}}{\alpha_n}\right\} \; ; \quad b = \sup \left\{\frac{\beta_{n-1}}{\beta_n}\right\} .$$

Then $0 < a, b < 1$. Define γ in ω by

$$\gamma_n = \max \{a^{-n+1} \; \alpha_n, \; b^{-n+1} \; \beta_n\}.$$

Then $\gamma_n \geqq \alpha_n, \beta_n$ for $n \geqq 1$. To prove $\sup \{\gamma_{n-1}/\gamma_n\} < 1$, we consider four cases (i) - (iv)

(i) Let $\gamma_n = a^{-n+1} \; \alpha_n$; $\gamma_{n-1} = a^{-n+2} \; \alpha_{n-1}$. Then $\gamma_{n-1}/\gamma_n \leqq a^2 < a$.

(ii) Let $\gamma_n = b^{-n+1} \; \beta_n$; $\gamma_{n-1} = b^{-n+2} \; \beta_{n-1}$ and proceed as in (i).

(iii) Let $\gamma_n = b^{-n+1} \; \beta_n$; $\gamma_{n-1} = a^{-n+2} \; \alpha_{n-1}$. Here $b^{-n+1} \; \beta_n \geqq a^{-n+1} \; \alpha_n > a^{-n+1} \; \alpha_{n-1}$ and so $\gamma_{n-1}/\gamma_n < a$.

(iv) Let $\gamma_n = a^{-n+1}\alpha_n$; $\gamma_{n-1} = b^{-n+2}\beta_{n-1}$ and proceed as in (iii).

To prove the infinite Köthe character of P, observe that if $\alpha \in P$, then $\alpha^k = \{\alpha_n^k\} \in P$ for every $k > 0$; in particular $\alpha^2 \in P$. □

PROPOSITION 15.6.2 The Köthe space $(\Lambda(P),T_p)$ is a complete nuclear space.

PROOF. It suffices to show the nuclearity of $(\Lambda(P),T_p)$. Consider α in P and let $\beta_n = a^{-n+1}\alpha_n$, $n \geq 1$, where $a = \sup\{\alpha_{n-1}/\alpha_n\}$. Then $\beta_n \geq \alpha_n$ and $\beta_{n-1}/\beta_n \leq a^2 < a$ for $n \geq 1$. Hence $\beta \in P$ and

$$\sum_{n\geq 1} \alpha_n/\beta_n = \sum_{i\geq 1} a^i = \frac{1}{1-a} .$$

By Theorem 1.4.14, $(\Lambda(P),T_p)$ is nuclear. □

NOTE. By using a result from [129], p. 290, it is easily seen that $\Lambda(P) \subset S$, the s.s. of all rapidly decreasing sequences.

It is readily verified that $\{e^n;e^n\}$ is a fully ℓ^1-base for $(\Lambda(P),T_p)$. Let $x^n = e^n$; $y^n = e^n + e^{n-1}$; $n \geq 1$, $e^0 = \{0\}$. Clearly $[y^n] = \Lambda(P)$ and since $\{y_n\}$ satisfies (2.1.4), $\{y^n\}$ is an S.b. for $(\Lambda(P),T_p)$ by Theorem 2.1.3. On the other hand, using Theorem 15.2.1, we also have

PROPOSITION 15.6.3 $\{y^n\}$ is an S.b. for the complete nuclear space $(\Lambda(P),\ T_p)$.

PROOF. Let $\alpha \in P$. Then, for arbitrary scalars $a_1,\ldots,a_m$,

$$\begin{aligned} p_\alpha\Big(\sum_{i=1}^{m} a_i(y^i-x^i)\Big) &= p_\alpha\Big(\sum_{i=2}^{m} a_i e^{i-1}\Big) \\ &\leq \lambda_\alpha\, p_\alpha\Big(\sum_{i=1}^{m} a_i x^i\Big), \end{aligned}$$

where $\lambda_\alpha = \sup\{\alpha_{i-1}/\alpha_i\}$. Hence, by Theorem 15.2.1, $\{y^n\}$ is an S.b. for $(\Lambda(P), T_p)$. □

NOTE. The sequence $\{y^n\}$ considered above is not $\{x^n\}$-triangular, since $\langle y^n, e^i\rangle = 0$ for $i \geqq n+1$ and so

$$y^n \neq x^n + \sum_{i\geqq n+1} \langle y^n, e^i\rangle\, x^i.$$

We finally come to a use of Propsoition 15.3.8. Attempts have been made in the past to expand analytic functions in terms of sequences of analytic functions (called Pincherle bases in [4]) other than the customary $\{z^n\}$ (cf. [4], [6], [20], [51]).

Let X_R be the space of all analytic functions in $D_R = \{z \in \mathbb{C}: |z| < R\}$ topologized by the norms $M_r(f) = \sup\{|f(z)| : |z| \leqq r\}$, $0 < r < R$. The space X_R is clearly complete (indeed, a Fréchet space). Corresponding to any fully ℓ^1-base $\{\alpha_n\}$ in X_R, let

$$\beta_n(z) = \alpha_n(z) + \sum_{i\geqq 1} A_{ni}\, \alpha_{n+i}(z),$$

where (A_{ni}) is an infinite matrix with entries in $\mathbb{C}$ such that the infinite series converges uniformly over compacta in D_R. Proposition 15.3.8 when applied to this particular situation, yields the following variant of Evagrafov's theorem ([51], p. 117); cf. also [9].

THEOREM 15.6.4 If

$$\limsup_{n\to\infty} \sum_{i\geqq 1} |A_{ni}| \frac{M_r(\alpha_{n+i})}{M_r(\alpha_n)} < 1, \quad 0 < r < R$$

then $\{\beta_n\}$ is a fully ℓ^1-base and $\{\alpha_n\} \approx \{\beta_n\}$.

As a special case, one derives the following result [20] on Pincherle bases.

COROLLARY 15.6.5 Let $\alpha_n(z) = z^n$ $(n \geqq 0)$ and

$$\beta_n(z) = z^n \left(1 + \sum_{i \geq 1} A_{ni} z^i\right),$$

where (A_{ni}) is as before. If

$$\limsup_{n \to \infty} \sum_{i \geq 1} |A_{ni}| r^i < 1, \ 0 < r < R$$

then $\{\beta_n\}$ is a fully ℓ^1-base for X_R and $\{\alpha_n\} \approx \{\beta_n\}$.

NOTE. One may similarly obtain results on Pincherle bases for analytic functions of several complex variables and for those represented by Dirichlet series.

16 Further theorems on stability

16.1 INTRODUCTORY REMARKS

In the last chapter, we exploited the classical Paley-Wiener stability theorem for $L^2[-\pi,\pi]$ to its full generality in complete l.c. TVS or even in complete TVS. By now, there are several stability conditions (cf. [187]) between an S.b. $\{x_n\}$ and a sequence $\{y_n\}$ which ensure the basis character of the latter. As rightly remarked in [187], these several stability theorems essentially split in two categories: 1) the "contractive operator" theorems and 2) the "compact operator" theorems. In fact, the entire set of stability theorems makes use of two basic lemmas: Lemma 15.2.2 (related to contractive mappings) and Lemma 16.3.4 (related to compact operators) and hence these are the two categories. The results of the last chapter fall into category 1).

It is not known whether all these stability theorems which have a strong bearing in Banach and Hilbert spaces (cf. [187]), are true in the general setting of locally convex spaces. However, some of these results are known to be true in spaces more general than Banach spaces and we will discuss these in subsequent sections of this chapter.

16.2 THEOREM OF KREIN-MILMAN-RUTMAN

After the Paley-Wiener stability criterion between two sequences, the next classical result in this direction is that of Krein, Milman and Rutman [150] whose slight variation and applications were established in [17] for Banach spaces. These results of [17] were subsequently extended to the setting of locally convex spaces by Kalton [96] whom we follow for the rest of this section.

Unless specified otherwise, $\{x_n;f_n\}$ denotes an arbitrary u.e-S.b. for the subspace $X_0 = [x_n]$ of an l.c. TVS (X,T); here $\{f_n\} \subset X_0^*$. Consider p_0 in D_T so that $|f_n(x)| \leq p_0(x)$ for $n \geq 1$; x in X_0 and call p_0 a *seminorm of equicontinuity* (s.o.s.) on X corresponding to $\{x_n;f_n\}$. Any sequence $\{y_n\}$ in X will be called a *deformation of the* u.e-S. *basic sequence* $\{x_n\}$ provided

$$\text{(K-M-R)} \qquad K_0 \equiv \sum_{i\geq 1} p_0(y_i-x_i) < 1;\ K_p \equiv \sum_{i\geq 1} p(y_i-x_i) < \infty,$$

$$\forall p \in D_T.$$

The following result also appears in [130], p. 93 and we prove it for the sake of completeness of our discussion on stability theorems.

PROPOSITION 16.2.1 If (X,T) is complete, then a deformation $\{y_n\}$ of $\{x_n\}$ is S. basic and $\{x_n\} \approx \{y_n\}$.

PROOF. For x in X_0 and p in D_T

$$\sum_{i\geq 1} |f_i(x)|\ p(y_i-x_i) \leq K_p p_0(x),$$

and since (X,T) is complete, $\sum_{i\geq 1} f_i(x)(y_i-x_i)$ converges and so does $\sum_{i\geq 1} f_i(x)y_i$. If $X_1 = [y_n]$, then the linear map $R : X_0 \to X_1$,

$$Rx = \sum_{i\geq 1} f_i(x)y_i$$

is well defined. For p in D_T and x in X_0, $p(Rx) \leq p(x) + K_p p_0(x)$ and so R is continuous; also

$$p_o(Rx) \geqq p_o(x) - \sum_{i \geqq 1} |f_i(x)| p_o(y_i - x_i) \geqq (1-K_o) p_o(x)$$

$$\Longrightarrow p(x) \leqq p(Rx) + K_p p_o(x) \leqq p(Rx) + K_p (1-K_o)^{-1} p_o(Rx).$$

Thus R is 1-1 and R^{-1} is continuous from $R[X_o]$ onto X_o. Observe that $R[X_o]$ is complete and hence closed in (X,T). Thus $X_1 = R[X_o]$. Hence $X_o \simeq X_1$ under R with $Rx_n = y_n$, $n \geqq 1$. □

NOTE. The preceding result belongs to category 1).

The next result is useful in completing a full discussion on an application of Proposition 16.2.1. Let us recall the definition of CSBS from section 2.1.

PROPOSITION 16.2.2 Let (X,T) be complete and $\{y_n\}$ a deformation of $\{x_n\}$. If $\{x_n\}$ is CSBS with P being a continuous projection from X onto X_o satisfying $K_o p_o(Px) \leqq \delta p_o(x)$, $0 < \delta < 1$, then $\{y_n\}$ considered above is also CSBS.

PROOF. We invoke both the notation and the proof of the preceding proposition.

Define $S : X \to X$, $Sx = Px - RPx$. Since for x in X_o and p in D_T, $p(x-Rx) \leqq K_p p_o(x)$, we have for p in D_T,

$$p(Sx) \leqq K_p p_o(Px) \leqq K_p K_o^{-1} \delta p_o(x).$$

Therefore by Lemma 15.2.2, $I-S$ is a topological isomorphism from (X,T) onto itself. Let R_1 be the inverse of $I-S$. Then $R_1(I-S) = (I-S)R_1 = I$. Put $Q = (I-S)PR_1$, then $Q^2 = Q$ and Q is continuous. Also $(I-S)P = (I-P+RP)P = RP$, and so $Q = RPR_1$. Hence $Q[X] \subset X_1$.

Let $y \in X_1$. There exists x in X_o with $Rx = y$ and so $y = RPx$ (here $Px = x$). But R_1 is an isomorphism and therefore $y = RPR_1 z$ for some z in X. Thus $X_1 = Q[X]$. □

Applications

Next we consider an application of Proposition 16.2.1; we begin with

THEOREM 16.2.3 Let (X,T) be a Fréchet space with $D_T = \{p_1 \leq p_2 \leq \ldots)$ and $\{y_n\}$ a sequence in X such that $y_n \not\to 0$ in T but $f_i(y_n) \to 0$ for each $i \geq 1$. Then there exists a subsequence $\{y_{n_k}\}$ of $\{y_n\}$ such that $\{y_{n_k}\} \approx \{u_k\}$ for some S.bl.s. $\{u_k\}$ of $\{x_n\}$ and $\{y_{n_k}\}$ is a regular S.b. for $[y_{n_k}]$.

PROOF. We may select $\{p_n\}$ so that $\sup\{p_m(S_n(x)) : n \geq 1\} = p_m(x)$ for each x in X and $m \geq 1$ (cf. [130], p. 27). Since the deletion of a finite number of seminorms in D_T does not alter the topology T, we may select, if necessary, a subsequence of $\{y_n\}$ which we denote by $\{y_n\}$ itself so that for some $\varepsilon > 0$, $p_1(y_n) > \varepsilon$ for $n \geq 1$.

Fix $n_1 = 1$ and $m_0 = 0$. Find $m_1 > 0$, then $n_2 > n_1$, thereafter $m_2 > m_1$ and proceed inductively to get $\{m_k\}$, $\{n_k\}$ in I such that

$$(+) \quad p_k\left(\sum_{i=1}^{m_{k-1}} f_i(y_{n_k})x_i\right),\ p_k\left(\sum_{i\geq 1+m_k} f_i(y_{n_k})x_i\right) < \varepsilon\left(\frac{1}{2}\right)^{k+4};\ k \geq 1.$$

Since

$$(*) \quad p_1\left(\sum_{i=m_{k-1}+1}^{m_k} f_i(y_{n_k})x_i\right) \geq p_1(y_{n_k}) - 2\varepsilon\left(\frac{1}{2}\right)^{k+4} \geq \frac{\varepsilon}{2},$$

the sequence $\{u_k\}$ is a bl.s. of $\{x_n\}$, where

$$u_k = \sum_{i=m_{k-1}+1}^{m_k} f_i(y_{n_k})x_i.$$

Then $\{u_k\}$ is an S.b. for $[u_k]$ (cf. the remark after Theorem 2.1.3); let $\{h_k\}$ be the s.a.c.f. corresponding to $\{u_k\}$.

In order to use the desired proposition, we have to show that $\{u_k;h_k\}$ is a u.e-S.b. for $[u_k] \equiv X_0$ (say). Let $u \in X_0$ and consider any block $[m_{k-1}+1,m_k]$. Then from $u = \sum_{k\geq 1} h_k(u)u_k$, $f_i(u) = h_k(u)f_i(u_k)$ for $m_{k-1} + 1 \leq i \leq m_k$ and this yields $h_k(u)u_k = S_{m_k}(u) - S_{m_{k-1}}(u)$. Therefore $p_1(h_k(u)u_k) \leq 2p_1(u)$ and consequently using (*), $|h_k(u)| \leq (4/\varepsilon)p_1(u)$. If $p_0 = (4/\varepsilon)p_1$, then p_0 is an s.o.e. on X corresponding to $\{u_k;h_k\}$. From (+),

$$\sum_{k\geq 1} p_0(y_{n_k}-u_k) \leq \frac{4}{\varepsilon} \sum_{k\geq 1} p_k(y_{n_k}-u_k) \leq 8 \sum_{k\geq 1} 2^{-k-4} = \frac{1}{2}.$$

Similarly, taking any p_j and using (+), one gets

$$\sum_{k\geq 1} p_j(y_{n_k}-u_k) \leq \sum_{k=1}^{j-1} p_j(y_{n_k}-u_k) + \sum_{k\geq 1} p_k(y_{n_k}-u_k) < \infty.$$

Hence $\{y_{n_k}\}$ is a deformation of the u.e-S. basic sequence $\{u_k\}$ and Propsoition 16.2.1 is applied to get the required result. □

<u>COROLLARY 16.2.4</u> If the hypothesis of Theorem 16.2.3 is fulfilled and $\{u_k\}$ is CSBS then so is $\{y_{n_k}\}$.

For the proof, it is enough to find an integer $j \geq 1$ such that $X_{1,j} \equiv [y_{n_k} : k \geq j]$ is complemented, X_1 being $[y_{n_k}]$. Let $P : X \to X_0 \equiv [u_k]$ be the continuous projection. The mappings $Q_n : X_0 \to X_0$ defined by

$$Q_n(\sum_{i\geq 1} h_i(u)u_i) = \sum_{i\geq n} h_i(u)u_i$$

form an equicontinuous family of projections, since X_0 is also a Fréchet space and so with p_0 as in the proof of the theorem, there exists q in D_T such that

$$p_0'(x) \equiv \sup_{n \geq 1} p_0(Q_n Px) \leq q(x), \quad \forall x \in X.$$

Then p_0'' is a T-continuous seminorm on X, where $p_o'' = \max(p_0, p_0')$. Therefore $\sum_{k\geq 1} p_0''(y_{n_k} - u_k) < \infty$ and so there exists $j \geq 1$ with

$$K_0' \equiv \sum_{k\geq 1} p_0''(y_{n_k} - u_k) < 1.$$

Observe that $Q_j P$ is a continuous projection of X onto $X_{o,j} \equiv [u_k : k \geq j] \simeq X_{1,j}$. Also

$$p_0''(Q_j Px) = \sup_{j\leq n<\infty} p_0\Big(\sum_{i\geq n} \alpha_i u_i\Big), \; Px = \sum_{i\geq 1} \alpha_i u_i$$

$$\leq p_0''(x)$$

$$\Longrightarrow K_0' p_0''(Q_j Px) \leq K_0' p_0''(x).$$

Hence from Proposition 16.2.3, $X_{1,j}$ is complemented in X.

Theorem 16.2.3 yields

<u>PROPOSITION 16.2.5</u> Let (X,T) be a Fréchet space with an S.b. $\{x_n; f_n\}$ and let X_o be a closed subspace of (X,T). If X_o is not a Montel space, then X_o contains a normalized S. basic sequence.

<u>PROOF</u>. Let T_1 be the l.c. topology on X generated by $\{f_n\}$. Then we have the notion of γ-convergence on the bi-l.c. TVS (X,T,T_1). Let $y_n \to 0$ (γ) imply that $y_n \to 0$ in T, where $\{y_n\} \subset X_o$ (cf. [G1], [G2] for details of γ-convergence). Since T and T_1 are metrizable, T and T_1 coincide on T-bounded subsets of X_o. Further, each T-bounded set is $\sigma(X,X^*)$-precompact ([180], p. 50) and hence T_1-precompact. Therefore, each T-bounded subset of X_o is T-relatively compact and this shows that X_o is a Montel space, a contradiction. Hence, there exists $\{y_n\}$ in X_o such that

$y_n \to 0$ (γ) but $y_n \not\to 0$ in T; that is, $\{y_n\}$ is T-bounded, $y_n \to 0$ in T_1 but $y_n \not\to 0$ in T. Now apply Theorem 16.2.3. □

NOTE. For another application and a variation of Proposition 16.2.1, one is referred to chapter 6 of [130].

16.3 OTHER STABILITY THEOREMS

In this section, we take up a few other results on stability which fall into category 2). However, let us first collect some lemmas. Following [153], we have

LEMMA 16.3.1 Let $\{u_n\}$ be a sequence in a quasi-complete l.c. TVS (X,T) and let $\{f_n\}$ be an equicontinuous sequence in X*. Assume that

$$\sum_{n\geq 1} u_n \text{ converges unconditionally} \tag{16.3.2}$$

in (X,T). Then for each α in ℓ^∞ with $|\alpha_n| \leq 1$, the linear operator $R \equiv R_\alpha : X \to X$ defined by

$$Rx = \sum_{n\geq 1} \alpha_n f_n(x) u_n \tag{16.3.3}$$

is a compact operator.

PROOF. Although this result is also given in [129], p. 182, we however prove it for the sake of completeness. There exists p in D_T with $|f_n(x)| \leq p(x)$ for all x in X and write v for $v_p = \{x \in X : p(x) \leq 1\}$. Clearly $v \in \mathcal{B}_T$. By Theorem 1.5.2, corresponding to each equicontinuous subset M of X*, there exists $N \equiv N(M)$ in $\mathbb{N}$ so that

$$\sum_{i=r}^{s} |f(u_i)| \leq 1; \ \forall r,s \geq N, \ \forall f \in M$$

$$\Longrightarrow \sum_{i=r}^{s} \alpha_i f_i(x) u_i \in M^o; \quad \forall r,s \geq N, \quad \forall x \in v.$$

Hence $\sum_{n\geq 1} \alpha_n f_n(x)u_n$ converges uniformly in v and so it converges at each point of X. Therefore, the map R of (16.3.3) is well defined. Let $R_n : X \to X$ be defined by

$$R_n x = \sum_{i=1}^{n} \alpha_i f_i(x)u_i .$$

Then $R_n x \to Rx$ unioformly on v. Now for x in v and q in D_T,

$$q(R_n x) \leq \sum_{i=1}^{n} q(u_i)$$

and so $R_n[v]$ is bounded in the finite dimensional space $R_n[X]$. Thus $R_n[v]$ is precompact. By Lemma 1.2.14, R[v] is precompact and hence relatively compact. □

The next result is an extension of Riesz's theorem on compact operators (cf. [140], p. 207).

<u>LEMMA 16.3.4</u> Let (X,T) be a TVS and $R : X \to X$ a compact operator. If α is not an eigenvalue of R, then $R-\alpha I$ is a topological isomorphism from (X,T) onto itself.

We are now in a position to state and prove the following theorem which extends the stability theorem of Weill [231] proved earlier for Banach spaces by Vĕic [228].

<u>THEOREM 16.3.5</u> Let $\{x_n;f_n\}$ be an equicontinuous S.b. for a quasi-complete l.c. TVS (X,T). Let $\{y_n\}$ be an ω-l.i. sequence in X such that

$$\sum_{n\geq 1} (y_n - x_n) \text{ is unconditionally convergent} \qquad (16.3.6)$$

in (X,T). Then $\{y_n\} \overset{\approx}{\sim} \{x_n\}$ and $\{y_n\}$ is an S.b. for (X,T).

<u>PROOF</u>. By putting $\alpha = e$ and $u_n = y_n - x_n$, $n \geq 1$, it follows from Lemma 16.3.1 that the operator $R : X \to X$, defined by $Rx = \sum_{n\geq 1} f_n(x)(y_n - x_n)$ is a compact operator.

Consequently the operator $S : X \to X$, $Sx = \sum_{n\geq 1} f_n(x)y_n$ is well defined and continuous with $S = R+I$. If $Rx = -x$, then $Sx = 0$ and so $f_n(x) = 0$ for $n \geq 1$. Thus -1 is not an eigenvalue of R. Therefore, by Lemma 16.3.4, $(X,T) \simeq (X,T)$ under S with $Sx_n = y_n$, $n \geq 1$. □

<u>COROLLARY 16.3.7</u> Let $\{x_n;f_n\}$ be a regular S.b. for a quasi-complete ω-barrelled space (X,T). Let $\{y_n\}$ be an ω-l.i. sequence in X satisfying (16.3.6), then $\{x_n\} \approx \{y_n\}$ and $\{y_n\}$ is an S.b. for (X,T).

Indeed, by Corollary 10.2.3, $\{f_n\}$ is equicontinuous.

Stability criteria in Banach spaces

There are several other stability theorems in Banach spaces whose locally convex analogues in general appear to be rather complicated and not of great interest. These different criteria are listed below and are respectively taken from [163], [228], [149], [150], [162] and [187]. Therefore, let $\{x_n;f_n\}$ be an S.b. for a Banach space $(X, \|\cdot\|)$ which is also supposed to have a sequence $\{y_n\}$ satisfying the following "nearness" conditions.

(M) $\|x_n-y_n\| \leq \alpha_n$, $n \geq 1$ and

$$\sup_{n\geq 1} \sup_{e_i=\pm 1} \left\| \sum_{i=1}^{n} \alpha_i e_i f_i \right\| < 1.$$

(V) $\inf \|x_n\| > 0$, $\{y_n\}$ is ω-l.i. and

$\sum_{n\geq 1} (y_n-x_n)$ is unconditionally convergent.

(K-L) $\{y_n\}$ is ω-l.i. and $\sum_{n\geq 1} \|f_n\| \, \|x_n-y_n\| < \infty$.

(K-M-R) $\sum_{n\geq 1} \|f_n\| \, \|x_n-y_n\| < 1$.

(M-R) $\sup\{\|f_n\| : n \geq 1\} \equiv N < \infty$ and

$$\sup_{\sigma \in \Phi} \left\| \sum_{i \in \sigma} (x_i - y_i) \right\| < \frac{1}{4N},$$

where σ is a finite set of indices.

(A) $\|x_n\| \to \infty$, $\{y_n\}$ is ω-l.i. and $\sum_{n \geq 1} (x_n - y_n)$ is unordered bounded in X

(B) $\sum_{n \geq 1} f_n$ is unordered bounded in X^*, $\{y_n\}$ is ω-l.i. and $\|x_n - y_n\| \to 0$

(C) $\sup \{\| \sum_{i=m}^{n} f(x_i - y_i) f_i \| : \|f\| \leq 1\} \to 0$ as $m,n \to \infty$.

(D) $\sup \{\| \sum_{n \geq 1} f(x_n - y_n) f_n \| : \|f\| \leq 1\} < 1$.

NOTE. The conditions (M), (K-M-R), (M-R) and (D) involve certain specific bounds whereas the rest are concerned with restrictions on $\{x_n\}$ and $\{y_n\}$.

We need further information before we come to the main result. The next lemma is a slight improvement on a similar result of [162]. It is also given in [129], p. 182 with a more restrictive hypothesis.

LEMMA 16.3.8 Let $\{u_n\}$ be a sequence in a quasi-complete l.c. TVS (X,T) and let $\{f_n\}$ be an equicontinuous sequence in X^*. Suppose that

$$\sum_{n \geq 1} u_n \text{ is unordered bounded} \qquad (16.3.9)$$

in (X,T). Then for each α in c_o, the linear operator $R \equiv R_\alpha : X \to X$ given by (16.3.3) is compact.

PROOF. Fix α in c_o. Consider an arbitrary β in ℓ^∞, $\varepsilon > 0$ and equicontinuous set M in X^*. Let $K = \sup |\beta_n|$. By Proposition 1.5.6, there exists K_M such that $\sum_{n \geq 1} |f(u_n)| \leq K_M$ for each $f \in M$. Determine N in $\mathbb{N}$ so that $|\alpha_i| \leq \varepsilon / 2KK_M$ for

$i \geqq N$. Thus for f in M and m, $n \geqq N$

$$|\langle \sum_{i=m}^{n} \beta_i \alpha_i u_i, f\rangle| \leqq \frac{\varepsilon}{2K_M} \sum_{i=m}^{n} |f(u_i)| < \varepsilon .$$

Hence $\sum_{n\geqq 1} \beta_n(\alpha_n u_n)$ converges for each β in ℓ^∞ and so $\sum_{n\geqq 1} \alpha_n u_n$ converges unconditionally in (X,T) (cf. Theorem 1.5.2). The result now follows from Lemma 16.3.1. □

The following is well known.

LEMMA 16.3.10 For a continuous operator R on a Banach space $(X, \|\cdot\|)$, if R is compact, then so is R* and if $\|R\| < 1$, then $\|R^*\| < 1$.

The next result is an extension of [187], p. 137.

LEMMA 16.3.11 Let $\{x_n; f_n\}$ be an S.b. for an l.c. TVS (X,T) which is ω-complete and such that both (X,T) and $(X^*, \beta(X^*,X))$ are barrelled. Let $y_n \in X$, $n \geqq 1$. (i) If $\sum_{n\geqq 1} f(x_n - y_n) f_n$ exists in $\beta(X^*,X)$ for each f in X*, then $S : X^* \to X^*$ defined by $Sf = \sum_{n\geqq 1} f(x_n - y_n) f_n$ is a continuous linear operator. (ii) If the series $\sum_{n\geqq 1} f(x_n - y_n) f_n$ converges uniformly on each equicontinuous subset of $(X^*, \beta(X^*,X))$, then $R : X \to X$, $Rx = \sum_{n\geqq 1} f_n(x)(x_n - y_n)$ is a continuous linear operator on (X,T), $R^* = S$ and $S^*Jx = Rx$, J being the canonical embedding, $J : X \to X^{**}$.

PROOF. (i) Here the result follows by the Banach-Steinhaus theorem.

(ii) Let B be bounded in (X,T), M an equicontinuous subset of X* and $\varepsilon > 0$. There exists $N \equiv N(M,B,\varepsilon)$ such that

$$\sup_{f\in M}\sup_{x\in B}\left|\sum_{i=m}^{n} f(x_i-y_i)f_i(x)\right| \leq \varepsilon \ ; \ \forall m,n \geq N$$

$$\Longrightarrow \sum_{i=m}^{n} f_i(x)(x_i-y_i) \in \varepsilon M^{o}; \ \forall m, \ n \geq N; \ x \in B.$$

Hence Rx is well defined, the continuity of R being a consequence of the Banach-Steinhaus theorem again. Observe that $\langle x,R^*f\rangle = \langle x,Sf\rangle$ and so $R^* = S$. Also

$$\langle f,R^{**}Jx\rangle = \langle R^*f,Jx\rangle = \langle x,Sf\rangle, \quad \text{or,}$$

$$\langle f,S^*Jx\rangle = \langle x,Sf\rangle = \langle x,R^*f\rangle. \quad \square$$

We also recall ([162]; cf. [129], p. 145)

LEMMA 16.3.12 Let a vector space X be equipped with a seminorm ρ. Corresponding to a finite subset σ of positive integers ($\sigma \in \Phi$), consider a subset $\{x_i : i \in \sigma\}$ of X and a subset $\{\alpha_i : i \in \sigma\}$ of $\mathbb{K}$. If $k = \sup\{|\alpha_i| : i \in \sigma\}$, then

$$\rho\Big(\sum_{i\in\sigma} \alpha_i x_i\Big) \leq 4k \sup_{\sigma_1\subset\sigma} \rho\Big(\sum_{i\in\sigma_1} x_i\Big).$$

Unless specified otherwise, we follow [177] for the rest of this section. The main result is

THEOREM 16.3.13 Let $\{x_n;f_n\}$ be an S.b. for a Banach space $(X, \|\cdot\|)$ containing a sequence $\{y_n\}$. If any of the conditions (M) to (D) stated earlier is satisfied, then $\{x_n\} \approx \{y_n\}$ and $\{y_n\}$ is an S.b. for $(X, \|\cdot\|)$.

PROOF. In view of Proposition 16.2.1 and Corollary 16.3.7, we need to prove the above result under all but the conditions (V) and (K-M-R).

(M) It is immediate that (M) yields

$$\sup_{n\geq 1} \sup \{ \sum_{i=1}^{n} \alpha_i |\phi(f_n)| : \phi \in X^{**}, \|\phi\| \leq 1\} < 1$$

$$\Longrightarrow \lambda \equiv \sup \{ \sum_{n\geq 1} \|x_n - y_n\| \; |\phi(f_n)| : \phi \in X^{**}, \|\phi\| \leq 1\} < 1.$$

Hence the operator $R : X \to X$, $Rx = \sum_{n\geq 1} f_n(x)(x_n - y_n)$ is well defined and continuous, since

$$\|Rx\| \leq \|x\| \sum_{n\geq 1} \left|\frac{Jx}{\|x\|}(f_n)\right| \; \|x_n - y_n\| \leq \lambda \|x\| .$$

By Lemma 15.2.2, $(X, \|\cdot\|) \simeq (X, \|\cdot\|)$ under $S = I - R$ and the result is proved in this case.

(K-L) By Lemma 1.2.14, the operator $R : X \to X$, $Rx = \sum_{n\geq 1} f_n(x)(x_n - y_n)$ is compact. Now proceed as in Theorem 16.3.5.

(M-R) We can find λ, $0 < \lambda < 1$ so that

$$\sup_{\sigma\in\Phi} \|\sum_{i\in\sigma} (x_i - y_i)\| \leq \frac{\lambda}{4N}.$$

Consider any finite set $\{\alpha_1, \ldots, \alpha_n\}$ of scalars and let $x = \sum_{i=1}^{n} \alpha_i x_i$. Then from Lemma 16.3.12

$$\|\sum_{i=1}^{n} \alpha_i (x_i - y_i)\| \leq 4 \sup |\alpha_i| \cdot \frac{\lambda}{4N} \leq \lambda \|\sum_{i=1}^{n} \alpha_i x_i\| ,$$

and the required result follows from Theorem 15.3.6.

(A) Here $\|f_n\| \to 0$ and $\{f_n / \|f_n\|\}$ is equicontinuous. Hence by Lemma 16.3.8, the operator $R : X \to X$ defined by

$$Rx = \sum_{n\geq 1} \|f_n\| \frac{f_n}{\|f_n\|}(x)(x_n - y_n) = \sum_{n\geq 1} f_n(x)(x_n - y_n)$$

is compact. Now proceed as in Theorem 16.3.5.

(B) Since $\{J(x_i - y_i)/ \|x_i - y_i\|\}$ is an equicontinuous sequence in X^{**} and $\|x_n - y_n\| \to 0$, the operator $S : X^* \to X^*$

defined by

$$Sf = \sum_{n\geq 1} \|x_n - y_n\| J\left(\frac{x_n - y_n}{\|x_n - y_n\|}\right)(f) f_n = \sum_{n\geq 1} f(x_n - y_n) f_n$$

is compact in $(X^*, \beta(X^*, X))$. Therefore the series $\sum_{i\geq 1}$ $f(x_i - y_i) f_i$ is uniformly Cauchy for $\|f\| \leq 1$ in X^* and so by Lemma 16.3.10, $S^*J : X \to X$ is compact. Now proceed as in (A).

(C) If $Sf = \sum_{n\geq 1} f(x_n - y_n) f_n$ and $S_n f = \sum_{i=1}^{n} f(x_i - y_i) f_i$, then $S_n f \to Sf$ uniformly for $\|f\| \leq 1$. Since S_n's are finite dimensional operators, S is compact. Now proceed as in (B).

(D) If $Sf = \sum_{n\geq 1} f(x_n - y_n) f_n$, then $\|s\| < 1$. By Lemma 16.3.11, S^*J is well defined and continuous on X. Further, by Lemma 16.3.10, $\|S^*J\| < 1$. Next proceed as in (M). □

NOTE. The stability theorems relating to the criteria (M), (K-M-R), (M-R) and (D) belong to the category 1) whereas the rest falls into the category 2).

16.4 SEVERAL STABILITY CONDITIONS: INTERRELATIONSHIP

This section is concerned with the possible interrelationship amongst various stability criteria (M) to (D) discussed in the last section and we follow [187] for most of the results.

THEOREM 16.4.1 Let $\{x_n; f_n\}$ be an S.b. for a Banach space $(X, \|\cdot\|)$ and let $y_n \in X$, $n \geq 1$. Then we have the following diagram

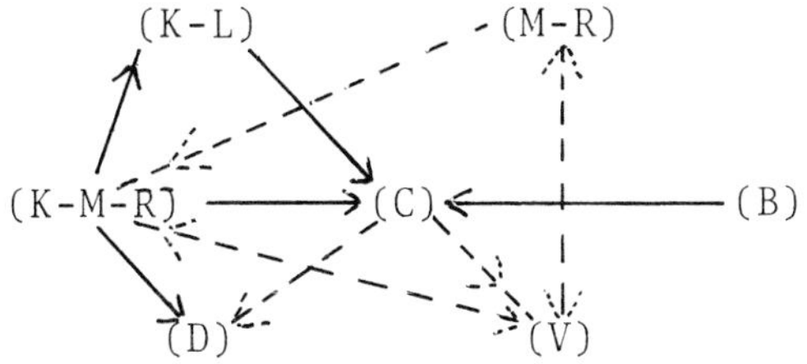

NOTE. In this diagram, whereas the gaps among various statements imply that they are independent of each other, a dotted line arrow between two statements means that the implication is not true in general and this fact is exhibited by a counter-example in the next subsection.

PROOF OF THEOREM 16.4.1 As far as the implications are concerned, we observe that (K-M-R) implies (K-L), (C) and (D), and (K-L) implies (C) are trivial.

(B) ==> (C). Let $K = \sup \{\|\sum_{i\in\sigma} f_i\| : \sigma \in \Phi\}$

Therefore, by Lemma 16.3.12,

$$\|\sum_{i=m}^{n} f(x_i - y_i) f_i\| \leqq 4K \sup \{|f(x_i - y_i)| : m \leqq i \leqq n\}$$

and this yields (C). In order to complete the discussion of the diagram, we need to construct counter-examples and this is carried out in the next subsection. □

Counter-examples

The negatives in the foregoing diagram are constructed with the help of the spaces $c_o = (c_o, \|\cdot\|_\infty)$ and $\ell^1 = (\ell^1, \|\cdot\|_1)$.

The first three examples are reproduced from [162]. For simplicity, we write $\|\cdot\|$ for both of the norms $\|\cdot\|_\infty$ and $\|\cdot\|_1$.

EXAMPLE 16.4.2 Here we show that (K-M-R) =≠> (V), (M-R). Consider the S.b. $\{x^n;f^n\}$ for c_o, where $x^n = ne^n$, $f_n = e^n/n$ and a sequence $\{y^n\}$ in c_o, where $y^n = \sum_{i=1}^{n-1} e^i/2n + (n-1/2n)e^n$, $n \geq 1$. Then $\|x^n-y^n\| = 1/2n$ and so

$$\sum_{n\geq 1} \|f^n\| \; \|x^n-y^n\| = \sum_{n\geq 1} \frac{1}{2n^2} = \frac{\pi^2}{12} < 1$$

and so (K-M-R) is satisfied. The first coefficient a_{1n} in $\sum_{i=1}^{n} (x^i-y^i)$ is $(2-h_n)/2$, $h_n = \sum_{i=1}^{n} (1/i)$. Since $\|\sum_{i=1}^{n} (x^i-y^i)\| \geq |a_{1n}|$, $\sum_{i\geq 1} (x^i-y^i)$ is not even unordered bounded and so (M-R) and (V) are not true.

EXAMPLE 16.4.3 (M-R) =≠> (K-M-R), (V). Consider the S.b. $\{x^n;f^n\}$ for c_o; $x^n = f^n = e^n$, $n \geq 1$. For any λ, $0 < \lambda < 1$, let $y^n = (1-\lambda/4)x^n$, $n \geq 1$. Hence

$$\|\sum_{i\in\sigma} (x^i-y^i)\| = \frac{\lambda}{4}$$

and so (M-R) holds. Since $\sum_{n\geq 1} (x^n-y^n)$ does not converge, (V) does not hold. Also $\|f^n\| \; \|x^n-y^n\| = \lambda/4$ and so (K-M-R) is not valid.

EXAMPLE 16.4.4 (V) =≠> (M-R), (K-M-R). Let $\{x^n;f^n\}$ and c_o be as in Example 16.4.3. Let $y^1 = 2x^1$ and $y^n = (n-1)x^n/n$, $n \geq 2$. Here $x^n-y^n = x^n/n$, $n \geq 2$ and $x^1-y^1 = -x^1$. For b in ℓ^∞, $\|\sum_{i=m}^{n} b_i x^i/i\| = \sup\{|b_i|/i : m \leq i \leq n\} \quad \|b\|_\infty$. By Theorem 1.5.2, $\sum_{i\geq 1} (x^i-y^i)$ is unconditionally convergent.

Since $\|f^i\| \ \|x^i-y^i\| = 1/i$, $i \geqq 1$ and $\|\sum_{i=1}^{n} (x^i-y^i)\| = 1$, (K-M-R) and (M-R) are not satisfied.

The next example is from [187].

EXAMPLE 16.4.5 (C) =/> (V). Consider the S.b. $\{x^n;f^n\}$ for ℓ^1; $x^n = f^n = e^n$, $n \geqq 1$. Put $y^n = (1-n^{-1})x^n$, $n \geqq 1$. Thus $x^n-y^n = x^n/n$ and, as $|\langle x^n-y^n,e\rangle| = 1/n$, by the theorem referred to above (V) is not satisfied. On the other hand, for $f \in \ell^\infty$

$$\sup_{\|f\| \leqq 1} \|\sum_{i=m}^{n} f(x^i-y^i)f^i\| = \sup_{\|f\| \leqq 1} \max_{m \leqq i \leqq n} |f(e^i/i)| = \frac{1}{m}$$

and so (C) holds.

NOTE. The above example also shows that (C) holds but (D) is not satisfied.

REMARK. If we drop any of the conditions (M) to (D) on the 'nearness' of a sequence $\{y_n\}$ to a given S.b. $\{x_n\}$, the former may cease to be an S.b. For instance, consider (c.f. [162])

EXAMPLE 16.4.6 Once again, consider the S.b. $\{x^n;f^n\}$ for c_o of Example 16.4.3. Let $y^1 = x^1$; $y^n = x^{n-1}-x^n$, $n \geqq 2$. Here $\|x^1-y^1\| = 0$, $\|x^n-y^n\| = 2$, $n \geqq 2$ and

$$\|\sum_{i=m}^{n} (x^i-y^i)\| = 2, \ n > m; \ \|\sum_{i=1}^{n} \varepsilon_i\alpha_i f^i\| = 2, \ n \geqq 2$$

where $\varepsilon_i = \pm 1$ and $\alpha_i = \|x^i-y^i\|$. Thus clearly (M), (K-L), (K-M-R), (A) and (B) are not satisfied. $\{y^n\}$ is ω-l.i., for if $\sum_{n \geqq 1} a_n y^n = 0$, then $a_1 + a_2 = 0$ and $a_{n-1} = a_n$, $n \geqq 3$. If $a = a_2 = a_3 = \ldots$, then

$$|a| \lim_{n\to\infty} \| \sum_{i=2}^{n} y^i - y^1 \| = 0.$$

But $\| \sum_{i=2}^{n} y^i - y^1 \| = 1$ for all $n \geqq 2$. Hence $a = 0$ and so $a_1 = a_2 = \ldots = 0$. Since

$$\sup_{\sigma\in\Phi} \| \sum_{i\in\sigma} x^i - y^i \| = 2,$$

(M-R) is not satisfied.

$\{y^n\}$ is not even minimal (otherwise we get an h in ℓ^1 with $h(y^1) = 1$, $h(y^n) = 0$, $n \geqq 2$ and this gives $h(x^{n-1}) = h(x^n)$, $n \geqq 2$, a contradiction) and so it cannot be S.basic.

16.5 STABILITY CRITERIA IN HILBERT SPACES

The results of the last section clearly remain valid for any separable Hilbert space. However, the rich norm structure of such spaces provides additional stability criteria which we discuss hereafter and we follow [187] for the rest of this section. Throughout, $X \equiv (X, \|\cdot\|)$ is assumed to be an arbitrary separable Hilbert space, the inner product between x and y in X being denoted by $\langle x,y\rangle$. Also, $\{x_n\}$ will henceforth stand for an orthonormal S.b. for X. For any sequence $\{y_n\}$ in X, consider the following set of conditions after [13], [14], [195] and [177]:

(B_1) $\sum_{n\geqq 1} \|x_n - y_n\|^2 = \lambda < 1$;

(B_2) $\{y_n\}$ is ω-l.i. and $\sum_{n\geqq 1} \|x_n - y_n\|^2 < \infty$,

(S) $\sup\{ \sum_{n\geqq 1} |\langle x, x_n - y_n\rangle|^2 : \|x\| \leqq 1\} \leqq \lambda < 1$

and

(E) If $g_n \to 0$ in $\sigma(X^*,X)$, then

$$\sum_{i\geqq 1} |\langle x_i - y_i, g_n\rangle|^2 \to 0 \text{ as } n \to \infty.$$

PROPOSITION 16.5.1 $(B_1) \Longrightarrow (B_2) \Longrightarrow (E)$; also $(B_1) \Longrightarrow (S)$.

PROOF. For $f \in X^*$ $(= X)$ with $\|f\| \leq 1$ and $n \geq m$, we have

$$(*) \quad \left\| \sum_{i=m}^{n} <x_i-y_i,f>f_i \right\|^2 = \sum_{i=m}^{n} |<x_i-y_i,f>|^2 \leq \sum_{i=m}^{n} \|x_i-y_i\|^2$$

(here $f_i = x_i$). Therefore, if (B_1) holds, proceeding as in Theorem 16.3.13 (D), we conclude that $\{y_n\}$ is an S.b. for X; in particular, $\{y_n\}$ is ω-l.i. Thus $(B_1) \Longrightarrow (B_2)$.

$(B_2) \Longrightarrow (E)$. Let $\varepsilon > 0$ and $K = \sup\ \{\|g_n\| : n \geq 1\}$. There exist N and M in $\mathbb{N}$ so that

$$\sum_{i \geq N+1} \|x_i-y_i\|^2 < \frac{\varepsilon}{2K}; \quad |<x_i-y_i,g_p>|^2 < \frac{\varepsilon}{2N},$$

for all $p \geq M$ and $1 \leq i \leq N$. Therefore $\sum_{i\geq 1} |<x_i-y_i,g_p>|^2 < \varepsilon$ for $p \geq M$.

The last implication follows from (*). □

PROPOSITION 16.5.2 If (S) holds, then $\{y_n\}$ is an S.b. for X.

PROOF. The existence of the operator $R : X \to X$, $Rx = \sum_{n\geq 1} <x,x_n-y_n>x_n$ is guaranteed by (S). Here $\|R\| < 1$ and so $\|R^*\| < 1$ with $R^*y = \sum_{n\geq 1} <x_n,y> (x_n-y_n)$, $y \in X^*$. Hence $I-R^*$ is a topological isomorphism on X.

PROPOSITION 16.5.3 If $\{y_n\}$ is ω-l.i. and (E) holds, then $\{y_n\}$ is an S.b. for X.

PROOF. Let $\{g_n\}$ be an arbitrary sequence in X with $g_n \to f$ in $\sigma(X^*,X)$ and consider the operator $S : X^* \to X^*$,

$Sf = \sum_{n\geq 1} \langle x_n - y_n, f\rangle f_n$ which is well defined by (E). Then

$$\| Sg_n - Sf \|^2 = \sum_{i\geq 1} |\langle x_i - y_i, g_n - f\rangle|^2 \to 0$$

as $n \to \infty$. Hence S is compact and so is the operator $S^* : X \to X$ defined by $S\, x = \sum_{n\geq 1} \langle x, x_n\rangle (x_n - y_n)$ and the result follows as before. □

Except for the appropriate counter-examples, we have therefore the diagram of implications below:

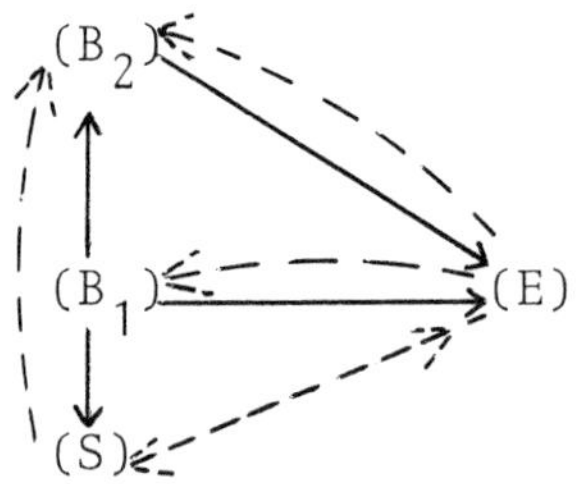

EXAMPLE 16.5.4 Here we show that (S) =≠> (B_1), (B_2) and (E). Consider the S.b. $\{x^n; f^n\}$ for the Hilbert space ℓ^2, $x^n = f^n = e^n$, $n \geq 1$. Let $y^n = e^n/2$, $n \geq 1$. Since $\|x^n - y^n\| = 1/2$, neither (B_1) nor (B_2) holds. Also $f^n \to 0$ in $\sigma(\ell^2, \ell^2)$ and $\sum_{i\geq 1} |\langle x^i - y^i, f^n\rangle|^2 = 1/4$, $\forall\, n \geq 1$, yield that (E) is not true. However, for $\|x\| \leq 1$,

$$\sum_{n\geq 1} |\langle x, x^n - y^n\rangle|^2 \leq \frac{1}{4} < 1$$

and so (S) is satisfied.

EXAMPLE 16.5.5 Here (E) =≠> (B_1), (B_2) and (S). Consider again ℓ^2 with its usual orthonormal S.b. $\{x^n\}$, $x^n = e^n$. Let $y^n = (1-n^{-1/2})x^n$, $n \geq 1$. Clearly (B_1) and (B_2) are not satisfied. Since $\sum_{n\geq 1} |\langle x^j, x^n - y^n\rangle|^2 = 1/j$, (S) is not satisfied. Now take any $\{g^n\}$ in ℓ^2 with $g^n \to 0$ in $\sigma(\ell^2,\ell^2)$. Observe that

$$\sum_{i\geq 1} |\langle g^n, x^i - y^i\rangle|^2 = \sum_{i\geq 1} |g_i^n|^2/i.$$

Let

$$K = \{u^\sigma \equiv \sum_{i\in\sigma} \frac{1}{i} x^i \in \ell^2 : \sigma \in \Phi\} .$$

By using the convergence of $\sum_{n\geq 1} 1/n^2$, it easily follows that K is precompact (for $\varepsilon > 0$, there exists N with $\sum_{i>N} 1/i^2 < \varepsilon$; let Φ_N be the family of all finite subsets of $\{1,\ldots,N\}$ and let $F_N = \{u^\sigma \in K : \sigma \in \Phi_N\}$, then $K \subset F_N + \varepsilon B$, $B = \{x \in \ell^2 : \|x\| \leq 1\}$) and therefore $g^n(x) \to 0$ uniformly on K (in fact, use the containment of K in $F_N + \varepsilon B$). Thus, if $M = \sup \{\|g^n\| : n \geq 1\}$ and $\varepsilon > 0$, there exists N so that $|g^n(x)| < \varepsilon/4M$ for $n > N$ and x in K; that is,

$$\left|\sum_{i\in\sigma} \frac{1}{i} g_i^n\right| < \frac{\varepsilon}{4M};\ \forall\sigma \in \Phi,\ n > N$$

$$\Longrightarrow \sup_{\sigma\in\Phi} \sum_{i\in\sigma} \frac{1}{i} M|g_i^n| \leq \varepsilon, \quad \forall n > N$$

by Lemma 16.3.12. Hence (E) is satisfied.

Concluding remarks

Concerning the S.b. character of a sequence $\{y_n\}$ in an l.c. TVS (X,T) containing an S.b. $\{x_n;f_n\}$, we have stated and proved several theorems by using one or another sufficient (stability) conditions in which the basic requirement has been to establish the contractive or compact character of the operator $R : X \to X$, $Rx = \sum_{n\geq 1} f_n(x)(x_n - y_n)$. In general, it appears difficult to ascertain whether the contractive or compact character of R yields one or another of the stability criteria mentioned before (of course, wherever necessary, we may additionally assume that $\{y_n\}$ is ω-l.i.). However, in the special case when X is a Hilbert space, $x_n = e^n$ and $f_n = e^n$, the situation is quite satisfactory for the conditions (S) and (E).

Indeed, let $\|R\| = \lambda < 1$, $Rx = \sum_{n\geq 1} \langle x, f_n\rangle (x_n - y_n)$. Then $\|R^*\| = \lambda$ and $R^*y = \sum_{n\geq 1} \langle y, x_n - y_n\rangle x_n$ and so $\|R^*y\|^2 = \sum_{n\geq 1} |\langle y, x_n - y_n\rangle|^2$. Hence (S) holds.

On the other hand, let $\{y_n\}$ be ω-l.i. and R be compact. Then R^* is compact. Consequently, if $g_n \to 0$ in $\sigma(X^*, X)$, then $R^*g_n \to 0$ in X and this yields (E).

Finally, we offer a comment about (B_1). If $0 < r \leqq 2$, then the truth of (B_{11}) : $\sum_{n\geq 1} \|x_n - y_n\|^r < 1$, implies $\sum_{n\geq 1} \|x_n - y_n\|^2 < 1$ and so (B_1) is weaker than the former ones (B_{11}) for $0 < r < 2$. However, for $r > 2$, (B_{11}) may not even yield the minimal character of $\{y_n\}$ as shown by

EXAMPLE 16.5.6 Let $\{e^n; e^n\}$ be the usual orthonormal S.b. for ℓ^2. Let $r = 2 + \varepsilon$, $\varepsilon > 0$ and $p = r/2 > 1$. There exists N such that $\sum_{n\geq N} (1/n)^p < 1$. Put $y^n = e^n$, $1 \leqq n \leqq N-1$; $y^n = e^n - (1/n)e^{(n)}$, $n \geqq N$; also, let $x^n = e^n/2$ and $u^n = y^n/2$. Then $\|x^n - u^n\|^2 = 0$ for $1 \leqq n \leqq N-1$ and $= 1/4n$ for $n \geqq N$. Therefore $\sum_{n\geq 1} \|x^n - u^n\|^r < 1$. However, $\{u^n\}$ is not minimal, for then there exists $\{f^n\}$ in ℓ^2 with $\langle u^n, f^m\rangle = \delta_{mn}$. Let $f^N = \{a_i\}$. Since $\langle u^n, f^N\rangle = 0$ for $1 \leqq n \leqq N-1$, $a_1, \ldots, a_{N-1} = 0$. Also $\langle u^N, f^N\rangle = 1$ yields $a_N = 2N/(N-1)$. Similarly, from $\langle u^n, f^N\rangle = 0$ for $n > N$ it follows that $a_{N+1}, \ldots = 2/(N-1)$. Thus $f^N \notin \ell^2$, a contradiction. However, $\{u^n\}$ is ω-l.i.

PART IV

17 Bases in λ-nuclear spaces

17.1 A HEURISTIC DESCRIPTION

The useful class of nuclear spaces is still enveloped by another important class of locally convex spaces, namely, λ-nuclear spaces. A related pertinent question is: how far has the theory of Schauder bases contributed to the growth of λ-nuclear spaces? Also, we may ask: are λ-bases related to λ-nuclear spaces in some way or another? In this chapter, we will ultimately answer the first question and we will take up the other problem in chapter 18.

However, we believe that λ-nuclear spaces are not as commonly known as nuclear spaces. As such, we will first develop an elementary and useful background of these spaces.

Nonetheless, it appears desirable, for our present purposes, to mention briefly the developments which culminated in the birth of λ-nuclear spaces.

When $\lambda = s$ (the space of rapidly decreasing sequences), Brudovskii ([21], [22]) was the first to characterize strongly nuclear spaces. Köthe [144] provided a correction to this and also introduced the concept of uniform s-nuclearity for certain sequence spaces. It was, hopefully, the paper [144] that motivated quite a number of workers to undertake the study of λ-nuclear spaces for different sequence spaces λ, e.g. [180], [193], [222], [191], etc. (cf. [112] for an exhaustive account in this direction). The problem of characterizing λ-nuclear spaces appears to be more pleasant than characterizing λ-nuclear operators; this observation is made in [112].

Unless stated otherwise, we follow [138] for all

subsequent sections of this chapter.

17.2 λ-NUCLEARITY

This section is exclusively devoted to the study of different types of 'λ-nuclear' maps and presents elementary facts about the associated nuclearity of an l.c. TVS. All unexplained notation is to be found in chapter 1. Throughout (X,T) or (Y,S) denotes an arbitrary l.c. TVS whereas λ is an arbitrary s.s.

DEFINITION 17.2.1 $R \in \mathcal{L}(X,Y)$ is called (i) *λ-nuclear* if there exist α in λ, an equicontinuous sequence $\{f_n\}$ in X^* and a sequence $\{y_n\}$ in Y with $\{g(y_n)\} \in \lambda^\times$ for each g in Y^* such that

$$Rx = \sum_{n\geq 1} \alpha_n f_n(x) y_n; \ \forall\, x \in X,$$

(ii) *pseudo-λ-nuclear* or *$\hat{\lambda}$-nuclear* if R satisfies all the conditions laid down in (i) with $\lambda^\times$ being replaced by ℓ^∞ and (iii) *quasi-λ-nuclear* if there exist α in λ and an equicontinuous sequence $\{f_n\}$ in X^* such that for each q in D_S,

$$q(Rx) \leq \sum_{n\geq 1} |\alpha_n f_n(x)|, \quad \forall x \in X.$$

The collection of all operators defined in (i) (resp. (ii), (iii)) will be denoted by $N_\lambda \equiv N_\lambda(X,Y)$ (resp. $N_\lambda^p \equiv N_\lambda^p(X,Y)$, $N_\lambda^q \equiv N_\lambda^q(X,Y)$).

DEFINITION 17.2.2 A seminorm p on X is called *quasi-λ-nuclear* if there exist α in λ and an equicontinuous sequence $\{f_n\}$ in X^* such that

$$p(x) \leq \sum_{n\geq 1} |\alpha_n f_n(x)|, \quad \forall x \in X.$$

The proof of the following proposition is straightforward

and hence omitted (cf. [216], p. 48 for details).

PROPOSITION 17.2.3 (a) $R \in \mathcal{L}(X,Y)$ is in N_λ^q if and only if there exists a quasi-λ-nuclear seminorm p on X such that the set $A_p = \{Rx : p(x) \leq 1\}$ is bounded in (Y,S). (b) The composition of any of the operators in Definition 17.2.1 with a continuous linear operator is of the same kind. (c) If λ is normal, then the composition of a continuous linear operator with an operator of λ-type (cf. Definition 1.4.18) yields an operator of the same kind.

We now return to the basic

DEFINITION 17.2.4 (X,T) is said to be *λ-nuclear* (resp. *$\hat{\lambda}$-nuclear*) if for each u in $\mathcal{B}_X$, there exists v in $\mathcal{B}_X$, with $v < u$ such that $\hat{K}_u^v : \hat{X}_v \to \hat{X}_u$ is λ-nuclear (resp. $\hat{\lambda}$-nuclear).

NOTE. If $\lambda = \ell^1$, ℓ^1-nuclear or $\hat{\ell}^1$-nuclear spaces are precisely nuclear spaces.

We now come to the main

THEOREM 17.2.5 For an l.c. TVS (X,T), the following statements are equivalent:

(i) X is $\hat{\lambda}$-nuclear (resp. λ-nuclear).

(ii) For each u in $\mathcal{B}_X$, $\hat{K}_u : X \to \hat{X}_u$ is $\hat{\lambda}$-nuclear (resp. λ-nuclear).

(iii) Each $R \in \mathcal{L}(X,F)$ is $\hat{\lambda}$-nuclear (resp. λ-nuclear) for any Banach space F.

PROOF. We prove the result for $\hat{\lambda}$-nuclearity and the result for the part relating to λ-nuclearity follows similarly.

(i) ==> (ii) follows from Proposition 17.2.3(b), since $\hat{K}_u = \hat{K}_u^v \circ \hat{K}_v$.

(ii) ==> (i). Let $u \in \mathcal{B}_X$. Then $\hat{K}_u : X \to \hat{X}_u$ is $\hat{\lambda}$-nuclear and so

$$(*) \quad \hat{K}_u x = \sum_{n \geq 1} \alpha_n f_n(x) x_u^{(n)} \quad ; \quad \forall x \in X,$$

where $\alpha \in \lambda$, $\{f_n\}$ is an equicontinuous sequence in X^* and $\sup \{\hat{p}_u(x_u^{(n)}) : n \geq 1\} < \infty$. Clearly $v_1 = \{x \in X : |f_n(x)| \leq 1, \forall n \geq 1\}$ is in $\mathcal{B}_X$ and set $v = u \cap v_1$. Let $\hat{f}_n \in (X_v)^*$ be defined by $\hat{f}_n(x_v) = f_n(x)$. Then $|\hat{f}_n(x_v)| \leq \hat{p}_v(x_v)$ for all $n \geq 1$ and x_v in X_v (see chapter 1). Thus from (*)

$$x_u = K_u^v(x_v) = \sum_{n \geq 1} \alpha_n f_n(x_v) x_u^{(n)}$$

and this shows the $\hat{\lambda}$-nuclearity of $\hat{K}_u^v$.

(ii) ==> (iii). Since $R \in \mathcal{L}(X,F)$, there exists p in D_T with $\|Rx\| \leq p(x)$ for all x in X. The set $v \in \mathcal{B}_X$, where $v = \{x \in X : p(x) \leq 1\}$ and so $\hat{K}_v : X \to \hat{X}_v$ is $\hat{\lambda}$-nuclear by (ii). Define $g_v : X_v \to F$ by $g_v(x_v) = Rx$; $x_v = x + \ker p$. Clearly $g_v \in \mathcal{L}(X_v, F)$ and $R = g_v \circ K_v$. Thus $R \in N_\lambda^p(X,F)$, by Proposition 17.2.3(b).

(iii) ==> (ii). This is trivial. □

In the sequel, we will also need

THEOREM 17.2.6 For an l.c. TVS (X,T), the following statements are equivalent:

(i) For each u in $\mathcal{B}_X$, there exists v in $\mathcal{B}_X$ with $v \prec u$ such that $\hat{K}_u^v \in N_\lambda^q(\hat{X}_v, \hat{X}_u)$.

(ii) For each u in $\mathcal{B}_X$, $\hat{K}_u \in N_\lambda^q(X, \hat{X}_u)$.

(iii) $\mathcal{L}(X,F) \subset N_\lambda^q(X,F)$ for each normed space F.

(iv) $\mathcal{L}_b(X,Y) \subset N_\lambda^q(X,Y)$, for each l.c. TVS (Y,S).

(v) $\mathcal{L}_p(X,Y) \subset N_\lambda^q(X,Y)$ for each l.c. TVS (Y,S).

PROOF. The proof of (i) <==> (ii) <==> (iii) follows on the lines of the proof of Theorem 17.2.5.

(iii) ==> (iv). If $R \in \mathcal{L}_b(X,Y)$, there exists u in $\mathcal{B}_X$ and B in $\mathcal{D}_Y$ (cf. section 1.2) such that $R[u] \subset B$. Further, $p_B(Rx) \leq p_u(x)$ for each x in X; therefore, the map $R_B^u : X_u \to Y_B$, $R_B^u(x_u) = Rx$, $x_u = x + \ker p_u$, is well defined and $R_B^u \in \mathcal{L}(X_u,Y_B)$, Y_B being equipped with the norm p_B. By the hypothesis, $K_u \in N_\lambda^q(X,X_u)$ and as $R = I_B \circ R_B^u \circ K_u$, where I_B is the inclusion map of Y_B into Y, R belongs to $N_\lambda^q(X,Y)$ by Proposition 17.2.3(b).

(iv) ==> (v) is trivially true.

(v) ==> (iii). Let $R \in \mathcal{L}(X,F)$ for any given normed space F. If $Y_\sigma = (F,\sigma(F,F^*))$, then $R \in \mathcal{L}_p(X,Y_\sigma)$ (cf. [190], p. 50). Therefore, $R \in N_\lambda^q(X,Y_\sigma)$. Hence by Proposition 17.2.3 (a), there exists a quasi λ-nuclear seminorm p in D_T so that $A_p = \{Rx : p(x) \leq 1\}$ is $\sigma(F,F^*)$-bounded and hence bounded in F. Thus $R \in N_\lambda^q(X,F)$ by the proposition referred to just now. □

17.3 $\hat{\Lambda}(P;\Psi)$-NUCLEARITY

Restrictions on the s.s. λ are likely to yield better and comparatively more convenient characterizations of λ-nuclear spaces and this is what we want to discuss in this section.

Throughout this section, we consider $\lambda = \Lambda(P;\Psi)$ introduced in section 1.3 where we assume without further reference that P is a *nuclear Köthe set of increasing type;* that is to say, each a in P is a strictly positive increasing sequence and $(\Lambda(P),T_p)$ is a nuclear space.

Before we come to the first main result of this section, let us collect certain relevant properties of $\Lambda(P;\Psi)$ in terms of a few lemmas.

LEMMA 17.3.1 If

$$\ell_\Psi = \{\alpha \in \omega: \sum_{n\geq 1} \Psi(|x_n|) < \infty\} ,$$

then $\ell_\Psi \subset \ell^1$.

PROOF. We follow [192]. The inverse $\Psi^{-1} : [0,\infty) \to [0,\infty)$ clearly exists and for $t_1, t_2 \in [0,\infty)$,

$$\Psi^{-1}(t_1+t_2) \geq \Psi^{-1}(t_1) + \Psi^{-1}(t_2).$$

Let $M = \sum_{i \geq 1} \Psi(|\alpha_i|)$ for α in ℓ_Ψ. If

$$x = \sum_{i=1}^{n} \Psi(|\alpha_i|), \quad y = \sum_{i \geq n+1} \Psi(|\alpha_i|),$$

then

$$\Psi^{-1}(M) = \Psi^{-1}(x+y) \geq \Psi^{-1}(x) \geq \sum_{i=1}^{n} \Psi^{-1}\Psi(|\alpha_i|),$$

for $n \geq 1$ and so $||\alpha||_1 \leq \Psi^{-1}(M)$. □

LEMMA 17.3.2 For $a \in P$, $\Lambda(a,\Psi) \subset \ell_\Psi$ and so $\Lambda(P,\Psi) \subset \ell_\Psi \subset \ell^1$, where

$$\Lambda(a;\Psi) = \{\alpha \in \omega : \sum_{n \geq 1} \Psi(|\alpha_n|)a_n < \infty\}.$$

PROOF. Indeed, if $\alpha \in \Lambda(a;\Psi)$, then

$$\sum_{n \geq 1} \Psi(|\alpha_n|) = \sum_{n \geq 1} \Psi(|\alpha_n|) \frac{a_n}{a_n} \leq \frac{1}{a_1} \sum_{n \geq 1} \Psi(|\alpha_n|)a_n < \infty.$$

Now apply Lemma 17.3.1. □

After [168], p. 32, we have

LEMMA 17.3.3 For each α in $\Lambda(P;\Psi)$, the sequence β belongs to $\Lambda(P;\Psi)$, $\beta_n = \sum_{i \geq n} |\alpha_i|$.

PROOF. Let $a \in P$. There exist b in P and c in ℓ^1 such that $a_n \leq b_n c_n$ (cf. Theorem 1.4.14). By Lemma 17.3.2, $\alpha \in \ell^1$. Hence, using subadditivity and continuity of Ψ,

$$\Psi(\sum_{i\geq n} |\alpha_i|) \leq \sum_{i\geq n} \Psi(|\alpha_i|)$$

and therefore

$$\sum_{n\geq 1} \Psi(\sum_{i\geq n} |\alpha_i|)a_n \leq \sum_{i\geq 1} \Psi(|\alpha_i|) \sum_{n=1}^{i} a_n$$

$$\leq \sum_{i\geq 1} \Psi(|\alpha_i|)b_i \sum_{n=1}^{i} c_n$$

$$\leq (\sum_{n\geq 1} c_n)(\sum_{i\geq 1} \Psi(|\alpha_i|)b_i) < \infty.$$

Thus $\beta \in \Lambda(P;\Psi)$. □

<u>LEMMA 17.3.4</u> Let X and Y be Hilbert spaces and $R \in \mathcal{L}(X,Y)$. Then the following statements are equivalent (here λ is equal to $\Lambda(P;\Psi)$):

(i) R is $\hat{\lambda}$-nuclear.

(ii) R is quasi-λ-nuclear.

(iii) R is of λ-type.

(iv) R^* is of λ-type.

(v) R^* is $\hat{\lambda}$-nuclear.

(vi) R^* is quasi-λ-nuclear.

<u>PROOF</u>. (i) ==> (ii). This is obvious.

(ii) ==> (iii). We have

$$\|Rx\| \leq \sum_{n\geq 1} |\alpha_n\langle x,f_n\rangle|; \quad \forall x \in X,$$

with the usual restrictions on $\{\alpha_n\}$ and $\{f_n\}$. If $M_n = \bigcap_{i=1}^{n} \{x \in X: \langle x,f_i\rangle = 0\}$, then M_n is a closed subspace of X with $\dim (X/M_n) \leq n$. If $M_n^{\perp}$ is the orthogonal complement of M_n, then $X = M_n \oplus M_n^{\perp}$. Since $X/M_n \simeq M_n^{\perp}$, $\dim M_n^{\perp} \leq n$.

Let u and v be closed unit balls of X and Y, respectively. If $x \in u \cap M_{n-1}$, then

$$\| Rx \| \leq \sum_{i \geq n} |\alpha_i| = \beta_n, \text{ say}$$

$$\Longrightarrow R[u \cap M_{n-1}] \subset \beta_n v.$$

Let $g \in R^*[v^o]$, then from the preceding inclusion, $g \in \beta_n (u \cap M_{n-1})^o = \beta_n \overline{(u^o + M_{n-1}^{\perp})}$, the closure being considered with respect to $\sigma(X^*,X)$ (cf. Lemma 1.2.16). Since u^o is $\sigma(X^*,X)$-compact (the Alaoglu-Bourbaki theorem), the set $u^o + M_{n-1}^{\perp}$ is $\sigma(X^*,X)$-closed ([190], p. 53). Hence

$$R^*[v^o] \subset \beta_n u^o + M_{n-1}^{\perp} \Longrightarrow \delta_n(R^*[v^o],u^o) \leq \beta, \; n \geq 1.$$

By Lemma 17.3.3, $\beta \in \lambda$. Using Proposition 1.4.8, we find that R^* is compact (since $\delta_n(R^*[v^o], u^o) \to 0$) and so from Proposition 1.4.19, $\alpha_n(R) = \alpha_n(R^*) = \delta_n(R^*) \equiv \delta_n(R^*[v^o],u^o)$. Hence $\{\alpha_n(R)\} \in \lambda \equiv \lambda(P,\Psi)$.

(iii) ==> (i). We have $\{\alpha_n(R)\} \in \lambda$. Since $\lambda \subset \ell^1$ by Lemma 17.3.2, $\{\alpha_n(R)\} \in \ell^1$ and so from Proposition 1.4.20 and Theorem 1.4.21,

$$Rx = \sum_{n \geq 1} \lambda_n \langle x,e_n \rangle h_n, \quad \forall x \in X$$

where $\lambda_n = \alpha_n(R)$; in particular, $R \in N_{\lambda}^{p}(X,Y)$.

(iii) <==> (iv). It follows from Proposition 1.4.19(ii).

(iv) <==> (v) <==> (vi) follow similarly. □

NOTE. If P is a nuclear Köthe set of increasing type as considered before, then the equivalence (i) <==> (iii) is also shown in [168], p. 33.

The preceding lemma immediately leads to

PROPOSITION 17.3.5 For an l.c. TVS (X,T) and the s.s.

$\lambda \equiv \Lambda(P;\Psi)$, the following statements are equivalent:

(i) (X,T) is $\hat{\lambda}$-nuclear.

(ii) For each u in $\mathcal{B}_X$, there exists v in $\mathcal{B}_X$ with $v < u$ such that $\hat{K}_u^v \in N_\lambda^q(\hat{X}_v,\hat{X}_u)$.

(iii) For each u in $\mathcal{B}_X$, $\hat{K}_u \in N_\lambda^q(X,\hat{X}_u)$.

(iv) $\mathcal{L}(X,F) = N_\lambda^q(X,F)$ for every normed space F.

(v) $\mathcal{L}_b(X,Y) = N_\lambda^q(X,Y)$ for each l.c. TVS (Y,S).

(vi) $\mathcal{L}_p(X,Y) = N_\lambda^q(X,Y)$ for each l.c. TVS (Y,S).

PROOF. (i) <==> (ii). In either of these cases, $\hat{X}_u$ is a Hilbert space (note that $\Lambda(P;\Psi) \subset \ell^1$ and make use of Proposition 1.4.1) and now apply Lemma 17.3.4.

(ii) <==> (iii) <==> (iv) <==> (v) <==> (vi). This follows from Theorem 17.2.6 and we need only to observe that each R in N_λ^q is precompact (e.g. [173], p. 59). □

NOTE. In (iv) above, F cannot be replaced by an arbitrary l.c. TVS (Y,S); indeed, if X is an infinite dimensional $\Lambda_\infty(P)$-nuclear space, where $\Lambda_\infty(P)$ is nuclear, then the identity map of X is not quasi-$\Lambda_\infty(P)$-nuclear (cf. [138], p. 173).

We come to the main

THEOREM 17.3.6 Let (X,T) be an l.c. TVS and $\lambda = \Lambda(P;\Psi)$. Then the following statements are equivalent:

(i) (X,T) is $\hat{\lambda}$-nuclear

(ii) For each u in $\mathcal{B}_X$, there exist $\xi \in \lambda$ and an equicontinuous sequence $\{f_n\}$ in X such that

$$p_u(x) \leq \sum_{n\geq 1} |\xi_n f_n(x)|, \ \forall\, x \in X. \qquad (17.3.7)$$

(iii) For each u in $\mathcal{B}_X$, there exists v in $\mathcal{B}_X$, $v < u$ such that $\{\delta_n(v,u)\} \in \lambda$.

PROOF. (i) ==> (ii). There exists v in $\mathcal{B}_X$ so that

$$(*) \qquad K_u^v(x_v) = \sum_{n\geq 1} \xi_n \hat{f}_n(x_v) y_u^n, \quad \forall x \in X$$

where $\hat{p}_u(y_u^n) \leq 1$ for $n \geq 1$; $|\hat{f}_n(x_v)| \leq \hat{p}_v(x_v)$, $n \geq 1$, $x \in X$ and $\xi \in \lambda$. (*) clearly yields (17.3.7) (cf. the discussion preceding Proposition 1.4.1).

(ii) ==> (iii). There exists v in $\mathcal{B}_X$ so that $|f_n(x)| \leq p_v(x)$ for all $n \geq 1$, $x \in X$ and we may choose p_v so that $p_u \leq p_v$. If $\hat{f}_n \in (X_v)^*$ is defined by $\hat{f}_n(x_v) = f_n(x)$, it follows from (17.3.7) that K_u^v is also quasi-nuclear and consequently, we may regard each $\hat{X}_u$ as a Hilbert space. Therefore, $\hat{K}_u^v$ is quasi-λ-nuclear from the Hilbert space $\hat{X}_u$ into the Hilbert space $\hat{X}_v$. Hence $\hat{K}_u^v$ is of λ-type (Lemma 17.3.4 (ii) ==> (iii)). Since $\delta_n(v,u) \leq \alpha_n(\hat{K}_n^v)$, we have finished.

(iii) ==> (i). Using Theorem 1.4.10 and Proposition 1.4.1, we may regard each of the spaces $\hat{X}_u$ as a Hilbert space. Then K_u^v being nuclear, is compact and so $\alpha_n(\hat{K}_u^v) = \delta_n(v,u)$ (cf. Proposition 1.4.19). Hence $\hat{K}_u^v$ is of λ -type and consequently $\hat{K}_u^v$ is λ-nuclear by Lemma 17.3.4. □

NOTE. It appears that the preceding theorem makes full use of the nuclear and increasing character of the Köthe set P. However, it is observed (without proof) in [222], p. 499 that the equivalence of (i) and (iii) above is true provided $\Lambda(P)$ is a nuclear G_∞-space, a condition stronger than ours.

Further restrictions on P

In this subsection, we assume P to be a *nuclear* G_∞*-Köthe set*; that is, P is a *Köthe set of infinite type* (cf. chapter 1) *and* $(\Lambda(P),T_p)$ *is a nuclear space*. Under these conditions, it can be easily verified (cf. [222], p. 498) that for each $k \geq 1$, there exist $b \in P$ and $M > 0$ such that

$$n^k \leqq Mb_n \quad \forall n \geqq 1. \tag{17.3.8}$$

With these restrictions on P and putting $\lambda = \Lambda(P;\Psi)$, we have

THEOREM 17.3.9 An l.c. TVS (X,T) is $\hat{\lambda}$-nuclear if and only if for each p in D_T, there exist α in λ and an equicontinuous sequence $\{f_n\}$ in X* such that

$$p(x) \leqq \sup_{n\geqq 1} \{|\alpha_n f_n(x)|\}, \forall x \in X. \tag{17.3.10}$$

PROOF. If (17.3.10) is satisfied, then so is (17.3.7) and the sufficiency part follows from the preceding theorem.

On the other hand, if (X,T) is $\hat{\lambda}$-nuclear, then by Theorem 17.3.6, we have (17.3.7). Let $a \in P$. We also have b in P satisfying (17.3.8). By the G_∞-character of P, we can find c in P with $a_n b_n \leqq c_n$, $n \geqq 1$. Thus

$$\sum_{n\geqq 1} \Psi(n^k|\xi_n|)a_n \leqq \sum_{n\geqq 1} n^k \Psi(|\xi_n|)a_n$$

$$\leqq M \sum_{n\geqq 1} \Psi(|\xi_n|)a_n < \infty.$$

Hence, if $\beta_n = n^k \xi_n (k > 1)$, (17.3.7) yields

$$p_u(x) \leqq \sup_{n\geqq 1} \{|\beta_n f_n(x)|\} \sum_{n\geqq 1} \frac{1}{n^k}$$

and this proves (17.3.10) with $\alpha_n = \beta_n(\sum_{n\geqq 1} 1/n^k)$, $n \geqq 1$. □

Relationship between λ and $\hat{\lambda}$-nuclearity

It is natural to enquire about the relationship between λ-and $\hat{\lambda}$-nuclear operators and the associated nuclearity of an l.c. TVS., where λ is an arbitrary s.s. If $\lambda \subset \ell^1$, then $N_\lambda^P(X,Y) \subset N_\lambda(X,Y)$. On the other hand, let $\lambda \subset \lambda \cdot \lambda$ and $R \in N_\lambda(X,Y)$, where R has the representation given in

Definition 17.2.1. Let us write $\alpha = \beta \cdot \gamma$ and $z_n = \gamma_n y_n$, β, $\gamma \in \lambda$. If $g \in Y^*$, then $\{g(z_n)\} = \gamma \cdot \{g(y_n)\} \in \lambda . \lambda^{\times} \subset \ell^1 \subset \ell^{\infty}$. Hence $R \in N_{\lambda}^{p}(X,Y)$. Specializing λ to Köthe spaces, we have

PROPOSITION 17.3.11 Let P be an arbitrary nuclear Köthe set of infinite type and Ψ associated with $\Lambda(P;\Psi)$ satisfy the additional property: there exist $M \geq 1$ and t_o in $[0,\infty)$ such that

$$\psi(\sqrt{t}) \leq \sqrt{\Psi(M\, t)}, \quad \forall t \in [0,t_o].$$

Then $\Lambda(P;\Psi) \subset \ell^1$ and $\Lambda(P;\Psi) \subset \Lambda(P;\Psi) . \Lambda(P;\Psi)$. Therefore, if $\lambda \equiv \Lambda(P;\Psi)$, then λ-and $\hat{\lambda}$-nuclear operators are the same and so are the associated nuclearity of spaces.

PROOF. Clearly, $\Lambda(P;\Psi) \subset \ell_{\Psi} \subset \ell^1$ (cf. Lemmas 17.3.1 and 17.3.2). For the other inclusion consider $\alpha \in \Lambda(P;\Psi)$. Then there exists N in $\mathbb{N}$ such that $|\alpha_n| \leq t_o$ for all $n \geq N$. Also $\Psi(M|\alpha_n|) \leq [M+1]\ \Psi(|\alpha_n|)$, $n \geq 1$. Thus $\{\Psi(M|\alpha_n|)\} \in \Lambda(P)$ and so $\{\sqrt{\Psi(M|\alpha_n|)}\} \in \Lambda(P)$ by Lemma 1.4.17. But

$$\Psi(\sqrt{|\alpha_n|}) \leq \sqrt{\Psi(M|\alpha_n|)}, \quad n \geq N.$$

Hence $\{\sqrt{|\alpha_n|}\} \in \Lambda(P;\Psi)$ and we are done with $\Lambda(P;\Psi) \subset \Lambda(P;\Psi) \cdot \Lambda(P;\Psi)$. The last part is a consequence of the discussion preceding this proposition. □

The kernel theorem

Theorem 17.3.6 appears to be the basic result to be used for studying bases in $\hat{\lambda}$-nuclear spaces. There are still a few more results on $\hat{\lambda}$-nuclear spaces for which we refer to [138]. However, for the sake of completeness, we reproduce from [138] the final result of this section on the characterization of λ-nuclear spaces, often referred to as the kernel theorem on λ-nuclear spaces.

We consider bilinear forms on $X \times Y$, where X and Y are two l.c. TVS and refer to [36] for their several stimulating properties. A bilinear form $B:X \times Y \to \mathbb{K}$ is said to be λ-*nuclear* (λ is an arbitrary s.s.) provided

$$B(x,y) = \sum_{n\geq 1} \alpha_n f_n(x) g_n(y),$$

where $\alpha \in \lambda$, $\{f_n\}$ is an equicontinuous sequence in X^* and $g_n \in Y^*$ ($n \geq 1$) such that $\{g_n(y)\} \in \lambda^\times$ for each y in Y.

In the next result, $\lambda = \Lambda(P;\Psi)$ where P and Ψ satisfy the conditions of Proposition 17.3.11.

<u>THEOREM 17.3.12</u> An l.c. TVS (X,T) is λ-nuclear if and only if for each Banach space F, every continuous bilinear form B on $X \times F$ is λ-nuclear.

<u>PROOF</u>. (Sufficiency). Let $u \in \mathcal{B}_X$ and consider the Banach space $(X^*(u^0), q_{u^0})$ (cf. section 1.2). If $B(x,f) = f(x)$ for $x \in X$ and $f \in X^*(u^0)$, then $|B(x,f)| \leq p_u(x) q_{u^0}(f)$ and so B is continuous. Thus

$$B(x,f) = \sum_{n\geq 1} \alpha_n f_n(x) g_n(f),$$

where $\alpha \in \lambda$, $\{f_n\}$ is an equicontinuous sequence in X^* and $g_n \in (X^*(u^0))^*$ with $\{g_n(f)\} \in \lambda^\times$ for each f in $X^*(u^0)$.

Write $\alpha_n = \beta_n \gamma_n; \beta, \gamma \in \lambda$ and put $\hat{f}_n = \gamma_n q^*_{u^0}(g_n) f_n$, where

$$q^*_{u^0}(g_n) = \sup_{f \in u^0} |g_n(f)|.$$

Thus for f in u^0,

$$|B(x,f)| \leq \sum_{n\geq 1} |\beta_n| \, |\hat{f}_n(x)|, \quad \forall x \in X$$

(*) $\Longrightarrow p_u(x) = \hat{p}_u(\hat{x}_u) \leq \sum_{n\geq 1} |\beta_n f_n(x)|$, $\forall x \in X$; (cf. (1.2.1)).

But $\gamma.\{g_n(f)\} \in \lambda.\lambda^\times \subset \ell^1 \subset \ell^\infty$, for each f in $X^*(u^o)$ and so by the Banach-Steinhaus theorem and the equicontinuity of $\{f_n\}$, $\{\hat{f}_n\}$ is an equicontinuous sequence in X^*. Thus (*) yields (17.3.7) and (X,T) is λ-nuclear by Theorem 17.3.6 and Proposition 17.3.11.

The converse is contained in [43], p. 87. □

17.4 BASES IN $\hat{\lambda}$-NUCLEAR SPACES

We begin this section with the assumption that P is a nuclear Köthe set of infinite type and $\lambda = \Lambda(P,\Psi)$. The presence of an S.b. in an l.c. TVS (X,T) is sometimes useful in characterizing the $\hat{\lambda}$-nuclearity of (X,T) as exhibited in (cf. [138])

THEOREM 17.4.1 Let an l.c. TVS (X,T) contain a u.e.-S.b. $\{x_n;f_n\}$. Then (X,T) is $\hat{\lambda}$-nuclear if and only if for each p in D_T, there corresponds a q in D_T with $p \leq q$ and an injection $\pi:\mathbb{N} \to \mathbb{N}$ with $\pi[\mathbb{N}] = \{n \in \mathbb{N}: p(x_n) \neq 0\}$ such that

$$\left\{\frac{p(x_{\pi(n)})}{q(x_{\pi(n)})}\right\} \in \lambda. \tag{17.4.2}$$

PROOF. Let (X,T) be $\hat{\lambda}$-nuclear and so it is nuclear. Therefore, by Theorem 17.3.6 and following the notation and proof of Theorem 12.4.6, we find p_∞ and q in $D_T (p_\infty \leq q)$ so that $\{\delta_n(u_q,u_p^\infty)\} \in \lambda$ and (combine (v), (vii) and (viii) of the proof of Theorem 12.4.6)

$$\frac{p(x_{\pi(n)})}{q(x_{\pi(n)})} \leq n\delta_n(u_q,u_p^\infty), \quad n \geq 1.$$

Using the inequality (17.3.8), we get $\{n\delta(u_q,u_p^\infty)\} \in \lambda$ and so (17.4.2) follows.

Conversely, let (17.4.2) hold for each p and the corresponding choice of q and π. Let $g_n(x) = p(x_n)f_n(x)$ and $\alpha_n = \sup\{|g_n(x)|: q_\infty(x) \leqq 1\}$, where $q_\infty(x) = \sup\{|f_n(x)|\ q(x_n): n \geqq 1\}$. Then

$$\alpha_{\pi(n)} \leqq \frac{p(x_{\pi(n)})}{q(x_{\pi(n)})}, \quad n \geqq 1,$$

and $\{\alpha_{\pi(n)}\} \in \lambda$. Let

$$\hat{g}_n(x) = \begin{cases} \dfrac{g_n(x)}{\alpha_n}, & \alpha_n \neq 0; \\ 0, & \alpha_n = 0. \end{cases}$$

Observe that $|\hat{g}_n(x)| \leqq q_\infty(x)$ for all x in X and $n \geqq 1$, and

$$p(x) \leqq \sum_{n\geqq 1} \alpha_{\pi(n)} |\hat{g}_{\pi(n)}(x)|, \quad \forall x \in X$$

The required result now follows after appealing to Theorem 17.3.6, (ii) ==> (i). □

The next result is generally referred to as the *Grothendieck-Pietsch-Köthe criterion* for $\hat{\lambda}$-nuclearity of an arbitrary Köthe space $(\Lambda(Q);T_Q)$. Thus, if P is as before and Q is an arbitrary Köthe set, we have

THEOREM 17.4.3 An arbitrary Köthe space $(\Lambda(Q),T_Q)$ is $\hat{\lambda}$-nuclear ($\lambda = \Lambda(P;\Psi)$) if and only if for each a in Q there exist b in Q with $a \leqq b$ and an injection $\pi:\mathbb{N} \to \mathbb{N}$, $\pi[\mathbb{N}] = \{n \in \mathbb{N};\ a_n \neq 0\}$ such that

$$\{\frac{a_{\pi(n)}}{b_{\pi(n)}}\} \in \lambda.$$

PROOF. It is clear that $\{e^n;e^n\}$ is a u-e-S.b. for

$(\Lambda(Q),T_Q)$, where $D_{T_Q} = \{p_a : a \in Q\}$, $p_a(x) = \sum_{n\geq 1} |x_n|a_n$. Since $p_a(e^{\pi(n)}) = a_{\pi(n)}$, the result follows from Theorem 17.4.1. □

NOTE. The preceding result is also observed in [43], p. 93 and [222], p. 499 without proof.

EXERCISE 17.4.4 Prove that for any s.s. μ, $(\mu,\eta(\mu,\mu^\times))$ is $\hat{\lambda}$-nuclear if and only if for each a in $\lambda^\times$ there exist b in $\lambda^\times$ with $|a_n| \leq |b_n|$, $n \geq 1$ and an injection $\pi:\mathbb{N} \to \mathbb{N}$, $\pi[\mathbb{N}] = \{n \in \mathbb{N} : a_n \neq 0\}$ such that

$$\left\{\frac{|a_{\pi(n)}|}{b_{\pi(n)}}\right\} \in \lambda.$$

NOTE. Theorem 17.4.3 is in fact an application of Theorem 17.4.1, the former therefore making full use of the nuclear infinite character of the Köthe set P. It is possible to obtain an extension of Theorem 17.4.3 for those Köthe sets P which are just nuclear set of increasing type (cf. the beginning of section 17.3) - an assumption which we make throughout the rest of this section. However, in doing so, we use the $\hat{\Lambda}(P:\Psi)$-nuclearity of diagonal maps. Given two arbitrary s.s. λ and μ, by a *diagonal map* $D: \lambda \to \mu$, we mean $D = \{d_n : n \geq 1\}$ such that $Dx = \{d_n x_n\} \in \mu$ for $x = \{x_n\} \in \lambda$. The following result (cf. [216] p. 67; see also [138], p. 182 where it is given without proof) extends similar results given in [137] and [144]. In fact, we let $\lambda = \Lambda(p;\Psi)$ with P as mentioned above, then we have

PROPOSITION 17.4.5 A diagonal map $D: \ell^1 \to \ell^1$ with $D = \{d_n\}$; $d_n \geq 0$, $n \geq 1$ is $\hat{\lambda}$-nuclear if and only if there exists an injection $\pi:\mathbb{N} \to \mathbb{N}$ with $\pi[\mathbb{N}] = \{n \in \mathbb{N} : d_n > 0\}$ such that $\{d_{\pi(n)}\} \in \lambda$.

PROOF. Let D be $\hat{\lambda}$-nuclear (observe that we already have

$d \in \ell^\infty$ and so $\|Dx\|_1 \leq \|d\|_\infty \|x\|_1$). By Lemma 17.3.2, D is nuclear and so it is compact; that is, D[u] is compact ($u = \{x \in \ell^1 : \|x\|_1 \leq 1\}$). Hence

$$d_n = |d_n| = \sum_{i \geq n} |d_i e_i^n| \leq \sup_{y \in D[u]} \sum_{i \geq n} |y_i|$$

$$\to 0 \text{ as } n \to \infty,$$

by a well known characterization of compact subsets of ℓ^1 (cf. [129], p. 108; [143], p. 282). Therefore, we find an injection $\pi : \mathbb{N} \to \mathbb{N}$ with $\pi[\mathbb{N}] = \{n \in \mathbb{N} : d_n > 0\}$ so that $d_{\pi(n)} \geq d_{\pi(n+1)}$, $n \geq 1$.

Let $x \in u$ and $R \in \mathcal{L}(\ell^1, \ell^1)$ with $\dim R[\ell^1] < n$. Invoking both the statement and notation of Lemma 1.2.17 where $X = Y = \ell^1$, one has

$$(*) \quad \|D-R\| \geq \|(D-R)x\|_1 = \|\{d_k x_k - \sum_{i=1}^{n-1} \alpha_i f^i(x) y_k^i\}_k\|_1 .$$

Consider the homogeneous system of (n-1) equations in n unknown $\zeta_1, \ldots, \zeta_n$:

$$\sum_{k=1}^{n} f^i_{\pi(k)} \xi_k = 0, \quad 1 \leq i \leq n-1.$$

This system has nontrivial solutions (cf. [151], p. 390) and we may determine the same with

$$\sum_{i=1}^{n} |\zeta_i| = 1.$$

Choose x in u with

$$x_i = \begin{cases} \zeta_k, & i = \pi(k),\ 1 \leq k \leq n; \\ 0, & \text{otherwise.} \end{cases}$$

Since

$$f^j(x) = \sum_{i \geq 1} f_i^j x_i = \sum_{k=1}^{n} f_{\pi(k)}^j \zeta_k = 0, \ 1 \leq j \leq n-1,$$

(*) yields

$$\| D-R \| \leq \| \{d_k x_k\} \|_1 \geq d_{\pi(n)}.$$

Thus $\alpha_n(D) \geq d_{\pi(n)}$, $n \geq 1$. However, $\{\alpha_n(D)\} \in \lambda$. In fact, truncating the $\hat{\lambda}$-nuclear expansion of D at the (n-1)th stage, we get

$$\| Dx - \sum_{i=1}^{n-1} \beta_i f^i(x) y^i \|_1 \leq (\sum_{i \geq n} |\beta_i|) \ \|x\|_1$$

where $\{\beta_i\} \in \lambda$, $\|f^i\|_\infty \leq 1$, $\|y^i\|_1 \leq 1$ and

$$Dx = \sum_{n \geq 1} \beta_n f^n(x) y^n, \quad x \in \ell^1.$$

Therefore $\alpha_n(D) \leq \gamma_n \equiv \sum_{i \geq n} |\beta_i|$ and now make use of Lemma 17.3.3. Consequently $\{d_{\pi(n)}\} \in \lambda$.

Conversely, let $\{d_{\pi(n)}\} \in \lambda$ and $M_d = \{n \in \mathbb{N}: d_n > 0\}$. Since $\{e^n;e^n\}$ is a u-S.b. for ℓ^1, it is easily seen that

$$Dx = \sum_{n \in M_d} \langle Dx, e^n \rangle e^n = \sum_{n \geq 1} d_{\pi(n)} \langle x, e^{\pi(n)} \rangle e^{\pi(n)}$$

and this proves the required converse. □

Now we have the main result promised in the preceding note.

<u>THEOREM 17.4.6</u> An arbitrary Köthe space $(\Lambda(Q), T_Q)$ is $\hat{\lambda}$-nuclear if and only if for each a in Q, there correspond a b in Q with $a \leq b$ and an injection $\pi: \mathbb{N} \to \mathbb{N}$ with $\pi[\mathbb{N}] = \{n \in \mathbb{N}: a_n \neq 0\}$ such that

$$\left\{\frac{a_{\pi(n)}}{b_{\pi(n)}}\right\} \in \lambda.$$

PROOF. For convenience, let $\Lambda = \Lambda(Q)$ and for a in Q, $\Lambda_a = \Lambda/p^{-1}(0)$; also $\hat{\Lambda}_a$ is the completion of Λ_a with respect to the norm $\hat{p}_a$ on Λ_a. Let us also write K_a for the quotient map, $K_a: \Lambda \to \Lambda_a$ and K_a^b for the canonical embedding, $K_a^b: \Lambda_b \to \Lambda_a$. Further, assume that $M_a = \{n \in \mathbb{N} : a_n \neq 0\}$ and ℓ_a^1 is the canonical preimage of ℓ^1 with respect to M_a given by $\{x \in \ell^1 : x_n = 0, \ \forall n \notin M_a\}$. Consider the natural map $\Psi_a: \Lambda_a \to \ell_a^1$, where for $\hat{x}$ in Λ_a, $\hat{x} = x + p_a^{-1}(0)$, $x \in \Lambda$, $\Psi_a(\hat{x}) = \{a_n x_n\}$. Then $\|\Psi_a(\hat{x})\|_{1,a} = p_a(x) = \hat{p}_a(\hat{x})$ and $\overline{\Psi_a[\Lambda_a]} = \ell_a^1$. Therefore $\hat{\Lambda}_a \overset{ism}{\simeq} \ell_a^1$ under the unique extension $\hat{\Psi}_a$ of Ψ_a.

With this background, $(\Lambda(Q), T_Q)$ is $\hat{\lambda}$-nuclear if and only if for each a in Q there exists b in Q, $a \leq b$ such that $\hat{K}_a^b$ is $\hat{\lambda}$-nuclear.

Put $\hat{D}_a^b = \hat{\Psi}_a \circ \hat{K}_a^b \circ \hat{\Psi}_b^{-1}$. Then $\hat{D}_a^b$ is a diagonal map from ℓ_b^1 onto ℓ_a^1 determined by $\{a_n/b_n\}$. Clearly $\hat{D}_a^b$ is $\hat{\lambda}$-nuclear if and only if $\hat{K}_a^b$ is $\hat{\lambda}$-nuclear. It only remains now to apply Proposition 17.4.5 to arrive at the desired conclusion both ways. □

Further consequences of Theorem 17.4.1

Theorem 17.4.3 which is a consequence of Theorem 17.4.1, makes an implicit use of the presence of the u.e-S.b. $\{e^n;e^n\}$ in $\lambda(Q)$ and after [144], the former motivates the following

DEFINITION 17.4.7 An arbitrary Köthe space $(\Lambda(Q), T_Q)$ is called *uniformly* $\hat{\lambda}$*-nuclear* ($\lambda = \Lambda(P;\Psi)$) if there exists π in P such that for each a in P, there exists b in P with $\{a_{\pi(n)}\} \in \{b_{\pi(n)}\}.\lambda$.

Clearly every uniformly $\hat{\lambda}$-nuclear Köthe space $\Lambda(Q)$ is $\hat{\lambda}$-nuclear. On the other hand, the next result which is an extension of [222], p. 499, speaks about the converse of the foregoing statement for $\Lambda(Q)$. In fact, we have (cf. [138])

THEOREM 17.4.8 Let Q be a Köthe set of infinite type and P is a nuclear Köthe set of infinite type. Suppose $\lambda = \Lambda(P,\Psi)$. Then the following statements are equivalent:

(i) $\Lambda(Q) \equiv (\Lambda(Q), T_Q)$ is $\hat{\lambda}$-nuclear.

(ii) For every a in Q, there exists c in Q, $c \geqq a$ such that $\{c_n^{-1}\} \in \lambda$.

(iii) There exists c in Q with $\{c_n^{-1}\} \in \lambda$.

(iv) $\Lambda(Q)$ is uniformly $\hat{\lambda}$-nuclear.

PROOF. (i) ==> (ii). Let $a \in Q$. By Theorem 17.3.6, there exists c in Q, $c \geqq a$, with $\{\delta_n(u_c, u_a)\} \in \lambda$, where $u_a = \{x \in \Lambda(Q): p_a(x) \leqq 1\}$ etc. Let

$$\Lambda_n = \text{sp}\,\{e^1, \ldots, e^n\}.$$

Then for x in Λ_n, $p_c(x) \leqq c_n p_a(x)/a_1$ and so for x in $u_a \cap \Lambda_n$, $p_c(a_1 x/c_n) \leqq 1$, giving $a_1/c_n \leqq \delta_n(u_c, u_a)$ (cf. Proposition 1.4.7). This proves (ii).

(ii) ==> (iii) and (iv) ==> (i) are trivial.

(iii) ==> (iv). Let $a \in Q$ and c be as in (iii).

Find d in Q with a, $c \leqq d$. Then find b in Q with $d_n^2 \leqq b_n$, $n \geqq 1$. Hence $a_n/b_n \leqq 1/d_n$ and as $\{d_n^{-1}\} \in \lambda$, we find $a/b \in \lambda$. Take π to be the identity map of $\mathbb{N}$ and (iv) follows. □

Finally, we have

THEOREM 17.4.9 Let Q be a Köthe set of finite type and let P be a nuclear Köthe set of infinite type with $\lambda = \Lambda(P,\Psi)$. Then the following statements are equivalent:

(i) $\Lambda(Q)$ is $\hat{\lambda}$-nuclear.

(ii) $Q \subset \lambda$.

(iii) $\Lambda(Q)$ is uniformly $\hat{\lambda}$-nuclear.

PROOF. (i) ==> (ii). Choose a in Q arbitrarily. After Theorem 17.3.6, find c in Q with $c \geqq a$ and $\{\delta_n(u_c,u_a)\} \in \lambda$. It is easily seen that $a_n/c_1 \leqq \delta_n(u_c,u_a)$ and so $a \in \lambda$.

(ii) ==> (iii). Let $a \in Q$. There exists b in Q with $a_n \leqq b_n^2$, $n \geqq 1$. But $b \in \lambda$ and so (iii) follows.

(iii) ==> (i) as before. □

Another result on uniform $\hat{\lambda}$-nuclearity

Throughout this subsection we write $\lambda = \Lambda(P)$, where P is a nuclear Köthe set of infinite type. Earlier we have given a characterization of uniform λ-nuclearity of a class of Köthe spaces $\Lambda(Q)$ when Q is a Köthe set of infinite or finite type. In some way, we relax the conditions on Q at the expense of P and characterize the uniform $\hat{\Lambda}(P)$-nuclearity of $\Lambda(Q)$.

Following [181], we have

THEOREM 17.4.10 Let $\lambda = (\Lambda(P),T_P)$ be stable and let $\Lambda(Q)$ be a metrizable Köthe space with $Q = \{c^k: k \geqq 1\}$. Then $(\Lambda(Q),T_Q)$ is uniformly $\hat{\lambda}$-nuclear if and only if it is $\hat{\lambda}$-nuclear.

PROOF. It suffices to prove the 'only if' part, as the other implication holds for an arbitrary Köthe space $\Lambda(Q)$. So, let $(\Lambda(Q),T_Q)$ be $\hat{\lambda}$-nuclear. Then by Theorem 17.4.3, for each k in $\mathbb{N}$, there exist j in $\mathbb{N}$ and an injection $\pi_k : \mathbb{N} \to \mathbb{N}$ with

$$(+) \quad c^k \leqq c^j,\ \pi_k[\mathbb{N}] = \{n \in \mathbb{N};\ c_n^k \neq 0\} \text{ and } \{c^k_{\pi_k(n)}/c^j_{\pi_k(n)}\} \in \lambda.$$

Further, for each n in $\mathbb{N}$, there exists k in $\mathbb{N}$ with $c_n^k \neq 0$; that is, $n \in \pi_k[\mathbb{N}]$ and so $n = \pi_k(m)$ for some m in $\mathbb{N}$. Besides, consider the bijection $\beta:\mathbb{N} \to \mathbb{N} \times \mathbb{N}$, $\beta(i) = (k,m)$ where i is uniquely expressed as, $i = 2^{k-1}(2m-1)$. Thus by Lemma 1.2.21, there exist an injection $\gamma:\mathbb{N} \to \mathbb{N} \times \mathbb{N}$ and a π in P satisfying (i) through (iv) of the lemma referred to.

Let us now fix k in $\mathbb{N}$, satisfying (+) and choose n in $\mathbb{N}$ with $\pi(n) \in \pi_k[\mathbb{N}]$ (obviously the interesting case is the one when such n's are countably infinite). Let $b_i^k = c_i^k/c_i^j$, $c_i^k \neq 0$; then $b_{\pi(n)}^k \neq 0$ and we have an unique m in $\mathbb{N}$ with $\pi(n) = \pi_k(m) = \pi_{\gamma_1(m)}(\gamma_2(n))$ (cf. (iv) of the lemma) and (cf. (ii) and (iii) of the lemma)

$$(*) \qquad n \leq \beta^{-1}(\gamma(n)) \leq \beta^{-1}(k,m).$$

Using Proposition 1.3.17 for $a \in P$, we find d in P and a positive constnat $M \equiv M(a,d)$ such that $a_{2n} \leq M\, d_n$, for each $n \geq 1$. As each member of P is increasing,

$$(**) \qquad a_{2^{k-1}(2r-1)} \leq M\, d_r, \quad \forall r \geq 1.$$

Thus

$$b_{\pi(n)}^k\, a_n = b_{\pi_k(m)}^k\, a_n \leq b_{\pi_k(m)}^k\, a_{\beta^{-1}(k,m)} \leq M\, b_{\pi_k(m)}^k\, d_m$$

by (*) and (**). But $\{b_{\pi_k(s)}^k : s \geq 1\} \in \lambda$ (cf. (+)) and m is uniquely determined by n, we find that $\{b_{\pi(n)}^k\} \in \lambda$. As λ is independent of k, the space $(\Lambda(Q),T_Q)$ is uniformly $\hat{\lambda}$-nuclear. □

18 λ-bases and λ-nuclearity

18.1 INTRODUCTION

In this chapter, we are concerned specifically with a possible solution of the second problem raised at the beginning of chapter 17, namely, can we obtain a possible relationship between a λ-base and the λ-nuclearity of the underlying space. A solution in this direction is achieved by introducing the concept of Q-fully λ-bases and Q*-fully λ-bases.

Throughout this chapter, we write (X,T) for an arbitrary l.c. TVS containing an S.b. $\{x_n;f_n\}$ and let λ denote an arbitrary s.s. equipped with its normal topology $\eta(\lambda,\lambda^{\times})$. If λ is given a specific structure, this will be mentioned at the beginning of the corresponding section or subsection. In addition, we also recall some notation, for example, Δ, γ, Ψ_p, etc. introduced in section 4.2. For most of the discussion we follow [133] and occasionally [138]. To begin with, let us introduce

DEFINITION 18.1.1 $\{x_n;f_n\}$ is said to be (i) Q-*semi* λ-*base* provided there exists π in *P* such that $\Psi_p^{\pi}(x) \in \lambda$ for each p in D_T and x in X, $\Psi_p^{\pi}(x) = \{f_{\pi(n)}(x)p(x_{\pi(n)})\}$, and (ii) Q-*fully* λ-*base* if it is Q-semi λ-base and for each p in D_T and α in $\lambda^{\times}$, there exists q in D_T so that for each x in X

$$Q_{p,\alpha}^{\pi}(x) \equiv \sum_{n\geq 1} |f_{\pi(n)}(x)p(x_{\pi(n)})\alpha_n| \leq q(x). \quad (18.1.2)$$

We will also need

DEFINITION 18.1.3 $\{x_n;f_n\}$ is said to be (i) Q^*-*semi* λ-*base* provided that for each p in D_T, there exists $\pi \equiv \pi(p)$ in P such that $\Psi_{p,\pi}(x) \in \lambda$ for each x in X, where $\Psi_{p,\pi}(x) = \{f_{\pi(n)}(x)p(x_{\pi(n)})\}$ and (ii) Q^*-*fully* λ-*base* if it is a Q^*-semi λ-base and for each p in D_T, the map $\Psi_{p,\pi}$ is T-$\eta(\lambda,\lambda^\times)$ continuous.

18.2 CHARACTERIZATIONS OF Q-FULLY λ-BASES

Our restrictions on (X,T), $\{x_n;f_n\}$ and λ are the same as those outlined in the beginning of the last section and, in addition, we assume that λ is perfect and that $(\lambda,\eta(\lambda,\lambda^\times))$ is nuclear. It will be useful to introduce

DEFINITION 18.2.1 A subset B of X is called *normal* if for $\{\alpha_n\}$ in ℓ^∞; $|\alpha_n| \leq 1$, $n \geq 1$, the element $\sum_{n\geq 1} \alpha_n f_n(x)x_n \in B$ whenever $x \in B$.

The concept of a normal set appears to be meaningful when $\{x_n;f_n\}$ is a b.m-S.b. (cf. chapter 6) and in the case $B \subset X$, let

$$\hat{B} = \{\sum_{n\geq 1} \alpha_n f_n(x)x_n : x \in B; |\alpha_n| \leq 1, n \geq 1\}. \quad (18.2.2)$$

LEMMA 18.2.3 Let $\{x_n;f_n\}$ be a b.m-S.b. for an S-space (X,T). Then for each bounded subset B of X, the set $\hat{B}$ given by (18.2.2) is bounded. Therefore, $\beta(X^*,X)$ is the topology generated by the polars of sets belonging to the family $\mathfrak{S}$ of all normal bounded sets in X.

PROOF. By Proposition 6.3.2, (ii), $\mu = \delta^\times$ and therefore from Theorem 3.3.5, $\hat{B}$ is bounded if and only if $F_1[\hat{B}] = \{\{\alpha_n f_n(x)\}: x \in B; |\alpha_n| \leq 1, n \geq 1\}$ is $\sigma(\delta,\delta^\times)$-bounded. Thus, using Proposition 1.3.6, $\hat{B}$ is bounded if and only if $F_1[\hat{B}]$ is $\eta(\delta,\delta^\times)$-bounded which follows clearly from the $\eta(\delta,\delta^\times)$-boundedness of $F_1[B]$. □

For an arbitrary s.s. λ, an l.c. TVS (X,T) with an S.b. $\{x_n;f_n\}$ and a π in P, let us introduce a formal map

$F_a^{\pi} : X \to X$ where for a in $\lambda_+^{\times}$ and x in X,

$$F_a^{\pi}(x) = \sum_{n\geq 1} a_{\pi(n)} f_n(x)x_n \qquad (18.2.4)$$

PROPOSITION 18.2.5 Let $\{x_n;f_n\}$ be a b.m-S.b. as well as a shrinking base for an S-space (X,T). Consider the following statements:

(i) $\{f_n;\Psi x_n\}$ is a Q-fully λ-base for $(X^*,\beta(X^*,X))$.

(ii) There exists π in P such that for every a in $\lambda_+^{\times}$, the mapping F_a^{π} is a well defined roughly bounded linear operator.

(iii) There exists π in P such that for each a in $\lambda_+^{\times}$, the mapping F_a^{π} is a continuous linear operator.

(iv) There exists π in P such that for every a in $\lambda_+^{\times}$, $F_a^{\pi} : X \to X$ is well defined.

Then the following conclusions (a) - (d) are valid:

(a) (iii) ==> (ii) ==> (i); (ii) ==> (iv);

(b) if (X,T) is barrelled also, then (iv) ==> (iii) and if (X,T) is bornological also, then (ii) ==> (iii);

(c) if (X,T) is semireflexive also, then (i) ==> (ii);

(d) if (X,T) is a Fréchet reflexive space, then (i) <==> (ii) <==> (iii) <==> (iv).

PROOF. (a) (iii) ==> (ii) ==> (iv). This is all trivial.

(ii) ==> (i). By Lemma 18.2.3, $\beta(X^*,X)$ is an $\mathcal{B}$-topology, where $\mathcal{B}$ is the collection of all normal bounded subsets of X. If $B \in \mathcal{B}$, then by (ii), $B_{\pi,a} \in \mathcal{B}$ for the given π in P and each a in $\lambda_+^{\times}$, where

$$B_{\pi,a} \equiv F_a^{\pi}[B] = \{\sum_{n\geq 1} a_{\pi(n)} f_n(x)x_n : x \in B\}.$$

For B in $\mathcal{B}$ and f in X*, let $p_B(f) = \sup\{|f(x)| : x \in B\}$.

Let $f \in X^*$ and $a \in \lambda_+^{\times}$. Then

$$p_{B_{\pi,a}}(f) = \sup_{x \in B} \left\{ \left| \sum_{n \geq 1} a_{\pi(n)} f_n(x) f(x_n) \right| \right\}.$$

$$= p_B\left(\sum_{n \geq 1} a_{\pi(n)} f(x_n) f_n \right),$$

where the infinite series clearly represents an element in X^* by the S-character of (X,T). Put $F_n = f(x_n) f_n$, $n \geq 1$. Then $p_{B_{\pi,a}}(F_n) = p_B(a_{\pi(n)} f(x_n) f_n)$.

Next, observe that if $f, g \in X^*$ with $|f(x_n)| \leq |g(x_n)|$, then $p_B(f) \leq p_B(g)$ for each B in $\mathcal{S}$. Indeed, for x in B,

$$\left| f\left(\sum_{n \geq 1} f_n(x) x_n \right) \right| \leq \sum_{n \geq 1} |f_n(x) g(x_n)| = \left| \sum_{n \geq 1} \alpha_n f_n(x) g(x_n) \right|,$$

where $|\alpha_n| = 1$, $n \geq 1$. So, if $y = \sum_{n \geq 1} \alpha_n f_n(x) x_n$ which is an element of B, we find that $|f(x)| \leq |g(y)| \leq p_B(g)$ for each x in B; thus we find that $p_B(f) \leq p_B(g)$.

Let us note that $|F_n(x_i)| \leq |f(x_i)|$ for all $i \geq 1$ and for each $n \geq 1$. Since $B_{\pi,a} \in \mathcal{S}$, we obtain

$$p_{B_{\pi,a}}(F_n) \leq p_{B_{\pi,a}}(f), \quad \forall n \geq 1$$

$$\implies \sup_{n \geq 1} \{ a_{\pi(n)} |f(x_n)| p_B(f_n) \} \leq p_{B_{\pi,a}}(f), \quad \forall f \in X^*$$

$$\implies \sup_{n \geq 1} \{ a_n |f(x_{\sigma(n)})| p_B(f_{\sigma(n)}) \} \leq p_{B_{\pi,a}}(f), \quad \forall f \in X^*,$$

where $\sigma = \pi^{-1}$. Let now a in $\lambda^{\times}$ be arbitrary. There exists b in $\lambda^{\times}$ such that

$$(*) \quad \sum_{n \geq 1} |f(x_{\sigma(n)}) p_B(f_{\sigma(n)}) a_n| \leq p_{B_{\pi,b}}(f) \sum_{n \geq 1} \left| \frac{a_n}{b_n} \right| < \infty$$

and since $\lambda = \lambda^{\times\times}$, $\{ f(x_{\sigma(n)}) p_B(f_{\sigma(n)}) \} \in \lambda$ for each f in X^* and B in $\mathcal{S}$. By using $(*)$ again (the first inequality), we conclude the truth of (i).

(b) Here (ii) ==> (iii) follows from Proposition 1.2.19.

(iv) ==> (iii). The continuous linear maps $F^{\pi}_{a,n} : X \to X$ defined by

$$F^{\pi}_{a,n}(x) = \sum_{i=1}^{n} a_{\pi(i)} f_i(x) x_i$$

are pointwise bounded and $F^{\pi}_{a,n}(x) \to F^{\pi}_{a}(x)$. Therefore, the required result follows from the Banach-Steinhaus theorem.

(c) Assume the truth of (i) and let $\mathfrak{S}$ be as in (a). There exists σ in P so that for each B in $\mathfrak{S}$ and a in λ_+, there exists B_1 in $\mathfrak{S}$ such that $\{\psi x_{\sigma(n)}(f) p_B(f_{\sigma(n)})\} \in \lambda^{\times}$ for each f in X* and

$$(*) \quad \sum_{n \geq 1} |f(x_n)| p_B(f_n) a_{\pi(n)} \leq p_{B_1}(f); \quad \forall f \in X^*,$$

where $\pi = \sigma^{-1}$.

By Theorem 9.4.2, $\{x_n; f_n\}$ is γ-complete and hence, in order to establish the existence of F^{π}_{a}, it suffices to establish the boundedness of $\{\sum_{i=1}^{n} a_{\pi(i)} f_i(x) x_i\}$ for each a in $\lambda^{\times}_{+}$ and x in X. Indeed, each x in X belongs to some B in $\mathfrak{S}$ and therefore for f in X*

$$\begin{aligned} |f(\sum_{i=1}^{n} a_{\pi(i)} f_i(x) x_i)| &\leq \sum_{n \geq 1} a_{\pi(n)} |f_n(x) f(x_n)| \\ &\leq \sum_{n \geq 1} a_{\pi(n)} |f(x_n)| p_B(f_n) \\ &\leq p_{B_1}(f), \end{aligned}$$

by (*) and the sequence in question is bounded. Thus the map $F^{\pi}_{a} : X \to X$ is well defined.

To prove the rough boundedness of F^{π}_{a}, let $B \in \mathfrak{S}$. Let f in X* and x in B be chosen arbitrarily. Then

$$|f(F_a^\pi x)| \leq \sum_{n\geq 1} a_{\pi(n)} |f(x_n) f_n(x)|$$

$$\leq p_{B_1}(f),$$

for some B_1 in $\mathcal{B}$ (cf. (*)). Hence $F_a^\pi[B]$ is bounded.

Finally, (d) is a consequence of (a) to (c). □

Next, we pass on to one of the main results of this section, namely,

THEOREM 18.2.6 Let $\{x_n;f_n\}$ be a b.m-S.b. for a Fréchet space (X,T). Then $\{x_n;f_n\}$ is a Q-fully λ-base for (X,T) if and only if there exists π in P so that the mapping $F_a^\pi : X \to X$ introduced in (18.2.4) is a well defined continuous linear operator for each a in $\lambda_+^\times$.

PROOF. Let $\{x_n;f_n\}$ be a Q-fully λ-base. Hence there exists σ in P such that for each a in $\lambda_+^\times$ and p in D_T, we find q in D_T with

$$\{f_{\sigma(n)}(x)\ p(x_{\sigma(n)})\} \in \lambda, \quad \forall x \in X;$$

$$(*) \qquad \sum_{n\geq 1} a_{\pi(n)} |f_n(x)| p(x_n) \leq q(x), \quad \forall x \in X,$$

where $\pi = \sigma^{-1}$. The ω-completeness of (X,T) forces, therefore the existence of F_a^π and since $p(F_a^\pi x) \leq q(x)$ by (*), the required necessity is proved.

For proving the converse, we first prove the existence of a fundamental system $\hat{\mathcal{B}}$ of neighbourhoods at the origin consisting of normal subsets of X.

To this end, define θ: $\ell^\infty \times X \to X$ by

$$\theta(\alpha,x) = \sum_{n\geq 1} \alpha_n f_n(x) x_n; \qquad \alpha \in \ell^\infty,\ x \in X.$$

Since $\{x_n;f_n\}$ is a b.m-S.b., θ is well defined. Let

$$F_{\alpha}^{(n)} : X \to X; \; K_{x}^{(n)} : \ell^{\infty} \to X$$

be defined by

$$F_{\alpha}^{(n)}(x) = \sum_{i=1}^{n} \alpha_i f_i(x) x_i \qquad (\alpha \text{ fixed})$$

$$K_{x}^{(n)}(\alpha) = \sum_{i=1}^{n} \alpha_i f_i(x) x_i \qquad (x \text{ fixed}).$$

Then using the continuity of these maps and the Banach-Steinhaus theorem, θ is separately continuous. Thus θ is jointly continuous (e.g. [201], p. 51). Therefore, for each u in $\mathcal{B}_X$, there exist v in $\mathcal{B}_X$ and $\varepsilon > 0$ such that

$$W \equiv W(v,\varepsilon) = \theta(S_{\varepsilon}, v) \subset u,$$

S_{ε} being the closed ε-ball in ℓ^{∞}. Since

$$W(v,\varepsilon) = \{ \sum_{n \geqq 1} \alpha_n f_n(x) x_n : x \in v; \; |\alpha_n| \leqq \varepsilon, \; n \geqq 1\},$$

$W(v,\varepsilon)$ is normal and as $\varepsilon v \subset W(v,\varepsilon)$, $W(v,\varepsilon)$ is also a T-neighbourhood at the origin. Let

$$\hat{\mathcal{B}} = \{\overline{\rceil W(v,\varepsilon)} : v \in \mathcal{B}_X, \; \varepsilon > 0\},$$

where $\rceil A$ denotes the balanced convex hull of $A \subset X$. We now show that each set in $\hat{\mathcal{B}}$ is normal. Let

$$z \in (\overline{\rceil W(v,\varepsilon)})^{\wedge}$$

$$\Longrightarrow z = \sum_{n \geqq 1} \alpha_n f_n(y) x_n; \; |\alpha_n| \leqq 1, \; n \geqq 1, \; y \in \overline{\rceil W(v,\varepsilon)}.$$

But

$$y = \sum_{i=1}^{N} \beta_i y^i; \quad \sum_{i=1}^{N} |\beta_i| \leqq 1, \; y^i \in W(v,\varepsilon), \; 1 \leqq i \leqq N.$$

Therefore

$$z = \sum_{n\geqq 1} \alpha_n \Big(\sum_{i=1}^{N} \beta_i f_n(y^i)\Big) x_n = \sum_{i=1}^{N} \beta_i \Big(\sum_{n\geqq 1} \alpha_n f_n(y^i) x_n\Big)$$

$$= \sum_{i=1}^{N} \beta_i u^i,$$

where $u^i \in W(v,\varepsilon)$ by the normal character of W. Thus

$$z \in \overline{\rceil W(v,\varepsilon)}$$

and hence $\overline{\rceil W(v,\varepsilon)}$ is normal; consequently $\mathcal{B}$ is a fundamental system of neighbourhoods at the origin of (X,T), consisting of normal subsets of X.

From the foregoing discussion, it follows that T is also given by $\{p_u : u \in \hat{\mathcal{B}}\}$, p_u being the Minkowski functional corresponding to u. Further, let $E(x,u) = \{\lambda > 0 : x \in \lambda u\}$; $x \in X$, $u \in \hat{\mathcal{B}}$. If $x,y \in X$ with $|f_i(x)| \leqq |f_i(y)|$, $i \geqq 1$; then $E(y,u) \subset E(x,u)$ and so $p_u(x) \leqq p_u(y)$.

Returning to the main proof, fix a in $\lambda_+^\times$. Since F_a^π is continuous, for each u in $\hat{\mathcal{B}}$ we find v in $\hat{\mathcal{B}}$ so that

$$p_u(F_a^\pi(x)) \leqq p_v(x), \quad \forall x \in X.$$

But for any choice of $n \geqq 1$ and x in X,

$$|f_i(a_{\pi(n)} f_n(x) x_n)| \leqq |f_i(F_a^\pi(x))|, \; i \geqq 1$$

and so for u in $\hat{\mathcal{B}}$,

$$p_u(a_{\pi(n)} f_n(x) x_n) \leqq p_u(F_a^\pi(x)) \leqq p_v(x)$$

and as $n \geqq 1$ is arbitrary, we get

$$(*)\quad \sup_{n\geq 1} \{a_n |f_{\sigma(n)}(x)| p_u(x_{\sigma(n)})\} \leq p_v(x); \quad \forall x \in X.$$

where $\sigma = \pi^{-1}$. The inequality (*) is valid for any choice of a in $\lambda_+^\times$, u in $\hat{\mathcal{B}}$ and with a corresponding choice of v depending upon u and a. Since $(\lambda, \eta(\lambda, \lambda^\times))$ is nuclear, to every a in $\lambda^\times$ we find b in $\lambda^\times$ with $\sum_{n\geq 1} |a_n|/|b_n| < \infty$.

Therefore, using (*),

$$\sum_{n\geq 1} |a_n f_{\sigma(n)}(x)| p_u(x_{\sigma(n)}) \leq (\sum_{n\geq 1} |a_n/b_n|)\, p_v(x),$$

where v depends upon u and b. However, this shows that $\{x_n;f_n\}$ is a Q-fully λ-base. □

The second main result of this section depends upon the (K)-property of λ (cf. (4.5.4)).

<u>THEOREM 18.2.7</u> Let λ satisfy the (K)-property also. Then an S.b. $\{x_n;f_n\}$ for a Fréchet space (X,T) is a Q-fully λ-base if and only if $\{f_n;\Psi x_n\}$ is a Q-fully λ-base for $(X^*, \beta(X^*,X))$.

<u>PROOF</u>. First, let $\{x_n;f_n\}$ be a Q-fully λ-base for (X,T). By the (K)-property of λ, $\lambda \subset \ell^1$. Further, there exists σ in P so that $\{f_{\sigma(n)}(x)p(x_{\sigma(n)})\} \in \lambda$ for each p in D_T and x in X. Hence by the ω-completeness of (X,T), $\sum_{n\geq 1} f_{\sigma(n)}(x)x_{\sigma(n)}$ converges in (X,T), say, to $y \in X$. Thus

Thus $f_{\sigma(n)}(y) = f_{\sigma(n)}(x)$, $n \geq 1$; that is, $x = y$. Therefore, $\{x_{\sigma(n)};f_{\sigma(n)}\}$ is a fully λ-base for (X,T) and so, from Proposition 4.5.12, (X,T) is nuclear. Hence (X,T) is reflexive also (cf. [178], p. 83) and, on using Theorem 9.4.2, we conclude that $\{x_n;f_n\}$ is shrinking and γ-complete. Also observe that $\sum_{n\geq 1} |f_n(x)|p(x_n) < \infty$ for every p in D_T and x in X (apply the (K)-property of λ and the fact that $\{f_{\sigma(n)}(x)p(x_{\sigma(n)})\} \in \lambda$) and therefore $\{x_n;f_n\}$ is also a b.m-S.b. Consequently, by Theorem 18.2.6, there exists π in P so that for each a in $\lambda_+^\times$, the map F_a^π defined by (18.2.4) is a continuous linear operator. Finally, use Proposition 18.2.5(a) to conclude the sufficiency of the result.

Conversely, let $\{f_n;\Psi x_n\}$ be a Q-fully λ-base for $(X^*,\beta(X^*,X))$. This, in particular, implies the shrinking character of $\{x_n;f_n\}$. Also, there exists σ in P so that for a in $\lambda_+^\times$ and $B \in \mathcal{D}_X$, there exists A in $\mathcal{D}_X$ such that $\{f(x_{\sigma(n)})p_B(f_{\sigma(n)})\} \in \lambda$ for each f in X* and

$$(*) \quad \sum_{n\geq 1} a_n|f(x_{\sigma(n)})|p_B(f_{\sigma(n)}) \leq p_A(f), \quad \forall f \in X^*.$$

Since $(X^*,\beta(X^*,X))$ is ω-complete ([80], p. 218), $\{f_{\sigma(n)};\Psi x_{\sigma(n)}\}$ is a fully λ-base (see the beginning of the proof of the first part). By Proposition 4.5.12, $(X^*,\beta(X^*,X))$ is nuclear and so is (X,T) (e.g. [178], p. 78). Thus (X,T) is reflexive and this yields the γ-complete character of $\{x_n;f_n\}$. It is now easily seen that if $|\alpha_n| \leq 1$, then $\{\sum_{i=1}^{n} \alpha_i f_i(x)x_i\}$ is bounded (use (*)) and hence $\{x_n;f_n\}$ is a b.m-S.b. Therefore, by Proposition 18.2.5, there exists π in P so that for each a in $\lambda_+^\times$, the map F_a^π is a continuous linear operator. Finally make use of Theorem 18.2.6 to obtain the desired result. □

NOTE. An inspection of the proofs of the preceding results suggests the following two theorems (for the first theorem, cf. also [35]) of which the proofs are obviously omitted.

THEOREM 18.2.8 Assume that λ satisfies the conditions of Theorem 18.2.7. Then an S.b. $\{x_n;f_n\}$ for a Fréchet space is a fully λ-base if and only if $\{x_n;\Psi x_n\}$ is a fully λ-base for $(X^*,\beta(X^*,X))$.

Replacing $(\lambda,\eta(\lambda,\lambda^\times))$ by $(\Lambda(P),T_p)$, where $\Lambda(P) \subset \ell^1$ and $(\Lambda(P),T_p)$ is nuclear, we have

THEOREM 18.2.9 Let $\{x_n;f_n\}$ be an S.b. for a Fréchet space (X,T). Then $\{x_n;f_n\}$ is a Q-fully $\Lambda(P)$-base (resp. a $\Lambda(P)$-fully base) for (X,T) if and only if $\{f_n;\Psi x_n\}$ is a Q-fully $\Lambda(P)$-base (resp. a fully $\Lambda(P)$-base) for $(X^*,\beta(X^*,X))$.

18.3 λ-BASES AND Λ-NUCLEARITY

We now take up the problem raised in the beginning of this chapter, namely, to seek a possible relationship of a λ-base with the Λ-nuclearity of the underlying space. This problem is also born out of the natural observation contained in the next two theorems.

THEOREM A A Fréchet space with an ℓ^1-base $\{x_n;f_n\}$ is ℓ^1-nuclear if and only if $\{f_n;\Psi x_n\}$ is an ℓ^1-base for $(X^*,\beta(X^*,X))$.

THEOREM B A Fréchet space X equipped with an S.b. is ℓ^1-nuclear if and only if each S.b. for X is an ℓ^1-base.

NOTE. The necessity in Theorem A is a direct consequence of Proposition 9.5.1 (we may overlook the ℓ^1-nuclearity of the space in question), whereas the converse is contained in Proposition 9.5.11.

Concerning Theorem B, the 'only if' part readily follows from Theorem 12.4.6, whereas the converse is contained in [235].

In general, Theorem A breaks down when ℓ^1 is replaced by another sequence. For, we have

EXAMPLE 18.3.1 Let $X = \Lambda(P)$, P being a countable nuclear Köthe set of infinite type. Then (X,T) is a nuclear Fréchet space, where $T = T_P$ (cf. Proposition 1.3.14). Using the infinite character of P, one can easily show that the S.b. $\{e^n;e^n\}$ for (X,T) is a semi $\Lambda(P)$-base and so it is a $\Lambda(P)$-base for (X,T).

To show that $\{e^n;e^n\}$ is a $\Lambda(P)$-base for $(X^*,\beta(X^*,X))$, it suffices to prove that for each bounded subset B of (X,T) and f in X*, $\{<f,e^n>p_B(e^n)\} \in \Lambda(P)$. The S.b. $\{e^n;e^n\}$ is already a shrinking base for (X,T) because (X,T), being a nuclear Fréchet space, is reflexive.

Now, let B be a T_p-bounded subset of $\Lambda(P)$. Hence by Lemma 1.4.16, there exists α in $\Lambda(P)$ with

$$\sup_{\gamma \in B} |\gamma_n| \leq |\alpha_n|, \; n \geq 1.$$

If $f \in (\Lambda(P),T_p)^*$, there exists $\rho > 0$ and c in P such that $|f(e^n)| \leq \rho c_n$, $n \geq 1$. Choose a in P arbitrarily. By the infinite character of P, we can find b in P with $a_n c_n \leq b_n$, $n \geq 1$. Thus

$$\sum_{n \geq 1} |<f,e^n>|p_B(e^n)a_n \leq \rho \sum_{n \geq 1} c_n|\alpha_n|a_n$$

$$\leq \rho \sum_{n \geq 1} |\alpha_n|b_n < \infty.$$

However, by Proposition 17.3.11 and Theorem 17.4.8, (X,T) is not $\Lambda(P)$-nuclear.

Relating our problem to Theorem B, we begin with

<u>PROPOSITION 18.3.2</u> Let P be a Köthe set of increasing type and P_o a nuclear Köthe set of infinite type. If an l.c. TVS (X,T) possesses a Q*-fully $\Lambda(P)$-base $\{x_n;f_n\}$ and $(\Lambda(p),T_p)$ is $\hat{\Lambda}(P_o;\Psi)$-nuclear, then (X,T) is $\hat{\Lambda}(P_o;\Psi)$-nuclear.

<u>PROOF</u>. Imitating the proof of Theorem 17.4.8, (i) ==> (ii), we find the existence of some a in P with $\{1/a_n\} \in \Lambda(P_o;\Psi)$. Further, for each p in D_T, there exist q in D_T and π in P such that

(*) $\sum_{n \geq 1} |f_{\pi(n)}(x)| \, p(x_{\pi(n)}) \, a_n \leq q(x), \quad \forall x \in X$

$$\Longrightarrow \frac{p(x_{\pi(n)})}{q(x_{\pi(n)})} \leq \frac{1}{a_n}, \quad \forall n \geq 1.$$

By (*) and the increasing character of P, $\{x_n;f_n\}$ is a u.e-S.b. for (X,T). Therefore, the result follows by Theorem 17.4.1. □

NOTE. The notion of a Q-fully λ-base is stronger than that of a Q*-fully λ-base. Thus the foregoing proposition remains valid if the base $\{x_n;f_n\}$ is taken to be a Q-fully λ-base as observed in [138].

Conversely, we have

PROPOSITION 18.3.3 Let P_0 be a nuclear Köthe set of infinite type. Let (X,T) be a $\hat{\Lambda}(P_0)$ nuclear space. Then each u.e-S.b. $\{x_n;f_n\}$ for (X,T) is a Q*-fully $\Lambda(P_0)$ base.

PROOF. By Theorem 17.4.1, for each p in D_T, there exist q in D_T and π in P so that $p(x_{\pi(n)}) = \alpha_n \, q(x_{\pi(n)})$, where $\alpha \in \Lambda(P_0)$. If $a \in P_0$ and $x \in X$, then

$$\sum_{n \geq 1} |f_{\pi(n)}(x)| \, p(x_{\pi(n)}) a_n \leq \sup_{n \geq 1} |f_n(x) q(x_n)| \sum_{n \geq 1} |\alpha_n| a_n$$

$$\leq M_a \, q_\infty(x); \quad \forall x \in X,$$

where $q_\infty(x) = \sup \{|f_n(x)| q(x_n) : n \geq 1\}$ is a continuous seminorm on X. □

PROPOSITION 18.3.4 Let P_1 and P_2 be two nuclear Köthe sets of infinite type. Then an l.c. TVS (X,T) having a u.e-S.b. $\{x_n;f_n\}$ is $\hat{\Lambda}(P_2;\Psi)$-nuclear whenever it is $\Lambda(P_1)$ nuclear and $\Lambda(P_1)$ is $\hat{\Lambda}(P_2;\Psi)$-nuclear.

PROOF. Indeed, by Proposition 18.3.3, $\{x_n;f_n\}$ is a Q*-

fully $\Lambda(P_1)$ base and now apply Proposition 18.3.2. □

The conclusion of Proposition 18.3.3 may still be strengthened at the expense of P_o and (X,T). Indeed, we have ([133]; [165])

PROPOSITION 18.3.5 Let P be a nuclear Köthe set of infinite type such that $(\Lambda(P),T_p)$ is stable. Then each S.b. $\{x_n;f_n\}$ for a $\Lambda(P)$-nuclear Fréchet space (X,T) is a Q-fully $\Lambda(P)$-base.

PROOF. Since $\Lambda(P) \subset \ell^1$, (X,T) is ℓ^1-nuclear by Proposition 17.3.11. Hence by Theorem 12.4.6, $\{x_n;f_n\}$ is a fully ℓ^1-base for (X,T) and therefore $(X,T) \simeq (\Lambda(P_o),T_{P_o})$ where $P_o = \{\{p_k(x_n)\} : k \geq 1\}$ (cf. Proposition 4.5.10). It now suffices to show that $\{e^n;e^n\}$ is a Q-fully $\Lambda(P)$-base for $\Lambda(P_o)$. Since $(\Lambda(P_o),T_{P_o})$ is $\hat{\Lambda}(P)$-nuclear, it is also uniformly $\hat{\Lambda}(P)$-nuclear by Theorem 17.4.10. Thus there exist π in P such that for each $k \geq 1$, one can find $m \geq 1$ $(m \equiv m(k))$ with

$$(*) \qquad \{p_k(x_{\pi(n)})/p_m(x_{\pi(n)})\} \in \Lambda(P).$$

To prove the required result, let $k \geq 1$ and a in P be chosen arbitrarily. Then for α in $\Lambda(P_o)$,

$$\sum_{n\geq 1} |\langle\alpha,e^{\pi(n)}\rangle| p_k(x_{\pi(n)}) a_n \leq \sup_{n\geq 1}\{|\alpha_n| p_m(x_n)\} \sum_{n\geq 1} \frac{p_k(x_{\pi(n)})}{p_m(x_{\pi(n)})} a_n$$

$$\leq K_a \sum_{n\geq 1} |\alpha_n| p_m(x_n) \equiv K_a \hat{p}_m(\alpha),$$

where

$$K_a = \sum_{n\geq 1} \frac{p_k(x_{\pi(n)})}{p_m(x_{\pi(n)})} a_n < \infty; \quad \hat{p}_m(\alpha) = \sum_{n\geq 1} |\alpha_n| p_m(x_n),$$

$\hat{p}_m$ being a member of the family $D_{T_{P_o}}$ generating the l.c. topology T_{P_o} on $\Lambda(P_o)$. Therefore

$$\sum_{n\geq 1} |<\alpha, e^{\pi(n)}>| \hat{p}_k(e^{\pi(n)}) a_n \leq K_a \hat{p}_m(\alpha), \quad \forall\alpha \in \Lambda(P_o)$$

and this completes the proof. □

19 Dragilev bases

19.1 MOTIVATION

We have had occasion to eulogize about the importance of the structure of the space X_r of analytic functions (for example, cf. section 15.1) and, again, let us single out a very interesting feature of this class which has, in fact, given rise to the concept of 'Dragilev bases' (often, these bases are called 'regular bases' - not the ones we have introduced in chapter 2).

Indeed, the Schauder bases we are going to take up in this chapter, essentially stem from the study of bases in X_r. More specifically, it was Dragilev [38] who observed that any S.b. in X_r after the former undergoes a suitable normalization and permutation is equivalent to the fundamental S.b. $\{\delta_n\}$. As a matter of fact, this observation, substantiated by its proof by Dragilev himself, is a step toward solving an equally difficult problem, namely, what are those spaces with bases such that all bases in the same space are equivalent in a certain sense. This natural problem is concerned with the relationship amongst all Schauder bases present in an l.c. TVS.

A few years later, Dragilev [39] himself abstracted his own ideas, laid down the foundation of 'regular bases' which we prefer to call Dragilev bases and solved the problem of 'equivalence' (in fact, quasi-equivalence) of such bases in nuclear Fréchet spaces.

This chapter is essnetially devoted to the basics of Dragilev bases, certain related notions and their extensions, including the solution of two fundamental problems on the theory of these bases, namely, (A) the Dragilev character of another S.b. present in a nuclear space already having a Dragilev base and (B) the quasi-equivalence of two Dragilev bases present in certain nuclear spaces. While there are many examples of spaces having Dragilev bases (e.g. Example 19.3.1), a nontrivial example of a product space having no Dragilev base will be taken up in section 19.4.

19.2 THE FIRST FUNDAMENTAL THEOREM

In this section we will precisely state and prove a result related to the problem (A) of the last section. Nevertheless, the present section is equally concerned with the basic facts on Dragilev bases and therefore, to begin with, let us introduce

DEFINITION 19.2.1 An S.b. $\{x_n; f_n\}$ for an l.c. TVS (X,T) is called a *Dragilev base* or a (d_o)-*base* provided there exists an equivalent family D of seminorms on X such that for each pair p,q in D whenever $p(x_i) \leq q(x_i)$, $i \geq 1$, then the close-up of $\{p(x_n)/q(x_n)\}$ is decreasing, it being understood that 0/0 is to be regarded as 0.

NOTE. In the above definition, if $q(x_i) = 0$ then $p(x_i) = 0$ and we regard the i-th term in $\{p(x_n)/q(x_n)\}$ as 0.

Whenever (X,T) is a metrizable l.c. TVS, it will be assumed throughout without further reference that D_T consists of a nondecreasing sequence $\{p_n\}$ of seminorms on X; in particular, this assumption applies to Fréchet spaces as well. The following is then obvious.

PROPOSITION 19.2.2 In a metrizable l.c. TVS (X,T), an S.b. $\{x_n; f_n\}$ is a d_o-base if and only if there exists an admissible family $D = \{p_1 \leq p_2 \leq \ldots\}$ of semi-norms such that for each $i \geq 1$,

$$p_i(x_n)/p_{i+1}(x_n) \geq p_i(x_{n+1})/p_{i+1}(x_{n+1}), \; n \geq 1.$$

Similarly, by redefining an equivalent topology in a metrizable l.c. TVS (X,T), we can rephrase Theorem 12.4.6 as follows.

THEOREM 19.2.3 A metrizable l.c. TVS (X,T) having a u.e-S.b. $\{x_n;f_n\}$ is nuclear if and only if for each $i \geq 1$

$$\sum_{n\geq 1} p_i(x_n)/p_{i+1}(x_n) < \infty .$$

In what follows, we assume throughout this chapter, unless the contrary is specified, that (X,T) is a Fréchet space with $D_T = \{p_1 \leq p_2 \leq \cdots \leq \ldots\}$, where each p_i is a norm. This is equivalent to the fact that an arbitrary Fréchet space possesses a continuous norm. This requirement on (X,T) facilitates the analysis involved in the proofs of most of the results on (d_o)-bases. Secondly, the majority of spaces of interest which have (d_o)-bases are of the kind specified above. Besides, in the definition of (d_o)-bases, we are always interested in the nonzero terms of the sequence $\{p(x_n)/q(x_n)\}$ and this is just possible when all the members of D_T are norms. In addition, we also assume that (X,T) is nuclear.

After [39], we pass on to the *first fundamental theorem* of this chapter.

THEOREM 19.2.4 Let $\{x_n;f_n\}$ be a (d_o)-base for (X,T) and let $\{y_n;g_n\}$ be another S.b. for (X,T). Then there exists π in *P* such that $\{y_{\pi(n)};g_{\pi(n)}\}$ is a (d_o)-base for (X,T).

PROOF. At the outset, let $D = \{p_1 \leq p_2 \leq \ldots\}$ denote an admissible sequence of norms on X such that

$$p_i(x_n)/p_{i+1}(x_n) \geq p_i(x_{n+1})/p_{i+1}(x_{n+1}), \; n \geq 1. \quad (19.2.5)$$

The crux of the proof lies in finding a sequence $\{k_n\}$ of integers and a π in P such that $k_{\pi(n)} < k_{\pi(n+1)} \to \infty$ as $n \to \infty$.

For convenience, let us write

$$a_{nk} = f_n(y_k);\ b_{nk} = g_n(x_k),\ n,k \geq 1.$$

Since

$$1 = g_n(y_k) = \sum_{k\geq 1} f_k(y_n)g_n(x_k) = \sum_{k\geq 1} a_{kn}b_{nk}$$

for every $n \geq 1$, we can find k_n in $\mathbb{N}$ with

$$(+)\quad |a_{k_n n}\ b_{nk_n}| > \frac{1}{\sigma\ k_n^2},\ \sigma = \frac{\pi^2}{5}.$$

Observe that (+) follows by using contradiction and the fact that $\sum_{k\geq 1} (1/k^2) = \pi^2/6$.

The S.b. $\{x_n\}$ and $\{y_n\}$ are u.e-S.b. and consequently $T \approx T^* \approx T^{**}$, where T^* (resp. T^{**}) is generated by $D^* = \{p_i^* : p_i \in D_T\}$ (resp. $D^{**} = \{p_i^{**} : p_i \in D_T\}$) with

$$p_i^*(x) = \sum_{n\geq 1} |f_n(x)|p_i(x_n) \qquad (19.2.6)$$

and

$$p_i^{**}(x) = \sum_{n\geq 1} |g_n(x)|p_i^*(y_n) \qquad (19.2.7)$$

(cf. Proposition 12.4.15). In particular, for each $i \geq 1$ there exists $j \geq 1$ such that $p_i^{**}(x) \leq K_1 p_j(x)$ and so in particular

$$p_i^{**}(x_k) = \sum_{n\geq 1} |g_n(x_k)|p_i^*(y_n) \leq K_1 p_j(x_k),\ k \geq 1.$$

Now

$$\sum_{k\geq 1} p_j(x_k)/p_{j+1}(x_k), \quad \sum_{k\geq 1} p_{j+1}(x_k)/p_{j+2}(x_k) < \infty$$

(cf. Theorem 19.2.3) and by the (d_o)-character of $\{x_n\}$, we conclude that for $k \geq 1$

$$p_j(x_k) < \frac{K_2}{k} p_{j+1}(x_k), \quad p_{j+1}(x_k) < \frac{K_3}{k} p_{j+2}(x_k),$$

where K_2 and K_3 are respectively the sums of the preceding infinite series. Thus

$$|g_n(x_k)| p_i^*(y_n) \leq K_1 K_2 K_3 \cdot \frac{1}{k^2} p_{j+2}(x_k); \quad \forall k,n \geq 1$$

$$(*) \qquad \Longrightarrow |b_{nk_n}| p_i^*(y_n) \leq K_1 K_2 K_3 \cdot \frac{1}{k_n^2} p_{j+2}(x_{k_n}), \quad n \geq 1.$$

On the other hand, for $n \geq 1$, we have

$$(**) \quad p_i^*(y_n) \geq |f_{k_n}(y_n)| \, p_i(x_{k_n}) = a_{k_n n} |p_i(x_{k_n}).$$

Let us write $\alpha_n = |a_{k_n n}|$ and $M_i = \sigma K_1 K_2 K_3$, then from (+), (*) and (**), we get

$$(++) \qquad \alpha_n p_i(x_{k_n}) \leq p_i^*(y_n) < M_i \ \alpha_n \ p_{j+2}(x_{k_n}),$$

valid for all $i,n \geq 1$ and some $j \geq 1$, $j \equiv j(i)$.

Next, we claim that each coordinate in $\{k_n : n \geq 1\}$ is repeated possibly only a finite number of times; for otherwise, there exists N in $\mathbb{N}$ such that $k_{n_r} = N$ for $r = 1,2,\ldots$. Hence for m in $\mathbb{N}$ with m depending upon i+1,

$$\sum_{n\geq 1} \frac{p_i^*(y_n)}{p_{i+1}^*(y_n)} \geq \frac{1}{M_{i+1}} \sum_{n\geq 1} \frac{p_i(x_{k_n})}{p_{m+2}(x_{k_n})}$$

$$\geq \frac{1}{M_{i+1}} \sum_{r\geq 1} (p_i(x_{k_{n_r}})/p_{m+2}(x_{k_{n_r}}))$$

$$= \infty .$$

However, on account of Theorem 19.2.3, the last inequality is not valid. Therefore we can find a π in P such that $k_{\pi(n+1)} \geqq k_{\pi(n)}$, $n \geqq 1$ and $k_{\pi(n)} \to \infty$. On account of Theorem 12.4.6, $\{y_n;g_n\}$ is a fully ℓ^1-base and since (X,T) is ω-complete, $\{y_n;g_n\}$ is also a u-S.b. In particular, $\{y_{\pi(n)};g_{\pi(n)}\}$ is an S.b. for (X,T). For each $i \geqq 1$, let

$$q_i(x) = \sum_{n\geqq 1} \alpha_{\pi(n)} |g_{\pi(n)}(x)| p_i(x_{k_{\pi(n)}}),\ x \in X.$$

Then by (**), (19.2.7) and (++)

$$q_i(x) \leqq \sum_{n\geqq 1} |g_{\pi(n)}(x)| p_i^*(y_{\pi(n)})$$

$$\leqq M_i \sum_{n\geqq 1} \alpha_{\pi(n)} |g_{\pi(n)}(x)| p_{j+2}(x_{k_{\pi(n)}})$$

$$= M_i\, q_{j+2}(x)$$

$$\Longrightarrow q_i(x) \leqq p_i^{**}(x) \leqq M_i\ q_{j+2}(x).$$

If $D_o = \{q_i\}$ and T_o is the l.c. topology on X generated by D_o, then $T_o \approx T^{**} \approx T$ and so D_o is admissible. Observe that on account of (19.2.5), the quotient

$$q_i(y_{\pi(n)})/q_{i+1}(y_{\pi(n)}) = p_i(x_{k_{\pi(n)}})/p_{i+1}(x_{k_{\pi(n)}})$$

decreases with the increase of n, and hence the S.b. $\{y_{\pi(n)},g_{\pi(n)}\}$ is a (d_o)-base for (X,T_o). □

Let (Y,S) be another Fréchet space with $D_S = \{q_1 \leqq q_2 \leqq \ldots\}$, each q_i being a norm on Y; then the definition of a (d_o)-base immediately yields

<u>PROPOSITION 19.2.8</u> Let (X,T) be as before (not necessarily nuclear) having a (d_o)-base $\{x_n;f_n\}$. If $(X,T) \simeq (Y,S)$ under R with $Rx_n = y_n$, then

$\{y_n\} \equiv \{y_n; f_n \circ R^{-1}\}$ is a (d_o)-base for (Y,S).

PROOF. Indeed, for each $i \geq 1$, let $p_i'(y) = p_i(R^{-1}y)$ for y in Y, where $D_T = \{p_1 \leq p_2 \leq \ldots\}$ for which (19.2.5) is satisfied. The family $D = \{p_1' \leq p_2' \leq \ldots\}$ is an admissible family for S since $(X,T) \simeq (Y,S)$. Further $p_i'(y_n) = p_i(x_n)$. □

COROLLARY 19.2.9 Let $\{x_n; f_n\}$ be a (d_o)-base for (X,T) [(X,T) is not necessarily nuclear]. Then every S.b. $\{y_n\} \equiv \{y_n; g_n\}$ such that $\{x_n\} \sim \{y_n\}$, is a (d_o)-base for (X,T).

Indeed, apply Theorem 13.6.6 to get R such that $(X,T) \simeq (X,T)$ under R with $Rx_n = y_n$, $n \geq 1$ and then apply Proposition 19.2.8.

An advantage of having a (d_o)-base for (X,T) lies in exact estimation of Kolmogorov diameters of neighbourhoods generated by certain admissible families of norms on X. In fact, we have (cf. also [39])

PROPOSITION 19.2.10 Let $\{x_n; f_n\}$ be a (d_o)-base for (X,T). Then there exists an admissible sequence $\{p_1 \leq p_2 \leq \ldots\}$ of norms on X such that for $i \leq j$

$$d_n(u_j, u_i) = p_i(x_n)/p_j(x_n), \tag{19.2.11}$$

where $n, i, j = 1,2,\ldots$.

PROOF. We may clearly assume that $i < j$, otherwise there is nothing to prove. By Proposition 19.2.2, $\{p_i(x_n)/p_j(x_n): n \geq 1\}$ is decreasing, where we may assume that

$$p_i(x) = \sum_{n\geq 1} |f_n(x)| p_i(x_n). \tag{19.2.12}$$

Indeed, $\{p_i^*\}$ is an admissible sequence of norms on X, where

p_i^* is defined by (19.2.6) and $p_i^*(x_n) = p_i(x_n)$, $n \geq 1$. The required estimate (19.2.11) is now a simple consequence of Corollaries 5.4.7 and 5.4.16. Indeed, the nuclearity and completeness of (X,T) force $\{x_n;f_n\}$ to be a fully ℓ^1-base (cf. Proposition 12.4.15). □

19.3 EQUIVALENCE OF (d_o)-BASES

Throughout this section, (X,T) stands for an arbitrary nuclear Fréchet space having an S.b. $\{x_n;f_n\}$, where $D_T = \{p_1 \leq p_2 \leq \ldots\}$. When (X,T) contains two (d_o)-bases, a natural question arises of their relationship : are these two bases similar? This is not true in general; indeed, it is due to the fact that normalization of a (d_o)-base does not change the (d_o)-character of the resulting S.b. Let us consider

EXAMPLE 19.3.1 Consider the nuclear Fréchet space $(\Lambda_\infty(\alpha),T_\infty)$, where $\alpha_n = n$, $n \geq 1$. If

$$p_i(x) = \sum_{n\geq 1} |x_n| i^n, \quad x \in \Lambda_\infty(\alpha);$$

then $\{p_i\}$ is an admissible sequence of norms on $\Lambda_\infty(\alpha)$. The S.b. $\{e^n;e^n\}$ for $(\Lambda_\infty(\alpha),T_\infty)$ is clearly a (d_o)-base corresponding to $\{p_i\}$ and so is the S.b. $\{n^n e^n;e^n/n^n\}$. By Theorem 14.3.20, $\{e^n\} \not\sim \{n^n e^n\}$. Indeed, if we assume the contrary, then for each i in N, there would exist j in N such that for some constant $K > 1$

$$n^n (i/j)^n \leq K, \quad \forall n \geq 1$$

which is, however, not true.

As the foregoing example suggests, we may therefore look for some weaker types of similarity between two (d_o)-bases. To formulate the results precisely, let us introduce the following

DEFINITION 19.3.2 Let (X_1,T_1) and (X_2,T_2) be two arbitrary l.c. TVS containing S.b. $\{u_n;g_n\}$ and $\{v_n;h_n\}$ respectively. Then $\{u_n\}$ and $\{v_n\}$ are said to be (i) *semisimilar* if there exist scalars $a_n > 0$, $n \geq 1$ such that $\{u_n\}$ and $\{a_n v_n\}$ are similar and (ii) *quasisimilar* if there exist π, σ in P and scalars $a_n > 0$, $n \geq 1$ such that $\{u_{\pi(n)}\}$ and $\{a_n v_{\sigma(n)}\}$ are similar S.b.

NOTE. We have: similarity ==> semisimilarity ==> quasisimilarity.

A good deal of attention has been paid in the literature to establishing the quasisimilarity of any two arbitrary (d_o)-bases in a nuclear Fréchet space (X,T). Indeed, this is a step in solving a much bigger

PROBLEM 19.3.3 What are those nuclear Fréchet spaces (X,T) having bases in which all S.b's are quasisimilar?

Several persons have tackled Problem 19.3.3 in specific circumstances; for instance, when (X,T) is either chosen to be a particular space or it is considered with some restrictions. Obviously, the latter situation is of more interest than the former and in order to appreciate this, let us recall a few more types of bases.

DEFINITION 19.3.4 An S.b. $\{x_n;f_n\}$ for (X,T) is called (i) a (d_1)-*base* provided it is a (d_o)-base and for the corresponding admissible sequence $\{p_i\}$ of norms, one has:

$$(d_1) \quad \exists k \ni \forall j \; \exists i \ni \quad \frac{p_j^2(x_n)}{p_k(x_n)\; p_i(x_n)} \to 0;$$

and (ii) a (d_2)-*base* provided it is a (d_o)-base and for the corresponding admissible sequence $\{p_i\}$ of norms, one has:

$$(d_2) \qquad \forall k\ \exists j \ni \forall i \quad \frac{p_j^2(x_n)}{p_k(x_n)\ p_i(x_n)} \to \infty \ .$$

<u>NOTE</u>. (d_1) and (d_2)-bases were introduced in [39]. The condition (d_1) and (d_2) are respectively equivalent to (δ_1) and (δ_2) where

$$(\delta_1) \qquad \exists k \ni \forall j\ \exists m \ni \sup \frac{p_j^2(x_n)}{p_k(x_n)\ p_m(x_n)} < \infty$$

and

$$(\delta_2) \qquad \forall k\ \exists\ m \ni \forall i \quad \sup \frac{p_k(x_n)\ p_i(x_n)}{p_m^2(x_n)} < \infty \ .$$

It is not difficult to see that $(d_1) \Longleftrightarrow (\delta_1)$ and $(d_2) \Longleftrightarrow (\delta_2)$. In fact, $(d_i) \Longrightarrow (\delta_i)$, $i = 1,2$, is obvious. Let (δ_1) be true. Then there exists i such that $p_m(x_n)/p_i(x_n) \to 0$ (use the Schwartz character of (X,T)) and so (d_1) follows. Similarly if (δ_2) holds, we can find $j \geq 1$ such that $p_m(x_n)/p_j(x_n) \to 0$ and this yields (d_2).

<u>NOTE</u>. Conditions (d_1) and (d_2) can also be equivalently stated in terms of diameters of sets; these are

$$(d_1') \qquad \exists u \ni \forall w\ \exists\ v \ni d_n(v,w)/d_n(w,u) \to 0$$

and

$$(d_2') \qquad \forall u\ \exists\ w \ni \forall v \quad d_n(w,u)/d_n(v,w) \to 0,$$

where it is understood that $\{x_n;f_n\}$ is a d_o-base for a nuclear Fréchet space. The equivalence of (d_i) and (d_i') is a simple consequence of Proposition 19.2.10 which is left to the reader as an exercise.

A nuclear Fréchet space (X,T) is said to belong to (d_i)-*class* if (X,T) possesses a (d_i)-base, $i = 1,2$.

Returning to Problem 19.3.3, the first classical solution in this direction was obtained in [39] which says that all S.b.'s in a (d_i)-class $(i = 1,2)$ are quasisimilar. A weaker version of this very result of [39] was subsequently obtained as a corollary of a more general result in [16] to which we shall return later. However, this result of [16] relaxed no further the class-classification hypothesis of (X,T) in solving Problem 19.3.3.

In the mid-seventies, a very satisfactory solution to Problem 19.3.3 was discovered by Crone and Robinson [29] and this is what we call the *second fundamental theorem* of this chapter. It runs as follows:

THEOREM 19.3.5 Let a nuclear Fréchet space contain a (d_o)-base $\{x_n\} \equiv \{x_n;f_n\}$. If $\{y_n\} \equiv \{y_n;g_n\}$ is any arbitrary S.b. for (X,T), then $\{x_n\}$ and $\{y_n\}$ are quasi-similar.

Djakov [37] has given a very simple and short proof of Theorem 19.3.5 and we will reproduce this here. Indeed, by virtue of Theorem 19.2.4, it is enough to prove

PROPOSITION 19.3.6 Let (X,T) be a nuclear Fréchet space having two arbitrary (d_o)-bases $\{x_n;f_n\}$ and $\{y_n;g_n\}$. Then $\{x_n\}$ and $\{y_n\}$ are semisimilar.

PROOF. Let $\{\mu_1 \leqq \mu_2 \leqq \ldots\}$ and $\{\nu_1 \leqq \nu_2 \leqq \ldots\}$ be admissible sequences of norms on X with respect to which $\{x_n\}$ and $\{y_n\}$ are (d_o)-bases. By Proposition 12.4.15, $T \approx T^*_\mu \approx T^*_\nu$, where T^*_μ (resp. T^*_ν) is generated by $\{\mu^*_i\}$ (resp. $\{\nu^*_i\}$) with

$$\mu^*_i(x) = [\sum_{n\geqq 1} |f_n(x)|^2 \ \mu^2_i(x_n)]^{1/2};$$

$$\nu^*_i(x) = [\sum_{n\geqq 1} |g_n(x)|^2 \nu^2_i(y_n)]^{1/2}.$$

Put $u_i = \{x \in X : \mu_i^*(x) < 1\}$ and $v_j = \{x \in X : \nu_j^*(x) < 1\}$. Without loss of generality, we may assume that $\mu_1^* \leq \nu_1^* \leq \mu_2^* \leq \cdots \leq \mu_i^* \leq \nu_i^* \leq \mu_{i+1}^*$, otherwise we pass on to suitable subsequences of norms with their respective positive multiples. The new set of sequences of norms together with their positive multiples satisfy all those conditions enjoyed by the original set of sequences of norms.

Let $i,j \in N$

(I) If $i \leq j$, $\mu_i^* \leq \nu_i^* \leq \nu_j^* \leq \mu_{j+1}^*$ and so from Proposition 1.4.4,

$$d_n(u_{j+1},u_i) \leq d_n(v_j,u_i) \leq d_n(v_j,v_i)$$

$$\Longrightarrow d_n(u_{j+1},u_i) \leq d_n(v_j,v_i)$$

$$\Longrightarrow \mu_i^*(x_n)/\mu_{j+1}^*(x_n) \leq \nu_i^*(y_n)/\nu_j^*(y_n),$$

by Proposition 19.2.10.

(II) Let $i > j$. Then $\nu_j^* \leq \mu_{j+1}^* \leq \mu_i^* \leq \nu_i^*$ and hence, as before,

$$d_n(v_i,v_j) \leq d_n(u_i,u_{j+1})$$

$$\Longrightarrow \nu_j^*(y_n)/\nu_i^*(y_n) \leq \mu_{j+1}^*(x_n)/\mu_i^*(x_n).$$

Therefore, in either of the cases

$$\mu_i^*(x_n)/\nu_i^*(y_n) \leq \mu_{j+1}^*(x_n)/\nu_j^*(y_n),$$

for all $i,j,n \geq 1$. Let

$$r_n = \sup_{i \geq 1} \{\mu_i^*(x_n)/\nu_i^*(y_n)\}.$$

Then

$$(+) \qquad \mu_i^*(x_n) \leqq r_n \, \nu_i^*(y_n) \leqq \mu_{i+1}^*(x_n),$$

for $i,n \geqq 1$. Note that for each n, r_n is well defined and positive. The sequence $\{r_n y_n\}$ is also an S.b. for (X,T). By (+) and Theorem 14.3.20, $\{x_n\} \sim \{r_n y_n\}$. □

19.4 SPACES HAVING NO (d_o)-BASE

As mentioned at the beginning of this chapter, the concept of a (d_o)-base has emerged from the base $\{\delta_n\}$ of the space X_r $(0 < r \leqq \infty)$ equipped with the compact-open topology T_r. However, surprisingly enough the product of the spaces X_r with $0 \leqq r < \infty$ and X_∞ behave differently; indeed, we will show in this final section that the product space $X_r \times X_\infty$ does not possess a (d_o)-base (cf. [39]).

In order to derive the desired conclusion from the space $X_r \times X_\infty$ in the form of Theorem 19.4.8, we need some preparatory information concerning the spaces X_r $(0 < r \leqq \infty)$ and their product and we collect this in the form of several results given below. At the outset, corresponding to a sequence $\{r_p\}$, $0 < r_1 < r_2 < \ldots < r_p \to r$, as $p \to \infty$ $(0 < r \leqq \infty)$, let us denote the countable family of norms generating the topology T_r by $\{|\cdot|_p : p \in \mathbb{N}\}$, where for $f \in X_r$,

$$|f|_p = \sup \{|f(z)| : |z| \leqq r_p\}. \qquad (19.4.1)$$

We now have

<u>LEMMA 19.4.2</u> Let $\{|\cdot|_p^1\}$ be an arbitrary increasing sequence of norms generating the topology T_r of X_r $(0 < r < \infty)$. Then there exist p_1, q_1 in $\mathbb{N}$ with $q_1 > p_1$ and constants α, a with $1 < \alpha < a < \infty$ such that for each $q > q_1$,

$$\alpha^n < \frac{|\delta_n|_q^1}{|\delta_n|_{p_1}^1} < a^n \qquad (19.4.3)$$

for sufficiently large n.

PROOF. For p = 1, there exist $K_1 > 0$ and p_1 in $\mathbb{N}$ such that

$$r_1^n < K_1 \, |\delta_n|_{p_1}^1 \; , \quad \forall n \geqq 0$$

and there exist K_{p_1} and s with

$$|\delta_n|_{p_1}^1 < K_{p_1} = r_s^n, \quad \forall n \geqq 0.$$

Let $r_t > r_s$. For this t, we find $q_1 (> p_1)$ and $K_t > 0$ such that

$$r_t^n < K_t \, |\delta_n|_{q_1}^1, \quad \forall n \geqq 0.$$

Define A and B so that

$$1 < A < \frac{r_t}{r_s} \; ; \; \frac{r}{r_1} \leqq B < \infty.$$

Then for $q > q_1$,

$$\left[\frac{A}{(K_{p_1} K_t)^{1/n}}\right]^n < \frac{|\delta_n|_q^1}{|\delta_n|_{p_1}^1}, \quad \forall n \geqq 1.$$

Choose λ such that $A^{-1} < \lambda < 1$ and write $\alpha = A\lambda > 1$. Then for some $n_1 \in \mathbb{N}$,

$$\lambda < \left(\frac{1}{K_{p_1} K_t}\right)^{1/n}, \quad \forall n \geqq n_1$$

Hence

$$(+) \qquad \alpha^n < \frac{|\delta_n|_q^1}{|\delta_n|_{p_1}^1}, \quad \forall n \geqq n_1.$$

On the other hand for $q \in \mathbb{N}$, there exists $n_q \in \mathbb{N}$ such that

$$|\delta_n|_q^1 < r^n, \quad \forall n \geq n_q.$$

Now choose $a = (\sup_n K_1^{1/n})B$. With this choice of a, we have

$$(++) \qquad \frac{|\delta_n|_q^1}{|\delta_n|_{p_1}^1} < a^n, \quad \forall n \geq n_q.$$

Combining (+) and (++), we get (19.4.3) valid for all $n \geq \max.(n_1, n_q)$. □

LEMMA 19.4.4 For an arbitrary increasing sequence $\{|\cdot|_p^2\}$ of norms generating the topology T_∞ of X_∞, there exists p_2 in $\mathbb{N}$ such that for any $\beta > 1$ and q_2 in $\mathbb{N}$, we can find $q > q_2$ and $b > 0$ such that

$$\beta^n < \frac{|\delta_n|_q^2}{|\delta_n|_{p_2}^2} < b^n \qquad (19.4.5)$$

for sufficiently large n.

PROOF. For $p = 1$, there exist p_2, $M_1 \geq 1$, and then ℓ, $M_{p_2} \geq 1$ such that

$$r_1^n < M_1 \, |\delta_n|_{p_2}^2 \; ; \; |\delta_n|_{p_2}^2 < M_{p_2} \, r_\ell^n, \quad \forall n \geq 0.$$

Let $\beta > 1$ be given and q_2 in $\mathbb{N}$ be chosen arbitrarily. Choose $r_s > \beta M_{p_2} r_\ell$. Then there exist $q > q_2$ and $M_s \geq 1$ such that

$$r_s^n < M_s \, |\delta_n|_q^2, \quad \forall n \geq 0.$$

For q, there exist t and $M_q > 0$ satisfying the inequality

$$|\delta_n|_q^2 < M_q\, r_t^n, \quad \forall n \geq 0.$$

If $n_o \in \mathbb{N}$ is such that $M_s/M_{p_2}^{n-1} \leq 1$ for $n \geq n_o$ and $b > \sup_n (M_1 M_q)^{1/n} (r_t/r_1)$, then

$$\beta^n < \frac{|\delta_n|_q^2}{|\delta_n|_{p_2}^2}, \quad \forall n \geq n_o; \quad \frac{|\delta_n|_q^2}{|\delta_n|_{p_2}^2} < b^n, \forall n \geq 1.$$

These inequalities together yield (19.4.5). □

REMARK. In the above lemma, we may find $q > p_2$; indeed, choose $q_2 = p_2$.

The next result which finds its application in the main theorem of this section, is of independent interest in the theory of Schauder bases in Fŕechet spaces.

PROPOSITION 19.4.6 Let (X,T) and (Y,S) be Fréchet spaces with S.b.'s $\{x_n;f_n\}$ and $\{y_n;g_n\}$ respectively. If $\{m_i\}$ and $\{n_i\}$ are two complementary subsequences of $\mathbb{N}$ with $\{m_i\} \cup \{n_i\} = \mathbb{N}$, then the sequence $\{z_k\}$ defined by

$$z_k = \begin{cases} (x_i,0); & k = m_i \\ & \qquad\qquad i,k = 1,2,\ldots \\ (0,y_i); & k = n_i, \end{cases} \tag{19.4.7}$$

is an S.b. for $X \times Y$ relative to its product topology. In addition, if $\{x_n\}$ and $\{y_n\}$ are u-S.b., then $\{x_k\}$ is an S.b. for $X \times Y$ for any two supplementary sequences $\{m_i\}$ and $\{n_i\}$ of $\mathbb{N}$.

PROOF. Let the families $\{p_n\}$ and $\{q_n\}$ of seminorms generate T and S respectively and $\{Q_n\}$ be the sequence of seminorms generating the product topology of $X \times Y$, where we assume

$$Q_n(z) = p_n(x) + q_n(y), \quad z = (x,y), \quad n \geq 1.$$

For $z \in X \times Y$ with $z = (x,y)$, define

$$\alpha_k(z) = \begin{cases} f_i(x) & k = m_i \\ g_i(y) & k = n_i \end{cases} \qquad i,k = 1,2,\ldots$$

Then α_k's are linear functionals on $X \times Y$ with $\alpha_k(z_\ell) = \delta_{k\ell}$, the Kronecker delta. Moreover, for p in $\mathbb{N}$, if s and t are the largest integers in $\mathbb{N}$ with $1 \leq m_s \leq p$ and $1 \leq n_t \leq p$, then we have

$$Q_n(z - \sum_{k=1}^{p} \alpha_k(z) z_k) = p_n(x - \sum_{i=1}^{s} f_i(x)x_i)$$

$$+ q_n(y - \sum_{i=1}^{t} g_i(y)y_i) \to 0 \quad \text{as } s,t \to \infty.$$

Thus $\{z_k\}$ is a t.b. for $X \times Y$ and hence an S.b. for $X \times Y$.

For the last statement, observe that for p in $\mathbb{N}$

$$z - \sum_{k=1}^{p} \alpha_k(z) z_k = (x - \sum_{i \in J_1^p} f_i(x)x_i, \ y - \sum_{i \in J_2^p} g_i(y)y_i)$$

where

$$J_1^p = \{i : k = m_i, \ 1 \leq m_i \leq p\}; \quad \text{and}$$

$$J_2^p = \{i : k = n_i, \ 1 \leq n_i \leq p\}.$$

Since $\{x_n\}$ and $\{y_n\}$ are unordered bases for X and Y; for given $\varepsilon > 0$ and $n \in \mathbb{N}$, there exist finite sets σ_1 and σ_2 of $\mathbb{N}$ such that

$$p_n(x - \sum_{i \in \sigma} f_i(x)x_i) < \frac{\varepsilon}{2}, \quad \forall \sigma \supset \sigma_1;$$

and

$$q_n(y - \sum_{i\in\sigma} g_i(y)y_i) < \frac{\varepsilon}{2}, \ \forall\sigma \supset \sigma_2.$$

Let $\sigma_0 = \sigma_1 \cup \sigma_2$ and $p_0 = \max \{m_i, n_i : i \in \sigma_1 \cup \sigma_2\}$. Then for $p \geq p_0$, $J_1^p \supset \sigma_1$; $J_2^p \supset \sigma_2$ and so for $p \geq p_0$,

$$Q_n(z - \sum_{k=1}^{p} \alpha_k z_k) < \varepsilon. \quad \square$$

We are now prepared to state and prove the main result of this section, namely,

THEOREM 19.4.8 The space $X = X_r \times X_\infty$ $(0 < r < \infty)$, possesses no (d_o)-base.

PROOF. Let the product space X have a (d_o)-base. Then for some supplementary sequences $\{m_i\}$ and $\{n_i\}$ of $\mathbb{N}$, the base

$$f_k = \begin{cases} (\delta_i, 0) & k = m_i \\ & \qquad\qquad i = 0,1,2,\ldots \\ (0, \delta_i) & k = n_i \end{cases} \tag{19.4.9}$$

is a (d_o)-base. Hence there exists a sequence $\{|\cdot|_p\}$ of increasing norms defining the topology of X such that

$$\frac{|f_k|_p}{|f_k|_{p+1}} \geq \frac{|f_k|_p}{|f_{k+1}|_{p+1}}, \ \forall k, p \geq 1$$

$$\Longrightarrow \frac{|f_k|_q}{|f_k|_p} \geq \frac{|f_{k+1}|_q}{|f_{k+1}|_p}, \ \forall q \geq p;\ k \geq 1. \tag{19.4.10}$$

Let us introduce increasing families $\{|\cdot|_p^1\}$ and $\{|\cdot|_p^2\}$ of norms generating the topologies of X_r and X_∞ respectively as follows:

$$|(x,0)|_p = |x|_p^1, \quad x \in X_r$$

and

$$|(0,y)|_p = |y|_p^2, \ y \in X .$$

With this choice of families of norms, we may assume without loss of generality that $p_1 = p_2 = p_o$, $q_1 = q_2 = q_o$ with $q_o > p_o$ in Lemmas 19.4.2 and 19.4.4.

Let us choose subsequences $\{m_{\ell_j}\}$ and $\{n_{s_j}\}$ so that

$$(*) \quad m_{\ell_{j-1}} \leqq m_{\ell_j - 1} < n_{s_j} < m_{\ell_j}, \quad \forall j \geqq 1.$$

Then the following two cases are possible:

(i) $\overline{\lim\limits_{j\to\infty}} s_j/\ell_j > \delta > 0$; or

(ii) $\lim\limits_{j\to\infty} s_j/\ell_j = 0$.

In case (i), for $\beta > a^{1/\delta}$, one can find $q > q_1$ such that for a subsequence of $\{j\}$,

$$(+) \quad \frac{|f_{m_{\ell_j}}|_q}{|f_{m_{\ell_j}}|_{p_o}} = \frac{|\delta_{\ell_j}|_q^1}{|\delta_{\ell_j}|_{p_o}^1} < a^{\ell_j} < \beta^{\delta\ell_j} < \beta^{s_j} <$$

$$\frac{|\delta_{s_j}|_q^2}{|\delta_{s_j}|_{p_o}^2} = \frac{|f_{n_{s_j}}|_q}{|f_{n_{s_j}}|_{p_o}} .$$

In case (ii), for $q > q_o$ and b satisfying Lemma 19.4.4, we have

$$(++) \quad \frac{|f_{m_{\ell_j - 1}}|_q}{|f_{m_{\ell_j - 1}}|_{p_o}} = \frac{|\delta_{\ell_j - 1}|_q^1}{|\delta_{\ell_j - 1}|_{p_o}^1} > \alpha^{\ell_j - 1} > b^{s_j} >$$

$$\frac{|\delta_{s_j}|_q^2}{|\delta_{s_j}|_{p_o}^2} = \frac{|f_{n_{s_j}}|_q}{|f_{n_{s_j}}|_{p_o}}$$

for sufficiently large j.

Since $q > p_o$, the inequalities (+) and (++) along with (*) contradict (19.4.10). Hence X has no (d_o)-base. □

20 Extension of Dragilev theory

20.1 INTRODUCTION

Attempts have been made in the past to strengthen the theory involved in the statements of Theorems 19.2.4 or 19.3.5 or both by way of either relaxing the basis hypothesis or the structure of the space in question. The first attempt in this direction was made in [16] which provides an extension of Theorem 19.2.4. Subsequently a study parallel to Theorem 19.3.5 was initiated in [1] and later both of these theorems were re-proved in [28] by weakening the (d_o)-character of the Schauder base. The present chapter which contains this development in chronological order, provides a dissection of analysis of the entire theory as related to the two theorems mentioned above.

20.2 THE FIRST THEOREM

In this section, we essentially consider a strengthening of the first fundamental theorem (Theorem 19.2.4) of the last chapter in the sense that here we do not take $\{x_n\}$ to be a (d_o)-base nor do we consider $\{y_n\}$ to be an S.b. for the whole of (X,T); we establish semisimilarity between $\{y_n\}$ and $\{x_{k_n}\}$ for a suitable $\{k_n\} \in I$. As before, we let (X,T) be an arbitrary nuclear Fréchet space and follow [16] throughout this section. At the outset, let us begin with

LEMMA 20.2.1 Let $\{x_n\} \equiv \{x_n;f_n\}$ be an S.b. for (X,T) with $D_T = \{p_1 \leq p_2 \leq \ldots\}$. Then there exists π in P such that for each i we can find $R_i > 0$ with

$$p_i(x_{\pi(n)}) \leqq R_i n^{-2} p_{i+2}(x_{\pi(n)}), \quad \forall n \geqq 1. \qquad (20.2.2)$$

PROOF. The beginning of the proof makes use of the arguments given in [40], p. 215. Put $c_{in} = p_i(x_n)/p_{i+1}(x_n)$. The claim is that for each $i \geqq 1$, we can find n_i in $\mathbb{N}$ such that $c_{in} \leqq \alpha_n$ for $n \geqq n_i$, where α_n is independent of i and $\{\alpha_n\} \in \ell^1$. Indeed, by Theorem 19.2.3, for each i, one finds n_i with

$$\sum_{n \geqq n_i} c_{in} \leqq \frac{1}{i^2}, \quad i \geqq 1.$$

We may clearly assume that $n_i < n_{i+1}$ and so $\{n_i\}$ is a fixed member of I. Let $n \geqq n_1$ be chosen arbitrarily. For this choice of n, there is a unique integer k with $n_k \leqq n < n_{k+1}$, and let $j_n = k$. Suppose that $\alpha_n = \max\{c_{in} : 1 \leqq i \leqq j_n\}$. Then

$$\sum_{n \geqq n_1} \alpha_n \leqq \sum_{n \geqq n_1} \sum_{i=1}^{j_n} c_{in} = \sum_{i \geqq 1} \sum_{n \geqq n_i} c_{in} < \infty$$

and this justifies the desired claim.

Now there exists π in P such that $\alpha_{\pi(n+1)} \leqq \alpha_{\pi(n)}$, $n \geqq 1$; thus

$$n\,\alpha_{\pi(n)} \leqq \sum_{i=1}^{n} \alpha_{\pi(i)} < \sum_{i \geqq 1} \alpha_i$$

$$\Longrightarrow K \equiv \sup\{n\alpha_{\pi(n)} : n \geqq 1\} \leqq \|\alpha\|_1.$$

Therefore, for each $i \geqq 1$, there exists $M_i \geqq K$ with $\sup\{nc_{i\pi(n)} : n \geqq 1\} \leqq M_i$; that is, for each $i \geqq 1$,

$$(+) \quad np_i(x_{\pi(n)}) \leqq M_i\, p_{i+1}(x_{\pi(n)}), \quad \forall n \geqq 1.$$

Similarly,

$$(++) \quad np_{i+1}(x_{\pi(n)}) \leqq M_i'\, p_{i+2}(x_{\pi(n)}), \quad \forall n \geqq 1.$$

From (+) and (++), the required inequality (20.2.2) follows, where $R_i = M_i M_i'$. □

LEMMA 20.2.3 Let $\{x_n;f_n\}$ be an S.b. for (X,T) and $\{y_n\}$ a CSBS in (X,T). Then there exist π in P and $\{k_n\} \subset N$ such that $f_{\pi(k_n)}(y_n) \neq 0$, $n \geq 1$ and for any p in D_T, there exists q in D_T so that

$$a_n p\,(x_{\pi(k_n)}) \leq p_\infty(y_n)\,;\; p(y_n) \leq a_n\, q(x_{\pi(k_n)});\; n \geq 1, \tag{20.2.4}$$

where

$$a_n = |f_{\pi(k_n)}(y_n)|,\; n \geq 1.$$

PROOF. By Proposition 12.4.1, there exists $\{g_n\}$ in X^* with $g_n(y_m) = \delta_{mn}$ such that for each p in D_T, there exists q_1 in D_T such that

$$p^\infty(x) \equiv \sup_{n\geq 1} \{|g_n(x)|\; p\,(y_n)\} \leq q_1(x),\; \forall x \in X.$$

Let $\sigma = 1 + \sum_{n\geq 1} 1/n^2$. By Lemma 20.2.1, we can find q_2 in D_T so that

$$p^\infty(x_{\pi(n)}) \leq \sigma^{-1}\, n^{-2}\, q_2(x_{\pi(n)}),\quad \forall n \geq 1,$$

where π in P does not depend upon p. Let $q = \max(q_1,q_2)$, $u_n = x_{\pi(n)}$ and $h_n = f_{\pi(n)}$. Then $\{u_n;h_n\}$ is an S.b. for (X,T) and

(i) $p^\infty(x) \equiv \sup_{n\geq 1} \{|g_n(x)|p\,(y_n)\} \leq q(x),\quad \forall x \in X;$

(ii) $p^\infty(u_n) \leq \sigma^{-1} n^{-2} q(u_n),\quad \forall n \geq 1.$

Since

$$1 = g_n(y_n) = \sum_{k \geq 1} h_k(y_n) g_n(u_k),$$

for each n, we can find k_n in $\mathbb{N}$ with

$$|g_n(u_{k_n}) h_{k_n}(y_n)| \geq \sigma^{-1} k_n^{-2}$$

(cf. the proof of Theorem 19.2.4). Thus

$$\text{(iii)} \quad |g_n(u_{k_n})| \geq a_n^{-1} \sigma^{-1} k_n^{-2}, \quad \forall n \geq 1.$$

By (i) and (ii)

$$|g_n(u_{k_n})| \; p\,(y_n) \leq p^{\infty}(u_{k_n}) \leq \sigma^{-1} k_n^{-2} \; q(u_{k_n}), \; \forall n \geq 1$$

and so (iii) leads to

$$\text{(iv)} \quad p\,(y_n) \leq a_n \; q(u_{k_n}), \quad \forall n \geq 1.$$

From the definition of p_∞,

$$\text{(v)} \quad p_\infty(y_n) \geq |h_{k_n}(y_n)| \; p(u_{k_n}), \quad \forall n \geq 1.$$

Inequalities (iv) and (v) result in (20.2.4). □

Before we come to the next crucial lemma, let us discuss some useful preliminaries.

DEFINITION 20.2.5 Let $\{u_n;h_n\}$ be an S.b. in an l.c. TVS (Y,S). A member p in D_T is called $\{u_n\}$-*normal* if $|h_n(y)|p(u_n) \leq p(y)$ for all y in Y and $n \geq 1$.

The existence of $\{u_n\}$-normal seminorms is guaranteed by

PROPOSITION 20.2.6 Let $\{u_n;h_n\}$ be an S.b. for a Fréchet space (Y,S) and for p in D_S, let

$$r_p(x) = \sup \{p(\sum_{i=m}^{n} h_i(x)u_i) : n, m \geq 1\}.$$

Then r_p is an $\{u_n\}$-normal seminorm and the topology generated by $\{r_p : p \in D_S\}$ is equivalent to S.

PROOF. The proof follows on the lines of Theorem 2.2.5 of [123] and the fact that $r_p(x) \geq |h_n(x)| \; r_p(u_n)$; $x \in Y$, $n \geq 1$. □

LEMMA 20.2.7 Let $\{x_n;f_n\}$ be an S.b. for a Fréchet nuclear space and $\{y_n\}$ a CSBS in (X,T). Then there exist π, σ in P and $\{k_n\} \subset \mathbb{N}$ with $\pi(k_{\sigma(n)}) \uparrow \infty$ such that $a_n \equiv |f_{\pi(k_n)}(y_n)| \neq 0$ and for each $\{x_n\}$-normal seminorm p in D_T, there exists r in D_T such that

$$a_n \; p(x_{\pi(k_n)}) \leq p(y_n) \leq a_n \; r(x_{\pi(k_n)}), \; n \geq 1. \quad (20.2.8)$$

PROOF. First of all note that $p_\infty(x) = \sup \{|f_n(x)p(x_n)| : n \geq 1\} \leq p(x)$. There exists $s \geq p$ in D_T such that

$$(*) \qquad \sum_{n\geq 1} p(y_n)/s(y_n) < \infty.$$

We now invoke both the notation and proof of Lemma 20.2.3. Then, if $r = \max(q,s)$, the inequality (20.2.8) is a simple consequence of (20.2.4) and from (*), $p(y_n)/r(y_n) \to 0$. Further, using the second inequality in (20.2.4), we can find $r_1 \geq r$ in D_T such that

$$r(y_n) \leq a_n \; r_1(x_{\pi(k_n)}).$$

Thus

$$(+) \quad \frac{p(y_n)}{r(y_n)} \geq \frac{a_n \; p(x_{\pi(k_n)})}{a_n \; r_1(x_{\pi(k_n)})} \implies p(x_{\pi(k_n)})/r_1(x_{\pi(k_n)}) \to 0.$$

Consider any $j \geq 1$. We can find an $\{x_n\}$-normal seminorm p for which $p(x_j) \neq 0$. Thus, the set $\{n : \pi(k_n) = j\}$ is finite; for otherwise, there is an infinite subsequence of

N on which $p(x_{\pi(k_n)})/r_1(x_{\pi(k_n)})$ remains a nonzero constant. Hence all the sets $\{n : \pi(k_n) = j\}$ with $j \geqq 1$ are finite; that is, $\pi(k_n) \to \infty$. If m_j is the cardinality of the sets $\{n : \pi(k_n) = j\}$, $j \geqq 1$, choose σ in P so that

$\pi(k_{\sigma(m_{j-1}+1)}) = \pi(k_{\sigma(m_{j-1}+2)}) = \ldots = \pi(k_{\sigma(m_{j-1}+m_j)}) = j$,

$j \geqq 1$, where $m_o = 0$. □

In a Fréchet space (Y,S) with an S.b. $\{u_n;h_n\}$, it is always possible to have an admissible sequence of $\{u_n\}$-normal seminorms on Y. Thus the following result is an immediate consequence of Lemma 20.2.7 and so its proof is omitted.

PROPOSITION 20.2.9 Let (X,T), $\{x_n;f_n\}$, $\{y_n\}$, π, σ and $\{k_n\}$ be as described in Lemma 20.2.7. Let $\{p_i\}$ be an admissible sequence of $\{x_n\}$-normal seminorms on X. Then for each $i \geqq 1$, there exist K_i and $j \geqq i$ such that

$$a_{\sigma(n)}p_i(x_{\pi(k_{\sigma(n)})}) \leqq p_i(y_{\sigma(n)}) \leqq K_i\ a_{\sigma(n)}p_j(x_{\pi(k_{\sigma(n)})}), \tag{20.2.10}$$

for $n \geqq 1$, where K_i is a positive constant depending upon p_i.

Finally we have

THEOREM 20.2.11 Let $\{x_n;f_n\}$ be an S.b. for a nuclear Fréchet space (X,T) and $D_T = \{p_1 \leqq p_2 \leqq \ldots\}$ an admissible sequence of $\{x_n\}$-normal seminorms on X. Further, let $\{y_n;g_n\}$ be a CSBS in (X,T). Then there exist π, σ in P, a sequence $\{k_n\} \subset \mathbb{N}$ with $\pi(k_{\sigma(n)}) \uparrow \infty$, scalars $a_n \equiv f_{\pi(k_n)}(y_n) \neq 0$, $n \geqq 1$ and an admissible sequence $\{q_1 \leqq q_2 \leqq \ldots\}$ of $\{y_n\}$-normal seminorms on Y such that

$$q_i(a^{-1}_{\sigma(n)}y_{\sigma(n)}) = p_i(x_{\pi(k_{\sigma(n)})});\ \forall i, n \geqq 1 \tag{20.2.12}$$

where $Y = [y_n]$.

PROOF. By Proposition 12.4.14, the sequence $D = \{p'_1 \leqq p'_2 \leqq \ldots\}$ is an admissible sequence of seminorms on Y, where

$$p'_i(y) = \sup_{n \geqq 1} \{|g_n(y)| p_i(y_n)\}, \quad y \in Y.$$

By Proposition 20.2.9, there exist π, σ in P, $\{k_n\} \subset \mathbb{N}$ and $\{a_n\}$ satisfying (20.2.10). Define

$$q_i(y) = \sup_{n \geqq 1} \{a_n |g_n(y)| p_i(x_{\pi(k_n)})\}; \; y \in Y, \; i \geqq 1.$$

By (20.2.10), $q_i(y) \leqq p'_i(y) \leqq K_i \, q_j(y)$ for each $i \geqq 1$ and $j \geqq i$. Hence $\{q_i\}$ is an admissible sequence of seminorms on Y. Since $|g_n(y)| q_i(y_n) = |g_n(y)| \; a_n \; p_i(x_{\pi(k_n)}) \leqq q_i(y)$, we have finished the proof. □

20.3 THE SECOND THEOREM

The discussion of this section is based on the results given in [1]. Before we proceed, we would like to offer a few comments on the structure of the space (X,T) which we are going to consider in this section. Essentially we confine ourselves to the class of those l.c. TVS (X,T) possessing an S.b. $\{x_n; f_n\}$ for which $p(x_i) \neq 0$, $i \geqq 1$ for any p in D_T and at the same time (X,T) is complete. All the familiar Köthe spaces having this property are indeed Fréchet spaces.

In the general setting of an l.c. TVS (X,T) having an S.b. $\{x_n; f_n\}$, a family D of seminorms on X will be called *complete* provided each p in D is $\{x_n\}$-normal and D generates an l.c. topology on X equivalent to T.

DEFINITION 20.3.1 An S.b. $\{x_n; f_n\}$ for an l.c. TVS (X,T) is called a (G_1) (resp. a (G_∞))-*base* if there exists an equivalent system D of seminorms on X such that $\{\{p(x_n)\}: p \in D\}$ is a Köthe set of finite (resp. infinite) type.

NOTE. The notion of a (G_1) (resp. (G_∞))-base is an extension of the (D_2) (resp. (D_1))-base introduced in [16] in metrizable spaces. In fact, let $\{x_n;f_n\}$ be a (d_o)-base corresponding to an admissible sequence $\{p_1 \leq p_2 \leq \ldots\}$ of norms on X; in addition, if the condition (Λ_1) (resp. (Λ_2)) is also satisfied, then $\{x_n;f_n\}$ is called a (D_1) (resp. (D_2))-base, where

$$(\Lambda_1) \quad \begin{cases} \text{(i)} & p_1(x_n) = 1, \ \forall n \geq 1; \\ \text{(ii)} & \forall k \ \exists \ i \ni p_k^2(x_n) \leq p_i(x_n) \ \ \forall n \geq 1. \end{cases}$$

and

$$(\Lambda_2) \quad \begin{cases} \text{(i)} & p_k(x_n) \to 1, \ \forall n \geq 1; \\ \text{(ii)} & \forall k \ \exists \ i \ni p_k(x_n)/p_i^2(x_n) \to 0. \end{cases}$$

The condition (i) in (Λ_1) and the (d_o)-character of $\{x_n;f_n\}$ shows that each (D_1)-base in a metrizable l.c. TVS is a (G_∞)-base. Also, from the (d_o)-character of $\{x_n;f_n\}$, for any $i \geq 1$,

$$\frac{p_i(x_{n+1})}{p_i(x_n)} \leq \frac{p_{i+1}(x_{n+1})}{p_{i+1}(x_n)} \leq \ldots \leq \frac{p_{i+j}(x_{n+1})}{p_{i+j}(x_n)} \to 1$$

and so a (D_2)-base is a (G_1)-base.

In the discussion which follows, we would require some properties of Köthe spaces of finite and infinite types. One of these which is easy to prove is contained in (17.3.8) and the other one is given in terms of

<u>LEMMA 20.3.2</u> Let $(\Lambda(P),T_P)$ be a nuclear Köthe set of

finite type, then for a in P

$$(*) \qquad \sum_{n\geqq 1} n^2 a_n < \infty.$$

PROOF. Let $\alpha \in \Delta(\Lambda(P))$. Then for each b in P, there exists c in P with $\alpha_n d_n(u_c,u_b) \to 0$, where $b \leqq c$. By (5.4.15),

$$d_n(u_c,u_b) \geqq \inf\{b_i/c_i : 1 \leqq i \leqq n\}$$

$$\geqq b_n/c_1.$$

Therefore $\alpha_n b_n \to 0$ for each b in P and hence by Theorem 1.4.10 (v), $n^4\ b_n \to 0$ for every b in P. Thus (*) follows. □

LEMMA 20.3.3 Let $\{x_n;f_n\}$ be a (G_1) or (G_∞)-u.e-S.b. for an ω-complete nuclear space (X,T) with D denoting an equivalent system of seminorms on X for which $\{x_n;f_n\}$ is a (G_1) or (G_∞)-base. Then for each p in D_T, there exists q in D_T such that

$$n^2 p(x_n) \leqq q(x_n), \quad \forall n \geqq 1$$

PROOF. Making use of Propositions 12.4.15 and 4.5.10 or alternatively Exercise 12.4.10, we find that $(X,T) \simeq (\Lambda(P),T_P)$ where $P = \{\{s(x_n)\}: s \in D\}$. In particular, $(\Lambda(P),T_P)$ is a nuclear Köthe space with $s(x_n) > 0$ for $n \geqq 1$ and for each s in D. If $p \in D_T$, we can find s in D with $p(x) \leqq s(x)$, for all x in X.

Let $\{x_n\} \equiv \{x_n;f_n\}$ be a (G_1)-base. We can find r in D such that $s(x_n) \leqq r^2(x_n)$. Hence by Lemma 20.3.2,

$$n^2 \sqrt{s(x_n)} \leqq n^2 r(x_n) \leqq \sum_{n\geqq 1} n^2 r(x_n) \equiv A < \infty$$

$$\Longrightarrow n^2 p(x_n) \leqq \sqrt{s(x_n)} \cdot n^2 \sqrt{s(x_n)} \leqq A r(x_n),$$

and the result is proved in this case.

Let $\{x_n\}$ be a (G_∞)-base. We find r in D and M > 0 with $n^2 \leqq Mr(x_n)$, $n \geqq 1$; cf. (17.3.8). Find t in D with $r(x_n), s(x_n) \leqq t(x_n)$; and k in D with $t^2(x_n) \leqq k(x_n)$ for $n \geqq 1$. Then

$$n^2 p(x_n) \leqq n^2 s(x_n) \leqq M\, t^2(x_n) \leqq Mk(x_n),$$

and we have finished the proof. □

<u>LEMMA 20.3.4</u> Let $\{x_n;f_n\}$ be a (G_1) or (G_∞) u.e-S.b. for an ω-complete nuclear space (X,T) with D denoting an equivalent system of $\{x_n\}$-normal seminorms for which $\{x_n\}$ is a (G_1) or (G_∞)-base. Let $\{y_n\}$ be a CSBS in (X,T) such that $\{y_n\}$ is an u.e-S.b. for $[y_n]$. Then there exist $\{k_n\} \subset \mathbb{N}$ with $k_n \to \infty$ and scalars $a_n \equiv |f_{k_n}(y_n)| > 0$ for $n \geqq 1$ such that for each p in D there exists q in D_T satisfying

$$a_n p(x_{k_n}) \leqq p(y_n) \leqq a_n q(x_{k_n}), \quad \forall n \geqq 1. \qquad (20.3.5)$$

<u>PROOF</u>. This follows exactly on the lines of the proof of Lemma 20.2.7 and in the proof here we make use of Lemma 20.3.3 in place of Lemma 20.2.1. □

<u>THEOREM 20.3.6</u> Let $\{x_n;f_n\}$, (X,T), D and $\{y_n\}$ be exactly the same as described in Lemma 20.3.4. Then there exist $\{k_n\} \subset \mathbb{N}$ with $k_n \to \infty$, scalars $a_n \equiv |f_{k_n}(y_n)| \neq 0$ and a system $D^* = \{H_p : p \in D\}$ of $\{y_n\}$-normal seminorms on $Y \equiv [y_n]$ such that

$$H_p(a_n^{-1} y_n) = p(x_{k_n}), \quad \forall n \geqq 1. \qquad (20.3.7)$$

<u>PROOF</u>. We again follow the earlier proof, namely, of Theorem 20.2.11. Here we introduce the seminorms I_p and

H_p on Y by means of the formulae:

$$I_p(y) = \sup_{n\geq 1} \{|g_n(y)|p(y_n)\};$$

$$H_p(y) = \sup_{n\geq 1} \{a_n|g_n(y)|p(x_{k_n})\},$$

where $p \in D$ and $\{g_n\}$ is the s.a.c.f. corresponding to the S.b. $\{y_n\}$ for Y. Now everything goes *verbatim* as in the proof of Theorem 20.2.11. □

<u>Corollary 20.3.8</u> Let (X,T) be an ω-complete nuclear space. If (X,T) admits a (G_1) or (G_∞)-u.e-S.b. corresponding to an equivalent family D of $\{x_n\}$-normal seminorms, then every CSBS $\{y_n\}$ which is also a u.e-S.b. for $Y = [y_n]$ is quasisimilar to a CSBS which is also a (G_1) or (G_∞)-u.e-S.b. for its closed linear span.

In fact, by Theorem 20.3.6, we have (20.3.7). Since $k_n \to \infty$, there exists π in P such that $\{k_{\pi(n)}\}$ is non-decreasing. Let $z_n = a^{-1}_{\pi(n)} y_{\pi(n)}$, then

$$H_p(z_n) = p(x_{k_{\pi(n)}}).$$

It is clear that $\{z_n\}$ is a CSBS in (X,T) since $\{y_n\}$ is also a u-S.b. and a u.e-S.b. for $Z = [z_n] = Y$. If $\{x_n\}$ is a (G_1) (resp. (G_∞))-base corresponding to D, then so is $\{z_n\}$ corresponding to $\{H_p : p \in D\}$. Finally, observe that

$$\text{conv. of } \sum_{n\geq 1} \alpha_n y_{\pi(n)} \iff \text{conv. of } \sum_{n\geq 1} a_{\pi(n)} \alpha_n z_n.$$

For the final theorem of this section we need some more information on the diametral dimensions of Köthe spaces. We reproduce these two lemmas for the sake of completeness. First, we have (cf. [218], p. 88)

<u>LEMMA 20.3.9</u> For a G_∞-space $(\Lambda(P),T_P)$ which is also Schwartz, $\Delta(\Lambda(P)) = \Lambda(P)^*$.

PROOF. Since $(\Lambda(P),T_P)$ is an AK-space, $\Lambda(P)^*$ can be taken to be the class of all sequences α in ω such that $|\alpha_n| \leq M\, a_n$, $n \geq 1$ for some a in P and a positive constant $M > 0$ (cf. Theorem 1.3.13).

We first prove that $P \subset \Delta(\Lambda(P))$ and since $\Delta(\Lambda(P))$ is normal, it follows that $\Lambda(P)^* \subset \Delta(\Lambda(P))$. Let $a \in P$. To prove that $a \in \Delta(\Lambda(P))$, let b be an arbitrary member of P. We can find d in P, $a \leq d$, $b \leq d$ such that a_n/d_n, $b_n/d_n \to 0$ and so $a_n b_n/d_n^2 \to 0$. Also, there exists e in P with $d_n^2 \leq e_n$, $n \geq 1$ and let $c = \max(b,e)$. Then $a_n b_n/c_n \to 0$. If $\varepsilon > 0$ is given, there exists k such that $b_n/c_n \leq \varepsilon/a_n \leq \varepsilon/\inf\{a_n : n \geq k\}$ for $n \geq k$, or, $\sup\{b_n/c_n : n \geq k\} \leq \varepsilon/a_k$. By Proposition 5.4.5, $\delta_k(u_c,u_b) \leq \varepsilon/a_k$, where $u_b = \{x \in \Lambda(P) : p_b(x) \leq 1\}$. Hence $a_n\delta_n(u_v,u_b) \to 0$; that is, $a \in \Delta(\Lambda(P))$.

On the other hand, let $\alpha \in \Delta(\Lambda(P))$. Then for a in P we can find b in P, $a \leq b$ such that $\alpha_n\delta_n(u_b,u_a) \to 0$. Let $M = \sup\{|\alpha_n|\delta_n(u_b,u_a) : n \geq 1\}$. By Lemma 5.4.14, $\delta_n(u_b,u_a) \geq \inf\{a_i/b_i : 1 \leq i \leq n\} \geq a_1/b_n$ and so $|\alpha_n| \leq (M/a_1)b_n$, $n \geq 1$. Hence $\alpha \in \Lambda(P)^*$. □

LEMMA 20.3.10 Let a G_1-space $(\Lambda(Q),T_Q)$ be nuclear. Then $\Delta(\Lambda(Q)) = \Lambda(Q)$.

PROOF. Let $\alpha \in \Delta(\Lambda(Q))$. Consider a in Q. Proceeding as in the last paragraph, $\{\alpha_n b_n\} \in \ell^\infty$ for each b in Q. Now choose b in Q with $a_n \leq b_n^2$, $n \geq 1$. Thus

$$\sum_{n\geq 1} |\alpha_n|\, a_n \leq \sup\{|\alpha_n| b_n\} \sum_{n\geq 1} b_n < \infty,$$

by Lemma 20.3.2 and hence $\alpha \in \Lambda(Q)$.

Conversely, let $\alpha \in \Lambda(Q)$. Choose any a in Q. Find b in Q with $a_n \leq b_n^2$. Then find c in Q such that

$$\sup_{i\geq n}\{b_i/c_i\} \leq \sum_{i\geq n}(b_i/c_i) \to 0 \text{ as } n \to \infty.$$

Let $d = \max(a,c)$. Then, using Proposition 5.4.5,

$$\delta_n(u_d,u_a) \leq \sup_{i\geq n} [\frac{a_i}{d_i}]$$

$$\leq b_n \sup_{i\geq n} [\frac{b_i}{c_i}]$$

$$\Longrightarrow \alpha_n\delta_n(u_d,u_a) \to 0.$$

Hence $\alpha \in \Delta(\Lambda(Q))$. □

We can now easily prove

PROPOSITION 20.3.11 Let $\{x_n;f_n\}$ and $\{y_n;g_n\}$ be two u.e-S.b. for an ω-complete nuclear space (X,T) such that both of them are either (G_1) or (G_∞)-bases. Then $\{x_n\} \sim \{y_n\}$.

PROOF. Let D_1 and D_2 be equivalent systems of seminorms on X under which $\{x_n\}$ and $\{y_n\}$ are both either (G_1) or (G_∞)-bases. Let

$$P_1 = \{\{p(x_n)\}: p \in D_1\};\ P_2 = \{\{q(y_n)\}: q \in D_2\}.$$

As mentioned in the proof of Lemma 20.3.3, $(X,T) \simeq (\Lambda(P_1), T_{P_1}) \simeq (\Lambda(P_2),T_{P_2})$. Since (X,T) is ω-complete, it is enough to prove in either of the cases that

(*) $\Lambda(P_1) = \Lambda(P_2)$.

(I) Let both of the bases be (G_1)-bases. Since $\Delta(\Lambda(P_1)) = \Delta(\Lambda(P_2))$, $\Lambda(P_1) = \Lambda(P_2)$ by Lemma 20.3.10.

(II) Let now $\{x_n\}$ and $\{y_n\}$ be (G_∞)-bases. Then by Lemma 20.3.9, $\Lambda(P_1)^* = \Lambda(P_2)^*$. Let $\alpha \in \Lambda(P_1)$. Consider any a in P_2, then $a \in \Lambda(P_2)^*$ and so for some $M > 0$ and b in P_1, $a_n \leq Mb_n$. This shows that $\alpha \in \Lambda(P_2)$, giving $\Lambda(P_1) \subset \Lambda(P_2)$. Similarly $\Lambda(P_2) \subset \Lambda(P_1)$. □

Summarizing the results of this section, finally we have

THEOREM 20.3.12 Let an ω-complete nuclear space (X,T) possess an u.e-S.b. $\{x_n;f_n\}$ which is also a (G_1) or (G_∞)-base corresponding to an equivalent family D of $\{x_n\}$-normal seminorms. Then each u.e-S.b. $\{u_n;h_n\}$ for (X,T) is quasisimilar to any other u.e-S.b. $\{y_n;g_n\}$.

PROOF. By Corollary 20.3.8, $\{u_n;h_n\}$ is quasisimilar to a u.e-S.b. $\{u_n';h_n'\}$ which is either a (G_1) or a (G_∞)-base depending on which $\{x_n;f_n\}$ is. Similarly, $\{y_n;g_n\}$ is quasisimilar to a u.e-S.b. $\{y_n';g_n'\}$ which is either a (G_1) or a (G_∞)-base depending on which $\{x_n;f_n\}$ is. By Proposition 20.3.11, $\{u_n'\} \sim \{y_n'\}$ and hence the result. □

20.4 PSEUDO-(d_o)-BASES

In this last section, we will deal with the possible extension of (d_o)-bases. In order to show that a given S.b. in a Fréchet space is not a (d_o)-base, it appears to be a very cumbersome exercise to disprove the inequality in Propsotiion 19.2.2 for every admissible sequence of norms. To overcome this difficulty and, at the same time to keep in view the obtaining of a possible solution to Problem 19.3.3, a notion of pseudo-(d_o)-bases was introduced in [28] which we follow hereafter in this section. Since we will be dealing only with nuclear Fréchet spaces having continuous norms, we assume hereafter in this section that (X,T) is an arbitrary nuclear Fréchet space whose topology T is always given by a countable sequence $D_T = \{p_1 \leq p_2 \leq \ldots\}$ of nondecreasing norms. Now we have

DEFINITION 20.4.1 An S.b. $\{x_n;f_n\}$ for (X,T) is called a *pseudo*-(d_o)-*base* provided there exists an admissible sequence $D_T = \{p_1 \leq p_2 \ldots\}$ of norms on X such that the following condition is satisfied:

$$\begin{cases} \forall i \ \exists j \ni \forall k > j \quad \exists \ell > i \text{ and } M > 0 \text{ with} \\ p_i(x_n)/p_\ell(x_n) \leq M\, p_j(x_m)/p_k(x_m), \quad \forall m \leq n. \end{cases} \tag{20.4.2}$$

NOTE. Comparing (20.4.2) with the inequality of Proposition 19.2.2, it is obvious that each (d_o)-base is a pseudo-(d_o)-base. Unlike possibly (d_o)-bases, once the inequality (20.4.2) is satisfied for one admissible family, it is satisfied for all admissible families. More precisely, we have

PROPOSITION 20.4.3 Let $\{x_n;f_n\}$ be a pseudo-(d_o)-base for (X,T) corresponding to an admissible sequence $D_1 = \{p_1 \leq p_2 \leq \ldots\}$ of norms on X. Then $\{x_n;f_n\}$ is a pseudo-(d_o)-base for any other admissible sequence $D_2 = \{q_1 \leq q_2 \leq \ldots\}$ of norms on X.

PROOF. Let $r \in \mathbb{N}$. There exists i in $\mathbb{N}$, $i > r$ such that $q_r \leq M_r p_i$ for some $M_r > 0$. Next we have integers, j,k,ℓ in $\mathbb{N}$ and a constant $M > 0$ satisfying (20.4.2). For j in $\mathbb{N}$ as described just now, choose $s > j$ in $\mathbb{N}$ and a constant $M_j > 0$ such that $p_j \leq M_j q_s$. Let $t > s$. We can find k_o in $\mathbb{N}$, $k_o \equiv k_o(t) > t(> j)$ and a constant $M_t > 0$ such that $q_t \leq M_t p_{k_o}$ Finally, pick up u in $\mathbb{N}$, $u > \ell$ and a constant $M_\ell > 0$ with $p_\ell \leq M_\ell\, q_u$. Observe that $u > r$ (indeed, $u > \ell > i > r$). Hence, for $m \leq n$

$$\frac{q_r(x_n)}{q_u(x_n)} \leq M_r\, M_\ell\, \frac{p_i(x_n)}{p_\ell(x_n)} \leq MM_rM_\ell\, \frac{p_j(x_m)}{p_{k_o}(x_m)}$$

$$\leq MM_rM_\ell\, M_jM_t\, \frac{q_s(x_m)}{q_t(x_m)}\,.$$

The proof is finished after putting $N = MM_rM_\ell\, M_jM_t$. □

The next result is comparable with Theorem 19.2.4.

PROPOSITION 20.4.4 Let $\{x_n;f_n\}$ be a pseudo-(d_o)-base for (X,T). If $\{y_n;g_n\}$ is another S.b. for (X,T), then tbere exists σ in P such that $\{y_{\sigma(n)};g_{\sigma(n)}\}$ is a pseudo-(d_o)-base for (X,T).

PROOF. Since $\{x_n;f_n\}$ is a pseudo-(d_o)-base, there exists an admissible sequence $\{p_i\}$ of norms on X satisfying (20.4.2). Let us invoke both the notation and statement of Theorem 20.2.11. Clearly $\{y_{\sigma(n)};g_{\sigma(n)}\}$ is an S.b. for (X,T). If $u_n = y_{\sigma(n)}$, then from (20.2.12) and (20.4.2), we obtain:

$$\forall i \ \exists \ j \ni \forall k > j \ \exists \ \ell > i \text{ and } M > 0 \text{ with}$$

$$q_i(u_n)/q_\ell(u_n) \leq Mq_j(u_m)/q_k(u_m), \ \forall m \leq n$$

and this proves the result. □

The main theorem

To prove the main theorem of this section, we first reproduce two lemmas from [29] which also provide an alternative proof of Theorem 19.3.5.

LEMMA 20.4.5 For each $p = 1,2,\ldots$, let a^p and $b^p \in \omega$ with $a^p_n, b^p_n > 0$ such that $a^p \cdot b^q \in \ell^\infty$ for all $p,q = 1,2,\ldots$. Then there exists $d \in \omega$, $d_n > 0$, $n \geq 1$ such that $a^p \cdot d$, $b^p/d \in \ell^\infty$ for all p.

PROOF. Let us first define A^p, B^p in ω for $p = 1,2,\ldots$. In fact, put $A^1 = a^1$ and $B^1 = c_1 b^1$ where $c_1 > 0$ is chosen so that $A^1_n B^1_n \leq 1$ for $n \geq 1$ (by the hypothesis). Assume the existence of A^i and B^i for $i = 1,\ldots,p-1$. Let $A^p = c'_p a^p$, where $c'_p > 0$ is chosen so that $A^p_n B^i_n \leq 1$, for $n \geq 1$ and $i = 1,\ldots,p-1$. Next, let $B^p = c_p b^p$, where $c_p > 0$ is chosen so that $A^i_n B^p_n \leq 1$, for $n \geq 1$ and $i = 1,\ldots,p$. Thus we have

$$A_n^p B_n^q \leqq 1; \quad \forall p,q,n \geqq 1.$$

Finally, let

$$d_n \equiv \sup_{q \geqq 1} B_n^q \leqq \inf_{p \geqq 1} \frac{1}{A_n^p},$$

and this yields the required sequence $\{d_n\}$. □

The proof of the next lemma depends upon the following result on sequence spaces.

PROPOSITION 20.4.6 Let $a^p \in \omega$ for $p = 1,2,\ldots$, where $a_1^{p+1} \geqq a_i^p > 0$; $i,p \geqq 1$. Then for each fixed $p \geqq 1$

$$[\bigcap_{q \geqq 1} \frac{a^p}{a^q} c_0]^\times = \bigcup_{q \geqq 1} \frac{a^q}{a^p} \ell^1.$$

PROOF. We have

$$\bigcup_{q \geqq 1} \frac{a^q}{a^p} \ell^1 \subseteq [\bigcap_{q \geqq 1} \frac{a^p}{a^q} c_0]^\times.$$

If the preceding inclusion is strict, then there exists y in ω such that

$$(*) \qquad y \in [\bigcap_{q \geqq 1} \frac{a^p}{a^q} c_0]^\times ; y \notin \frac{a^q}{a^p} \ell^1, \quad \forall q \geqq 1.$$

The second estimate yields a strictly increasing sequence $\{n_q\}$ such that

$$\sum_{i=n_{q-1}+1}^{n_q} (a_i^p/a_i^q)|y_i| > 2^{2q}, \; q = 1,2,\ldots, n_o = 0.$$

Define x in ω by

$$x_i = \frac{a_i^p}{a_i^q} \cdot \frac{1}{2^q}; \; n_{q-1} < i \leqq n_q, \; q \geqq 1$$

and let $z^q = (a^q/a^p)x$. Fix q_o in N. Since

$$z_i^{q_o} = \frac{a_i^{q_o}}{a_i^q} \cdot \frac{1}{2^q} \; ; \; n_{q-1} < i \leqq n_q, \; q \geqq 1$$

we find that $z_i^{q_o} \leqq 1/2^q$ for $n_{q-1} < i \leqq n_q$, $q \geqq q_o$ and so $z^{q_o} \in c_o$. As $q_o \geqq 1$ is arbitrary,

$$x \in \bigcap_{q \geqq 1} \frac{a^p}{a^q} c_o.$$

But the inequality

$$\sum_{i \geqq 1} |x_i y_i| \geqq \sum_{i=n_{q-1}+1}^{n_q} \frac{a_i^p}{a_i^q} \cdot \frac{1}{2^q} |y_i| > 2^q, \; q \geqq 1$$

violates the first condition in (*) and so the result is proved. □

Let $\lambda = \Lambda(P_1)$ be the Köthe space with $P_1 = \{a^p : p \geqq 1\}$, where a^p is defined as above. Similarly, if $b^p \in \omega$, $p \geqq 1$ with $b_i^{p+1} \geqq b_i^p > 0$, we put $\mu = \Lambda(P_2)$, where $P_2 = \{b^p\}$. Now we have

LEMMA 20.4.7 Let

$$\bigcup_{p \geqq 1} \bigcap_{q \geqq 1} \frac{a^p}{a^q} c_o = \bigcup_{r \geqq 1} \bigcap_{s \geqq 1} \frac{b^r}{b^s} c_o.$$

Then there exists d in ω with $d_n > 0$, $n \geqq 1$ such that $\lambda = d.\mu$.

PROOF. We have

$$\bigcap_{q \geqq 1} \frac{a^p}{a^q} c_o \subset \bigcup_{r \geqq 1} \bigcap_{s \geqq 1} \frac{b^r}{b^s} c_o, \; \forall p \geqq 1$$

$$\Longrightarrow \forall p \geqq 1 \; \exists \; r(p) \in \mathbb{N} \ni \bigcap_{q \geqq 1} \frac{a^p}{a^q} c_o \subset \bigcap_{s \geqq 1} \frac{b^{r(p)}}{b^s} c_o.$$

Therefore, using Proposition 20.4.6, we conclude that

$$\forall p \geqq 1 \ \exists \ r(p) \ni \forall s \quad \frac{b^s}{b^{r(p)}} \ell^1 \subset \bigcup_{q \geqq 1} \frac{a^q}{a^p} \ell^1$$

$$\Longrightarrow \forall p \ \exists \ r(p) \ni \forall s \ \exists \ q(p,s) \ni \frac{a^p}{b^{r(p)}} \ell^1 \subset \frac{a^{q(p,s)}}{a^p} \ell^1,$$

(*) or, $\forall p \ \exists \ r(p) \ni \forall s \ \exists \ q(p,s) \ni \dfrac{a^p}{b^{r(p)}} \cdot \dfrac{b^s}{a^{q(p,s)}} \in \ell^\infty .$

Similarly, we find that

(+) $\forall p \ \exists \ r'(p) \ni \forall s \ \exists \ q'(p,s) \ni \dfrac{b^p}{a^{r'(p)}} \cdot \dfrac{a^s}{b^{q'(p,s)}} \in \ell^\infty .$

For $p \geqq 1$, let us write

$$R(p) = \max \{r(p), q'(1,p), q'(2,p), \ldots, q'(p,p)\};$$

$$R'(p) = \max \{r'(p), q(1,p), q(2,p), \ldots, q(p,p)\}.$$

Then

$$\frac{a^p}{b^{R(p)}} \cdot \frac{b^s}{a^{R'(s)}} \in \ell^\infty ; \quad \forall p, s \geqq 1.$$

Indeed, if $p \geqq s$ then $R(p) \geqq q'(s,p)$, $R'(s) \geqq r'(s)$. If $s \geqq p$, then $R(p) \geqq r(p)$, $R'(s) \geqq q(p,s)$. Hence

$$\frac{a^p}{b^{R(p)}} \ \frac{b^s}{a^{R'(s)}} \leqq \frac{a^p}{b^{q'(s,p)}} \ \frac{b^s}{a^{r'(s)}} ;$$

$$\frac{a^p}{b^{R(p)}} \ \frac{b^s}{a^{R'(s)}} \leqq \frac{a^p}{b^{r(p)}} \ \frac{b^s}{a^{q(p,s)}} ,$$

and make use of (*) and (+).

Finally, make use of Lemma 20.4.5 to get d in ω, $d_n > 0$ ($n \geqq 1$) such that

$$\frac{a^p}{b^{R(p)}}\cdot d \in \ell^\infty \quad ; \forall p \geq 1, \quad \frac{b^s}{a^{R'(s)}} / d \in \ell^\infty ; \forall s \geq 1.$$

The last expressions clearly imply that $\lambda = d.\mu$. In fact, let x in λ and s in $\mathbb{N}$ be chosen arbitrarily. Then

$$\sum_{i\geq 1} b_i^s \frac{|x_i|}{d_i} \leq \sup_{i\geq 1} \left(\frac{b_i^s}{a_i^{R'(s)} d_i}\right) \sum_{i\geq 1} a_i^{R'(s)} |x_i| < \infty$$

$$\Longrightarrow \frac{x}{d} \in \mu \Longrightarrow \lambda \subset d.\mu.$$

Similarly $d.\mu \subset \lambda$. □

We require a few more computations on sequence spaces in the form of

PROPOSITION 20.4.8 Let $a^p \in \omega$ with $a_i^{p+1} \geq a_i^p > 0$ for $i, p \geq 1$. If $\lambda = \cap \{(1/a^p)\ \ell^1 : p \geq 1\}$, then $\lambda^\times = \cup\{a^p\ \ell^\infty : p \geq 1\}$. If in addition, $\sum_{n\geq 1} (a_n^p/a_n^q) < \infty$ whenever $p < q$, then

$$(*) \qquad \lambda \cdot \lambda^\times = \bigcup_{q\geq 1} \bigcap_{r>q} \left(\frac{a^q}{a^r}\right) c_o.$$

PROOF. For the first part, let us observe that

$$\bigcup_{p\geq 1} a^p\ \ell^\infty \subset \lambda^\times.$$

Let $y \in \lambda^\times$ but $y \notin \cup\{a^p\ \ell^\infty : p \geq 1\}$. Then there exists $\{n_p : p \geq 1\} \in I$ such that $(a_{n_p}^p)^{-1} |y_{n_p}| > 2^{2p}$, $p \geq 1$. Let

$$x_i = \begin{cases} 1/2^p\ a_i^p, & i = n_p,\ p \geq 1; \\ 0\ , & \text{otherwise.} \end{cases}$$

For any p_o in $\mathbb{N}$,

$$\sum_{i\geq 1} a_i^{p_o}|x_i| = \Big(\sum_{p<p_o} + \sum_{p\geq p_o}\Big) a_{n_p}^{p_o} \cdot \frac{1}{2^p\, a_{n_p}^p} < \infty ,$$

and so $x \in \lambda$. But

$$\sum_{i\geq 1} |x_i y_i| = \sum_{p\geq 1} \frac{1}{2^p\, a_{n_p}^p} |y_{n_p}| \geq \sum_{p\geq 1} 2^p = \infty,$$

and hence a contradiction.

For the second part, let x be an element of the right-hand s.s. in (*) above. Then $x \in (a^q/a^r)c_o$, for some $q \geq 1$ and each $r > q$. Since c_o is normal, $x \in (a^q/a^r)c_o$ for each $r \geq 1$. We can find y^r in c_o and a sequence $\{n_r\}$ $(n_{r-1} < n_r$, $n_o = 0)$ so that

$$x = \frac{a^q}{a^r} y^r,\ r \geq 1;\ |y_n^r| < \frac{1}{2^r},\ \forall n > n_r,\ n \geq 1.$$

Define u,v in ω with

$$u_n = \frac{1}{2^r a_n^r}\ ;\ v_n = 2^r y_n^r a_n^q,\ n_r+1 \leq n \leq n_{r+1},\ r \geq 1$$

and for $1 \leq n \leq n_1$, choose u_n, v_n so that $u_n v_n = x_n$.

The first claim is that $a^p u \in \ell^1$, for $p \geq 1$. In fact, for p in $\mathbb{N}$, choose R in $\mathbb{N}$ so that $a^p/a^R \in \ell^1$. Then

$$\sum_{n\geq 1} \frac{a_n^p}{a_n^r} \leq \sum_{n\geq 1} \frac{a_n^p}{a_n^R} \equiv B_{p,R},\ \forall r \geq R$$

$$\Longrightarrow \sum_{n=n_r+1}^{n_{r+1}} \frac{a_n^p}{a_n^r} \leq B_{p,R},\ \forall r \geq R$$

$$\Longrightarrow \sum_{n\geq 1} a_n^p u_n \leq \sum_{n=1}^{n_R} a_n^p u_n + B_{p,R} \sum_{r\geq R} \frac{1}{2^r} < \infty .$$

Secondly, $v \in a^q\, \ell^\infty \subset \lambda^\times$. Finally, $u_n v_n = x_n$, for $n > n_1$. Thus $x = u.v$ with $u \in \lambda$ and $v \in \lambda^\times$ and this shows that

$$\bigcup_{q\geq 1} \bigcap_{r>q} \frac{a^q}{a^r} c_o \subset \lambda.\lambda^{\times}.$$

The reverse inclusion is obviously true. □

PROPOSITION 20.4.9 Let $\{x_n;f_n\}$ be an S.b. for (X,T). Then, corresponding to $D_T = \{p_i\}$ given by (19.2.12), the inverse diametral dimension $\Delta^*(X)$ is given by

$$\Delta^*(X) = \bigcup_{i\geq 1} \bigcap_{j>i} \left(\frac{a^i_\pi}{a^j_\pi}\right) c_o,$$

where $a_i = \{p_i(x_n) : n \geq 1\}$, $\pi \in P$ and $a^i_\pi/a^j_\pi = \{a^i_{\pi(1)}/a^j_{\pi(1)} \geq a^i_{\pi(2)}/a^j_{\pi(2)} \geq \ldots\} \in \ell^1$, π being dependent upon i,j.

PROOF. We obviously have (for u,v in $\mathcal{B}_X$)

$$\Delta^*(X) = \{\alpha \in \omega: \exists\ u\ \forall v < u,\ \alpha \in \delta(v,u)c_o\}.$$

where $\delta(u,v) = \{\delta_n(v,u)\}$. Since T is also generated by the countable norms given by (19.2.12), we have

$$\Delta^*(X) = \{\alpha \in \omega: \exists\ u_i\ \forall u_j < u_i,\ \alpha \in \delta(u_j,u_i)c_o\}.$$

$u_i = \{x \in X : p_i(x) \leq 1\}$, where p_i is given by (19.2.12). But $i < j \iff p_i < p_j \iff u_j \subset u_i$ and so

$$\Delta^*(X) = \{\alpha \in \omega: \exists\ i\ \forall j > i,\ \alpha \in \delta(u_j,u_i)c_o\}.$$

Also $\{p_i(x_n)/p_j(x_n)\} \in \ell^1$ and so for some π in P, $p_i(x_{\pi(n+1)})/p_j(x_{\pi(n+1)}) \leq p_i(x_{\pi(n)})/p_j(x_{\pi(n)})$, $n \geq 1$. Following the proof of Corollaries 5.4.7 and 5.4.16, we find

$$\delta_n(u_j,u_i) = \frac{p_i(x_{\pi(n)})}{p_j(x_{\pi(n)})},$$

and we have completed the proof. □

PROPOSITION 20.4.10 Let $\{x_n; f_n\}$ be a pseudo-(d_o)-base for (X,T) corresponding to an admissible sequence $D_T = \{p_1 \leq p_2 \leq \ldots\}$ of norms on X satisfying (19.2.12). Let $a^i = \{p_i(x_n)\}$ and $\lambda = \cap \{(1/a^i)\ell^1 : i \geq 1\}$. Then

$$\Delta^*(X) = \lambda \cdot \lambda^\times .$$

PROOF. By Theorem 19.2.3, Propositions 20.4.8 and 20.4.9,

$$\lambda \cdot \lambda^\times = \bigcup_{j \geq 1} \bigcap_{k > j} \left(\frac{a^j}{a^k}\right) c_o ; \quad \Delta^*(X) = \bigcup_{i \geq 1} \bigcap_{\ell > i} \left(\frac{a^i_\pi}{a^\ell_\pi}\right) c_o ,$$

where π in P is the one mentioned in Proposition 20.4.9.

Since $\{x_n; f_n\}$ is a pseudo-(d_o)-base,

$$\forall i \; \exists \; j \ni \forall k > j \; \exists \; \ell > i \text{ and } M > 0 \text{ with}$$

$$(+) \qquad \frac{a^i_n}{a^\ell_n} \leq M \frac{a^j_m}{a^k_m} , \quad \forall m \leq n.$$

There exists π in P for which a^i_π / a^ℓ_π is a decreasing sequence and

$$(*) \qquad \frac{a^i_{\pi(n)}}{a^\ell_{\pi(n)}} \leq M \frac{a^j_n}{a^k_n} , \quad \forall n \geq 1.$$

(*) clearly holds for $n = 1$. Let $n > 1$. If $\pi(n) \geq n$, (*) trivially follows from (+). If $\pi(n) < n$, there exists $m < n$ with $\pi(m) \geq n$. Since $a^i_{\pi(n)}/a^\ell_{\pi(n)} \leq a^i_{\pi(m)}/a^\ell_{\pi(m)}$, we have (*) again by virtue of (+). Now observe that (*) easily yields the inclusion $\Delta^*(X) \subset \lambda.\lambda^\times$.

Similarly, there exists σ in P such that $\{a^j_{\sigma(n)}/a^k_{\sigma(n)}\}$ decreases and

$$(**)\qquad \frac{a_n^i}{a_n^\ell} \leq M\, \frac{a^j_{\sigma(n)}}{a^k_{\sigma(n)}}, \quad \forall n \geq 1.$$

(**) easily leads to $\lambda.\lambda^\times \subset \Delta^*(X)$. □

PROPOSITION 20.4.11 Any two pseudo-(d_o)-bases $\{x_n;f_n\}$ and $\{y_n;g_n\}$ for (X,T) are semisimilar.

PROOF. Let $D_1 = \{p_1 \leq p_2 \leq \ldots\}$ and $D_2 = \{q_1 \leq q_2 \leq \ldots\}$ be the admissible sequences of norms on X for which $\{x_n\}$ and $\{y_n\}$ are pseudo-(d_o)-bases, it being understood that p_i satisfies (19.2.12) for $i \geq 1$ and similarly for $i \geq 1$

$$q_i(x) = \sum_{n\geq 1} |g_n(x)|q_i(x_n), \ \forall x \in X.$$

Put $a^i = \{p_i(x_n) : n \geq 1\}$ and $b^j = \{q_j(y_n) : n \geq 1\}$; also, let

$$\lambda = \bigcap_{i\geq 1} \left(\frac{1}{a^i}\right) \ell^1; \quad \mu = \bigcap_{j\geq 1} \left(\frac{1}{b^j}\right) \ell^1.$$

By using the normal character of c_o (as in the proof of the second part of Proposition 20.4.8) and Lemma 20.4.7 as well as Proposition 20.4.10, there exists d in ω with $d_n > 0$ for $n \geq 1$ such that $\lambda = d.\mu$. Hence

$$\text{conv. of } \sum_{n\geq 1} a_n x_n \iff \text{conv. of } \sum_{n\geq 1} a_n d_n^{-1} y_n,$$

and this completes the proof. □

Finally, we have the following theorem which is analogous to Theorem 19.3.5 and provides a solution to Problem 19.3.3

THEOREM 20.4.12 Let (X,T) possess a pseudo (d_o)-base. Then all S.b's for (X,T) are quasisimilar.

PROOF. Make use of Propositions 20.4.4 and 20.4.11. □

21 ℓ^1-bases and nuclearity: Wojtyński's theorem

21.1 INTRODUCTION

In this last chapter, we intend to present one of the deepest results of the basis theory (Theorem 21.1.1) on the existence of conditional bases in certain non-nuclear Köthe-Fréchet spaces. In particular, this result yields another theorem (Theorem 21.1.2), giving sufficient conditions on a Fréchet space with an S.b. for the space to be a nuclear space and, as we will see later, these conditions also turn out to be necessary ones (Theorem 21.7.1). Thus, we are in the pleasant situation of having another characterization of nuclearity of a class of spaces in terms of bases, the only other one we have discussed so far being contained in Theorem 12.4.6 (cf. also Theorem 17.4.1).

We follow [235] hereafter to prove the main results of this chapter contained in

THEOREM 21.1.1 Corresponding to $p \geq 1$, each non-nuclear Köthe space $\ell^p[a_{mn}]$ possesses an S.b. which is not an u-S.b.

THEOREM 21.1.2 If all bases of a Fréchet space X with an S.b. are ℓ^1-bases, then X is nuclear.

REMARKS. The proof of the basic theorem is quite cumbersome and requires several intermediary results. In the meantime, we may sum up the outline governing the desired proof as follows: (a) estimation of (i) upper bounds of the base constants K and (ii) lower bounds of the unconditional base constants K_u of different bases of ℓ_n^p, $n \in \mathbb{N}$; $1 \leq p \leq \infty$, (b) establishing sufficient conditions for the nuclearity of vector-valued Köthe-Fréchet (v.v.K-F) spaces and (c) construction of non-nuclear v.v.K-F spaces from the non-nuclear Köthe-Fréchet spaces. Finally, the basic theorem is proved by using a contradiction method, thereby constructing an S.b. in a certain v.v.K-F space isomorphic to a non-nuclear subspace of $\ell^p[a_{mn}]$ with the help of (a) (i) and (c). The required contradiction is arrived at by assuming the u-S.b. nature of this base and making use of (b) with the help of (a) (ii).

Part of (a) (i) is, however, also discussed in the next section while the rest of the outline will be taken up in subsequent sections in the order (a), (b) and (c).

21.2 PREPARATORY BACKGROUND

This section is devoted to providing ourselves with the immediate prerequisites, including terminology which will be needed in proving several lemmas and a few theorems in subsequent sections, that will ultimately help us prove the basic theorem, namely, Theorem 21.1.1.

Spaces $\ell^p[P]$ and $\ell^p[P;X_n]$

Without further reference, $X \equiv (X,T)$ will denote an arbitrary Fréchet space with $D_T = \{p_1 \leq p_2 \leq \ldots\}$ and P, Q, etc. are arbitrary countable Köthe sets; for instance, we write $P = \{a^m : m \geq 1\}$ with $a_{m,n} \equiv a_{mn} = a_n^m$, m, n ≥ 1 and in such a situation P will henceforth be called a *Köthe* matrix. If $a_{m+1,n} \geq a_{mn}$, for m,n ≥ 1, then $P \equiv [a_{mn}]$ is called a *monotone Köthe matrix* (m.K.m.). It is understood that each column has at least one non-zero term otherwise we discard this column.

If J is a finite subset of $\mathbb{N}$, we write

$$\ell^p_J = \{x = \{x_i : i \in J\} : \|x\|_{J,p} \equiv$$

$$[\sum_{i \in J} |x_i|^p]^{1/p} < \infty\} , \quad 1 \leqq p < \infty;$$

and

$$\ell^\infty_J = \{x = \{x_i : i \in J\} : \|x\|_{J,\infty} \equiv \sup \{|x_i| : i \in J\} < \infty\} .$$

When J = $\mathbb{N}$, we merely write ℓ^p for $\ell^p_{\mathbb{N}}$. Also, if the omission of the suffix J does not create any ambiguity in the context, we will interchangeably write ℓ^p for ℓ^p_J and $\|\cdot\|_p$ for $\|\cdot\|_{J,p}$.

Recall the space $\Lambda_p[P]$ introduced in Proposition 5.4.12, where $P = \{\{p(x_n)\} : p \in D_T\}$. In the special case when P is a Köthe matrix, that is, replacing $\{p(x^n)\}$ by $\{a_{mn} : n \geqq 1\}$, we write $\ell^p[P]$ for $\Lambda_p[P]$. There are other reasons for doing so; for instance, $\Lambda_p[P]$ is isomorphic to ℓ^p in certain special cases. Further, the topology T_p of $\ell^p[P]$ is generated by $\{\|\cdot\|_m\}$, where for $1 \leqq p < \infty$

$$||\alpha||_m = [\sum_{n \geqq 1} a^p_{mn} |x_n|^p]^{1/p}; \ m \geqq 1, \qquad (21.2.1)$$

and if $p = \infty$, the topology T_∞ of $\ell^\infty[P]$ is generated by $\{||\cdot||_m\}$, where

$$||\alpha||_m = \sup_{n \geqq 1} \{a_{mn}|x_n|\}, \ m \geqq 1. \qquad (21.2.2)$$

<u>LEMMA 21.2.3</u> Let P be an m.K.m. and $p \geqq 1$. Then $(\ell^p[P], T_p)$ is nuclear if and only if for each $i \geqq 1$ and $s > 0$ there exists $j > i$ such that

$$\sum_{n \geqq 1} [a_{in}/a_{jn}]^s < \infty. \qquad (21.2.4)$$

Corresponding to a Köthe matrix P, any matrix $P^* = [a_{m_k n_j}]$ where $\{m_k\}, \{n_j\} \subset \mathbb{N}$ is called a *submatrix* of P.

An m.K.m. P is called *nuclear* (resp. *non-nuclear*) according to whether (21.2.4) is satisfied (resp. is not satisfied).

Hereafter, we write X_n for a Banach space with $\dim X_n < \infty$, $n \geq 1$, the norm on each X_n being designated by the same symbol $\|\cdot\|$. For $1 \leq p \leq \infty$, let us write $\ell^p[P;X_n] \equiv \ell^p([a_{mn}];X_n) = \ell^p_{\mathbb{N}}([a_{mn}]; \{X_n : n \in \mathbb{N}\})$ for the collection of all vector valued sequences $\bar{x} = \{x^n\}$ with $x^n = \{x^n_1,\ldots,x^n_{\dim X_n}\} \in X_n$, $n \geq 1$ such that for each $m \geq 1$

$$\|\bar{x}\|_m \equiv \begin{cases} [\sum\limits_{n\geq 1} (a_{mn}\|x^n\|)^p]^{1/p} < \infty, \text{ if } 1 \leq p < \infty; \\ \sup\limits_{n\geq 1} \{a_{mn}\|x^n\|\} < \infty, \text{ if } p = \infty, \end{cases} \tag{21.2.5}$$

and endow this space with the topology $\tilde{T}_p$ generated by $\{\|\cdot\|_m\}$. The earlier arguments and terminology for ℓ^p_J also apply to $\ell^p_J[P^*;X_n] \equiv \ell^p_J([a_{mn} : n \in J]; \{X_n : n \in J\})$ where $P^* = [a_{mn} : m \geq 1, n \in J]$, $J \subset \mathbb{N}$.

The spaces $\ell^p[P]$ and $\ell^p[P;X_n]$ are Fréchet spaces and all subspaces of these spaces will henceforth be assumed to be equipped with the topologies induced by T_p and $\tilde{T}_p$ respectively.

We write $\ell^p_0[P]$ for the dense subspace ϕ of $\ell^p[P]$. Similarly, $\ell^p_0[P;X_n]$ is defined to be the dense subspace of $\ell^p[P;X_n]$, consisting of $\bar{x} = \{x^n\}, x^n$ being the zero vector of X_n for all but a finite number of indices n.

<u>LEMMA 21.2.6</u> If P is an m.K.m. and $\ell^p[P]$ is nuclear, $p \geq 1$, then $\ell^1[P] \simeq \ell^p[P]$ under the identity map and a *fortiori*, $\ell^1_0[P] \simeq \ell^p_0[P]$ under the same map.

<u>PROOF</u>. Fix $p \geq 1$ and find q with $1/p + 1/q = 1$. Now apply Hölder's inequality and (21.2.4) with $s = q$. □

Next, let $A_n : X_n \to X_n$ be a linear operator and define for a given $p \geq 1$ and an m.K.m. P, the operator

$$\begin{cases} \bigoplus_{n\geq 1} A_n : \ell^p[P;X_n] \to \ell^p[P;X_n], \\ \bigoplus_{n\geq 1} A_n(\bar{x}) = \{A_n(x^n) : n \geq 1\}. \end{cases} \tag{21.2.7}$$

It is easily seen that if $\|A_n\| \leq K$ for all $n \geq 1$, then $\bigoplus_{n\geq 1} A_n$ is continuous. Let I be the identity operator of X_n, $n \geq 1$ and $\alpha > 0$. Write D^α for $\bigoplus_{n\geq 1} A_n$ if $A_n = (n \dim X_n)^\alpha I$ and if $\dim X_{n+1} \geq \dim X_n \geq n$ and $A_n = (\dim X_n)^\alpha I$, let us write B^α for $\bigoplus_{n\geq 1} A_n$.

Auerbach bases

Every finite dimensional space $(E, \|\cdot\|)$ has an S.b., but the well-known Auerbach lemma suggests (e.g. [178], p. 135) that we can find an S.b. $\{x_n;f_n\}$ for E, called an *Auerbach base* such that $\|x_n\| = \|f_n\| = 1$ for $n = 1,2,\ldots,N \equiv N_E = \dim E$. For x in E, let

$$\|x\|_1 = \sum_{i=1}^{N} |f_i(x)| \; ; \; \|x\|_\infty = \sup_{1\leq i\leq N} |f_i(x)|. \tag{21.2.8}$$

Then

$$\begin{cases} \|x\|_\infty \leq \|x\| \leq \|x\|_1 : \\ \|x\|_1 \leq N \|x\|_\infty, \end{cases} \tag{21.2.9}$$

and as expected, all these three norms on E are equivalent. Write $\tilde{E}$ for E equipped with the norm $\|\cdot\|_1$

LEMMA 21.2.10 Let $P \equiv [a_{mn}]$ be an m.K.m. and $K > 0$ be a constant such that to every $i \geq 1$ there exists $j \geq 1$ with

$$a_{in}(\dim X_n) \leq K\, a_{jn}, \quad \forall n \geq 1 \tag{21.2.11}$$

Then $\ell^1[P;X_n] \simeq \ell^1[P;\tilde{X}_n]$ under the identity map I_p.

PROOF. By (21.2.9), for each x^n in X_n and $n \geqq 1$,

$$a_{in} \|x^n\| \leqq a_{in} \|x^n\|_1 \leqq K\, a_{jn} \|x^n\|$$

and so

$$\|\bar{x}\|_i \leqq \|I_p(\bar{x})\|_{1,i} \leqq K \|\bar{x}\|_j . \quad \square$$

Bases in ℓ^p and in its truncations

In this subsection we construct bases in ℓ_n^p $(1 < p < \infty)$ and find upper bounds for the corresponding base constants (b.c.). The case $p = 2$ is irrelevant here, for this is replaced when we confine ourselves to Babenko bases in ℓ^2 – in fact in L^2. An upper bound of the b.c. in ℓ_n^1 is covered in the discussion on a specific S.b. $\{d^k;f^k\}$ for ℓ^1 in a subsequent section.

To begin with, let us rephrase some of the basic ideas on bases in the setting of normed spaces. Let us therefore consider an S.b. $\{x_k;f_k\}$ for a Banach space $(B, \|\cdot\|)$. For brevity, denote by σ_n the set $\{1,2,\ldots,n\}$ and let

$$\begin{aligned} K^n\{x_k\} &= \sup\,\{\,\|S_i\| : 1 \leqq i \leqq n\}; \\ K_u^n\{x_k\} &= \sup\,\{\,\|S_\sigma\| : \sigma \subset \sigma_n\} \end{aligned} \qquad (21.2.12)$$

[If $N \equiv \dim B < \infty$, then $x_i = 0$ for $i > N$ and so

$$K^n\{x_k\} \equiv K^n\{x_k : 1 \leqq k \leqq N\} = \sup\{\|S_i\| : 1 \leqq i \leqq n\};$$

$$K_u^n\{x_k\} \equiv K_u^n\{x_k : 1 \leqq k \leqq N\} = \sup\{\|S_\sigma\| : \sigma \subset \sigma_n\},$$

where $1 \leqq n \leqq N$ and $K^n\{x_k\} = K^N\{x_k\}, K_u^n\{x_k\} = K_u^N\{x_k\}$, $n \geqq N$.]
Since $\|x\| \leqq \sup_{i \geqq 1} \|S_i(x)\|$, we find $i \leqq K^n\{x_k\}$; indeed

$$\|S_n(x)\| \leqq \sup_{1\leqq i\leqq n} \|S_i(S_n(x)\| \Longrightarrow 1 \leqq \sup_{1\leqq i\leqq n} \frac{\|S_i(S_n(x))\|}{\|S_n(x)\|}$$

$$\leqq \sup_{1\leqq i\leqq n} \|S_i\|.$$

Also, by Theorem 2.1.3,

$$K\{x_k\} \equiv \sup \{K^n\{x_k\} : n \geqq 1\} = \sup \{\|S_n\| : n \geqq 1\} < \infty ,$$

thus

$$1 \leqq K^n\{x_k\} \leqq K^{n+1}\{x_k\}; \lim_{n\to\infty} K^n\{x_k\} = K\{x_k\}. \quad (21.2.13)$$

On the other hand, we always have

$$K_u^{n+1}\{x_k\} \geqq K_u^n\{x_k\} \geqq K^n\{x_k\} \geqq 1, \quad \forall n \geqq 1 \quad (21.2.14)$$

and if $\sup \{K_u^n\{x_k\} : n \geqq 1\} \equiv K_u\{x_k\} < \infty$, then

$$K_u\{x_k\} = \lim_{n\to\infty} K_u^n\{x_k\}. \quad (21.2.15)$$

By Theorem 6.1.1, $K_u\{x_k\}$ exists if and only if $\{x_k;f_k\}$ is an u-S.b., whereas $K\{x_k\}$ alwyas exists. The constants $K\{x_k\}$ and $K_u\{x_k\}$ are respectively called the *base constant* (b.c.) and the *unconditional base constant* (u.b.c.).

Let $D = \{z \in \mathbb{C} : |z| = 1\}$ and equip D with the usual Lebesgue measure. Suppose that for $1 \leqq p < \infty$, $L^p \equiv L^p(D)$ denotes the space of all complex-valued measurable functions f on D with the norm $\|\cdot\|_p$, where

$$\|f\|_p = [\frac{1}{2\pi} \int_D |f|^p \, d\mu]^{1/p}.$$

Writing a trignometric polynomial in its complex form and using a result from [158], p. 50-51, we conclude that $\{\delta_n : n = 0, \pm 1, \pm 2, \ldots\}$ is an S.b. for L^p, $1 < p < \infty$ such that it is a u-S.b. for $p = 2$, where $\delta_n(z) = z^n$. Thus, in any case, by Theorem 2.1.3, for each p, $1 < p < \infty$, there exists $N_p \geqq 1$ so that

$$K\{\delta_n\} \leqq N_p. \qquad (21.2.16)$$

Next we construct finite dimensional function spaces similar to $\ell_n^p \equiv \ell_J^p$, $J = \{1,\ldots,n\}$. For this purpose, consider points on the x-axis which are *distinct modulo* 2π; thus these points are contained in any interval of length 2π and for simplicity, let us confine ourselves to the interval $[0,2\pi]$. We now consider n points x_j $(n \geqq 1)$

$$x_j = (2\pi/n)j, \; 1 \leqq j \leqq n. \qquad (21.2.17)$$

The points x_j or their images $\varepsilon_{n,j}$ under the map $t \to \exp(it)$, $0 \leqq t \leqq 2\pi$ are called the *nodal points* of $[0,2\pi]$ or D, where $\varepsilon_{n,j} = \exp\{(2\pi i/n)j\}$; $1 \leqq j \leqq n$, $n \geqq 1$. Let $D_n = \{\varepsilon_{n,1},\ldots,\varepsilon_{n,n}\}$ and $L_n^p = \{f : D_n \to \mathbb{C}\}$ with the norm $\|f\|_p$, where

$$\|f\|_p = [\frac{1}{n} \sum_{j=1}^{n} |f(\varepsilon_{n,j})|^p]^{1/p}. \qquad (21.2.18)$$

LEMMA 21.2.19 For each n in $\mathbb{N}$ and $1 \leqq p < \infty$,

$$L_n^p \overset{iso}{\cong} \ell_n^p,$$

under the map $R_n : L_n^p \to \ell_n^p$, where for f in L_n^p

$$R_n(f) = \{n^{-1/p} f(\varepsilon_{n,j}) : 1 \leqq j \leqq n\} \text{ and } \|R_n(f)\|_p = \|f\|_p.$$

For $k = 0,1,\ldots,$ consider the nodal points $x_j = [2\pi/(2k+1)]j$, $1 \leqq j \leqq 2k+1$. The element e_n^{2k+1} in ℓ_{2k+1}^p is given by $\{(e_n^{2k+1})_j\} = \{\exp(inx_j) : 1 \leqq j \leqq 2k+1\}$. Put $g_n^{2k+1} = e_n^{2k+1}/(2k+1)^{1/p}$. Let f_n^{2k+1} and h_n^{2k+1} be in L_{2k+1}^p, where

$$\begin{cases} f_n^{2k+1}(\varepsilon_{2k+1,j}) = (2k+1)^{1/p}\exp(inx_j) \equiv (2k+1)^{1/p}(e_n^{2k+1})_j; \\ h_n^{2k+1}(\varepsilon_{2k+1,j}) = (e_n^{2k+1})_j, \ 1 \leq j \leq 2k+1. \end{cases} \tag{21.2.20}$$

Then

$$\begin{cases} R_{2k+1}(f_n^{2k+1}) = e_n^{2k+1}, \ R_{2k+1}(h_n^{2k+1}) = g_n^{2k+1}; \\ \| h_n^{2k+1} \|_p = \| g_n^{2k+1} \|_p = 1, \ n \geq 1. \end{cases} \tag{21.2.21}$$

<u>LEMMA 21.2.22</u> The sequence $\{h_n^{2k+1} : 1 \leq n \leq 2k+1\}$ (resp. $\{g_n^{2k+1} : 1 \leq n \leq 2k+1\}$) is an S.b. for L^p_{2k+1} (resp. ℓ^p_{2k+1}) with their norms equal to unity, where $1 \leq p < \infty$.

<u>PROOF</u>. In view of Lemma 21.2.19, (21.2.20) and (21.2.21), it suffices to prove the result for $\{h_n^{2k+1}\}$.

Consider 2k+1 arbitrary scalars $\alpha_1,\ldots,\alpha_{2k+1}$ with

$$\sum_{n=1}^{2k+1} \alpha_n h_n^{2k+1} = 0 \implies \sum_{n=1}^{2k+1} \alpha_n \exp(inx_j) = 0,$$

for $j = 1,\ldots,2k+1$. The determinant Δ of these 2k+1 equations in the unknowns $\alpha_1,\ldots,\alpha_{2k+1}$ is given by

$$\Delta = e^{2(k+1)i\pi} \prod_{\mu>\nu} (e^{ix_\mu} - e^{ix_\nu}) \neq 0$$

(cf. [242], Vol. II, p.1). Hence $\alpha_1 = \ldots = \alpha_{2k+1} = 0$. □

Let $L^{p,2k+1} = \mathrm{sp}\,\{\delta_n : 1 \leq n \leq 2k+1\}$ be the (2k+1)-dimensional subspace of L^p. Define $A_{2k+1}: L^{p,2k+1} \to L^p_{2k+1}$ by $A_{2k+1}(\delta_n) = h_n^{2k+1}$; $n = 1,\ldots,2k+1$ so that if

$$F = \sum_{n=1}^{2k+1} \alpha_n \delta_n, \tag{21.2.23}$$

then

$$A_{2k+1}(F) = \sum_{n=1}^{2k+1} \alpha_n h_n^{2k+1}. \tag{21.2.23)'$$

Hence from (21.2.21), (21.2.23) and (21.2.23)',

$$G_{2k+1}(F) \equiv R_{2k+1} \circ A_{2k+1}(F) = \alpha \equiv \sum_{n=1}^{2k+1} \alpha_n g_n^{2k+1}. \tag{21.2.24}$$

We now pass on to the crucial

LEMMA 21.2.25 Given p, $1 < p < \infty$, there exists $C_p > 0$ such that

$$\|A_{2k+1}\|, \|A_{2k+1}^{-1}\| \leqq C_p.$$

PROOF. Fix arbitrary positive constants $M_o, \ldots, M_{2k}$ such that $M_j - M_{j-1} = 2\pi/(2k+1)$, $1 \leqq j \leqq 2k$. Define $\omega_{2k+1} : [0, 2\pi] \to \mathbb{R}$ with $\omega_{2k+1}(t) = M_{j-1}$, for $x_{j-1} \leqq t < x_j$, $1 \leqq j \leqq 2k+1$ and $\omega_{2k+1}(2\pi) = M_{2k} + 2\pi/(2k+1)$, where $x_o = 0$. Put $S(t) = F(e^{it})$. Now

$$\|A_{2k+1}(F)\|_p = [\frac{1}{2k+1} \sum_{j=1}^{2k+1} |S(x_j)|^p]^{1/p}$$

$$= [\frac{1}{2\pi} \int_0^{2\pi} |S(t)|^p \, d\omega_{2k+1}(t)]^{1/p}$$

$$\leqq M[\frac{1}{2\pi} \int_0^{2\pi} |F(e^{it})|^p \, dt]^{1/p},$$

where M is an absolute constant (cf. [242], Vol. II, p. 30). Hence $\|A_{2k+1}\| \leqq M$.

On the other hand,

$$\| A_{2k+1}^{-1}(\delta) \|_p = \| F \|_p = [\frac{1}{2\pi} \int_0^{2\pi} |F(e^{it})|^p]^{1/p},$$

where $\delta \in L_{2k+1}^p$ and

$$\delta = \sum_{n=1}^{2k+1} \alpha_n h_n^{2k+1}.$$

Therefore, from [241], Vol. II, p. 30, there exists $H_p > 0$ so that

$$\| A_{2k+1}^{-1}(\delta) \|_p \leqq H_p [\frac{1}{2\pi} \int_0^{2\pi} |F(e^{it})|^p \, d\omega_{2k+1}(t)]^{1/p}$$

$$\Longrightarrow \| A_{2k+1}^{-1}(\delta) \|_p \leqq H_p \| \delta \|_p \Longrightarrow \| A_{2k+1}^{-1} \| \leqq H_p.$$

Put $C_p = \max(M, H_p)$ and we get the result. □

<u>LEMMA 21.2.26</u> Let $1 < p < \infty$ and consider the base $\{g_n^{2k+1} : 1 \leqq n \leqq 2k+1\}$ for ℓ_{2k+1}^p. Then there exists $M_p > 0$ such that

$$K\{g_n^{2k+1}\} \equiv K\{g_n^{2k+1}: 1 \leqq n \leqq 2k+1\} \leqq M_p, \ \forall k \geqq 1. \tag{21.2.27}$$

<u>PROOF</u>. Make use of (21.2.16), (21.2.24) and Lemmas 21.2.19, 21.2.25. □

<u>NOTE</u>. Lower bounds of the corresponding u.b.c. of $\{g_n^{2k+1}\}$ in ℓ_{2k+1}^p, $1 < p < \infty$, $p \neq 2$ will be obtained in Lemma 21.3.4 and 21.3.5. The case $p = 2$ is not of interest here since $\{\delta_n\}$ is a u-S.b. for L^2.

Babenko bases

We briefly touch upon the first example of a conditional base in L^2 due to Babenko [237] and pass on to a lemma in preparation for finding a lower bound of the u.b.c. for ℓ_n^2 in a subsequent section. The b.c. is clearly finite in this case.

Fix α, $0 < \alpha < 1/2$ and define $f_{\alpha,k}$ in $L^2 \equiv L^2[-\pi,\pi]$ by $f_{\alpha,k}(s) = |s|^\alpha \exp(iks)$, $k = 0, \pm 1, \pm 2, \ldots$. Then $\{f_{\alpha,k}\}$ is a conditional S.b. for L^2, the corresponding s.a.c.f. $\{t_{\alpha,k}\}$ being given by

$$t_{\alpha,k}(f) = \frac{1}{2\pi} \int_{-\pi}^{\pi} f(t)\ |t|^{-\alpha} \exp(-ikt)dt$$

(cf. [215], p. 428 or [154], p. 73-74 also).

LEMMA 21.2.28 Let $g_\alpha \in L^2$ with $g_\alpha(s) = |s|^{-\alpha}$, $\frac{1}{4} < \alpha < \frac{1}{2}$. Then there exists $A_\alpha > 0$ such that

$$t_{\alpha,k}(g_\alpha) \geqq A_\alpha\ k^{2\alpha-1}.$$

PROOF. We have

$$t_{\alpha,k}(g_\alpha) = \frac{k^{2\alpha-1}}{\pi} \int_0^{k\pi} \frac{\cos t}{t^{2\alpha}}\, dt.$$

Let

$$b_n \equiv \int_{n\pi}^{(n+1)\pi} \frac{\cos t}{t^{2\alpha}}\, dt =$$

$$\int_0^{\pi/2} (-1)^n\ [(n\pi+t)^{-2\alpha} - ((n+1)\pi - t)^{-2\alpha}]\cos t\ dt; \quad \text{and}$$

$$s_k = \sum_{n=0}^{k} b_n;\ k,n = 0,1,\ldots\ .$$

By Leibnitz's test, $\{s_k\}$ converges and

$$s_n \to \int_0^\infty \frac{\cos t}{t^{2\alpha}}\, dt = \frac{\pi}{2\,\Gamma(2\alpha)\cos \pi\alpha} \equiv B_\alpha > 0$$

(cf. [241], p. 107). Therefore $s_n > B_\alpha/2$ for all $n > N$. If $A_\alpha = \min\{(1-2\alpha)^{-1}\pi^{-2\alpha}, \pi^{-1} s_o, \ldots, \pi^{-1} s_N, \pi^{-1} B_\alpha/2\}$, we get the required result. □

21.3 LOWER BOUNDS OF u.b.c. IN ℓ_n^p

The proof of the basic theorem depends, besides other factors, upon the estimation of lower bounds of the u.b.c. of suitable bases in ℓ_n^p for all integers $n \geq 1$ and $1 \leq p < \infty$. As such we consider different cases : (A) $p = 1$, (B) $p = 2$, (C) $1 < p < 2$ and (D) $2 < p < \infty$, construct bases and obtain lower bounds of the corresponding u.b.c. in question.

Case (A): lower bound of u.b.c. in ℓ_n^1

We have

LEMMA 21.3.1 There exists an S.b. $\{b^k; f^k\}$ for ℓ^1 such that

(i) $Y_n \equiv sp\{b^1, \ldots, b^n\} \overset{iso}{\cong} \ell_n^1$ for every $n \in \mathbb{N}$;

(ii) For every $0 < \varepsilon < 1/2$, there exists $K \equiv K(\varepsilon)$ with

$$K_u^n\{b^k\} \geq K n^{1/2-\varepsilon}, \quad \forall n \geq 1.$$

In particular, for each $n \geq 1$, there exists an S.b. $\{b^1, \ldots, b^n\}$ for ℓ_n^1 such that

$$K\{b^i : 1 \leq i \leq n\} \leq K_1;\ K_u\{b^i : 1 \leq i \leq n\} \geq K_2\, n^{\alpha(1)},$$

where K_1, K_2 and $\alpha(1)$ are positive constants independent of n.

PROOF. Let us define $\{b^k\}$ in ℓ^1 and $\{f^k\}$ in ℓ^∞ by

$$b^1 = e^1;\ b^k = e^k - e^{k-1},\ k \geq 2;\ f^k = \sum_{i \geq k} e^i,\ k \geq 1.$$

For x in ℓ^1,

$$(*) \qquad \langle x, f^k \rangle = \sum_{i \geq k} x_i$$

and so

$$S_n(x) = \sum_{i=1}^{n} f^i(x) b^i = \{x_1, \ldots, x_{n-1}, \sum_{i \geq n} x_i, 0, 0, \ldots\}$$

$$\Longrightarrow \| S_n(x) - x \|_1 = \Big| \sum_{i \geq n+1} x_i \Big| + \sum_{i \geq n+1} |x_i| \to 0 \text{ as } n \to \infty.$$

Since $f^k(b^j) = \delta_{jk}$, $\{b^k; f^k\}$ is an S.b. for $\ell^1 \equiv (\ell^1, \|\cdot\|_1)$.

(i) It is easily verified that Y_n and ℓ^1_n are the same set - theoretically and so the identity map works as an isometric isomorphism between Y_n and ℓ^1_n.

(ii) Write $\gamma = 1/2 + \varepsilon$ and define x^o in ℓ^1 by $x^o = \{n^{-\gamma} - (n+1)^{-\gamma} : n \geq 1\}$. Then

$$(+) \qquad f^n(x^o) = \frac{1}{n^\gamma};\quad n \geq 1.$$

For $m \geq 1$, put $\sigma_m = \{1, 3, \ldots, 2m-1\}$. Then using (+), we easily compute, for $m \geq 2$

$$\langle S_{\sigma_m}(x^o), e^i \rangle = \begin{cases} 0, & i > 2m-1; \\ i^{-\gamma}; & i \text{ is odd}, \leq 2m-1; \\ -(i+1)^{-\gamma}, & i \text{ is even}, < 2m-1 \end{cases}$$

$$\Longrightarrow \| S_{\sigma_m}(x^o) \|_1 > \sum_{j=2}^{2m} \frac{1}{j^\gamma} > \int_2^{2m} \frac{dt}{t^\gamma}$$

$$==> \| S_{\sigma_m}(x^0) \|_1 > \frac{1}{1-\gamma} [(2m)^{1-\gamma} - 2^{1-\gamma}]$$
$$\geq (2m)^{\frac{1}{2} - \varepsilon} D(\varepsilon),$$

where $D(\varepsilon) = (1-1/2^{\frac{1}{2} - \varepsilon})/(\frac{1}{2} - \varepsilon)$.

Observe that $\|x^0\|_1 = 1$ and so $\|S_{\sigma_m}\| \geq \|S_{\sigma_m}(x^0)\|_1$, $m \geq 2$. On the other hand, $\|S_{\sigma_1}(x^0)\|_1 = \|f^1(x^0)e^1\|_1 = |\sum_{i\geq 1} x_i^0| = 1$ and so $\|S_{\sigma_1}\| \geq 1$. Hence if $K(\varepsilon) = \min\{1, 1/2^{\frac{1}{2}-\varepsilon}, D(\varepsilon)\}$, we obtain the inequality in (ii).

The last statement is a consequence of (i), (ii) and Theorem 2.1.3 (conclude an inequality similar to (21.2.16) for the S.b. $\{b^n\}$ in ℓ^1). □

Case (B): Lower bound of u.b.c. in ℓ_n^2

First we prove a general result in the form of

LEMMA 21.3.2 Let $\{x^n; f^n\}$ be an S.b. for $\ell^p\}(1 \leq p \leq 2)$ with $\|x^n\|_p = 1$ for $n \geq 1$. Then there exists $D_p > 0$ such that

$$K_u^n\{x^k\} \geq D_p \sup \{[\sum_{i=1}^n |f^i(x)|^2]^{1/2} / \|x\|_p : x \in \ell^p\},\ n \geq 1$$

PROOF. Let $G_n = \{\{\varepsilon_i\}: \varepsilon_i = 1,\ i > n;\ \varepsilon_i = \pm 1,\ 1 \leq i \leq n\}$ and G be the union of G_n's, $n \geq 1$.

For g in G, let $A_g : \ell^p \to \ell^p$ with

$$A_g(x) = \sum_{i=1}^n \varepsilon_i f^i(x) x^i + \sum_{i>n} f^i(x) x^i,$$

where $g \in G_n$ for some $n \geq 1$. The ultimate aim of introducing A_g is to express a lower bound of $K_u^n\{x^k\}$ in terms of A_g. So define $P_g = (I-A_g)/2$, I being the identity map of ℓ^p. It is easily seen that P_g is at most an n-dimensional projection of ℓ^p into $Y_n \equiv sp\{x^1,\ldots,x^n\}$.

Given g in G_n, if $\sigma_g = \{i:\varepsilon_i = -1, \varepsilon_i \in g\}$, then $P_g = S_{\sigma_g}$. Similarly, if $\sigma \subset \{1,\ldots,n\}$, let $g_\sigma = \{\varepsilon_i\}$ with $\varepsilon_i = -1$ for $i \in \sigma$ and $\varepsilon_i = 1$ for $i \in \mathbb{N}\setminus\sigma$. Then $P_{g_\sigma} = S_\sigma$. Therefore

$$(*) \quad K_u^n\{x^k\} = \sup_{g\in G_n} \|P_g\| \geq \frac{1}{2}[\sup_{g\in G_n} \|A_g\| - 1].$$

By Theorem 1.6.5, there exists $M_p \equiv M_p(\ell^p) > 0$ such that for any finite sequence $u^1,\ldots,u^n$ in ℓ^p,

$$[\sum_{i=1}^n \|u^i\|_p^2]^{1/2} \leq M_p \sup_{\varepsilon_i=\pm1} \|\sum_{i=1}^n \varepsilon_i u^i\|_p,$$

where $1 \leq p \leq 2$. Let $x \in \ell^p$ and $u^i = f^i(x)x^i$, $1 \leq i \leq n$. Hence

$$\begin{aligned}[\sum_{i=1}^n |f^i(x)|^2]^{1/2} &\leq M_p \sup_{g\in G_n} \|A_g(S_n(x))\|_p \\ &\leq M_pK_p \|x\|_p \sup_{g\in G_n} \|A_g\|, \; K_p = K\{x^k\}.\end{aligned}$$

Thus, with $N_p = M_pK_p$, we have (cf. (*))

$$(*) \quad K_u^n\{x^k\} \geq \frac{1}{2}[N_p^{-1} \sup\{[\sum_{i=1}^n |f^i(x)|^2]^{1/2}/ \|x\|_p : x \in \ell^p\} - 1]$$

For simplicity, put $A_n = \sup \{[\sum_{i=1}^n |f^i(x)|^2]^{1/2}/ \|x\|_p : x \in \ell^p\}$.

If n_2 is the smallest integer such that $A_{n_2}/N_p > 1$, then there exists $\beta \equiv \beta(n_2,p) > 0$ with $N_p^{-1} - A_{n_2}^{-1} \geq 2\beta$ and hence

$$(**) \qquad \frac{1}{2}\left(\frac{A_n}{N_p} - 1\right) \geq BA_n, \; n \geq n_2.$$

On the other hand, if $n_2 > 1$ we let $n_1 = n_2 - 1$, then $1 \geqq A_n/N_p$ for $1 \leqq n \leqq n_1$. But $K_u^n\{x^k\} \geqq 1$ (cf. (21.2.14)). Hence

$$(++) \qquad K_u^n\{x^k\} \geqq \frac{A_n}{N_p}, \quad 1 \leqq n \leqq n_1.$$

Let $D_p = \min(N_p^{-1}, \beta)$ and make use of (+), (**) and (++) to get the required inequality. □

We now come to the main lemma of this subsection, namely,

LEMMA 21.3.3 For $1/4 < \alpha < 1/2$, the Babenko base $\{f_{\alpha,k}; t_{\alpha,k}\}$ for $L^2 \equiv L^2[-\pi,\pi]$ satisfies the following inequality

$$K_u^n\{f_{\alpha,k}\} \geqq D_\alpha n^{2\alpha-\frac{1}{2}}; \quad \forall n \geqq 1,$$

where D_α is a constant independent of $n \geqq 1$.

In particular, for each $n \geqq 1$, there exists an S.b. $\{h^1,\ldots,h^n\}$ for ℓ_n^2 such that

$$K\{h^i : 1 \leqq i \leqq n\} \leqq K_2'; K_u\{h^i : 1 \leqq i \leqq n\} \geqq K_2 n^{\alpha(2)},$$

where K_2', K_2 and $\alpha(2)$ are positive constants independent of $n \geqq 1$.

PROOF. Multiplying each $f_{\alpha,k}$ by $\sqrt{(2\alpha+1)}/\pi^\alpha$ and writing the resulting term again as $f_{\alpha,k}$, we may assume that $\|f_{\alpha,k}\|_2 = 1$ for all k. In this case the members of the new s.a.c.f. become $(\pi^\alpha/\sqrt{(2\alpha+1)})\, t_{\alpha,k}$ which we write as $t_{\alpha,k}$.

Further L^2 is separable and so

$$(*) \qquad (L^2, \|\cdot\|_2) \overset{\text{iso}}{\cong} (\ell^2, \|\cdot\|_2).$$

Hence by Lemmas 21.2.28 and 21.3.2

$$(+)\qquad K_u^n\{f_{\alpha,k}\} \geqq D_2 A_\alpha \, \|g_\alpha\|_2^{-1} \Big[\sum_{i=1}^{n} i^{4\alpha-2}\Big]^{1/2}, \quad \forall n \geqq 1.$$

If $n \geqq 3$, then

$$\sum_{I=1}^{n} i^{4\alpha-2} \geqq \sum_{i=2}^{n-1} \int_i^{i+1} s^{4\alpha-2} ds \geqq \frac{1}{4\alpha-1}\Big[1-\Big(\frac{2}{3}\Big)^{4\alpha-1}\Big] n^{4\alpha-1}$$

and so from (+), for $n \geqq 3$,

$$K_u^n\{f_{\alpha,k}\} \geqq \bar{A}_\alpha n^{2\alpha-1/2}, \quad \bar{A}_\alpha = \frac{D_2 \pi^\alpha \sqrt{1-2\alpha}}{\sqrt{4\alpha-1}}\Big[1-\Big(\frac{2}{3}\Big)^{4\alpha-1}\Big]^{1/2} A_\alpha.$$

Let $D_\alpha = \min(\bar{A}_\alpha, B_\alpha, C_\alpha)$ where $B_\alpha = 1$ and $C_\alpha = 2^{1/2-2\alpha}$, then (as $K_u^n \geqq 1$)

$$K_u^n(f_{\alpha,k}) \geqq D_\alpha n^{2\alpha-\frac{1}{2}}, \quad \forall n \geqq 1,$$

and we are finished with the proof of the major part. The last statement follows with the help of (*). □

<u>Case (C): Lower bound of u.b.c. in ℓ_n^p, $1 < p < 2$</u>

The major result is

<u>LEMMA 21.3.4</u> For each p, $1 < p < 2$, there exist positive constants K_p', K_p and $\alpha(p)$ independent of $k \geqq 1$ such that for the base $\{g_n^{2k+1}: 1 \leqq n \leqq 2k+1\}$ of ℓ_{2k+1}^p as in Lemma 21.2.22,

$$\text{(i)}\quad K_u\{g_n^{2k+1}: 1 \leqq n \leqq 2k+1\} \geqq K_p(2k+1)^{\alpha(p)}; \ \forall k \geqq 1$$

and also

$$\text{(ii)}\quad K\{g_n^{2k+1}: 1 \leqq n \leqq 2k+1\} \leqq K_p', \ \forall k \geqq 1.$$

<u>PROOF</u>. In view of Lemma 21.2.26, we have only to prove

(i). However, (i) is proved with the help of Lemma 21.3.2 by suitably choosing x in ℓ^p. As such, let

$$x = \left[\frac{1}{2k+1}\right]^{1-\frac{1}{p}} \sum_{n=1}^{2k+1} g_n^{2k+1}.$$

Since

$$\sum_{n=1}^{2k+1} (e_n^{2k+1})_j = \begin{cases} 2k+1, & j = 2k+1; \\ 0, & 1 \leq j \leq 2k, \end{cases}$$

we get $x_{2k+1} = 1$ and $x_j = 0$, $1 \leq j \leq 2k$. Therefore $\|x\|_p = 1$. If $\{F_n^{2k+1}\}$ is the s.a.c.f. corresponding to $\{g_n^{2k+1}\}$, then

$$F_n^{2k+1}(x) = (2k+1)^{1/p-1}; \quad 1 \leqq n \leqq 2k+1,$$

and hence by Lemma 21.3.2, for $k \geqq 1$, we have

$$K_u\{g_n^{2k+1}\} \geqq D_p(2k+1)^{1/p-1/2},$$

and we have completed the proof. □

Case (D): Lower bound of u.b.c. in ℓ_n^p, $2 < p \leqq \infty$

We now deal with the final case of determining the u.b.c. (and also b.c.) of suitable bases in ℓ_{2k+1}^q, where $2 < q < \infty$. Given ℓ_{2k+1}^q, we have ℓ_{2k+1}^p for a unique p, $p^{-1}+q^{-1} = 1$ so that $(\ell_{2k+1}^p)^* = \ell_{2k+1}^q$ and $\{F_n^{2k+1} : 1 \leqq n \leqq 2k+1\}$ defined above is an S.b. for ℓ_{2k+1}^q. Further, by a known result, $\|S_\sigma\| = \|S_\sigma^*\|$ where S_σ corresponds to the expansion operator of $\{g_n^{2k+1}\}$ and S_σ^* is the dual expansion operator for the base $\{F_n^{2k+1}\}$. Hence from the proof of the preceding lemma,

$$K_u\{F_n^{2k+1}\} = K_u\{g_n^{2k+1}\} \geq D_p(2k+1)^{1/p-1/2}$$

$$\Longrightarrow K_u\{F_n^{2k+1}\} \geq D_q(2k+1)^{1/2-1/q}$$

Thus, using Lemma 21.2.26, we have proved

LEMMA 21.3.5 For each q, $2 < q < \infty$, there exist positive constants K_q', K_q and $\alpha(q)$ independent of $k \geq 1$ such that for the base $\{F_n^{2k+1}\}$ of ℓ_{2k+1}^q one has:

(i) $K_u\{F_n^{2k+1}: 1 \leq n \leq 2k+1\} \geq K_q(2k+1)^{\alpha(q)}$, $k \geq 1$;

(ii) $K\{F_n^{2k+1}: 1 \leq n \leq 2k+1\} \leq K_q'$, $k \geq 1$.

Similarly, with the help of Lemma 21.3.1, we can derive

LEMMA 21.3.6 There exist positive constants K_1, K_2 and α such that for the base $\{f^i: 1 \leq i \leq n\}$ of ℓ_n^∞,

$$K_u\{f^i: 1 \leq i \leq n\} \geq K_2 n^\alpha;\ K\{f^i: 1 \leq i \leq n\} \leq K_1,\ \forall n \geq 1.$$

The main lemma

Summarizing the lemmas of this section dealing with the cases (A), (B), (C) and (D), the resulting lemma is the one given below, in which we are ultimately interested in proving the basic theorem of this chapter.

LEMMA 21.3.7 For each p, $1 \leq p < \infty$ and every n in $\mathbb{N}$, there exist bases $\{e_k^n: 1 \leq k \leq n\}$ in ℓ_n^p and positive constants $C_1(p)$, $C_2(p)$ and $\alpha(p)$ independent of $n \geq 1$ such that

(i) $K\{e_k^n\} \equiv K\{e_k^n: 1 \leq k \leq n\} \leq C_1(p)$, $n \geq 1$;

(ii) $K_u\{e_k^n\} \equiv K_u\{e_k^n: 1 \leq k \leq n\} \geq C_2(p)n^{\alpha(p)}$, $n \geq 1$.

21.4 NUCLEARITY OF $\ell^p[P;X_n]$

The aim of this section is to provide a set of sufficient conditions yielding the nuclearity of the v.v. K-F space $\ell^p[P;X_n]$ discussed in section 21.1, namely, in the form of

THEOREM 21.4.1 Let $P \equiv [a_{mn}]$ be an m.K.m. and $A_n: X_n \to X_n$ be linear operators $(n \geq 1)$, where for some $\alpha > 0$,

$$\|A_n\| \geq (\dim X_n)^{\alpha};\ n \geq 1 \qquad (21.4.2)$$

with

$$\dim X_{n+1} \geq \dim X_n \geq n,\ n \geq 1. \qquad (21.4.3)$$

If $\bigoplus_{n\geq 1} A_n: \ell_0^p[P;X_n] \to \ell_0^p[P;X_n]$ is continuous for any p with $1 \leq p \leq \infty$, then $\ell^p[P;X_n]$ is nuclear.

We first need an elementary

PROPOSITION 21.4.4 Let $P \equiv [a_{mn}]$ be an m.K.m. For some $\alpha > 0$, let $A^{\alpha}: \ell_0^p[P] \to \ell_0^p[P]$ be continuous where $A^{\alpha}(x) = \{n^{\alpha}x_n\}$. Then $\ell^p[P]$ is nuclear, where $1 \leq p \leq \infty$.

PROOF. Using composition of maps A^{α} if necessary, we may assume that $\alpha > 1$. For each $m \geq 1$, there exist $n \geq m$ and $C_m > 0$ with $k^{\alpha}\, a_{mk} \leq C_m\, a_{nk}$, for $k \geq 1$. Now apply Lemma 21.2.3. □

PROOF OF THEOREM 21.4.1 Let $1 \leq p < \infty$ and we similarly proceed for $p = \infty$. If necessary, we may consider $(\dim X_n)^2 A_n$ in place of A_n and thus we may let $\alpha > 2$. For each $n \geq 1$, we can find x_0^n in X_n with $\|A_n(x_0^n)\| = \|A_n\|$ and $\|x_0^n\| = 1$. For ξ in $\ell_0^p[P]$, let $\bar{x}_0 = \{\xi_n x_0^n\}$ be an element of $\ell_0^p[P;X_n]$. Thus using the continuity of $\bigoplus_{n\geq 1} A_n$, to every m we can find

$k \geq m$ and $C_m > 0$ with

$$(*) \qquad \sum_{n\geq 1} a_{mn}^p \|A_n\|^p |\xi_n|^p \leq C_m^p \sum_{n\geq 1} a_{kn}^p |\xi_n|^p; \ \forall \xi \in \ell_o^p[P].$$

By (*) we conclude the continuity of A^α on $\ell_o^p[P]$ and hence from Proposition 21.4.4, $\ell^p[P]$ is nuclear where $1 \leq p < \infty$. Proceeding as in Lemma 21.2.6, we find that (under the identity map)

$$(**) \qquad \ell^p[P;X_n] \simeq \ell^1[P;X_n]; \ \ell_o^p[P;X_n] \simeq \ell_o^1[P;X_n].$$

In view of the preceding discussion, the problem is now reduced to establishing the nuclearity of $\ell^1[P;X_n]$.

We next reduce the problem to ascertaining the nuclearity of $\ell^1[P;\ell_{N_n}^1]$ by way of proving

$$(+) \qquad \ell^1[P;X_n] \simeq \ell^1[P;\ell_{N_n}^1]; \ N_n = \dim X_n, \ n \geq 1.$$

In this direction, observe that (*) yields (21.2.11) and so by Lemma 21.2.10, $\ell^1[P;X_n] \simeq \ell^1[P;\tilde{X}_n]$, under the identity map $\tilde{I}_p$, where $\tilde{X}_n$ is the space X_n equipped with the norm $\|\cdot\|_1$ (cf. (21.2.8)).

A seminorm on $\ell^1[P;\tilde{X}_n]$ will be denoted by $\|\cdot\|_{m,1}$ when it is generated with the help of the norm $\|\cdot\|_1$ on X_n in a manner similar to $\|\cdot\|_m$ which is obtained with the help of the norm $\|\cdot\|$ on X_n. For each $n \geq 1$, determine an Auerbach base $\{X_i^{N_n}; f_i^{N_n}, 1 \leq i \leq N_n\}$ for $\{X_n, \|\cdot\|\}$; then under the map $R_{N_n} : \tilde{X}_n \to \ell_{N_n}^1$, $R_{N_n}(x^{N_n}) = \{f_i^{N_n}(x^{N_n}) : 1 \leq i \leq N_n\}$,

$$(\tilde{X}_n, \|\cdot\|_1 \overset{iso}{\cong} (\ell_{N_n}^1; \|\cdot\|_1),$$

the right-hand norm $\|\cdot\|_1$ being the usual norm on $\ell_{N_n}^1$. Next, if $\bar{x} = \{x^{N_n}\} \in \ell^1[P,\tilde{X}_n]$, then

$$\| (\bigoplus_{n\geq 1} R_{N_n}) (\bar{x}) \|_{m,1} = \| \bar{x} \|_{m,1},$$

where the firstnorm on the left relates to a norm on $\ell^1[P;\ell^1_{N_n}]$. Hence, under $\bigoplus_{n\geq 1} R_{N_n}$,

$$\ell^1[P;\tilde{X}_n] \overset{iso}{\cong} \ell^1[P,\ell^1_{N_n}],$$

and therefore we get (+) under $\bigoplus_{n\geq 1} R_{N_n} \circ I_P$.

It now remains to prove the nuclearity of $\ell^1[P;\ell^1_{N_n}]$. This is accomplished by constructing a nuclear K-F space $\ell^1[Q]$ and showing

$$(++) \quad \ell^1[P;\ell^1_{N_n}] \overset{iso}{\cong} \ell^1[Q].$$

Indeed, let

$$s_n = \sum_{i=1}^{n-1} N_i; \; s_1 = 0, \; N_o = 0, \qquad (21.4.5)$$

and put $b_{mk} = a_{mn}$ for $m \geq 1$ and $s_n + 1 \leq k \leq s_{n+1}$, $n \geq 1$. For $\bar{\alpha} = \{\alpha^{N_1}, \alpha^{N_2}, \ldots, \alpha^{N_n}, \ldots\}$ in $\ell^1[P;\ell^1_{N_n}]$, where $\alpha^{N_N} = \{\alpha_i^{N_n}: 1 \leq i \leq N_n\}$, $n \geq 1$, define $\beta_k = \alpha^{N_n}_{k-s_n}$ with $s_n + 1 \leq k \leq s_{n+1}$, $n \geq 1$. Then $\beta = \{\beta_k\}$ belongs to $\ell^1[Q]$, $Q = [b_{mk}]$. Hence there is a linear map $R:\ell^1[P;\ell^1_{N_n}] \to \ell^1[Q]$ such that $R(\bar{\alpha}) = \beta$ with $\bar{\alpha}$ and β as mentioned above; R is also seen to be 1-1 and onto. Indeed R is an isometry from $\ell^1[P;\ell^1_{N_n}]$ onto $\ell^1[Q]$ and so (++) is proved.

The only part which we wish to prove finally, is the nuclearity of $\ell^1[Q]$. Since $\bigoplus_{n\geq 1} A_n$ is continuous, using (**) we find that (*) is also valid for $p = 1$ and so if ξ in $\ell^1_o[P]$ is defined by $|\xi_n| = \|\alpha^{N_n}\|_1$ where $\alpha^{N_n} \in \ell^1_{N_n}$, we get

$$(***)\qquad \sum_{n\geq 1} a_{mn} N_n^{\alpha} \|\alpha^{N_n}\|_1 \leq C_m \sum_{n\geq 1} a_{kn} \|\alpha^{N_n}\|_1,$$

the inequality being valid for any choice of α^{N_n} in $\ell^1_{N_n}$, $n \geq 1$. Now, let $\beta \in \ell^1_o[Q]$ and define α^{N_n} in $\ell^1_{N_n}$ with $\alpha_i^{N_n} = \beta_{i+s_n}$, $1 \leq i \leq N_n$, $n \geq 1$. Then (***) yields

$$\sum_{n\geq 1} N_n^{\alpha} \sum_{i=1+s_n}^{s_{n+1}} b_{mi}|\beta_i| \leq C_m \sum_{n\geq 1} \sum_{i=1+s_n}^{s_{n+1}} b_{ki}|\beta_i|.$$

But $N_n^2 \geq i$ for $1 + s_n \leq i \leq s_{n+1}$ and so if $\alpha = 2 + 2\varepsilon$, we find that

$$\|A^{1+\varepsilon}(\beta)\|_m \leq C_m \|\beta\|_k, \quad \forall\ \beta \in \ell^1_o[Q].$$

By Proposition 21.4.4, $\ell^1[Q]$ is nuclear. □

21.5 NON-NUCLEAR SPACES $\ell^p[P]$, $1 \leq p < \infty$

The final major result which helps prove the basic theorem and deals with the outline (c) of the introductory remarks is the following

THEOREM 21.5.1 Let $P = [a_{mn}]$ be a non-nuclear m.K.m. Then there exists a non-nuclear submatrix $Q = [a_{m_k n_j}]$ such that for each p, $1 \leq p \leq \infty$

$$\ell^p[Q] \simeq \ell^p[D;\ell^p_{q(n)}]$$

for some m.K.m. $D = [d_{mn}]$ and for a sequence $\{q(n)\} \subset \mathbb{N}$ with $q(n+1) \geq q(n) \geq n$.

Several lemmas are involved before we actually come to the proof of this theorem.

LEMMA 21.5.2 For each non-nuclear m.K.m. $P = [a_{mn}]$, there exists a non-nuclear m.K.m. $Q \equiv [b_{mn}] = [a_{m_k n_j}]$ such that $b_{mn} \neq 0$ for $m,n \geq 1$.

PROOF. Since (21.2.4) is not satisfied for $s = 1$, there exists m_0 such that for each $k \geq 1$,

$$\sum_{n\geq 1} a_{m_0 n}/a_{m_0+k,n} = \infty. \tag{21.5.3}$$

Let $b_{kn} = a_{m_0+k,n}$, $k,n \geq 1$. Then $[b_{kn}]$ is a non-nuclear m.K.m. By (21.5.3), $a_{m_0 n} \neq 0$ for $n = \{n_j\}$ in I. Rewrite $[b_{mn}]$ for $[b_{kn_j}]$. □

LEMMA 21.5.4 Let P be an m.K.m. with $a_{mn} \neq 0$ for $m,n \geq 1$. Then for each p, $1 \leq p \leq \infty$ and k in $\mathbb{N}$, $\ell^p[P] \simeq \ell^p[Q]$ where $Q = [b_{mn}]$, $b_{mn} = a_{m+k-1,n}/a_{kn}, m,n \geq 1$.

PROOF. Define $T_k : \ell^p[P] \to \ell^p[Q]$, $T_k(x) = \{a_{kn}x_n\}$. Then $\|T_k(x)\|_m = \|x\|_{m+k-1}$ and $\|T_k^{-1}y\|_m \leq \|y\|_m$. □

LEMMA 21.5.5 Let P be an m.K.m. with (i) $a_{1n} = 1$, $n \geq 1$ and

(ii) $\liminf_{n\to\infty} a_{mn} < \infty$, $\forall\, m \geq 1$.

Then one of the following conclusions 1° or 2° holds good:

1°. There exists a submatrix $Q = [a_{m_k n_j}]$ with

$$\sum_{j\geq 1} 1/a_{m_k n_j} = \infty, \ \forall\, k \geq 1 \tag{21.5.6}$$

(and *a fortiori* that Q is non-nuclear) and

$$\lim_{j\to\infty} a_{m_k n_j} = \infty, \ \forall\, k \geq 1. \tag{21.5.7}$$

2°. There exists a submatrix $Q = [a_{mn_j}]$ such that

$$\limsup_{j \to \infty} a_{mn_j} < \infty , \ \forall\, m \geq 1. \qquad (21.5.8)$$

PROOF. Let $A_m = \{\sigma \subset \mathbb{N}: \sup_{n\in\sigma} a_{mn} < \infty\}$,

$$\mathcal{L}_m = \{\sigma \subset \mathbb{N}: \sum_{n\in\sigma} 1/a_{mn} < \infty\} \text{ and } \rho_{km} = \{n \in \mathbb{N}: a_{mn} < k\}.$$

Since P is an m.K.m.,

$$(*)\quad \begin{cases} A_1 \supset A_2 \supset A_3 \supset \ldots; \\ \mathcal{L}_1 \subset \mathcal{L}_2 \subset \mathcal{L}_3 \subset \ldots; \\ \Phi \subset A_m, \mathcal{L}_m, \ m \geq 1; \\ A_{m_1} \cap \mathcal{L}_{m_2} \subset \Phi, \ m_1 \geq m_2. \end{cases}$$

Let us write

$$A = \cap\{A_m : m \geq 1\}; \ \mathcal{L} = \cup \{\mathcal{L}_m : m \geq 1\}.$$

Coming to the proof, first observe that $\Phi \subset A$ is always true. Further, 2° holds if and only if $A \neq \Phi$.

Let, therefore, $A = \Phi$ and we will show that 1° holds. The proof runs in several steps.

Let E_m denote the family of subsets of $\mathbb{N}$ generated by $\mathcal{L} \cup A_m$ for each $m \geq 1$. The claim is that there exists m_0 such that E_{m_0} is proper; for otherwise $\mathbb{N} \in E_m$ for $m \geq 1$, yields $\mathbb{N} = B_m \cup A_{m_1}$ with $B_m \in A_m$ and $A_{m_1} \in \mathcal{L}_{m_1}$ where we may take $m_1 > m$ (cf. (*)). By hypothesis (ii) each A_m contains infinite sequences (members of I) and so $\mathbb{N}$ or any set of the form $\mathbb{N}\setminus F$, $F \in \Phi$ cannot belong to any $\mathcal{L}_m$ (cf. the second and fourth statements of (*)). Hence we may assume that each B_m is infinite. By induction we get $\{k_n\} \in I$ so that $k_1 = 1$ and

$$\mathbb{N} = B_n \cup A_{k_{n+1}} ; B_n \in A_{k_n}, A_{k_{n+1}} \in \mathcal{L}_{k_{n+1}}, n \geq 1.$$

By (*), $B_{n+1} \cap A_{k_{n+1}} \in \Phi$ and so $F_n \equiv B_{n+1} \setminus B_n \in \Phi$. This shows that there are infinitely many points in $B_n \cap B_{n+1}$ for each $n \geq 1$. If $F_0 = \emptyset$, then

$$B_i \setminus (\bigcup_{j=1}^{i-1} F_j) = \bigcap_{j=1}^{i} B_j \equiv \{n_1^i, n_2^i, \ldots\}, \ i \geq 1$$

and if $\sigma = \{n_1^1, n_2^2, \ldots\}$, then $\sigma \in I$ and $\sigma \setminus B_i \in \Phi$ for each $i \geq 1$. Clearly $\sigma \in A_m$ for $m \geq 1$ and this contradicts the fact that $A = \Phi$.

Therefore there exists an infinite subset B of $\mathbb{N}$ so that for some m_0, $B \notin E_{m_0}$ and it follows that $B \setminus \rho_{km_0} \notin \mathcal{L}$ for each $k \geq 1$. Thus

$$B \setminus \rho_{km_0} \notin \mathcal{L}_m, \forall k,m \geq 1 \Longrightarrow B \setminus \rho_{km_0} \notin \mathcal{L}_k, \forall k \geq 1$$

Hence for each $k \geq 1$ there exists a finite subset σ_k of $B \setminus \rho_{km_0}$ such that

$$(+) \qquad \sum_{i \in \sigma_k} 1/a_{ki} > k, \ \forall\ k \geq 1.$$

Without loss of generality we may arrange the members of σ_k and that of δ in an increasing order where

$$\delta = \bigcup \{\sigma_k : k \geq 1\}$$

and it follows that

$$(**) \qquad \delta \cap \rho_{km_0} \in \Phi, \forall k \geq 1.$$

Let us write $\delta = \{n_j\}$. Since assumption of $\lim_{j \to \infty} a_{m_0 n_j} < \infty$ contradicts (**), we have

(+ *) $\lim_{j\to\infty} a_{mn_j} = \infty$, $\forall\, m \geq m_o$.

On the other hand, by (+), for $k \geq m \geq m_o$

$$\sum_{j\geq 1} \frac{1}{a_{mn_j}} \geq \sum_{i\in\sigma_k} \frac{1}{a_{ki}} > k$$

(*+) $\Longrightarrow \sum_{j\geq 1} 1/a_{mn_j} = \infty$.

If $Q = [a_{m_o+k,n_j} : j,k \geq 1]$, then with the help of (*+) and (+ *), Q satisfies the requirements (21.5.6) and (21.5.7) respectively. The remaining part concerning non-nuclearity of Q follows with the help of (i) and (*+). □

<u>LEMMA 21.5.9</u> Let $a,b \in \omega$ with

(i) $0 < b_n \leq a_n$, $n \geq 1$; (ii) $a_n \to 0$

Further, let $\rho_i \subset N$, $i \geq 1$ with $\rho_i \cap \rho_j = \emptyset$ $(i \neq j)$ and

(iii) $\sum_{i\geq 1} \sum_{n\in\rho_i} b_n = \infty$; (iv) $b_n < C/2^{i^\alpha}$, $n \in \rho_i$

where $C > 0$ and $0 < \alpha \leq 1$ are constants independent of $i \geq 1$. Then there exists a sequence $\{\sigma_j\} \subset \Phi$ with $\sigma_i \cap \sigma_j = \emptyset$ $(i \neq j)$ and

(1) $\forall j\ \exists i \ni \sigma_j \subset \rho_i$; (2) $\sum_{j\geq 1} \sum_{n\in\sigma_j} b_n = \infty$;

(3) $\#(\sigma_j) \geq j$; (4) $b_n \leq D/2^{j^\beta}$, $n \in \sigma_j$;

where $D > 0$ and $0 < \beta \leq 1$ are constants independent of $j \geq 1$.

<u>PROOF</u>. Put $C_1 = \sup\{a_n : n \geq 1\}$; $\delta_k = \{n \in N : C_1/2^k < a_n \leq C_1/2^{k-1}\}$, $k \geq 1$. Each $\delta_k \in \Phi$ and $\delta_k \cap \delta_j = \emptyset$, $k \neq j$.

Let

$$L_{i,k} = \rho_i \cap \delta_k.$$

Then $\{L_{i,k} : i,k \geq 1\}$ is a pairwise disjoint sequence of finite subsets of N.

By considering max $(i,k) = i$ or k separately, we find that if $n \in L_{i,k}$, then (cf. (iv))

$$b_n < \max\,(C, 2C_1)/2^{[\max(i,k)]^{\alpha}}.$$

In order to arrange $\{L_{i,k}\}$ into a sequence $\{V_s\}$, let us write $V_s = L_{i,k}$, where $s : N \times N \to N$ with

$$s \equiv s(i,k) = \frac{(i+k-1)(i+k-2)}{2} + k,$$

and the arrangement of this 1-1 onto correspondence is shown below:

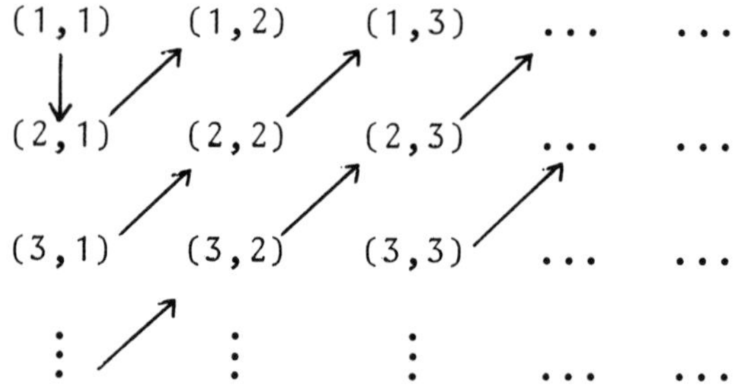

It is the sequence $\{V_s\}$ which we will ultimately arrange as $\{\sigma_j\}$ satisfying the required statements. In the meantime, let us observe some of the useful properties of $\{V_s\}$.

Since $s(i,k) < 4[\max\,(i,k)]^2$, we find that whenever $n \in L_{i,k} = V_s$

$$b_n < \max\,(C, 2C_1)/2^{s^{\beta}}$$

for all $s \geq 64$ and so we can find $D > 0$ so that for n in

V_s $(s \geq 1)$,

$$(*) \qquad b_n \leq D/2^{s^\beta},$$

where $\beta = \alpha/3$.

Let $R = \{s \in N: \#(V_s) < s\}$, $V_s = V_s^*$ if $s \in \mathbb{N} \setminus R$ and $V_s = \bar{V}_s$ if $s \in R$. Then

$$(+) \qquad \sum_{s \in R} \sum_{n \in V_s} b_n \leq D \sum_{s \geq 1} s(1/2)^{s^\beta} < \infty.$$

Further

$$\rho_i = \bigcup_{k \geq 1} L_{i,k}, \ \forall\, i \geq 1$$

and so

$$\sum_{i \geq 1} \sum_{n \in \rho_i} b_n \leq \sum_{s \in \mathbb{N} \setminus R} \sum_{n \in V_s^*} b_n + \sum_{s \in R} \sum_{n \in \bar{V}_s} b_n$$

$$(**) \qquad \Longrightarrow \sum_{s \in \mathbb{N} \setminus R} \sum_{n \in V_s} b_n = \infty,$$

by (iii) and (+).

Let us write $\{\sigma_j : j \geq 1\}$ for $\{V_s : s \in \mathbb{N} \setminus R\}$ enumerated in the same order. Thus $\sigma_j = V_s$ for $s \geq j$ and so (3) and (4) follow. (2) is a consequence of (**), whereas (5) results from the definition of δ_k. The rest is obvious. □

<u>LEMMA 21.5.10</u> Let $P = [a_{mn}]$ be an m.K.m. with (i) $a_{1n} = 1$, $n \geq 1$; (ii) $a_{mn} \to \infty$ as $n \to \infty$ for every $m \geq 2$ and (iii) $\sum_{n \geq 1} (1/a_{mn}) = \infty$ for $m \geq 1$. Then there exists a sequence $\{\sigma_k\} \subset \Phi$ with $\sigma_i \cap \sigma_j = \emptyset$, $i \neq j$ and

(a) $\#(\sigma_{k+1}) \geq \#(\sigma_k) \geq k$;

(b) $\forall m\ \exists\ K(m) \ni \forall k;\ n_1, n_2 \in \sigma_k \Longrightarrow \dfrac{a_{mn_1}}{a_{mn_2}} \leq K(m)$;

(c) $\sum_{k\geq 1} \sum_{n\in\sigma_k} 1/a_{mn} = \infty, \forall m \geq 1.$

PROOF. Put $d_{mn} = 1/a_{mn}$. In view of (iii), we can find $\{n_m\}$, $n_m \uparrow \infty$, $n_1 = 1$ such that

$$\sum_{j=n_m}^{n_{m+1}-1} d_{mj} > 1, \forall m \geq 1.$$

Define c_n so that $c_n = d_{mn}$ for $n_m \leq n \leq n_{m+1}-1$, $m \geq 1$; thus $c_n \leq d_{mn}$ for all $n \geq n_m$, $m \geq 1$. Let $D_{1,1} = \mathbb{N} \setminus \{1,2,...,n_2-1\}$, $D_{1,j} = \emptyset$ for $j > 1$ and let

$$a_n = \begin{cases} d_{2n}, & n \geq n_2 \\ c_n, & 1 \leq n \leq n_2-1 \end{cases} ; \; b_n = c_n, \; n \geq 1.$$

Then $\{a_n\}$, $\{b_n\}$ and $\{D_{1,j}\}$ satisfy conditions (i)-(iv) of Lemma 21.5.8 and as such we get a sequence $\{D_{2,j}: j \geq 1\}$ of mutually disjoint finite subsets of $\mathbb{N}$ with

(1)' $\forall j \; \exists \; i \ni D_{2,j} \subset D_{1,i}$; (2)' $\sum_{j\geq 1} \sum_{n\in D_{2,j}} c_n = \infty$;

(3)' $\#(D_{2,j}) \geq j$; (4)' $c_n \leq D/2^{j^\beta}$, $n \in D_{2,j}$, $j \geq 1$;

(5)' $d_{2n}/d_{2m} < 2$; $m,n \in D_{2,j}$, $j \geq 1$.

It is clear that we can find j_1 such that for n in $D_{2,j}$ with $j > j_1$, $c_n \leq d_{3n}$ and use (2)' to get $k_1 > j_1$ so that

$$\sum_{j=1+j_1}^{k_1} \sum_{n\in D_{2,j}} c_n > 1.$$

Let $\tilde{D}_{2,j} = D_{2,j+k_1}$, $j \geq 1$; $b_n = c_n$; $n \geq 1$, $a_n = d_{3n}$, $n \geq n_3$

and $a_n = c_n$, $1 \leq n < n_3$. Thus $\{a_n\}$, $\{b_n\}$ and $\{\tilde{D}_{2,j}\}$ satisfy the hypothesis of Lemma 21.5.9 and so we get $\{D_{3,j}\}$ satisfying conditions similar to (1)' - (5)' above. Observe that $\tilde{D}_{2,i} \subset \mathbb{N} \setminus \{1,\ldots,n_3-1\}$ for $i \geq 1$.

Choose $j_2 > \#(D_{2,k_1}) \geq k_1$ such that $c_n \leq d_{4n}$ for $n \in D_{3,j}$, $j > j_2$ and find $k_2 > j_2$ with

$$\sum_{j=1+j_2}^{k_2} \sum_{n \in D_{3,j}} c_n > 1.$$

Continuing this process indefinitely, we get sequences of finite subsets $\{D_{m,j} : j \geq 1\}$, $m \geq 1$ and natural numbers j_i and k_i with $j_{m+1} > \#(D_{m+1,k_m}) \geq k_m$, $m \geq 1$ and $1 < j_1 < k_1 < j_2 < k_2 < \ldots$ such that

(I) $c_n \leq d_{m+2,n}$, for $n \in D_{m+1,j}$, $j > j_m$;

(II) $\{D_{m,j}\}$ are pairwise disjoint for $j_m < j \leq k_m$, $m \geq 1$

and $\#(D_{m+1,j}) \geq j$, $m \geq 1$;

(III) $a_{kp}/a_{kq} < 2$; $p,q \in D_{k,j}$; $k \geq 1$, $j \geq 1$;

(IV) $\displaystyle\sum_{j=1+j_m}^{k_m} \sum_{n \in D_{m+1,j}} c_n > 1$, $\forall\, m \geq 1$.

The sequence $\{D_{m+1,j}\}$ is suitably arranged as a sequence $\{\sigma_k\}$ to yield the required (a), (b) and (c).

Indeed, let $s_0 = 0$ and $s_m = \sum_{p=1}^{m} (k_p - j_p)$, $m \geq 1$.

Then for each $k \geq 1$ we can find $m \geq 1$ so that $k = s_{m-1}+j-j_m$ for j in $\mathbb{N}$ satisfying the inequality $j_m < j \leq k_m$ and let $\sigma_k = D_{m+1,j}$. Observe that $1+s_{m-1} \leq k \leq s_m$ if and only if $j_m+1 \leq j \leq k_m$, $m \geq 1$.

By (II), $\#(\sigma_k) \geq j$ and it is easily seen that $k < j$ and so $\#(\sigma_k) > k$. If necessary, we may reorder the sets σ_k so that $\#(\sigma_{k+1}) \geq \#(\sigma_k)$ and we get (a)

(b) trivially follows when $m = 1$ and so fix $m \geq 2$. Consider any $k > s_m$ so that for some $t = 0,1,\ldots,$ $1+s_{m+t} \leq k \leq s_{m+t+1}$. Since the sets σ_j, $1 \leq j \leq s_{m+t}$ contain only a finite number of indices p,q and $\sigma_k = D_{m+t+2,j} \subset D_{m,i}$, an application of (III) yields (b).

Finally, fix $i \geq 1$ and find $m \geq 1$ with $m \geq i$. Then

$$\sum_{k\geq 1} \sum_{n\in\sigma_k} d_{in} \geq \sum_{m\geq i} \sum_{j=1+j_m}^{k_m} \sum_{n\in D_{m+1,j}} d_{mn} = \infty$$

by (I) and (IV). Hence (c) is proved. □

<u>LEMMA 21.5.11</u> Let P satisfy the assumptions of Lemma 21.5.10. Then there exist a non-nuclear submatrix $Q = [b_{mn}]$, a sequence $\{k_n\}$ in I with $k_n > n$ and an m.K.m. $R = [d_{mn}]$ such that for any p, $1 \leq p \leq \infty$

$$\ell^p[Q] \simeq \ell^p[R;\ell^p_{k_n}].$$

<u>PROOF</u>. By Lemma 21.5.10, we have a sequence $\{\sigma_n\}$ of mutually disjoint finite subsets of $\mathbb{N}$ satisfying (a), (b) and (c), it being understood that indices of $\{\sigma_n\}$ are arranged in the increasing order. Let us put $k_n = \#(\sigma_n)$ and $\delta = \cup \{\sigma_n : n \geq 1\}$

Define $b_{mn} = a_{mn}$; $n \in \delta$, $m \geq 1$ and $d_{mn} = \inf\{a_{mi} : i \in \sigma_n\}$. We set $Q = [b_{mn} : m \geq 1, n \in \delta]$ and $R = [d_{mn} : m \geq 1, n \geq 1]$. Corresponding to each x in $\ell^p_\delta[Q]$, define u in $\ell^p[R;\ell^p_{k_n}]$ by grouping coordinates of x according to their indices in σ_n, $n \geq 1$. Thus there is a 1-1 linear map $A:\ell^p_\delta[Q] \to \ell^p[R;\ell^p_{k_n}]$ and $\|Ax\|_m \leq \|x\|_m$. By the conclusion (b), the map A is also shown to be onto and if u is in $\ell^p[R;\ell^p_{k_n}]$, then $\|A^{-1}(u)\|_m \leq K(m)\|u\|_m$. Hence these spaces are isomorphic. Use (c) to conclude the non-nuclearity of Q. □

PROOF OF THEOREM 21.5.1 By Lemma 21.5.2, we find a non-zero non-nuclear submatrix $Q = [a_{m_k n_j}]$. Hence there exists k_0 such that

$$(*) \qquad \sum_{j\geq 1} a_{m_{k_0} n_j}/a_{m_k n_j} = \infty, \ \forall\ k \geq k_0.$$

If $b_{m_k n_j} = a_{m_{k+k_0-1} n_j}/a_{m_{k_0} n_j}$, then by Lemma 21.5.4, for each p, $1 \leq p \leq \infty$, $\ell^p[Q] \simeq \ell^p[R]$, where $R = [b_{m_k n_j}]$ and so it is enough to prove the result for $\ell^p[R]$.

Thus, in order to prove the required theorem, we may assume without loss of generality that P satisfies the conditions 1° or 2° of Lemma 21.5.5 and we establish the result under these conclusions separately.

Suppose that we have 2°. Then there exists a sequence $\delta = \{n_j\}$ such that the submatrix $P_1 \equiv [a_{mn_j}]$ satisfies (21.5.8) and therefore $\ell^p_\delta[P_1] \simeq \ell^p_\delta$ under the identity map. But

$$\ell^p_\delta[[c_{mn_j}];\ \ell^p_{n_j}] \simeq \ell^p_\delta,$$

where $c_{mn_j} = 1$ for j, $m \geq 1$ and we have finished.

Next, let us assume 1°. Then we have an m.K.m. (submatrix) $P_2 = [a_{m_k n_j}]$ satisfying (21.5.6) and (21.5.7). Let P_3 be the matrix P_2 with the first row of the latter replaced by a_{1n_j}, $j \geq 1$. Then P_3 satisfies the hypothesis (i), (ii) and (iii) of Lemma 21.5.10. Hence from Lemma 21.5.11, there exist a non-nuclear submatrix $P_4 = [a_{m_k n_{j_s}}]$, a sequence $\{q_{n_{j_s}}\}$ in I with $q_{n_{j_s}} > s$, $s \geqq 1$ and an m.K.m. $M = [d_{m_k n_j}]$ such that for each p, $1 \leqq p < \infty$

$$\ell^p_\lambda[P_4] \simeq \ell^p_\lambda[M; \ell^p_{q_{n_{j_s}}}],$$

where $\lambda = \{n_{j_s}\}$. □

21.6 PROOF OF THE BASIC THEOREM

With all the necessary background covered in the last few lemmas, paragraphs and theorems, we are finally ready to pass on to the proof of the basic theorem, namely,

PROOF OF THEOREM 21.1.1 We first reduce the proof to a comparatively easy situation. So, use Theorem 21.5.1 to get Q,D and $\{q(n)\}$ as mentioned there. Let $R = [b_{mn}]$, where $b_{mn} \equiv a_{m'_r n'_s}$ with $\{m'_r\} = \mathbb{N} \setminus \{m_k\}$ and $\{n'_s\} = \mathbb{N} \setminus \{n_j\}$. Then

$$\ell^p[P] \simeq \ell^p[D; \ell^p_{q(n)}] \oplus \ell^p[R],$$

and it suffices to prove the result for $\ell^p[D; \ell^p_{q(n)}]$.

Write k_n for $q(n)$ for convenience and construct bases $\{e_i^{k_n} : 1 \leqq i \leqq k_n\}$ for $\ell^p_{k_n}$, $n \geqq 1$ so that conditions (i) and (ii) of Lemma 21.3.7 are satisfied (hereafter, whenever we mention (i) or (ii), they will mean with respect to this lemma).

Let $k_o = 0$ and $s_m = \sum_{n=1}^{m-1} k_n$, $m \geqq 1$, $s_1 = 0$. For $j = s_m+i$; $1 \leqq i \leqq k_m$, $m \geqq 1$, let

$$d^j = \{\underbrace{0,\ldots,0}_{(m-1)\text{-times}}, e_i^{k_m}, 0, 0, \ldots\},$$

where the entries 0 designate the zero vectors of $\ell^p_{k_n}$, $n \geqq 1$.

Clearly $d^j \in \ell^p[D;\ell^p_{k_n}]$ for $j \geqq 1$ and we proceed to show that $\{d^j\}$ is an S.b. for this space. Let $M, r \in \mathbb{N}$ and $\alpha_1,\ldots,\alpha_M,\ldots,\alpha_{M+r}$ be arbirrary scalars. There exists $m \geqq 1$ such that $s_m + 1 \leqq M \leqq s_{m+1}$ and so $M = s_m+t$, $1 \leqq t \leqq k_m$. Thus

$$\left\| \sum_{j=1}^{M} \alpha_j d^j \right\|_s^p = \sum_{i=1}^{m-1} d^p_{si} \left\| \sum_{j=1+s_i}^{s_{i+1}} \alpha_j e_j^{k_i} \right\|_p^p$$

$$+ d^p_{sm} \left\| \sum_{j=1+s_m}^{s_m+t} \alpha_j e_j^{k_m} \right\|_p^p .$$

Considering two cases when $t + r \leqq k_m$ and $t + r > k_m$, proceeding as above and using (i), we get

$$\left\| \sum_{j=1}^{M} \alpha_j d^j \right\|_s \leqq \max\,(1,C_1) \left\| \sum_{j=1}^{m+r} \alpha_j d^j \right\|_s,$$

and now apply Theorem 2.1.3 to conclude the S.b. character of $\{d^j\}$.

Rewrite the base elements of $\ell^p_{k_n}$ as $\{e_i^{k_n} ; i \in F_n\}$ where $F_n = \{i : 1 + s_n \leqq i \leqq s_{n+1}\}$. If x is an arbitrary element of $\ell^p[D;\ell^p_{k_n}]$, then

$$(*) \quad x = \sum_{j \geqq 1} \alpha_j d^j = \{x_1^{k_1}, x_2^{k_2},\ldots\};\; x_n^{k_n} = \sum_{i=1+s_n}^{s_{n+1}} \alpha_i e_i^{k_n},$$

where $x_n^{k_n} \in \ell^p_{k_n}$, $n \geqq 1$. As usual, let S_σ be the expansion operator on $\ell^p_{k_n}$ corresponding to the base $\{e_i^{k_n} : i \in F_n\}$ with $\sigma \subset F_n$. There exists $\sigma_n \subset F_n$ so that $\|S_{\sigma_n}\| = K_u\{e_i^{k_n} : i \in F_n\}$ and so if $A_n = S_{\sigma_n}$, then from (ii)

$$\|A_n\| \geq C_2 k_n^\alpha = C_2 \ (\dim \ell_{k_n}^p)^\alpha$$

Observe that (following the notation in (*))

$$\underset{n\geq 1}{\oplus} A_n(x) = \{ \sum_{i\in\sigma_n} \alpha_i e_i^{k_n} : n \geq 1\}.$$

If $\{d^j\}$ is a u-S.b., it follows with the help of (6.1.3) that $\underset{n\geq 1}{\oplus} A_n$ is continuous and hence using Theorem 21.4.1, $\ell^p[D;\ell_{k_n}^p]$ turns out to be nuclear, a contradiction. Therefore, $\{d^j\}$ is a conditional S.b. for $\ell^p[D;\ell_{k_n}^p]$. □

21.7 ANOTHER CHARACTERIZATION OF NUCLEAR SPACES

Chapters 5 and 12 deal with several results concerning the impact of different types of bases present in a space on its nuclearity. In particular, we draw attention to Theorem 12.4.6 which characterizes nuclearity of an arbitrary l.c. TVS having a u.e-S.b. By using (12.4.7), it follows that each u.e-S.b. in an arbitrary nuclear space X is a semi ℓ^1-base and in addition, if X is also ω-complete then this base turns out to be an ℓ^1-base. Thus, if a Frećhet space is nuclear, then each of its S.b. (if present at all) is an ℓ^1-base. The converese of this statement was first investigated and answered by Wojtynski in [235] in the form of Theorem 21.1.2 and, as such, let us now turn to

PROOF OF THEOREM 21.1.2 Let (X,T) be non-nuclear. By the hypothesis, X contains an S.b. $\{x_n;f_n\}$ which is an ℓ^1-base. Indeed, $\{x_n;f_n\}$ is 1-Köthe (Definition 5.4.11 and use Proposition 5.4.1). Therefore, $(X,T) \simeq (\ell^1[P],T_1)$, where $P = [a_{mn}]$ with $a_{mn} = p_m(x_n)$ and $D_T = \{p_1 \leq p_2 \leq \ldots\}$. By Theorem 21.1.1, (X,T) therefore contains a conditional base, a contradiction. □

Summing up the discussion of this section, we have finally proved

THEOREM 21.7.1 Let a Fréchet space X contain an S.b. Then X is nuclear if and only if all bases on X are ℓ^1-bases.

REMARK. If each seminorm p_m of a Fréchet space (X,T) with $D_T = \{p_1 \leqq p_2 \leqq \dots\}$ is Hilbertian, then (X,T) is called a countably-Hilbert (c-H) space. There is an interesting characterization of nuclearity of c-H spaces in terms of unconditional bases, namely,

THEOREM 21.7.2 Let a c-H space X contain an S.b. Then X is nuclear if and only if all its bases are unconditional.

Whereas the necessity of the above result is a consequence of Theorem 21.7.1, the converse depends upon showing the isomorphism between X and a Köthe space $\ell^2[a_{mn}]$ and then making use of Theorem 21.1.1. The reader is, however, referred to [235] for its proof.

References

[1] Alpseyman, M. A generalization of Dragilev's theorem; Jour. reine u. angew. Math., 276 (1975), 124-129.

[2] Alpseyman, M. Basic sequences in some nuclear Köthe sequence spaces; Dissertation, University of Michigan, Ann Arbor, 1978.

[3] Arsove, M.G. Proper bases and automorphisms in the space of entire functions, Proc. Amer. Math. Soc., 8 (2) (1957), 264-271.

[4] Arsove, M.G. The Pincherle basis problem and a theorem of Boas; Math. Scand., 5 (1957), 271-275.

[5] Arsove, M.G. Proper bases and linear homeomorphisms in the spaces of analytic functions; Math. Ann., 135 (1958), 235-243.

[6] Arsove, M.G. Proper Pincherle bases in the space of entire functions; Quart. Jour. Math., 9 (33) (1958), 40-54.

[7] Arsove, M.G. Similar bases and isomorphisms in Fréchet spaces; Math. Ann., 135 (1958), 283-293.

[8] Arsove, M.G. Bases semblables et isomorphismes dans les espace de Fréchet; Extrait des Comptes Rend. Sc. Aca. Scs., 246 (1958), 1143-1145.

[9] Arsove, M.G. The Paley-Wiener theorem in metric linear spaces; Pac. Jour. Math., 10 (1960), 365-379.

[10] Arosve, M.G. and Edwards, R.E. Generalized bases in topological linear spaces; Pac. Jour. Math., 19 (1960), 95-113.

[11] Banach, S. Sur une propriété caractéristique des fonctions orthogonales; Comp. Rend. Aca. Scs., 180 (1925), 1637-1640.

[12] Banach, S. Theorie des opérations linéaires; Chelsea Pub. Co., New York, 1955.

[13] Bary, N.K. Sur la stabilité de certaines propriétés des systemes orthogonaux; Mat. Sbornik, 12 (1943), 3-27.

[14] Bary, N.K. Sur les systèmes complets de fonctions orthogonales; Mat. Sbornik; 14 (1944), 51-108.

[15] Bennett, G. Some inclusion theorems for sequence spaces; Pac. Jour. Math., 46 (1973), 17-30.

[16] Bessaga, C. Some remarks on Dragilev's theorem; Studia Math., 31 (1968), 307-318.

[17] Bessaga, C. and Pelczynski, A. On bases and unconditional convergence of series in Banach spaces; Studia Math., 17 (1958), 151-164.

[18] Bessaga, C. and Pelczynski, A. A generalization of results of R.C. James concerning absolute bases in Banach spaces; Studia Math., 17 (1958), 165-174.

[19] Bessaga, C. and Pelczynski, A. Wlasnosci baz w przestrzeniach typu B_0 (Polish); Prace Mat., 3 (1959), 123-142.

[20] Boas, Jr., R.P. Expansions of analytic functions; Trans. Amer. Math. Soc. 48 (1940), 467-487.

[21] Brudovskii, V.S. Associated nuclear topology, mappings of type s and strongly nuclear spaces; Soviet Math. Dokl., 9 (1968), 61-63.

[22] Brudovskii, V.S. s-type mappings of locally convex spaces; Soviet Math. Dokl., 9 (1968), 572-574.

[23] Cắc, N.P. On symmetric Schauder bases in a Fréchet space; Studia Math., 32 (1969), 95-98.

[24] Cook, T.A. Weakly equicontinuous Schauder bases; Proc. Amer. Math. Soc., 23 (3) (1969), 536-537.

[25] Cook, T.A. Schauder decompositions and semi-reflexive spaces; Math. Ann., 182 (1969), 232-235.

[26] Cook, T.A. On normalized Schauder bases; Amer. Math. Monthly, 77 (1970), 167.

[27] Cooke, R.G. Infinite matrices and sequence spaces; MacMillan, London, 1950.

[28] Crone, L., Dubinsky, Ed., and Robinson, W.B. Regular bases in products of power series spaces; Jour. Fnl. Anal., 24(3) (1977), 211-222.

[29] Crone, L. and Robinson, W.B. Every nuclear Fréchet space with a regular basis has the quasi-equivalence property; Studia Math., 52 (1975), 203-207.

[30] Daoud, D. Bases in the space of entire Dirichlet functions of two complex variables; Collect. Math., (1985), 35-42.

[31] Daoud, D. On the class of entire Dirichlet functions of several complex variables having finite order point; to appear in Portug. Math.

[32] Das, N.R. Bi-locally convex spaces and Schauder decompositions; Ph.D. Dissertation, Ind. Inst. Techno, Kanpur, 1982.

[33] Davis, W.J. and Dean, D.W. The direct sum of Banach spaces with respect to a basis; Studia Math., 28 (1967), 209-219.

[34] De-Grande-De Kimpe, N. Criteria for nuclearity in terms of generalized sequence spaces; Arch. Math. 28 (1977), 644-651.

[35] De-Grande-De Kimpe, N. On Λ-bases; Jour. Math. Anal. Appl., 53 (1976), 508-520.

[36] Dineen, S. Complex analysis in locally convex spaces; NM 83, North-Holland Pub. Co. Amsterdam, 1981.

[37] Djakov, P.B. A short proof of the theorem of Crone and Robinson on quasi-equivalence of regular bases; Studia Math., 53 (1975), 269-271.

[38] Dragilev, M.M. Standard form of basis for the space of analytic functions (Russian); Uspehi Mat. Nauk; 15 (1960), 181-188.

[39] Dragilev, M.M. On regular bases in nuclear spaces; Amer. Math. Soc. Transl. (2), 93 (1970), 61-82 [Russian Mat. Sb., 68 (1965), 153-173].

[40] Dragilev, M.M. On special dimensions defined on some classes of Köthe spaces; Math. USSR Sbornik, 9 (1969), 213-228.

[41] Dubinsky, Ed. Basic sequences in s; Studia Math., 59 (1977), 283-293.

[42] Dubinsky, Ed. The structure of nuclear Fréchet spaces; LN720, Springer-Verlag, Berlin, 1979.

[43] Dubinsky, Ed. and Ramanujan, M.S. λ-Nuclearity; Mem. Amer. Math. Soc., No. 128, Rhode Island, 1972.

[44] Dubinsky, Ed. and Retherford, J.R. Schauder bases and Köthe sequence spaces; Bull. Aca. Scs. (Ser. Sc. Math. Astr. Phy.), 16 (9) (1966), 497-501.

[45] Dubinsky, Ed. and Retherford, J.R. Bases in compatible topologies; Studia Math., 28 (1967), 221-226.

[46] Dubinsky, Ed. and Retherford, J.R. Schauder bases and Köthe sequence spaces; Trans. Amer. Math. Soc., 130 (1968), 265-280.

[47] Dunford, N. and Morse, A.P. Remarks on the preceding paper of James A. Clarkson; Trans. Amer. Math. Soc., 40 (1936), 415-420.

[48] Dvoretzky, A. and Rogers, C.A. Absolute and unconditional convergence in linear normed spaces; Proc. Nat. Aca. Sc., U.S.A., 36 (1950), 192-197.

[49] Dynin, A. and Mityagin, B.S. Criterion for nuclearity in terms of approximative dimension (Russian); Bull. Aca. Polo. Sc. (Ser. Sc. Math. Astr. Phy.), 88 (1960), 535-540.

[50] Ellis, H.W. and Holperin, I. Haar functions and the basis problem; Jour. Lond. Math. Soc., 31 (1956), 28-39.

[51] Evgrafov, M.A. The method of approximating systems in the space of analytic functions and its applications to interpolation (Russian); TrudyMosk. Mat. Obsch, 5 (1956), 89-201.

[52] Fenske, G. and Schock, E. Über die diametrale Dimension von lokalkonvexen Räumen, BMBW-GMD, Nr. 10, Bonn, (1969), 13-22.

[53] Fenske, G. and Schock, E. Nuklearität und lokale Konvexität von Folgenräume; Math. Nachr., 45 (1978), 327-335.
[54] Gross, F. Generalized series and orders and types of entire functions of several complex variables; Trans. Math. Soc., 120 (1) (1965), 124-144.
[55] Fullerton, R.E. Geometric structure of absolute basis system in a linear topological space; Pac. Jour. Math., 12 (1962), 137-147.
[56] Garling, D.J.H. On the symmetric sequence spaces; Proc. Lond. Math. Soc., 16(3) (1966), 85-106.
[57] Garling, D.J.H. The β- and γ-duality; Proc. Camb. Phil. Soc., 63 (1967), 963-981.
[58] Garling, D.J.H. Symmetric bases of locally convex spaces; Studia Math., 30 (1968), 163-181.
[59] Gautam, S.K.S. Bases in certain spaces of analytic Dirichlet transformations; Ph.D. Dissertation, Ind. Inst. Techno., Kanpur, 1972.
[60] Gelbaum, B.R. Expansions in Banach spaces; Duke Math. Jour., 17 (1950), 187-196.
[61] Gelbaum, B.R. A nonabsolute basis for Hilbert spaces; Proc. Amer. Math. Soc., 2 (5) (1951), 720-721.
[62] Gelfand, I.M. and Shilov, G.E. Generalized functions, Vol. 2; Academic Press, New York, 1968.
[63] Gelfand, I.M. and Vilenkin, N.Y. Generalized functions, Vol. 4; Academic Press, New York, 1964.
[64] Grothendieck, A. Resume de la theorie metrique des products tensciels topologiques, Bol. Soc. Mat. Sao Păulo 8, 1-79 (1956).
[65] Grothendieck, A. Topological vector spaces; Gordon and Breach, Science Pub., New York, 1975.
[66] Gupta, Manjul, A topological study of spaces of analytic functions of several variables; Ph.D. Dissertation, Ind. Inst. Techno., Kanpur, 1973.
[67] Gupta, Manjul and Das, N.R. Bi-locally convex spaces, Portugaliae Math., 42 (4) (1983-84), 417-433.

[68] Gupta, Manjul and Kamthan, P.K. Space of entire functions of several variables in a non-Archimedean field; Acta Mex. Cien. Techno., 8 (1,2,3) (1974), 19-33.

[69] Gupta, Manjul and Kamthan, P.K. Dominating sequences and functional equations; Period.Math. Hungar., 15 (3) (1984), 219-231.

[70] Gupta, Manjul and Kamthan, P.K. λ-similar bases; Internat. Jour. Math. and Math. Sc.; 10 (2), (1987), 227-232.

[71] Gupta, Manjul and Kamthan, P.K. Paley-Wiener type stability theorems in locally convex spaces; preprint.

[72] Gupta, Manjul, Kamthan, P.K. and Das, N.R. Bi-locally convex spaces and Schauder decompositions, Ann. Mat. Pura ed Appl., 23, Ser. IV (1983), 267-284.

[73] Gupta, Manjul, Kamthan, P.K. and Deheri, G. αμ-duals and holomorphic (nuclear) mappings; Collect. Math., 36 (1986), 33-71.

[74] Gupta, Manjul, Kamthan, P.K. and Rao, K.L.N. Generalized Köthe sequence spaces and decompositions, Ann. Mat. Pura ed Appl., 113, Ser. IV (1977), 287-301.

[75] Gupta, Manjul, Kamthan, P.K. and Ruckle, W.H. Symmetric sequence spaces, bases and applications; Jour. Math. Anal. Appl., 113 (1) (1986), 210-229.

[76] Gupta, Manjul and Patterson, J. Generalized sequence spaces and matrix transformations; Houston Jour. Math., 10 (3), (1984), 387-398.

[77] Gurevic, L.A. On unconditional bases (Russian); Uspehi Mate Nauk (N.S.) 8 (1953), 153-156.

[78] Harvey, J.R. Sequence spaces and the basis concept in Banach spaces; Dissertation, Uni. Texas, Austin, 1969.

[79] Holding, S.H. Note on completeness theorems of Paley-Wiener type; Ann. Math., 49 (2) (1948), 953-955.

[80] Horváth, J. Topological vector spaces and distributions I; Addison-Wesley, Reading, Mass., 1966.
[81] Hussain, T. and Kamthan, P.K. Spaces of entire functions represented by Dirichlet series; Collect. Math., 19(3) (1968), 203-216.
[82] Hutton, G. On the approximation numbers of an operator and its adjoint, Math. Ann., 210 (1974), 277-280.
[83] Iyer, V.G. On the space of integral functions; Jour. Ind. Math. Soc., 12 (1948), 13-40.
[84] Iyer, V.G. On the space of integral functions II; Quart. Jour. Math., Oxford Series (2), 1 (1950), 86-96.
[85] Iyer, V.G. On the space of integral functions (V); Jour. Ind. Math. Soc., 24 (1960), 269-278.
[86] Jacob, Jr., R.T. Matrix transformations involving simple sequence spaces; Pac. Jour. Math., 70 (1977), 179-187.
[87] James, R.C. Bases and reflexivity of Banach spaces; Ann. Math., 52 (1950), 518-527.
[88] James, R.C. Bases in Banach spaces; Amer. Math. Monthly, 89 (9) (1982), 625-640.
[89] Jarchow, H. Locally convex spaces; B.G. Tuebner, Stuttgart, 1981.
[90] Johnson, W.B. Markushevich bases and duality theory; Trans. Amer. Math. Soc., 149 (1970), 171-177.
[91] Jones, O.T. and Retherford, J.R. On similar bases in barrelled spaces; Proc. Amer. Math. Soc., 18 (4) (1967), 677-680.
[92] Kalton, N.J. Schauder decompositions in locally convex spaces; Proc. Camb. Phil. Soc., 68 (1970), 377-392.
[93] Kalton, N.J. Schauder bases and reflexivity; Studia Math., 38 (1970), 255-266.

[94] Kalton, N.J. Schauder decompositions and completeness; Bull. Lond. Math. Soc., 2 (1970), 34-36.

[95] Kalton, N.J. Unconditional and normalized bases; Studia Math., 38 (1970), 243-253.

[96] Kalton, N.J. Normalization properties of Schauder bases; Proc. Lond. Math. Soc., 22 (1971), 91-105.

[97] Kalton, N.J. Some forms of the closed graph theorem; Proc. Camb. Phil. Soc., 70 (1971), 401-408.

[98] Kalton, N.J. Mackey duals and almost shrinking bases; Proc. Camb. Phil. Soc., 74 (1973), 73-81.

[99] Kalton, N.J. On absolute bases; Math. Ann., 200 (1973), 209-225.

[100] Kamthan, P.K. Proximate order (R) of entire functions represented by Dirichlet series; Collect. Math., 14 (3) (1962), 275-278.

[101] Kamthan, P.K. FK-space for entire Dirichlet functions; Collect. Math., 20 (3) (1969), 272-280.

[102] Kamthan, P.K. Various topologies on the space of entire functions; Labdev. Jour. Sc. Tech., 9 (1971), 143-151.

[103] Kamthan, P.K. A study on the space of entire functions of several complex variables; Yokohama Math. Jour., 21 (1973), 11-20.

[104] Kamthan, P.K. Normalized bases in topological vector spaces; Portug. Math., 38 (3-4) (1979), 55-65.

[105] Kamthan, P.K. Monotone bases and orthogonal systems in Fréchet spaces; Yokohama Math. Jour., 21 (1973), 5-10.

[106] Kamthan, P.K. Regular and bounded bases; Tamkang Jour. Math., 7 (1976), 203-205.

[107] Kamthan, P.K. Bases in a certain class of a Fréchet space; Tamkang Jour. Math., 7 (1) (1976), 41-49.

[108] Kamthan, P.K. Kolmogorov diameters and their applications; Bull. Aligarh Mus. Uni., 7 (1977), 9-21.

[109] Kamthan, P.K. A lemma on the convergence of linear maps; preprint.
[110] Kamthan, P.K. The behaviour of trnasformations on sequence spaces; Collect. Math., 30 (1) (1982), 77-87.
[11] Kamthan, P.K. A nuclearity criterion for spaces having a Schauder base; Acta. Math. Viet., 7 (2) (1982), 41-45.
[112] Kamthan, P.K. λ-nuclear operators and applications; Jour. Bihar Math. Soc. (Silver Jubilee No.), 9 (1985), 1-16.
[113] Kamthan, P.K. and Gautam, S.K.S. Bases in the space of analytic Dirichlet transformations; Collect. Math. 23 (1) (1972), 9-16.
[114] Kamthan, P.K. and Gautam, S.K.S. Bases in a certain space of functions analytic in half-plane; Ind. Jour. Pure and Appl. Math., 6 (9) (1975), 1066-1075.
[115] Kamthan, P.K. and Gautam, S.K.S. Restricted double automorphisms in the space of analytic Dirichlet functions; Jour. Korean Math. Soc., 12 (2) (1975), 79-87.
[116] Kamthan, P.K. and Gautam, S.K.S. Bases in a certain space of Dirichlet entire transformations; Ind. Jour. Pure and Appl. Math., 6(8) (1975), 856-863.
[117] Kamthan, P.K. and Gautam, S.K.S. On bases in the space of Dirichlet entire trnasformations of finite order; Portug. Math., 38 (3-4), (1979), 67-78.
[118] Kamthan, P.K. and Gupta, Manjul, Expansion of entire functions of several complex variables; Trans. Amer. Math. Soc., 192 (1974), 371-382.
[119] Kamthan, P.K. and Gupta, Manjul, Analytic functions in bicylinders; Ind. Jour. Pure and Appl. Math., 5 (12) (1974), 1119-1126.
[120] Kamthan, P.K. and Gupta, Manjul, Space of entire functions of several complex variables having finite order point; Math. Japon., 20 (1) (1975), 7-19.

[121] Kamthan, P.K. and Gupta, Manjul, Characterization of bases in topological vector spaces; Tamkang Jour. Math., 7 (1) (1976), 51-55.

[122] Kamthan, P.K. and Gupta, Manjul, Schauder bases and sequential duals; Bull. Roy. Soc. Scs. Liège, 46 (1977), 153-155.

[123] Kamthan, P.K. and Gupta, Manjul, Uniform bases in locally convex spaces; Jour. reine u. angew. Math., 295 (1977), 208-213.

[124] Kamthan, P.K. and Gupta, Manjul, Weak Schauder bases and completeness; Proc. Roy. Irish Aca., 78A (1978), 51-54.

[125] Kamthan, P.K. and Gupta, Manjul, Schauder decompositions and their applications to continuity of maps; Jour. reine u. angew. Math., 298 (1978), 104-107.

[126] Kamthan, P.K. and Gupta, Manjul, Several notions of absolute bases; Jour. reine u. angew Math., 307/308 (1979), 79-83.

[127] Kamthan, P.K. and Gupta, Manjul, Addendum and corrigendum: Several notions of absolute bases; Jour. reine u. angew. Math.; 317 (1980), 220.

[128] Kamthan, P.K. and Gupta, Manjul, Weak unconditional Cauchy series; Rendi. Circo. Mat. Palermo, 29 (1980), 364-368.

[129] Kamthan, P.K. and Gupta, Manjul, Sequence spaces and series; Marcel Dekker, New York, 1981.

[130] Kamthan, P.K. and Gupta, Manjul, Theory of bases and cones; Pitman, London, 1985.

[131] Kamthan, P.K. and Gupta, Manjul, A lemma on Schauder bases; Note di Mate., 5 (1985), 83-89.

[132] Kamthan, P.K. and Gupta, Manjul, Representation theory of operators, a monograph under preparation.

[133] Kamthan, P.K., Gupta, Manjul and Sofi, M.A. λ-bases and their applications; Jour. Math. Anal. Appl., 88 (1982), 76-99.

[134] Kamthan, P.K. and Ray, S.K. On decompositions in barrelled spaces; Colloq. Math., 34 (1) (1975), 73-79.

[135] Kamthan, P.K. and Ray, S.K. Equivalence of sequences of subspaces in topological vector spaces; Math. Japon., 20 (4) (1976), 333-340.

[136] Kamthan, P.K. and Ray, S.K. Schauder bases and the Köthe structure of associated sequence spaces; Boll. U.M.I., 18A (5) (1981), 299-303.

[137] Kamthan, P.K. and Sofi, M.A. Operators of type s_ϕ and associated nuclearity; Tamkang Jour. Math., 12 (2) (1981), 233-244.

[138] Kamthan, P.K. and Sofi, M.A. λ-bases and λ-nuclearity; Jour. Math. Anal. Appl. 99 (1984), 164-188.

[139] Karlin, S. Bases in Banach spaces; Duke Math. Jour., 15 (1948), 971-985.

[140] Kelley, J.L. and Namioka, I. Linear topological vector spaces; D. Van Nostrand, Princeton, 1963.

[141] Knowles, R.J. and Cook, T.A. Incomplete reflexive spaces without Schauder bases; Proc. Camb. Phil. Soc., 74 (1973), 83-86.

[142] Köthe, G. Über nukleare lineare Räume; Studia Math., 31 (1968), 267-271.

[143] Köthe, G. Topological vector spaces I; Springer-Verlag, Berlin, 1969.

[144] Köthe, G. Stark nukleare Folgenräume; Jour. Fac. Sci. Uni. Tokyo, Sect. IA Math. (1970), 291-296.

[145] Köthe, G. Nukleare (F)-und (DF)-Folgenräume; Theory of sets and topology (1972), 327-332.

[146] Kozlov, V.Ya. On bases in the space $L_2[0,1]$ (Russian); Mat. Sb., 26 (68) (1950), 85-102.

[147] Krishnamurthy, V. On the spaces of certain classes of entire functions; Jour. Austr. Math. Soc. I (2) (1960), 147-170.

[148] Krishnamurthy, V. On the continuous endomorpshisms in the spaces of certain classes of entire functions; Proc. Nat. Inst. Sci. (India), Prt. A, 26 (6) (1960), 642-655.

[149] Krein, M.G. and Lusternik, L.A. Functional analysis, mathematics in the SSSR for 30 years (1917-1947) (Russian), Moscow-Leningrad (1947), 608-697.

[150] Krein, M.G., Milman, D. and Rutman, R. On a property of a basis in a Banach space (Russian); Kharkov. Zap. Mat. Obsh., 16 (4) (1940), 182.

[151] Kreyszig, E. Advanced engineering mathematics; John-Wiley and Sons, Inc., New York, 1968.

[152] Lawrence, C. and Robinson, W.B. Every nuclear Fréchet space with a regular basis has the quasi-equivalence property; Studia Math. 52 (1975), 203-207.

[153] Lindenstrauss, J. and Pelczynski, A. Absolutely summing operators in $\mathcal{L}_p$ spaces and their applications; Studia Math., 29 (1968), 275-326.

[154] Lindenstrauss, J. and Tzafriri, L. Classical Banach spaces I, sequence spaces; Springer-Verlag, Berlin, 1977.

[155] Mandelbrojt, S. Dirichlet series; Rice Inst. Pamphlet, Vol. 31, Houston, 1944.

[156] Markushevich, A.I. Sur les bases dans l'espace des fonctions analytiques (Russian); Mat. Sb., 17 (59) (1945), 211-252. [Eng. Transl. Amer. Math. Soc., 22 (1962), 1-42.]

[157] Markushevich, A.I. Theory of functions of a complex variable, Vol. II; Prentice-Hall, Englewood Cliffs, 1968.

[158] Marti, J.T. Theory of bases; Springer-Verlag, Berlin, 1969.

[159] McArthur, C.W. Convergence of monotone nets in ordered topological vector spaces; Studia Math., 34 (1970), 1-16.

[160] McArthur, C.W. Developments in Schauder basis theory; Bull. Amer. Math. Soc., 78 (6) (1972), 877-908.

[161] McArthur, C.W. and Retherford, J.R. Uniform and equicontinuous Schauder bases of subspaces; Cand. Jour. Math., 17 (1965), 207-212.

[162] McArthur, C.W. and Retherford, J.R. Some applications of an inequality in locally convex spaces; Trans. Amer. Math. Soc., 137 (1969), 115-123.

[163] Milman, V.D. Perturbations of sequences of elements of a Banach space (Russian); Sibirsk. Math. Z., 6 (1965), 398-412.

[164] Mityagin, B.S. Approximative dimension and bases in nuclear spaces; Russian Math. Surveys, 16 (1961), 59-127.

[165] Mori, Y. On bases in λ(P)-nuclear spaces; Math. Semi. Notes, Kobe Uni., 5 (1977), 49-58.

[166] Nagy, B.Sz. Expansion theorems of Paley-Wiener type; Duke Math. Jour., 14 (1947), 975-978.

[167] Narumi, S. A theorem on the expansion of analytic functions by infinite series; Tôhku Math. Jour., 30 (1928), 441-444.

[168] Nalimarkka, E. On operator ideals and locally convex A-spaces with applications to λ-nuclearity, Dissertation; Ann. Aca. Sci. Fenn., 1977.

[169] Newns, F. On the representation of analytic functions by infinite series; Phil. Trans. Roy. Soc. London (A), 245 (1953), 429-468.

[170] Nikol'skii, V.N. The best approximation and a basis in a Fréchet space (Russian); Dokl. Akad. Nauk SSSR, 59 (1948), 639-642.

[171] Orlicz, W. Über unbendingte Konvergenz in Funktionräumen I. Studia Math., 4 (1933), 33-37.

[172] Orlicz, W. Über unbendingte Konvergenz in Funktionräumen II; Studia Math., 4 (1933), 41-47.

[173] Paley, R.E.A.C. and Wiener, N. Fourier transforms in the complex domain; Colloquium Pub., AMS, New York, 1934.

[174] Pelczynski, A. and Singer, I. On non-equivalent bases and conditional bases in Banach spaces; Studia Math., 25 (1964), 5-25.

[175] Pelczynski, A. and Szlenk, W. An example of a non-shrinking base; Rev. Roum. Math. Pures et Appl., 10 (1965), 961-966.

[176] Pietsch, A. (F)-Räume mit absoluter basis; Studia Math., 26 (1966), 233-238.

[177] Pietsch, A. Absolute p-summierende Abbildungen in normierten Räumen; Studia Math., 28 (1967), 333-353.

[178] Pietsch, A. Nuclear locally convex spaces; Springer-Verlag; Berlin, 1972.

[179] Pollard, H. Completeness theorems of Paley-Wiener type; Ann. Math., 45 (2) (1944), 738-739.

[180] Ramanujam, M.S. Power series spaces and the associated nuclearity; Math. Ann. 189 (1970), 161-168.

[181] Ramanujam, M.S. and Terzioğlu, T. Diametral dimensions of cartesian products, stability of smooth sequence spaces and applications; Jour. reine u. angew., 280 (1971), 163-171.

[182] Ray, S.K. Decompositions of topological vector spaces, Ph.D. Dissertation; Ind. Inst. Techno., Kanpur, 1974.

[183] Retherford, J.R. Basic sequences and the Paley-Wiener criterion; Pac. Jour. Math., 14 (3) (1964), 1019-1027.

[184] Retherford, J.R. Bases, basic sequences and reflexivity of linear topological spaces; Math. Ann. 164 (1966), 280-285.

[185] Retherford, J.R. A semishrinking basis which is not shrinking; Proc. Amer. Math. Soc., 19 (3) (1968), 766.

[186] Retherford, J.R. The Paley-Wiener criterion, preprint.

[187] Retherford, J.R. and Holub, J.R. The stability of bases in Banach and Hilbert spaces; Jour. reine u. angew. Math., 246 (1971), 136-148.

[188] Retherford, J.R. and McArthur, C.W. Some remarks on bases in linear topological spaces; Math. Ann., 164 (1966), 38-41.

[189] Ritt, J.F. On certain points in Dirichlet series; Amer. Jour. Math., 50 (1928), 73-86.

[190] Robertson, A.P. and Robertson, W. Topological vector spaces; Cambridge Univ. Press, Cambridge, 1966.

[191] Robinson, W.B. On $\Lambda_1(\alpha)$-nuclearity; Duke Math. Jour., 40 (3) (1973), 541-546.

[192] Rosenberger, B. F-Normideale von Operatoren in Normierten Räumen; GMBW-GMD-44, Bonn, (1971).

[193] Rosenberger, B. ϕ-nukleare Räume; Math. Nachr., 52 (1972), 147-160.

[194] Rosenberger, B. and Schock, E. Über nukleare (F)-Räume mit Basis; Compo. Math., 25 (2) (1972), 207-219.

[195] Ruckle, W.H. The infinite sum of closed subspaces of an F-space, Duke Math. Jour., 31 (1964), 543-554.

[196] Ruckle, W.H. Infinite matrices which preserve Schauder bases; Duke Math. Jour., 33 (1966), 547-549.

[197] Ruckle, W.H. On the characterization of sequence spaces associated with Schauder bases; Studia Math., 28 (1967), 279-288.

[198] Ruckle, W.H. Symmetric coordinate spaces and symmetric bases; Can. Jour. Math., 19 (1967), 828-838.

[199] Ruckle, W.H. Topologies on sequence spaces; Pacific Jour. Math., 42 (1) (1972), 235-249.

[200] Ruckle, W.H. The strong ϕ-topology on symmetric sequence spaces; preprint.

[201] Rudin, W. Functional analysis; Tata McGraw-Hill Pub. Co. Ltd., New Delhi, 1976.

[202] Russo, J.P. Monotone and e-Schuader bases of subspaces, Dissertation; Florida St. Uni., Tallahassee, 1965.

[203] Russo, J.P. Monotone and e-Schauder bases of subspaces; Can. Jour. Math., 20 (1968), 233-241.

[204] Saxon, S. Basis cone basis theory, Dissertation; Florida St. Uni., Tallahassee, 1965.

[205] Schaefer, H.H. Topological vector spaces; Macmillan, New York, 1966.

[206] Schafke, F.W. Über einige unendliche lineare Gleichungssysteme; Math. Nachr., 3 (1949), 40-58.

[207] Schafke, F.W. Das Kriterium von Paley-Wiener im Banachschen Räumen; Math. Nachr., 3 (1949), 59-61.

[208] Schock, E. (F)-Räume mit absoluter Basis; BMBW-GMD-10, Bonn, (1969), 23-28.

[209] Schock, E. Diametrale Dimension approximative Dimension und Anwendungen; BMBW-GMD-43, Bonn, (1971), 1-48.

[210] Singer, I. On Banach spaces with symmetric bases (Russian); Rev. Roum. Math. Pures et Appl., 6 (1961), 159-166.

[211] Singer, I. Basic sequences and reflexivity of Banach spaces; Studia Math., 21 (1962), 351-369.

[212] Singer, I. Some characterizations of symmetric bases in Banach spaces; Bull. Aca. Sc., Ser. Scs. Math. Astr. Phy., 10 (4) (1962), 185-192.

[213] Singer, I. On a theorem of N.K. Bari and I.M. Gelfand; Arch. Math., 19 (1968), 508-510.

[214] Singer, I. Some remarks on domination of sequences; Math. Ann., 184 (1970), 113-132.

[215] Singer, I. Bases in Banach spaces I; Springer-Verlag, Berlin, 1970.

[216] Sofi, M.A. λ-bases and λ-nuclearity, Ph.D. Dissertation; Ind. Inst. Techno., Kanpur, 1981.

[217] Taylor, A.E. Introduction to functional analysis; John-Wiley and Sons, Inc., New York, 1963.

[218] Terzioğlu, T. Die diametrale Dimension von lokalkönvexen Räumen; Collect. Math., 20 (1969), 49-99.

[219] Terzioğlu, T. On Schwartz spaces; Math. Ann., 182 (1969), 236-242.

[220] Terzioğlu, T. Linear operators on nuclear spaces and their extension properties: Rev. Fac. Sc., l'Uni. d'Istanbul, Ser. A, 37 (1972), 1-21.

[221] Terzioğlu, T. Symmetric bases of nuclear spaces; Jour. reine u. angew. Math., 252 (1972), 200-204.

[222] Terzioğlu, T. Smooth sequence spaces and associated nuclearity, Proc. Amer. Math. Soc., 37 (2) (1973), 497-504.

[223] Terzioğlu, T. Stability of smooth sequences; Jour. reine u. angew. Math., 276 (1975), 184-187.

[224] Tikhomirov, V.M. Diameters of sets in function spaces and the theory of best approximation; Russian Math. Surveys, 15 (1960), 75-112.

[225] Tseitlin, Ya.M. On spaces with absolute basis; Siberia Math. Jour., 12 (5) (1971), 838-841.

[226] Tumarkin, Ju.B. Locally convex spaces with basis; Soviet Math. Dokl., 11 (1970), 1672-1675.

[227] Urysohn, P. Sur une problème de M. Fréchet relatif aux classes des fonctions holomorphes; C.R. Congr. Soc. Sav. Sci., 1924.

[228] Věic, B.E. Some stability properties of bases; Soviet Math., 5 (1964), 1141-1144.

[229] Webb, J.H. Sequential convergence in locally convex spaces; Proc. Camb. Phil. Soc., 64 (1968), 341-364.

[230] Weill, L.J. Unconditional bases in locally convex spaces, Dissertation; Florida St. Uni., Tallahassee, 1966.

[231] Weill, L.J. Stability of bases in complete barrelled spaces; Proc. Amer. Math. Soc., 18(6) (1967), 1045-1050.

[232] Weill, L.J. Unconditional and shrinking bases in locally convex spaces; Pac. Jour. Math., 29 (1969), 467-483.

[233] Wilansky, A. Functional analysis: Blaisdell Pub. Co., New York, 1964.

[234] Wilansky, A. Modern methods in topological vector spaces; McGraw-Hill Int. Book Co., New York, 1978.

[235] Wojtyński, W. On conditional bases in non-nuclear Fréchet spaces; Studia Math., 35 (1970), 77-96.

[236] Woods, P.C. Markuschevich basis and semi-reflexivity; Proc. Amer. Math. Soc., 37 (1) (1973), 212-220.

ADDITIONAL REFERENCES

[237] Babenko, K.I. On conjugate functions (Russian): Doklady Akad. Nauk, SSSR, 62 (1948), 157-160.

[238] Diestel, J. Sequence and series in Banach spaces; Springer-Verlag, Berlin, 1984.

[239] Jain, P.K. and Jain, D.R. On bases in a space of entire Dirichlet transforms of several complex variables; Tamkang Jour. Math., 8 (2) (1977), 183-195.

[240] Kalton, N.J. Subseries convergence in topological groups and vector spaces; Israel Jour. Math., 10 (4) (1971), 402-412.

[241] Titchmarsh, E.C. Theory of functions, Oxford University Press, Oxford, 1939.

[242] Zygmund, A. Trigonometric series, Vols. I and II; Cambridge University Press, Cambridge, 1968.

Index